Ibn al-Haytham's Geometrical Methods and the Philosophy of Mathematics

This fifth volume of *A History of Arabic Sciences and Mathematics* is complemented by four preceding volumes which focused on the main chapters of classical mathematics: infinitesimal geometry, theory of conics and its applications, spherical geometry, mathematical astronomy, etc.

This book includes seven main works of Ibn al-Haytham (Alhazen) and of two of his predecessors, Thābit ibn Qurra and al-Sijzī:

- The circle, its transformations and its properties;
- Analysis and synthesis: the founding of analytical art;
- A new mathematical discipline: *the Knowns*;
- The geometrisation of place;
- Analysis and synthesis: examples of the geometry of triangles;
- Axiomatic method and invention: Thābit ibn Qurra;
- The idea of an *Ars Inveniendi*: al-Sijzī.

Including extensive commentary from one of the world's foremost authorities on the subject, this fundamental text is essential reading for historians and mathematicians at the most advanced levels of research.

Roshdi Rashed is one of the most eminent authorities on Arabic mathematics and the exact sciences. A historian and philosopher of mathematics and science and a highly celebrated epistemologist, he is currently Emeritus Research Director (distinguished class) at the Centre National de la Recherche Scientifique (CNRS) in Paris, and is the former Director of the Centre for History of Medieval Science and Philosophy at the University of Paris (Denis Diderot, Paris VII). He also holds an Honorary Professorship at the University of Tokyo and an Emeritus Professorship at the University of Mansourah in Egypt.

J. V. Field is a historian of science, and is a Visiting Research Fellow in the Department of History of Art and Screen Media, Birkbeck, University of London, UK.

Culture and Civilization in the Middle East

General Editor: Ian Richard Netton
Professor of Islamic Studies, University of Exeter

This series studies the Middle East through the twin foci of its diverse cultures and civilisations. Comprising original monographs as well as scholarly surveys, it covers topics in the fields of Middle Eastern literature, archaeology, law, history, philosophy, science, folklore, art, architecture and language. While there is a plurality of views, the series presents serious scholarship in a lucid and stimulating fashion.

For a full list of books in the series, please go to https://www.routledge.com/middleeaststudies/series/SE0363

PUBLISHED BY ROUTLEDGE

Ibn al-Haytham's Geometrical Methods and the Philosophy of Mathematics

A History of Arabic Sciences and Mathematics

Volume 5

Roshdi Rashed

Translated by
J. V. Field

Routledge
Taylor & Francis Group

LONDON AND NEW YORK

مركز دراسات الوحدة العربية
CENTRE FOR ARAB UNITY STUDIES

First published 2017
by Routledge

2 Park Square, Milton Park, Abingdon, Oxfordshire OX14 4RN
52 Vanderbilt Avenue, New York, NY 10017

Routledge is an imprint of the Taylor & Francis Group, an informa business

First issued in paperback 2019

Publisher's Note

This book has been prepared from camera-ready copy provided by the
author

British Library Cataloguing in Publication Data
A catalogue record for this book is available from the British Library

Library of Congress Cataloging in Publication Data
The Library of Congress has cataloged volume 1 of this title under the
LCCN: 2011016464

ISBN: 978-0-415-58219-3 (hbk)
ISBN: 978-0-367-86529-0 (pbk)

Typeset in Times New Roman
by Taylor & Francis Books

CONTENTS

FOREWORD

This book is a translation of *Les Mathématiques infinitésimales du IXe* *au XIe siècle*, vol. IV: *Méthodes géométriques, transformations ponctuelles et philosophie des mathématiques*. The French version, published in London in 2002, also included critical editions of all the Arabic mathematical texts that were the subjects of analysis and commentary in the volume.

It also included: 1. *Les emprunts d'Ibn Hūd aux* Connus *et à* L'Analyse et la Synthèse (*Ibn Hūd's borrowings from* The Knowns *and* On Analysis and Synthesis); 2. *Al-Baghdādī critique d'Ibn al-Haytham* (*Al-Baghdādī as a critic of Ibn al-Haytham*), with the critical edition of the Arabic text, French translation and mathematical commentaries.

The whole book, apart from these two appendices, has been translated, with great scholarly care, by Dr J. V. Field. The translation of the primary texts was not simply made from the French; I checked a draft English version against the Arabic. This procedure converged to give an agreed translation. The convergence was greatly helped by Dr Field's experience in the history of the mathematical sciences and in translating from primary sources. I should like to take this opportunity of expressing my deep gratitude to Dr Field for this work.

Very special thanks are due to Aline Auger (Centre National de la Recherche Scientifique), who helped me check the English translations against the original Arabic texts, prepared the camera ready copy and compiled the indexes.

Roshdi Rashed

PREFACE

In his magisterial *Aperçu historique*, Michel Chasles, having given the titles of books and referred to the achievements of the great geometers of the Hellenistic period, writes:

> [...] then, for two or three centuries more, there came the writers of commentaries, who have passed on to us the works and the names of the geometers of Antiquity; then finally the centuries of ignorance, during which Geometry slumbered among the Arabs and the Persians until the Renaissance of learning in Europe.[1]

In this peremptory judgement, Chasles, whose good faith cannot be doubted, was setting out what was known by historians in the mid nineteenth century rather than describing historical facts. At the time, investigations into the history of geometry in classical Islam were few and scattered, and it is not surprising that Chasles' judgement became the received opinion. And indeed we find it repeated tirelessly in the historical introductions to manuals of geometry, such as that by Robert Deltheil and Daniel Caire,[2] as well as in writings penned by historians of geometry, even sometimes up to the present day. Nevertheless, a little later a better, though still far from satisfactory, grasp of the historical facts struck the first blows against this general prejudice. Today, for the majority of historians of mathematics, the opinion for which Chasles was the spokesman has given way to a different one, less absolute without however being more accurate, which could be summarised as follows: Arab geometers, while they never reached the high level of the geometry of the classical Greeks, did at least have the merit of recognising the importance of their work and of having preserved both its spirit and its matter, even going so far as to add several notable details. For this, the names that are mentioned are those of Thābit ibn Qurra and of Naṣīr al-Dīn al-Ṭūsī. Although it is more nuanced, but also more eclectic, this way of looking at things in fact derives from the same logic: to stop at the threshold of questions, without indicating the criteria and setting out the reasoning that could have led to this modest contribution to geometry. It is not clear why, according to the proponents of this theory, the geometers of classical Islam should have thus confined themselves to the role of

---

[1] M. Chasles, *Aperçu historique sur l'origine et le développement des méthodes en géométrie*, 3rd ed., Paris, 1889, p. 23.

[2] R. Deltheil and D. Caire, *Géométrie et compléments*, Paris, 1989.

conscientious preservers of the Hellenistic geometrical heritage, while they were making major advances in all the other disciplines: algebra, number theory, trigonometry, and so on. It is inexplicable that the notable developments in these latter disciplines and in mathematised disciplines such as astronomy and optics should have had so little effect on work in geometry. It is not clear why the single exception made by the historians of mathematics should be for the development of the theory of parallel lines.

To understand how such an opinion came into being, we may point to historians' ideology, the failings of historical research in this area and the huge size of the field of investigation, which is often examined only in part and in a narrowly focussed way; geometers are considered one at a time and their individual contributions are often divided up, which makes it difficult to perceive the underlying mathematical rationality, the more so because the development of geometry in classical Islam may appear in a somewhat paradoxical light.

The geometers of classical Islam are the heirs of the Greek geometers and, one might say, of them alone. Geometry, from the ninth century onwards, is incomprehensible without a knowledge of the works of Euclid, Archimedes, Apollonius, Menelaus and others, which were translated into Arabic. But to understand the linkage between the two phases of the history we first need to make a critical examination of how, from the ninth century on, the geometers took possession of this immense heritage.

This task is enormous; it amounts to finding the relation between Greek and Arabic geometrical work. We hope that the volumes of this book will contribute to carrying out this task, because establishing this relationship is not only necessary for grasping the history of geometry from the Greeks to at least the eighteenth century, but also we cannot manage without it if we wish to make a rigorous assessment of what was contributed by Arabic geometry. This is also the method that must be adopted if we wish to avoid writing in the very worst style for history, namely that of eclecticism: in this style, work written in Arabic is reduced to versions of Greek geometry, or again we discern in it the seeds of future geometry, but always in small pieces and in particular cases.

To look at the wider picture, it seems to be best to look back from the twelfth century, when geometrical research had already been carried out in Arabic for three centuries. Now this picture, while very different from that of the third century BC, is also much larger. In the twelfth century the domain of geometry includes all the area of Greek geometry, but we also find territory that is almost virgin: algebraic geometry represented by the works of al-Khayyām and Sharaf al-Dīn al-Ṭūsī; Archimedean geometry given renewed vigour by more substantial use of arithmetical sums and the employment of geometrical transformations; and it extended into domains hardly glanced at earlier: solid angles, lunes, and so on; the geometry of projections, that is to say the study of projections as a complete subject area within geometry, as it is presented in the works of al-Qūhī and Ibn Sahl; trigonometry (for example, in al-Bīrūnī); the theory of parallel lines, and so on. Some of these subject areas were known to Greek geometers, others were hardly suspected and there were others whose very existence was inconceivable for the Greeks. But it is difficult to draw a map of such a continent. One risks going astray if one proceeds author by author and relies on the books that are available. We need to begin with the research traditions: first to identify them, to make rough reconstructions of them, accepting that the description will need to be filled out later, and to recognise variations and individual styles. Without employing this method, the historian cannot find the patterns of reasoning which govern research work in geometry. If we do not know how to formulate things, the history is obscured and it is impossible to recognise the lines of division that run through it. So we do not see epistemological analysis as an optional luxury: it provides our only means of identifying traditions and styles. This is the task we set ourselves in the volumes of this series. In the first two volumes, we tried to reconstruct the subject area of 'infinitesimal geometry', with its dominant type of reasoning. We shall not repeat here, in summary form, what we set out there in detail; we shall merely note that these mathematicians combined infinitesimal arguments with projection, and infinitesimal arguments with point-to-point geometrical transformations. Moreover, they brought together geometry of position and geometry of measurement, to an incomparably greater extent than had been done in the past. In other works, we have considered Ibrāhīm ibn Sinān, al-Qūhī and Ibn Sahl, who all lived in the tenth century, and we have noted the same things in relation to projections, transformations, geometry of position and geometry of measure. In the third volume of the present series, we have proceeded in the same way: reconstructing the tradition that led to the opening up of a new area in geometry, 'geometrical constructions by means of the conic sections', new criteria for constructability and new means for carrying out constructions (notably the use of transformations).

The introduction of the concepts of transformation and projection as concepts proper to geometry, and (*a fortiori*) the concept of motion, the use of motion in definitions and proofs encouraged geometers to make more extensive use of transformations – which is what Ibn al-Haytham later does in his treatise *On the Properties of Circles*, translated in this volume – and to examine methods for discovery and proof and also to give justifications for making use of these concepts, particularly that of motion. This again is exactly what Ibn al-Haytham turns to in his treatise *Analysis and Synthesis* and in his book *The Knowns*, and it is what explains why he needed to geometrise the concept of place; which he did.

But, to assess these works, and all the others that the reader will find here translated into English for the first time and the subject of commentaries, it was better not to isolate the works from their context and the other writings in the tradition to which they belong. So the reader will find here two texts – one by Thābit ibn Qurra, the other by al-Sijzī – to which Ibn al-Haytham's treatise *Analysis and Synthesis* is related. These two texts were already published, in an unsatisfactory edition, so we made critical editions of them, as rigorously as possible, as well as a French translation that was as precise as possible. Here the two works appear in English for the first time. In the same spirit, we have also included another text by al-Sijzī.

Ibn al-Haytham's *On Place* was the target of thunderbolts hurled by the Aristotelian philosopher 'Abd al-Laṭīf al-Baghdādī, who devoted a complete book to his criticisms. We gave the *editio princeps* and the first translation of this text in *Les Mathématiques infinitésimales* (vol. IV, 2002).

As has been the rule in this series of volumes, Christian Houzel, Directeur de recherche at the Centre National de la Recherche Scientifique, reread the analyses and historical and mathematical commentaries that I wrote to accompany all these texts.

Pascal Crozet, Chargé de recherche at the Centre National de la Recherche Scientifique, did the same for the analysis and commentaries on the texts by al-Sijzī. Badawi El-Mabsout, Professor at the University of Paris VI, has read our commentary on the geometry of triangles. I thank them all very much for their comments and criticisms, which have been of considerable benefit to this work.

I am also grateful to Aline Auger, Ingénieur d'Études at the Centre National de la Recherche Scientifique, who has prepared this book for printing and compiled the indexes.

I also thank Professors S. Demidov and M. Rozhanskaya who helped to arrange for me to visit St Petersburg where I was able to work on the manuscript of Ibn al-Haytham's book *On the Properties of Circles*; Professor B. Rosenfeld who, more than forty years ago, courteously sent me photographs of a very significant part of this St Petersburg manuscript; the Nabī Khān family and Obaidur-Rahman Khān for having allowed me to work on the manuscript of al-Sijzī's texts; and finally Professor Y. T. Langermann for having sent me a microfilm of the text by al-Baghdādī.

Roshdi RASHED
Bourg-la-Reine, December 2001

INTRODUCTION

MOTION AND TRANSFORMATIONS IN GEOMETRY

From the mid ninth century onwards, mathematicians were readier than before to make use of geometrical transformations. The works of al-Farghānī, of the brothers Banū Mūsā – particularly those of the younger one, al-Ḥasan – and those of Thābit ibn Qurra provide the most striking examples. A century later, geometrical transformations have even acquired a group name: *al-naql*, as it is written by al-Sijzī.[1] A careful reading of Ibn Sahl, al-Qūhī and al-Sijzī, for example, shows that geometers were not concerned solely with studying figures but also with investigating relationships between them. Transformations do, of course, appear before the ninth century: for instance they are used by Archimedes and Apollonius.[2] But in the ninth century they are used much more frequently and applied much more widely. There is a noticeable difference between the ancients and the moderns: among the former certain transformations arise in the course of proofs – as can be seen in Archimedes – whereas among the latter a new point of view emerges: transformations are used directly in geometrical investigations. We have had several occasions to draw attention to the emergence of this new attitude, this changed perception of geometrical objects. We have also presented it as one of the consequences of research in geometry becoming more active from the ninth century onwards – but

---

[1] See Appendix, Text 2.

[2] We may, indeed, note that in *On Conoids and Spheroids* Archimedes makes use of an orthogonal affinity; but this book was not known to Arabic mathematicians. On Archimedes' use of this technique, see *Founding Figures and Commentators in Arabic Mathematics*, A History of Arabic Sciences and Mathematics, vol. 1, Culture and Civilization in the Middle East, London, 2012, pp. 347–9. As for Apollonius, it is possible that he made use of some transformations in *On Plane Loci*. All that we know about this book comes from Pappus, and we are not sure what method Apollonius may have used. All the same, later commentators, such as Fermat, recognised, when they were 'reconstructing' it, that the text employed some transformations, including inversion (see R. Rashed, 'Fermat and Algebraic Geometry', *Historia Scientiarum*, 11.1, 2001, pp. 24–47). Ninth- and tenth-century mathematicians certainly did not have copies of this book by Apollonius. Perhaps they had indirect knowledge of the statements of some of its theorems.

not an element in causing the revival. Let us now look at the areas covered by the revival.

The first field in which this new orientation of geometry became apparent rapidly acquired the name 'the science of projection (*'ilm al-tasṭīḥ*)'. This part of geometry separated off from astronomy when it became necessary to establish a firm basis for providing an exact representation of the sphere in order to construct astrolabes. We need to recall two significant historical facts. In the mid ninth century, questions regarding projection were already matters of discussion, or of controversy, in which contributions were made by, among others, mathematicians such as the Banū Mūsā, al-Kindī, al-Marwarrūdhī (astronomer to Caliph al-Ma'mūn) and al-Farghānī.[3] Moreover, insufficient emphasis has been given to the fact that these questions concerning projection were raised and debated by mathematicians who knew the then-recent translation of Apollonius' *Conics*. This intersection of research on projections with the geometry of conic sections can be seen clearly in al-Farghānī's book *The Perfect* (*al-Kāmil*). Al-Farghānī devotes a complete chapter of his book to the geometry of projections, a chapter called 'Introduction to the geometrical propositions by which the figure used for the astrolabe is demonstrated'. In this chapter he presents the first truly geometrical study of conical projections.[4] From al-Farghānī to al-Bīrūnī in the eleventh century, notably through the work of al-Qūhī and Ibn Sahl,[5] we see an increase in the scope and vigour of this geometrical work. We have, in short, the opening up of a

---

[3] See *Géométrie et dioptrique au Xe siècle. Ibn Sahl, al-Qūhī et Ibn al-Haytham* Paris, 1993, pp. CIII–CIV; English trans. *Geometry and Dioptrics in Classical Islam*, London, 2005, p. 337.

[4] In fact, in this chapter, the author undertakes a purely geometrical study of conical projections. Al-Farghānī first proves *a lemma*: the conical projection with pole *P* onto the tangent to a circle at the point diametrically opposite of a chord is a segment of the tangent such that the endpoints of the chord and of the segment lie on a circle that is invariant in the inversion with the same pole *P* that transforms the given circle into the tangent. In the two propositions that follow, al-Farghānī establishes that the projection of a sphere with pole at the point *P* of the sphere, onto the plane tangent to the sphere at the point diametrically opposite or onto a plane parallel to that plane, is a stereographic projection. See *al-Kāmil*, ms. Kastamonu 794, fols 90–94; and R. Rashed, 'Les mathématiques de la terre', in G. Marchetti, O. Rignani and V. Sorge (eds), *Ratio et superstitio*, Essays in Honor of Graziella Federici Vescovini, Textes et études du Moyen Âge, 24, Louvain-la-Neuve, 2003, pp. 285–318.

[5] In these authors we find a purely geometrical investigation of conical projections from an arbitrary point, as well as cylindrical projections. See *Géométrie et dioptrique au Xe siècle*, pp. CVII–CXXV, the treatise by al-Qūhī, pp. 190–230 and the commentary by Ibn Sahl, pp. 65–82. See also R. Rashed, 'Ibn Sahl et al-Qūhī: Les projections. Addenda & corrigenda', *Arabic Sciences and Philosophy*, vol. 10.1, 2000, pp. 79–100.

new field of geometry, an area in which Ptolemy's *Planisphaerium* is, at best, a distant ancestor. And in this new field important contributions are made by the geometrical research on sundials carried out by many mathematicians, among them Thābit ibn Qurra and his grandson, Ibn Sinān.

The second area in which we see the development of the use of transformations also became more active in the mid ninth century, as a result of another encounter, again between the *Conics* – or the tradition stemming from Apollonius – and the Archimedean tradition; that is between a geometry of position and forms and a geometry of measurement. Al-Ḥasan ibn Mūsā and his brothers, as well as their pupil Thābit ibn Qurra, make use of transformations from the start, either in the statements of certain propositions, or in the course of the proof. The Banū Mūsā applied a homothety in their *Book for Finding the Area of Plane and Spherical Figures*,[6] thereby departing from the method used by Archimedes, and they used an orthogonal affinity in their text *On the Cylinder and on Plane Sections*, transmitted by Ibn al-Samḥ.[7] Thābit ibn Qurra employed a cylindrical projection, an orthogonal affinity, and a homothety, as well as a combination of these last two transformations, in his *Book on the Sections of the Cylinder and on its Curved Surface*.[8] In the following century,

---

[6] See *Les Mathématiques infinitésimales du IXᵉ au XIᵉ siècle*, vol. I, pp. 36–7; English trans. *Founding Figures and Commentators*, pp. 42–3.

[7] In his treatise, whose content has been transmitted to us by Ibn al-Samḥ, al-Ḥasan ibn Mūsā defines an orthogonal affinity in relation to the minor axis and an orthogonal affinity in relation to the major axis, and obtains the ellipse as the image of the circle (Propositions 6, 7, 8). From the property of the orthogonal affinity, al-Ḥasan ibn Mūsā shows that for any $n > N$, the ratio, P_n/P'_n, of the areas of the two homologous inscribed polygons, one inscribed in the ellipse of area S and the other in the ellipse of areas S', is equal to the ratio k of the affinity. In other terms the ratio of the areas is preserved when it tends to the limit (*Les Mathématiques infinitésimales du IXᵉ au XIᵉ siècle*, vol. I, p. 885; English trans. *Founding Figures and Commentators*, p. 615). It almost goes without saying that the style is novel and that thereafter it was considered important to give definitions of certain transformations, to investigate their properties and to make use of them in the course of the proofs. This treatise a had a considerable effect on its successors, beginning with the Banū Mūsā's own pupil, Thābit ibn Qurra.

[8] In fact, following the example of his teachers, al-Ḥasan and Muḥammad – two of the Banū Mūsā brothers – Thābit ibn Qurra considerably develops the use of transformations. Thus, in his important treatise *On the Sections of the Cylinder and its Lateral Surface*, he makes extensive use of transformation: orthogonal affinities, homotheties and cylindrical projections. Furthermore, he works by combining transformations – affinity and homothety. We should emphasise that Thābit ibn Qurra does not content himself with merely using these transformations, but also makes a point of establishing certain of their properties. For example, Proposition 10 of this book consists of

Ibrāhīm ibn Sinān makes considerably increased use of geometrical transformations in order to reduce the number of lemmas in his text on *The Measurement of the Parabola*.[9]

Finally, we see increasingly frequent use of geometrical transformations in a third field: geometrical constructions that employ conic sections, as well as constructions for generating conics. For example, this was the procedure adopted in many studies of the regular heptagon, in which we often find an appeal to a similarity.[10] We see the same thing in treatises on the generation of conics, such as the one in which Ibn Sinān uses an

---

establishing that cylindrical projection of a circle onto a plane that is not parallel to the plane of the circle is a circle or an ellipse. To prove this proposition, Thābit ibn Qurra combines projections (see *Les Mathématiques infinitésimales du IXᵉ au XIᵉ siècle*, vol. I, p. 458; English trans. *Founding Figures and Commentators*, p. 333). It is again Thābit ibn Qurra who explicitly introduces motion in his attempt to prove Euclid's fifth postulate. See B. A. Rosenfeld, *A History of Non-Euclidean Geometry. Evolution of the Concept of a Geometric Space*, Studies in the History of Mathematics and Physical Sciences, 12, New York, 1988, pp. 49–56 and C. Houzel, 'Histoire de la théorie des parallèles', in R. Rashed (ed.), *Mathématiques et philosophie de l'antiquité à l'âge classique: Hommage à Jules Vuillemin*, Paris, 1991, pp. 163–79.

[9] Ibrāhīm ibn Sinān follows the example of his grandfather and continues to increase the use of transformations. Thus, he has made masterful use of an equi-affinity (a bijective affinity) in his treatise on *The Measurement of the Parabola* (see *Les Mathématiques infinitésimales du IXᵉ au XIᵉ siècle*, vol. I, p. 675; English trans. *Founding Figures and Commentators*, p. 459). Ibn Sinān shows that this transformation preserves the ratio of *areas* for triangles and for polygons. Then he shows that the same holds for areas with curved boundaries. He also shows that this affinity transforms an arc of a parabola into an arc of a parabola. In another, equally important treatise he makes use of affinities and projections in order to draw the parabola and an ellipse. To draw a hyperbola, he introduces a projective transformation – one that is no longer affine or linear – designed to transform the circle into a hyperbola whose *latus rectum* is equal to the transverse diameter; see *Fī rasm al-quṭūʿ al-thalātha* (*On the Drawing of the Three Sections*), in R. Rashed and H. Bellosta, *Ibrāhīm ibn Sinān: Logique et géométrie au Xᵉ siècle*, Leiden, 2000, pp. 245–62. Geometrical transformations are also frequently used in his other works, such as his book *On Sundials* or his *Anthology of Problems* (*ibid.*, Chapters IV and V).

[10] Mathematicians started by constructing a triangle of one of the types (1, 2, 4), (1, 5, 1), (1, 3, 3) or (2, 3, 2) and then transformed it to inscribe it in a circle. See *Les Mathématiques infinitésimales du IXᵉ au XIᵉ siècle*, vol. III: *Ibn al-Haytham. Théorie des coniques, constructions géométriques et géométrie pratique*, London, 2000, Chapter III; English translation: *Ibn al-Haytham's Theory of Conics, Geometrical Constructions and Practical Geometry*. A History of Arabic Sciences and Mathematics, vol. 3, Culture and Civilization in the Middle East, London, 2013.

orthogonal affinity and an oblique affinity.[11] Abū al-Wafā' follows his example in making a wider use of transformations.[12]

So, after a century, that is in the mid tenth century, transformations were being used more often and in different fields of geometry. And in fact, as we shall see, from this time onwards we find transformations being used in *Anthologies of Problems*, such as those of Ibn Sinān[13] and al-Sijzī,[14]

[11] See note 9.

[12] O. Neugebauer and R. Rashed, 'Sur une construction du miroir parabolique par Abū al-Wafā' al-Būzjānī', *Arabic Sciences and Philosophy*, 9.2, 1999, pp. 261–77.

[13] See R. Rashed and H. Bellosta, *Ibrāhīm ibn Sinān: Logique et géométrie au Xe siècle*.

[14] Like his contemporaries, al-Sijzī often makes use of transformations; see Appendix (p. 529, n. 15), as well as his treatise *Fī taḥṣīl al-qawānīn al-handasiyya al-maḥdūda* (*On Obtaining Determinate Geometrical Theorems*), ms. Istanbul, Reshit 1191, fols 70^r–72^v. In Proposition 3, we notice the use of an inversion, though without its being identified as such by al-Sijzī. He shows that if from a point A of a circle, we draw a chord AB, the tangent AC and the straight line AD such that $D\hat{A}B = C\hat{A}B$, then the straight line AD is the inverse of the circle in the inversion (B, BA^2). We have

$$BA^2 = BH \cdot BE = BD \cdot BG.$$

Now $B\hat{G}A = B\hat{A}C$ (angle enclosed between the chord BA and the tangent AC).

But $B\hat{A}C = B\hat{A}D$, so $B\hat{G}A = B\hat{A}D$; the triangles BGA and BAD are similar, hence $\dfrac{BA}{BD} = \dfrac{BG}{BA}$, so

$$BA^2 = BG \cdot BD.$$

Similarly,

$$BA^2 = BH \cdot BE.$$

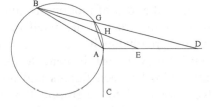

In Proposition 12, he shows that if in a given circle we consider a chord CD with mid point E, AB the diameter through E and two secants through C, we have

$$BA \cdot BE = BG \cdot BI = BC^2$$
$$= BK \cdot BH = BM \cdot BL.$$

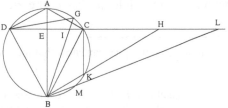

Here we can see that the proposition does indeed deal with the transformation of the circle into a straight line.

This result also corresponds to the inversion (B, BC^2) in which the image of the circle is the straight line DC.

This proposition appears again in the *Anthology of Problems* (fol. 45^{r-v}). Al-Sijzī's proof can be summarised as follows: $H\hat{D}B + C\hat{K}B = 2$ right angles and $D\hat{C}B$ and $L\hat{C}B = 2$ right angles; but $B\hat{D}C = D\hat{C}B$, so $C\hat{K}B = L\hat{C}B$; so the two triangles HCB and KCB are similar, $\dfrac{CB}{BH} = \dfrac{BK}{CB}$ so $BK \cdot BH = CB^2$.

as well as in several others of their writings, or in various works by al-Qūhī – for example, his *Two Geometrical Problems* (*Mas'alatayn handasiyyatayn*).[15] Thus, the use of conical and cylindrical projections, of affinities, of homothety, of translation, of similarity, and sometimes of inversion, was standard practice in the second half of the tenth century. Ibn al-Haytham was to continually extend this practice into further areas of geometry.

One of the major consequences of this new 'transformationist' approach was a *de facto* involvement of motion in statements and proofs of geometrical theorems. This is not a matter of kinematic motion, but of geometrical motion, that is, no account is taken of the time in which the motion takes place. This involvement of motion was confirmed by works that appeared in another area – because it was needed there – the area centred on attempts to prove Euclid's fifth postulate. Here again it was Thābit ibn Qurra who took the decisive step.[16] Ibn al-Haytham was to follow his example, but the basis of his work was more kinematic.[17]

The result was that, at the end of the tenth century, two types of questions were unavoidable. The first type amounted to demanding a justification for geometrical transformations in themselves, starting by trying to find what characterised them. Arguments were required to legitimate the use of transformations, arguments deriving explicitly from geometrical fundamentals. The questions of the second type related directly to the introduction of the idea of motion: asking how it can be admitted in definitions and in proofs, when it has itself never been defined. These two types of questions are obviously closely connected, and in their turn are given additional force by a third: if, henceforth, we are to concern ourselves with the relationships between figures and no longer only study the figures themselves, we need to determine the 'place' for these relationships. So it becomes impossible to leave the question of 'place' undecided and to

---

In the same way, we can show that the triangles *LCB* and *MCB* are similar, hence

$$BL \cdot BM = CB^2.$$

The equation $BA \cdot BE = BC^2$ expresses a property of the right-angled triangle *ACE*. Because triangles *ICB* and *GCB* are similar we have $BG \cdot BI = BC^2$.

In Proposition 5 of the same treatise *On Obtaining Theorems*, al-Sijzī makes use of a homothety; the proposition concerns two circles that touch one another externally. In the *Anthology of Problems* (fols 58v–59r), al-Sijzī again employs a homothety, as he also does in other propositions (for example, on fol. 49v).

[15] See commentary on Proposition 3 of the first part of *The Knowns*, pp. 310–12.

[16] See B. A. Rosenfeld, *A History of Non-Euclidean Geometry*, pp. 50 ff. and C. Houzel, 'Histoire de la théorie des parallèles'.

[17] See B. A. Rosenfeld, *A History of Non-Euclidean Geometry*, pp. 59 ff. and C. Houzel, 'Histoire de la théorie des parallèles'.

accept the Aristotelian notion of 'place' as envelope. These ideas are fundamental to classical geometry, and from the end of the tenth century, specifically with the work of Ibn al-Haytham, the questions become significant sources for reflection and invention. Let us look at the question of motion.

It is a fact that any consideration of motion was banned *de jure* in discussing the elements of geometry. This was the view of Platonic geometers, an opinion dictated by the theory of Ideas; it was also the view of Aristotelian geometers, because they saw mathematical entities as abstractions from physical objects. The true reason for this attitude may well be, rather, that there was little need for a concept of motion in a geometry that was essentially concerned with studying figures. But even when the need is felt, though as yet only slightly, it is not uncommon to avoid the use of motion *de jure* while nevertheless introducing it surreptitiously or involuntarily. This is, indeed, what Euclid did in the *Elements*. He avoids motion, but allows it, in disguised form, when he uses superposition. Superposition cannot actually take place without displacement, even if the displacement is only as seen by the mind's eye. And it is well known that, when Euclid defines the sphere, he opens up the possibility of motion, though as it were against his inclination. Outlawing motion nevertheless remained the watchword for a long time thereafter. We may, for instance, note al-Khayyām's criticism of Ibn al-Haytham's use of motion in his attempt to prove the fifth postulate.[18]

---

[18] *Fī sharḥ mā ashkala min muṣādarāt Kitāb Uqlīdis* (*Commentary on the Difficulties of Some of Euclid's Postulates*), in R. Rashed and B. Vahabzadeh, *Al-Khayyām mathématicien*, Paris, 1999; English translation: *Omar Khayyam. The Mathematician*, Persian Heritage Series no. 40, New York, 2000. In this important text he writes: 'But this is a statement which has no relation whatsoever to geometry for several reasons. Notably, how can the line be moved on the two lines while remaining perpendicular? And how can one demonstrate that this is possible? And notably, what is the relation between geometry and motion? And what is the meaning of motion?' (p. 310; English translation, p. 219).

Al-Khayyām launches a counterattack against those who defend motion in geometry by referring to Euclid's definition of the sphere in *Elements*, XI. He writes: 'The true and obvious definition of the sphere is known, namely, that it is a solid figure which is contained by a single surface, inside of which is a point such that all the straight lines drawn from it to the containing surface are equal. And Euclid deflected from this definition towards what he said by a lack of discernment and by negligence; for he is indeed very negligent in these Books in which he mentions the solids, relying on the student's skill when he would reach them. But if this definition had a meaning, one would have defined the circle by saying, That the circle is a plane figure generated by the rotation of a straight line in a plane surface, in such a manner that one of its two extremities remains fixed in its place while the other gets ultimately to the starting point

It is another matter to make use of motion *de facto*, without worrying about the legitimacy of introducing it. By not mentioning the question of legitimacy, one avoids contradicting earlier opinion – hence the success of this, so to say, practical or even pragmatic use of motion – in transformations by ancient geometers and by those of the tenth to eleventh centuries. In any case, this usage is what prevailed among geometers who worked on transcendental or algebraic curves in Antiquity; and later in Archimedes' work *On Conoids and Spheroids*, as in his *Spirals*; in Apollonius in his *Conics*, and so on. Such use of motion became even more common in the course of the ninth and tenth centuries.

It is a different matter, again, to introduce motion as one of the fundamental notions in geometry. That is to take a positive attitude to motion and its place in definitions and proofs. But to take such a step requires a reformulation of geometrical ideas, or at least of a certain number of them, to re-express them in terms of motion, and also to put new effort into refining the notion of geometrical 'place'. Now, to make such a modification we of course require new assumptions, new modes of thought and a new method, in short a new mathematical discipline. Ibn al-Haytham is, as far as I know, the first to have attempted to carry out this reformulation, introducing the concept of a discipline of 'knowns', by devising an *ars analytica* and redefining the notion of 'geometrical place'. After that we have to wait until the second half of the seventeenth century, with the *analysis situs* of Leibniz, before we find anything similar.

Ibn al-Haytham's geometrical writings can easily be divided into several coherent groups. We have already distinguished some of these: works on infinitesimal geometry; studies of conics, and studies of geometrical constructions carried out with the help of conics; as well as a group of writings that give a theoretical treatment of practical problems in geometry.[19] These groupings of works are not the only ones, and others must also be distinguished. We may note that the firm coherence within each of these groups of writings derives from a substantial tradition of research that Ibn al-Haytham wanted to bring to completion, that is to take it as far as the underlying logical possibilities allowed: to pursue research work jointly in Archimedean geometry and in the geometry of Apollonius; to establish more and more connections between a metrical geometry and a

---

of the motion. However, as people have deflected from that kind of definition because of the existence of motion, and have taken into consideration what should not enter into the art as a principle, we must follow their tracks and be consistent with the demonstrative rules and the universal laws mentioned in the works on logic' (*ibid.*, pp. 310–12; English translation, pp. 219–20).

[19] See *Les Mathématiques infinitésimales du IXe au XIe siècle*, vol. III, Chap. IV.

geometry of position and forms. From the Banū Mūsā to al-Qūhī, Ibn al-Haytham's predecessor, there had been many contributions to this new tradition that is a distinguishing characteristic of Arabic work in geometry. This movement towards a unification of geometry, which Ibn al-Haytham saw himself as carrying forward, obviously could not progress without modifications to inherited ideas and methods, or without giving rise to new types of problem. One of these relates to the idea of motion, in its various forms, and to the legitimacy of the use of motion in geometry. We have noted that, from the time of the Banū Mūsā in the ninth century onwards, the idea of motion was present and was employed either in its own right, as in the work of Thābit ibn Qurra on the fifth postulate, or was used in the form of a geometrical transformation. In the work of Ibn al-Haytham, the part played by motion has become so large, and motion is used so repeated-ly, that it was no longer possible to simply accept it de facto without asking questions about its legitimacy. Ibn al-Haytham devoted a substantial group of writings to such questions.

The first text is his *Commentary on the Postulates in the Book of Euclid*.[20] This *Commentary* is not only valuable for what it tells us about Euclid, it also sheds light on Ibn al-Haytham's intentions by providing us with information about his novel project. This is no less than to establish the principles of geometry, with a view to their including transformations and motion – which means, first of all, providing a justification of Euclid, whose purpose is to free the ideas in the *Elements* from the doubts that were attached to them. So the *Commentary* is not at all an 'Against Euclid', but rather a 'Meta-Euclid'. To justify the ideas of the *Elements* is to look for the ideas that form the basis of the work on the level of epistemology and of ontology. Indeed one would expect no less, given the part the *Elements* play in geometry, not only in the time of Ibn al-Haytham but even as late as the eighteenth century.

So in this *Commentary* Ibn al-Haytham is consciously undertaking a task of providing mathematical explanations. Such an undertaking required him to re-examine the definitions, the postulates and many of the propo-sitions, so as to provide answers to several closely interlinked questions, none of which had been formulated either by Euclid or by his ancient successors. Only 'moderns' had glanced at some of them, and that only in passing. The kinds of question that arise are: What means do we have to establish that a geometrical notion is what it is? Why is it as it is? How do we recognise that it exists? We may note that in these questions 'how' takes precedence over 'why', which allows us to enquire about methods,

---

[20] Ibn al-Haytham, *Sharḥ muṣādarāt Kitāb Uqlīdis*, ms. Istanbul, Feyzullah 1359, fols 150ʳ–237ᵛ.

and that all three questions direct us, whether we will or no, into a difficult process of raising questions about fundamentals. So Ibn al-Haytham the mathematician, necessarily committed to his discipline, that of mathematical research and its development, must double as a philosopher if he wants to succeed in his metamathematical task.

In his *Commentary*, Ibn al-Haytham proceeds in an orderly manner. He begins by setting out the different definitions for each geometrical concept, and takes up the one given by Euclid, which he tries to justify in the following way: he gives a definition of the same concept in which motion is explicitly involved, proves that this is the most appropriate definition, and with the help of this definition goes on to interpret and justify Euclid's definition by going beyond it. On examining his procedure more closely, we may note that the idea of motion enters into Euclid's concept, most notably by providing a new basis for its existence. Motion in fact appears at the heart of a theory of abstraction that has been fundamentally remodelled because of a belief in the independent existence of forms, as the best means of replying to the questions described above, notably to the question concerning existence. Let us briefly examine this point.

We have had occasion to remark upon a demand that seemed to become more and more pressing in the second half of the tenth century, and became a necessity in the work of Ibn al-Haytham: to provide a proof of existence even if one can employ a construction. Thus, when Ibn al-Haytham uses intersections of conic sections to solve solid problems, he takes care to prove the existence of the point of intersection.[21] Moreover, it was not long before he generalised this demand by extending it to definitions. It thus became inevitable to return to geometrical objects themselves, in order to make sure of their existence. This return to geometrical entities themselves necessarily led to questions about their nature: this time what was required was a philosophical account.

The theory that *mathemata* are the result of abstraction, which no one questioned at the time, not even Ibn al-Haytham, can indeed account for them as concepts, but it is certainly not capable of establishing their existence. According to this theory a mathematical object – a triangle, a circle, an angle and so on – is an entity in the mind conceived as separate

---

[21] R. Rashed, 'L'analyse et la synthèse selon Ibn al-Haytham', in *Mathématiques et philosophie de l'Antiquité à l'âge classique*. Études en hommage à Jules Vuillemin, éditées par R. Rashed, Paris, 1991, pp. 131–62. See also *Les Mathématiques infinitésimales du IXe au XIe siècle*, vol. III; English translation: *Ibn al-Haytham's Theory of Conics, Geometrical Constructions and Practical Geometry.*

from matter.[22] A straight line, like a circle, does not exist in the perceptible world – natural or artificial – even though in order to conceive of it we need to start by 'separating' it from the edges of perceptible surfaces. As Ibn al-Haytham notes, we can even provide a perceptible model of it by means of a very fine thread, pulled taut. But this model merely helps the 'imagination' to conceive of the straight line, without in any way establishing that it exists. So the question is how to recognise existence, and to answer it we have to modify the theory of abstraction. That is what Ibn al-Haytham sets out to do.

In the *Commentary* as in another slightly later text – the *Book for Resolving Doubts in the Book of Euclid* – Ibn al-Haytham develops a theory that we may summarise as follows: if the act of 'separating' *mathemata* from the perceptible is necessary for conceiving them, it cannot of itself alone cause them to be grasped as ideal objects, that is as 'invariable intellectual forms', as Ibn al-Haytham expresses it, nor can this act establish that the *mathemata* exist. In other words, abstraction allows us to conceive the straight line as a particular line, the boundary conceived from a whole class of surfaces of perceptible bodies, but not as 'a line that is placed in the same way for all its points',[23] as is laid down in Euclid's definition, with all the disputes it has aroused over translations and interpretations.[24] Next, Ibn al-Haytham invokes the performance of another act, the act of 'imagination'. He himself never says what he meant by this term (which at best is equivocal), a term philosophers had been using since the mid ninth century in a multiplicity of senses. Putting together the various uses Ibn al-Haytham makes of this term, one may deduce the following definition: this is an act by which thought, working on the traces left by natural or artificial objects in the *sensus communis*, within itself isolates from them invariable intellectual forms. So, when given this slant by Ibn al-Haytham, the term 'imagination (*takhayyul*)' seems to be peculiar to thought, as a kind of intellectual vision that works on the traces perceptible objects leave in the *sensus communis*. With this act, *mathemata* are thereafter assured of an existence in thought, and imagination itself takes on double dimensions: those of knowledge and those of existence.

---

[22] This is the theory widely known at the time, accepted by the majority of commentators on Aristotle. See I. Mueller, 'Aristotle's Doctrine of Abstraction in the Commentators', in R. Sorabji (ed.), *Aristotle Transformed: the Ancient Commentators and their Influence*, London, 1990, pp. 463–84.

[23] *Fī ḥall shukūk Kitāb Uqlīdis*, ms. Istanbul, University 800, fol. 4ʳ.

[24] See for example, M. Federspiel, 'Sur la définition euclidienne de la droite', in R. Rashed (ed.), *Mathématiques et philosophie de l'Antiquité à l'âge classique*, pp. 115–30.

Ibn al-Haytham, moreover, lays heavy emphasis upon the ontological dimension when he writes:

> Certainly, things that exist are divided into two classes: those that exist by sensation and those that exist by imagination and judgement. A thing that truly exists is one that does so by imagination and judgement. A thing that exists by sensation is not a thing that truly exists, for two reasons. The first is that the senses often err; and if the senses err, then the person who perceives does not perceive his error, and if the senses err and the person who perceives does not perceive his error, then one cannot be certain of the existence of the truth of what exists by sensation. A thing that exists of which one cannot be certain does not truly exist. That is one of the two reasons. The second reason is that perceptible things are entities subject to corruption, continually changing and not stable; and if they have not got a stable truth, they do not truly exist. In any case, nothing among perceptible things exists, in the strict sense. A thing that exists by the imagination is a thing that exists in the strict sense, because the form grasped in the imagination is imagined according to its true nature and does not change or vary except by the variation of the person who imagines it.[25]

This could not be clearer: ideal mathematical entities, distinct and invariable intellectual forms, exist independently of the subject who apprehends them, even if that person apprehends them in a particular way. In Ibn al-Haytham's eyes, this somewhat Platonic theory, of limited scope (it is true) and with some difficulties, justifies the existence of mathematical forms. A justification is required, but this one is too sweeping to satisfy a working mathematician. So it needs to be supplemented by other causative mechanisms that can demonstrate how these forms are produced in the 'imagination'. Only then can the existence of the forms in thought acquire an operational dimension that allows the forms to be discussed as causative agents.

As a preliminary to the task he is engaged upon, Ibn al-Haytham considers motion in geometry. In this respect he is part of a tradition. About a century earlier, Thābit ibn Qurra had, in fact, firmly and explicitly introduced motion into his writing on the fifth postulate. Thābit is in particular concerned with displacement, which is required if one wishes to discuss superposition. Accordingly he defines the disc as resulting from rotating a segment of a straight line, one of whose two endpoints is fixed.[26] Thābit brings in motion not only in his definitions but also in his proposed proof

---

[25] *Fī ḥall shukūk Kitāb Uqlīdis*, ms. Istanbul, University 800, fols 10ᵛ–11ʳ.

[26] Thābit ibn Qurra, *Fī anna al-khaṭṭayn idhā ukhrijā 'alā aqall min zāwiyatayn qā'imatayn iltaqayā*, ms. Paris, BN 2457, fol. 157ʳ.

of the fifth postulate. To this we should add his use of transformations, to which we have already referred.

As we have said, in his *Commentary* Ibn al-Haytham follows each of Euclid's definitions with another definition involving motion. Thus, after Euclid's definition of the straight line, he writes:

> The most specific and the most perfect among the definitions of the straight line is: if we fix two points and cause it to turn, its position does not change; because by that definition of the straight line we rid ourselves of any doubt that could occur in it.[27]

Now it is precisely by this rotatory motion about an axis or about itself – *infitāl* – that the straight line is distinguished from all other lines that, for their part, change position if they undergo this type of rotation. As Ibn al-Haytham sees it, this definition by motion underpins and justifies Euclid's definition, and this rotatory motion is the means available to the 'imagination' to assure itself that the straight line exists in thought, and to recognise it for what it is. So this definition, later called 'genetic', is 'the most specific and the most perfect', insofar as it gives knowledge of the straight line, not by one or other of its properties, but by its efficient cause. The same procedure is followed for all the other definitions – angle, circle and so on. For example, the circle is defined as the figure generated by the rotation of a straight line about one fixed endpoint, with the second endpoint moving. This rotation also ensures that this object of thought, the circle, exists since it is the efficient cause.

Once Ibn al-Haytham has introduced motion in the definitions, he continues the process by introducing it into the postulates, to justify them or, as he hopes, to prove the fifth one. In this last case, he draws inspiration from Ibn Qurra but, unlike him, chooses a kinematic conception of motion.[28] He proceeds in the same way in many of the propositions.

In the *Commentary*, as in the *Book for Resolving Doubts in the Book of Euclid*, Ibn al-Haytham introduces motion – rotation, displacement and so on – as a fundamental notion in geometry, with the intention of providing justification for Euclid's choice of concepts, to give them a new basis for their existence (that is a new mathematical ontology), so as, in the end, to free them from the doubts and ambiguities that might affect them. In other words, he is writing a commentary on the *Elements*. We shall return to this commentary later. However, introducing motion into geometry imposes two further, complementary, tasks. On the one hand we need to carry out new mathematical research on geometrical transformations and, on the

---

[27] *Sharḥ muṣādarāt Kitāb Uqlīdis*, ms. Istanbul, Feyzullah 1359, fol. 155ᵛ.
[28] *Ibid.*, see in particular fol. 162ᵛ.

other hand, we need to find ways to think systematically through geometry as a whole starting from the concept of motion. This project, conceived and set in train by Ibn al-Haytham, was taken up in the seventeenth century by several mathematicians such as Fermat, La Hire and Leibniz. However, we have to wait two more centuries before the project is fully realised, that is with the introduction of the idea of a group into geometry in the course of the last third of the nineteenth century.

To accomplish the first of these tasks, Ibn al-Haytham wrote several studies of transformations; see, among them his book *On the Properties of Circles*, in which he investigates affine properties and homothety. It seems to have been in the same spirit that he composed his book *On the Properties of Conic Sections*. To make it possible to think systematically about motion in geometry, Ibn al-Haytham invents a new geometrical discipline: The Knowns, and proposes an *Ars analytica* of them in another book, *Analysis and Synthesis,* that he himself presents as exemplifying the method employed in this discipline. Here again, Ibn al-Haytham is in fact following in a tradition, one started by Thābit ibn Qurra in his text *On the Means of Arriving at Determining the Construction of Geometrical Problems*, a tradition that was taken up by Ibn Sinān in his book *The Method of Analysis and of Synthesis in Geometrical Problems*, and then followed by al-Sijzī in an attempt to devise an *ars inveniendi*.

But motion and transformations are obviously not compatible with a conception of 'place' as the limit of the surrounding body. Another, more abstract representation is necessary in order to take account of the fact that a body can change form while retaining the same quantity of volume. We need the concept of place as homogeneous, unaffected by the forms an object can take, and, moreover, a concept that lends itself to mathematical treatment. That is precisely what Ibn al-Haytham is trying to provide in his text on place, and, as far as I know, for the first time.

So, apart from the two books concerned with Euclid's *Elements*, this fifth volume, dealing with transformations and geometrical methods as well as the philosophy of mathematics – not that of the philosophers but that of the mathematicians – includes all of Ibn al-Haytham's writings on the philosophy of mathematics that have come down to us. Thus, we have:

1. *The Properties of Circles*
2. *The Knowns*
3. *Analysis and Synthesis*
4. *A Geometrical Problem*
5. *The Properties of the Triangle*
6. *On Place*.

In the original French edition of the present volume, *Les Mathématiques infinitésimales*, vol. IV, we gave the *editio princeps* of these writings as well as the first translation of them and the first historical and mathematical commentary on them. Here, we present the first translation into English. Notes and references have been brought up to date.

But, to place Ibn al-Haytham's writings in their context, we also include the writings of Thābit ibn Qurra and of al-Sijzī that were mentioned above.

CHAPTER I

THE PROPERTIES OF THE CIRCLE

INTRODUCTION

The long list of mathematical works written by Ibn al-Haytham includes several that are still missing. Among these are three that, in the mathematics of infinitesimals, speak for themselves: *The Greatest Line that can be Drawn in a Segment of a Circle*, a *Treatise on Centres of Gravity* and a *Treatise on the Qaraṣṭūn*. These treatises are all concerned with the geometry of measure. Their absence not only deprives historians of mathematics of facts that would have helped them to appreciate more clearly the range of Ibn al-Haytham's œuvre, but also, more seriously, it makes it absolutely impossible for them to understand the structures of this œuvre and the network of meanings that they carry. If the first of the treatises cited above had been at our disposal, we should have a better understanding of the distance the author of a treatise on problems of figures with equal perimeters, on figures with equal areas and on the solid angle, travelled along the road of what was later to be called the calculus of variations.

This state of affairs is not peculiar to the geometry of measure; it is found also in the other type of geometry developed by Ibn al-Haytham and his predecessors: the geometry of position and forms. Among the books that until very recently were still missing we have one with the title *On the Properties of Circles*. A book with such a title is of course intriguing and surprising.[1] We ask ourselves what Ibn al-Haytham might deal with in a book whose title appears so strikingly modern. His predecessors, his contemporaries and Ibn al-Haytham himself wrote books and papers on one or another aspect of a geometrical figure, for example triangles, but rarely on all its properties taken together as a whole. Furthermore, Ibn al-Haytham had written more than once on the circle, on finding its perimeter and finding its area. We may ask what reasons he might have had to return to the subject of the circle.

---

[1] Another book by Ibn al-Haytham on conic sections, unfortunately missing, has an analogous title: *Fī khawāṣṣ al-quṭū'* (*On the Properties of Conic Sections*).

These were the kinds of questions that could have been asked, until I was able to produce a copy of the treatise, and establish a text of it, though a rather damaged one. Ibn al-Haytham's short introduction could not fail to sharpen the reader's curiosity and raise questions. The author indeed proposes to investigate the properties of the circle, or at least a certain number of them, since 'the properties of circles are numerous, and their number is almost infinite' (p. 87). He promises not to include in this treatise properties that have already been discovered. There is even a request to the reader that if, in the course of his reading, he happens to come upon a result that is already obtained elsewhere, he will see this as no more than a coincidence produced without the author's knowledge. Thus, Ibn al-Haytham explicitly lays claim to novelty and originality.

Thus, the question becomes: where does Ibn al-Haytham see this novelty? A mathematician of his standing, of his universally inventive genius, could not describe as *new* a result that was secondary or partial: only an idea he considered fundamental could be called 'new'. This last statement is not a *petitio principi* on our part, but the conclusion of a sufficiently long analysis of similar situations in the mathematical and optical works of Ibn al-Haytham. If it does sometimes happen that Ibn al-Haytham makes a mistake when proving a result, he always has a sharp eye in relation to the value of his programme of research.

And we shall in fact show that in this book Ibn al-Haytham did not confine himself to dealing with metrical properties of the circle, but also considered affine properties. It is as if he had intended to carry out a systematic exploration of the properties of the circle, and to classify them; which then has led him to investigate harmonic division and above all to devote about a third of the book to affine properties – similar ranges and, in particular, homothety. As far as I know, this is the first treatise in which this last form of geometrical transformation is studied in its own right.

The book sheds light on an important characteristic of Ibn al-Haytham's research in geometry: his interest in geometrical transformations. As we have already pointed out, it is precisely this research that Ibn al-Haytham is pursuing in his book *The Knowns*. Four propositions from the treatise *On the Properties of Circles* reappear in *The Knowns*, and it is in any case very likely that the latter treatise was composed after the former one. *The Knowns* is closely connected[2] with another of Ibn al-Haytham's treatises, *On Analysis and Synthesis*, and his interest in transformations must have encouraged him to return to the concept of place – which is what Ibn al-Haytham worked on in a short paper that has survived – so it seems

---

[2] See below, p. 231.

that *On the Properties of Circles* belongs to a substantial and homogeneous group of treatises.

The above statements are concerned with facts, titles and names, and are thus verifiable. It seems they might shed light on the novelty to which Ibn al-Haytham lays claim. But perhaps we should be wary of distorting the account of his thought by using the term homothety when he himself does not use such a word in his writings. We might be committing the cardinal sin of anachronism. The case would be further aggravated if we take into account that the term is not yet in use even at the end of the eighteenth century – it is indeed not to be found either in the *Encyclopédie* of d'Alembert and Diderot, or in the writings of mathematicians of the time – for example those of Euler and Clairaut. It is not until we come to Michel Chasles that the word 'homothety (homothétie)' appears, employed to designate a similarity of both form and position.[3] All the same, it would hardly be reasonable to deny that mathematicians working earlier than the 1830s had any knowledge of homothety. In fact, in the history of mathematical concepts, such exclusive attitudes are often adopted at the cost of rather harsh simplifications; and, one might say, here as elsewhere, it does not greatly matter if we are taxed with anachronism, whatever judgment one wishes to convey by the term. On the other hand, it seems as important as it is difficult to discern what degree of rational awareness Ibn al-Haytham might have had of this concept, in work that followed on from that of Euclid, Pappus, the Banū Mūsā, Ibn Qurra, Ibn Sinān, al-Būzjānī, al-Qūhī, al-Sijzī and others, and preceded the work of Fermat and many later mathematicians. So the best approach is to examine the final group of propositions in his book, which deal with this concept, before setting about making comparisons with the works of his predecessors.

1. *The concept of homothety*

As we have already noted, in his book *On the Properties of Circles*, Ibn al-Haytham considers similar divisions, homothetic triangles and harmonic divisions and pencils, before again addressing homothety in his last ten propositions. A consecutive commentary on all the propositions, and in particular on the last ten, will be given below. Here we merely wish to point out the salient characteristics of this research on homothety, so as to

---

[3] M. Chasles, *Aperçu historique sur l'origine et le développement des méthodes en géométrie*, Paris, 1889, p. 597, note. This is his memoir on the two general principles of geometry: duality and homography.

get a better grasp of the idea Ibn al-Haytham may have had of this transformation.

So let us begin with Proposition 32. Ibn al-Haytham has two circles that touch one another – internally or externally (in the latter case equal or unequal circles) – and he intends to prove that certain elements are the transformed versions of others. In precise terms, let AC and CE be the two diameters from the point of contact C, and let CBD be a secant; we have

(1) the arcs BC and DC are similar,
(2) the arcs AB and DE are similar,
(3) $\dfrac{CB}{CD} = \dfrac{CA}{CE}$.

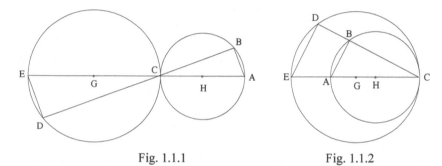

Fig. 1.1.1   Fig. 1.1.2

The reasoning emphasises that AB and ED are parallel, and the results in the statement follow immediately.

Ibn al-Haytham thus proves that to every secant straight line passing through C there correspond points B, D such that relation (3) holds. Now this ratio corresponds to the homothety with centre C and ratio $k = \pm\dfrac{R_H}{R_G}$ (where R_H et R_G are the radii of the two circles).

What we should note here is that Ibn al-Haytham's procedure does not simply involve using homothetic triangles. He starts with two tangent circles, both given, and is seeking to prove that one is the transform of the other in a homothety, in order to deduce from this some correspondences between arcs. This procedure is different from that of his predecessor, al-Sijzī, who seems to make no deductions about the circles and the arcs.[4] But it is also obviously different from the approach that starts from a single figure and finds another as its transform. In this last case, the homothety can be used heuristically, which is not so in the previous case.

[4] See p. 5, n. 14.

The novelty of Ibn al-Haytham's procedure consists, at least partly, in his identifying the elements of the homothety, its centre and its ratio. Thus, in Proposition 35, he again starts with two circles, but they are unequal and tangent externally. He draws the common exterior tangent and forms two homothetic triangles. He then characterises the position of the point of intersection of that tangent and the line of centres by means of a ratio. This gives the centre and the ratio of the homothety. He then deduces that the homologous radii are parallel and the homologous arcs are similar. Ibn al-Haytham does not stop there: once he has defined these concepts, he applies them to the other cases of the figure, taking as his starting point the point that divides the line segment joining the centres in the ratio of the radii, externally or internally; a point and a ratio that are none other than the centre and the ratio of one or other of the two homotheties in which one of the circles appears as the transform of the other.

Let us briefly return to Ibn al-Haytham's method of proceeding; we run the risk of repeating ourselves, but our purpose is to understand his approach better. In Proposition 35, as later in Propositions 39 and 40, he continues to start with two circles $\mathscr{C}_1$ and $\mathscr{C}_2$, tangent externally or separate and unequal, as in Proposition 39. Let us consider – Proposition 35 – EE' as a common tangent to the two circles; it cuts the line of centres HI in a point K beyond H. Ibn al-Haytham's first concern is to determine the property of the point K.

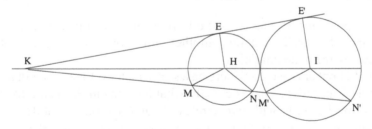

Fig. 1.2

From the property of the tangent EE', he deduces that $EH \parallel E'I$, from which he deduces that the two triangles KEH and $KE'I$ are homothetic; hence

(1) $$\frac{KI}{KH} = \frac{R_I}{R_H}.$$

Ibn al-Haytham then proves that any secant passing through K to cut $\mathscr{C}_2$ in M and N and that cuts $\mathscr{C}_1$ in M' and N', the points homologous to them, defines similar arcs MN and $M'N'$. His reasoning proceeds as follows: for a point K, centre of the positive homothety $h\left(K, \dfrac{R_I}{R_H}\right)$, the following two properties are equivalent.

1) K is the point of intersection of the line of centres with the external common tangent;

2) K lies on the straight line HI such that $\dfrac{\overline{KI}}{\overline{KH}} = \dfrac{R_I}{R_H}$.

Ibn al-Haytham's reasoning is the same in Propositions 39 and 40, where the circles are separate; the point K on the extended part of HI is defined by (1). In Proposition 39, Ibn al-Haytham proves that if KE is a tangent to $\mathscr{C}_2$, then it is a tangent to $\mathscr{C}_1$; that is $E' \in \mathscr{C}_1$ and KE' is a tangent to $\mathscr{C}_1$ at E'. Here, as in the following proposition, he proves that to any element of $\mathscr{C}_2$ (a point, an arc, a radius, a tangent, an angle and so on) there corresponds a homologous element of $\mathscr{C}_1$.

We may note that Ibn al-Haytham deals only with circles that touch one another, externally or internally, and separate circles, but never circles that intersect. This might have been a restriction designed to make the procedure easy. But it is not. The study of the homothety $h\left(K, +\dfrac{R_I}{R_H}\right)$ carried out in Propositions 35 and 39 would apply in an identical manner in the case where the circles cut one another; this application could not pass unnoticed by the mathematician.

Ibn al-Haytham also considers cases in which, in today's language, the homothety is negative. That is exactly what happens in Propositions 36 and 37. Here too he starts with two circles $\mathscr{C}_1$ and $\mathscr{C}_2$, one separate from the other, equal or unequal, and a point K on the segment IH such that

$$\frac{\overline{KI}}{\overline{KH}} = -\frac{R_I}{R_H};$$

he proves that if KE touches $\mathscr{C}_2$ at E, then it touches $\mathscr{C}_1$ at E' such that $h(E) = E'$. He proves that any secant passing through K cuts off two similar arcs on $\mathscr{C}_2$ and $\mathscr{C}_1$. In short, the two homotheties of ratio $\pm\dfrac{R_I}{R_H}$ are studied for circles that touch externally and for circles that are separate from each other. As for circles that touch internally, in Proposition 32 Ibn al-Haytham studies the positive homothety. He returns to this study for circles touching internally in Proposition 43, but does not refer to the negative homothety for these last cases.

More generally, in all the research work incorporated into *On the Properties of Circles*, Ibn al-Haytham is much concerned with the properties of common tangents and proves that such tangents pass though one of the centres of homothety; he also emphasises the fact that homologous radii are parallel, the fact that angles at the centre are equal and repeatedly refers to homologous angles, whose homologous arcs he deduces are similar. That is, he emphasises the properties of the homothety, which itself now becomes the object of study. He was no doubt able to deduce from this that the two chords of the arcs were parallel and to find the ratio between them, one chord being that joining two arbitrary points on one of the circles, the other chord being that joining the homologous points of the second circle. Noticing that such chords were parallel would have simplified the investigation of the fact that certain straight lines are at right angles to one another, which plays a part in Propositions 35, 38, 41 and 42.

2. *Euclid, Pappus and Ibn al-Haytham: on homothety*

Ibn al-Haytham's contribution to formulating the concept of a homothety is not confined to what can be found in his book *On the Properties of Circles*. But, before examining the corrections and generalisation he later introduced, we should take a brief look at the work of Euclid and of Pappus, to try to find possible relationships with the concepts they employ. Such investigation is required because it has been suggested that the same concept is indeed to be found in the work of these mathematicians. For Euclid, the reference is to Propositions 2, 5 and 6 of the sixth book of the *Elements*. These propositions consider two straight lines cut by two parallel straight lines.

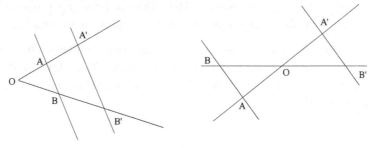

Fig. 1.3

In Proposition VI.2 we have

$$\frac{OA'}{OA} = \frac{OB'}{OB},$$

and in Propositions VI.5 and VI.6 we have

$$\frac{OA'}{OA} = \frac{OB'}{OB} = \frac{A'B'}{AB}.$$

Similar triangles such as OAB and $OA'B'$ are what we call homothetic triangles. All the same, it is clear that Proposition VI.2 – which, moreover, serves as a basis for other propositions – is a special case of what is known as Thales' theorem for two parallel straight lines. So we cannot identify this case with the one that appeals to homothetic triangles, with the centre and the ratio of a homothety, as in Propositions 11 and 26 of the treatise by Ibn al-Haytham. And it is precisely when we have the idea of the latter that we recognise the former; it is once we have an idea of that transformation, or at least of the correspondence between the two figures, that we identify Euclid's results as an application of the transformation, but surely not conversely. Moreover, a knowledge of the property of homothetic triangles allowed Ibn al-Haytham to deduce the property of similar divisions on parallel straight lines, which are indeed called homothetic divisions. This is to say that the new property is fertile, and it is the property Ibn al-Haytham uses in Propositions 4 and 6. In short, Euclid's work does not anticipate homothety, but rather is included in homothety.

One might ask whether the situation is different in regard to Pappus' *Mathematical Collection*. It has been said this is so, at least for Propositions 102, 106 and 118 of the seventh book.

In the first two of these propositions, Pappus reasons in the same way. So it is enough to take Proposition 102. This is how the proposition is presented in the wording employed by the Alexandrian mathematician:

Let there be two circles $AB\Gamma$, ΔEB touching one another at the point B; let us draw through the point B straight lines $\Gamma B\Delta$, ABE, and let us draw the straight lines that join $A\Gamma$, ΔE; I say that the straight lines $A\Gamma$, ΔE are parallel.[5]

[5] Pappus d'Alexandrie, *La Collection mathématique*, French trans. P. Ver Eecke, Paris/Bruges, 1933, p. 638.

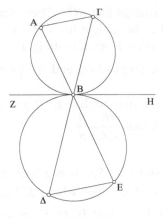

Fig. 1.4

To prove this proposition, Pappus begins from *Elements* III.32; if *HBZ* is the common tangent at the point *B*, he obtains

$$A\hat{B}Z = A\hat{\Gamma}B \text{ and } H\hat{B}E = B\hat{A}E;$$

but $A\hat{B}Z = H\hat{B}E$, so $A\hat{\Gamma}B = B\hat{A}E$, hence $A\Gamma \parallel \Delta E$.

We have seen that Ibn al-Haytham established a proposition close to this one – but not identical with it – employing a different approach. Unlike Pappus, he immediately turns his attention to the homothetic triangles *BAL* and *BEK*. Better still, he determines the centre and the ratio of homothety.

Proposition 118 of Book VII is the one most often cited in connection with the question of homothety. The statement is:

> Let there be two circles *AB*, *ΓΔ*; let us extend the straight line *AΔ* and let us make it such that the straight line *EH* is to the straight line *HZ* as the radius of the circle *AB* is to the radius of the circle *ΓΔ*; I say that, if a straight line drawn from the point *H*, to cut the circle *ΓΔ*, is extended, it also cuts the circle *AB*.[6]

It has already been pointed out that the statement is not perfectly precise.[7] Pappus in fact begins his proof by saying:

> [...] let us draw from the point *H* the straight line *HΘ* tangent to the circle *ΓΔ*; let us draw the straight line joining *ZΘ*, and let us draw the straight line

[6] Pappus, *La Collection mathématique*, p. 657.
[7] *Ibid.*, p. 657, note 3 by Ver Eecke.

EK parallel <to the straight line *ZΘ*>. Then, since the straight line *EK* is to the straight line *ZΘ* as the straight line *EH* is to the straight line *HZ*, the line passing through the points *H*, *Θ*, *K* is straight.[8]

Let us return to the proposition and try to make the most of Pappus' text.

What we are given is two circles (E, R_E) and (Z, R_Z) and a point *H* on the straight line *EZ* such that $\dfrac{HE}{HZ} = \dfrac{R_E}{R_Z}$.

a) If *HΘ* is a tangent to (Z, R_Z), it is a tangent to (E, R_E). To prove this statement, Pappus draws $EK \parallel Z\Theta$, where *K* lies on the circle (E, R_E). So we have $\dfrac{EK}{Z\Theta} = \dfrac{HE}{HZ}$; it follows that *H*, *Θ*, *K* are collinear, and that the angle *K* is a right angle.

b) A secant drawn to the circle (Z, R_Z) cuts it between *Δ* and *Θ*; if we extend it, it passes between *B* and *K*; now *HK* is a tangent to (E, R_E), so the secant also cuts the circle (E, R_E).

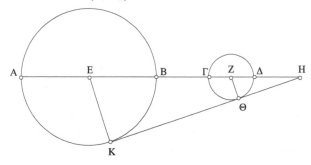

Fig. 1.5

So we see that Pappus starts with two parallel radii, and from the equality of two ratios, given as equal by hypothesis, and that he draws his conclusion without appealing to the homothetic triangles. In contrast, in a closely similar case (see Propositions 39 and 40) Ibn al-Haytham draws attention to these triangles and uses the property of the homothety.

---

[8] Pappus, *La Collection mathématique*, pp. 657–8.

3. Ibn al-Haytham and homothety as a point by point transformation

All in all, it is very difficult to read Pappus' texts as an application of homothety. So it seems Ibn al-Haytham might have been the first to have used homothety, before then going on to study it in its own right. This too turns out to be far from the truth. If we confine our attention simply to Ibn al-Haytham's predecessors from the ninth century onwards, notably those who worked on geometrical transformations, we find they did indeed genuinely make use of homothety in their writings on infinitesimal mathematics as well as in those on geometrical analysis. For example, in the ninth century the Banū Mūsā made use of homothety in their study of concentric circles and regular polygons.[9] Similarly, Thābit ibn Qurra used homothety for concentric circles and ellipses in his work on plane sections of the cylinder.[10] Similarly, others in the tenth century, such as Ibn Sinān, al-Qūhī and al-Sijzī,[11] in considering problems of geometrical analysis, used homothety before Ibn al-Haytham did. In this respect, we can again cite al-Qūhī and al-Sijzī, to look no further. Here, as elsewhere, Ibn al-Haytham's work appears as the final stage in a tradition of research that is already a century and a half old. So, at least in historical terms, it is understandable why Ibn al-Haytham included this transformation and its

---

[9] R. Rashed, *Les Mathématiques infinitésimales du IXe au XIe siècle*, vol. I: *Fondateurs et commentateurs: Banū Mūsā, Thābit ibn Qurra, Ibn Sinān, al-Khāzin, al-Qūhī, Ibn al-Samḥ, Ibn Hūd*, London, 1996, Chapter I, p. 37; English translation: *Founding Figures and Commentators in Arabic Mathematics*, A History of Arabic Sciences and Mathematics, vol. 1, Culture and Civilization in the Middle East, London, 2012, p. 43.

[10] *Ibid.*, for example Chapter II, pp. 475–6; English trans. p. 352.

[11] We have mentioned several times that Ibn Sinān frequently employs geometrical transformations, both in his works on infinitesimal mathematics and in his research on conics. Among these numerous transformations, we also find homothety. See R. Rashed and H. Bellosta, *Ibrāhīm ibn Sinān: Logique et géométrie au Xe siècle*, Leiden, 2000, for example pp. 486–7, 551–2, 719–20. Ibn Sinān makes use of homothety, but without really investigating its properties as Ibn al-Haytham was to do. Ibn Sinān's successor al-Qūhī, who took research on projections much further than his predecessors, also concerned himself with transformations and with homothety. Thus, in the first three propositions of his paper called *Two Geometrical Problems*, he gives the result stated and proved by Ibn al-Haytham in Proposition 3 of *The Knowns* (*al-Mas'alatayn al-handasiyyatayn*, ms. Istanbul, Aya Sofya 4832, fols 123^v–124^v; see the note on Proposition 3, p. 310). Al-Qūhī's younger contemporary, Aḥmad ibn 'Abd al-Jalīl al-Sijzī, in turn continues the use of transformations. Better still, he isolates the concept of transformation in its own right as an auxiliary method in analysis and in synthesis (see Appendix, Text 2). In various places he employs homothety, similarity, and even a primitive form of inversion (for homothety, see below, Proposition 32, note 22).

applications in his treatise *On the Properties of Circles*. In this book, we see homothety being put to use as a technique for studying the correspondences between two figures; but it is much more significant that we are also witnessing the first known investigation into certain properties of the transformation: under homothety an arc becomes an arc, a radius becomes a radius, an angle between two straight lines becomes an angle between the two homologous straight lines; for two arcs related by homothety the tangents at homologous points are parallel, and so on.

This, it seems, is where the novelty of Ibn al-Haytham's work is to be found. It would, however, be a mistake to ignore a limitation that is intrinsic to his conception that as yet – in this book on *The Properties of Circles* – prevents him from seeing homothety as a true point by point transformation. We have already noted that Ibn al-Haytham starts with two circles in order to prove that one is the transform of the other. Moreover, there is no proposition in his book in which he starts with the centre of homothety, the ratio of the homothety and a circle in order to then find another circle as the image of the first one. But if we do not look beyond this limitation, we are forgetting the place this book by Ibn al-Haytham occupies among his other works, and underestimating the intrinsic dynamics of on-going research in mathematics. Ibn al-Haytham's treatise, as we now know, forms part of a group of writings in which he concerns himself with the geometry of transformations. Composing these works seems to be a necessary response to needs arising from various changes that affected the internal relationships between mathematical disciplines, and from the new outlook in some disciplines. We may note here, without elaborating further, that there was an increasingly close interpenetration between an Archimedean tradition of geometry and a tradition of the geometry of position and form. We may also take note that there was an awareness of algebra, direct or indirect but always huge. Even Ibn al-Haytham, a geometer par excellence, wrote on algebra.[12] It is as these researches unfold that geometrical transformations appear more and more as being a new field of geometry; and it is at the end of this development that Ibn al-Haytham writes this group of books, to which his treatise *The Knowns* also belongs.

To attempt to illustrate this conceptual relationship, simply in regard to homothety, we shall return to a work by Ibn al-Haytham in which this affine transformation plays a part, not only in plane geometry but also in

---

[12] *Les Mathématiques infinitésimales du IXe au XIe siècle*, vol. II: *Ibn al-Haytham*, London, 1993, Tableau récapitulatif, no. 90, p. 532; English trans. *Ibn al-Haytham and Analytical Mathematics*. A History of Arabic Sciences and Mathematics, vol. 2, Culture and Civilization in the Middle East, London, 2012, List of Ibn al-Haytham's works, p. 420.

geometry in three dimensions; we shall then conclude with his study of this transformation in *The Knowns*.

In his work on figures with equal perimeters and equal surfaces (isoperimetric and isepiphanic figures), in which he presents the first account of a solid angle, Ibn al-Haytham employs homothety to obtain a sphere from another sphere. This is not the place to return to Ibn al-Haytham's proof, of which we have already given an analysis.[13] Here we shall confine ourselves to looking at some elements relating to the way homothety works.

Ibn al-Haytham starts from a sphere with centre A, and in this sphere two pyramids with vertex A, whose bases are similar regular polygons. We can take the planes of these polygons to be parallel; in this case the centres of their circumcircles, B and E, lie on a straight line through A. In these circles we can also take the radii corresponding to the vertices of the polygons as parallel two by two; this we have $BC \parallel EG$ and $BD \parallel EH$. In this case triangles CBD and GEH are similar. So if $BF \perp CD$ and $EM \perp GH$, we have

$$\frac{BF}{EM} = \frac{BC}{EG}.$$

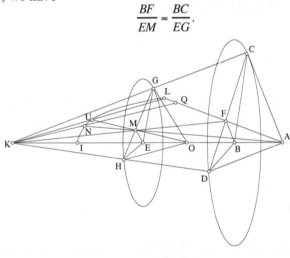

Fig. 1.6

Ibn al-Haytham proves that the point K, the point of intersection of FM and AE, is the centre of the homothety with ratio $\dfrac{KB}{KE}$. This homothety is

---

[13] *Les Mathématiques infinitésimales du IXe au XIe siècle*, vol. II, pp. 374–6 and vol. III; English trans. *Ibn al-Haytham and Analytical Mathematics*, pp. 289–95. The proof is long and complicated: five large pages. The figure has seventeen distinct points, eighteen solids, eight curves and thirty-five different straight lines.

then used throughout the proposition. He proves that this homothety transforms B, C, D into E, G, H respectively; that the planes (BCD) and (EGH) are homologous, as are the chords CD and GH. He then proves that the two angles CAD and GOH are equal if O is the transform of A. These angles are angles at the centre in two different spheres (A, AC) and (O, OG); so the planes (CAD) and (GOH) cut these spheres in similar arcs CLD and GUH. To show this, he proves that the two spheres and the two planes are homologous in this same homothety, $h\left(K, \dfrac{KB}{KE}\right)$.

In fact, Ibn al-Haytham starts from the sphere (A, AC) using the previous homothety, and obtains the sphere (O, OG), then he proves that the plane (ACD) is homologous to the plane (OGH) and that the arc CLD is homologous to the arc GUH.

Thanks to widening its application to include plane figures as well as three-dimensional ones and its explicit use as a geometrical transformation, the status of homothety as a point by point transformation seems to be unambiguous. This is precisely what is confirmed by the study that Ibn al-Haytham carries out in *The Knowns*.

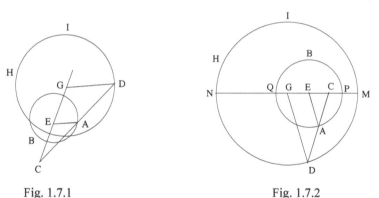

Fig. 1.7.1   Fig. 1.7.2

The book followed naturally from this new research in geometry, and to some extent serves to complete it. In this book, Ibn al-Haytham studies the variability of elements of the figures and their transformations. It is lines that provide the basis for the theoretical treatment he puts forward in his work. In this connection, he returns to homothety in at least seven propositions in the first chapter of the book, as well as in the second chapter. So let us consider a single example, the first one, to illustrate his ideas. In Proposition 3 of the first chapter, he starts with a circle $\mathscr{C}(E, R)$, an arbitrary point C distinct from E, and a point A on the circumference of the circle. With the point A he associates a point D on the extended part of

CA and such that $\dfrac{CA}{AD} = k$; he then proves that D lies on another known circle. Thus, he proves that D is the image of A in the homothety $h\left(C, \dfrac{k+1}{k}\right)$. That is to say that D lies on the circle with centre $G = h(E)$, $CG = \dfrac{k+1}{k} CE$, and radius $R_1 = \dfrac{k+1}{k} R$.

It is hardly necessary to repeat that here, as in the preceding example, homothety appears as a point by point transformation. Indeed Ibn al-Haytham seems to want to confirm this idea, by proving what is more or less the converse in the following proposition – the fourth: if a straight line from the centre of the homothety, C, cuts the first circle in a point A, it will cut the second circle in D, and we shall have $\dfrac{CA}{AD} = k$.

This is, moreover, the form in which the concept of a homothety appears after Ibn al-Haytham, for example in the work of Fermat. In the first proposition of his book *The Reconstruction of the Two Books on Plane Loci by Apollonius of Perga* (*Apollonii Pergaei libri duo de locis planis restituti*), Fermat proves that the homothetic image of a straight line is a parallel straight line and that the homothetic image of a circle is a circle.[14]

The history of the concept of a homothety from Euclid to Ibn al-Haytham and then to Fermat cannot be written as that of the prefiguration of a concept; it is rather the history of a double transition, a matter of gradual progression on the technical level, but somewhat abrupt on the theoretical one: from a correspondence between figures to the transformation of a figure, from technical use in the course of a proof to the study of the properties of the transformation. But, if we want to understand this double development, we must widen our scope beyond the narrow frame of the history of a concept. While not allowing ourselves to be led astray by the romantic notion of a complete history, we need, in this case, to situate homothety within the geometry of transformations, of which certain traces can be seen in the work of Archimedes and Apollonius, before it becomes a defined area of geometry from the mid ninth century onwards, and develops further in far distant climes in the seventeenth century. We need to remember that for Ibn al-Haytham homothety appears at the same time as other affine and projective transformations; and that later, in the book by Fermat that we have just cited, homothety is connected with similarity –

---

[14] *Œuvres de Fermat*, publiées par les soins de MM. Paul Tannery et Charles Henry, Paris, 1896, vol. III, pp. 3–5.

notably with inversion. To the educated eye, the character of this intellectual landscape is not at all Hellenistic.

To this conclusion we may add another: although establishing the textual tradition was a necessary condition for tracing the evolution of the concept of homothety in Ibn al-Haytham's works in the eleventh century, it is nevertheless the conceptual relationship that provided answers to the questions that arose in investigating the history of the text: it is highly likely that the text *On the Properties of Circles* preceded *The Knowns*.

4. *History of the text*

Ibn al-Haytham's treatise *On the Properties of Circles* (*Fī khawāṣṣ al-dawā'ir*) appears in the list of his writings established by Ibn Abī Uṣaybi'a.[15] This important treatise was thought to be lost until the recent discovery of the manuscript in the V.I. Lenin Library in Kuibychev. Along with some writings by al-Bīrūnī, by Kamāl al-Dīn al-Fārisī, by al-Khafrī and by al-Kāshī, this manuscript includes several treatises by Ibn al-Haytham, one of them being the one that interests us here. This valuable collection has been transferred to St Petersburg and is now in the National Library with pressmark no. 600, Arabic new series.

The whole collection was copied on thin and transparent paper, slightly grey in colour. Because of the transparency of the paper, it often happens that the words of the text on the verso of the page show through on the recto, and vice versa, which sometimes makes reading awkward. Damp, decay of part of the folios and a tear in the lower left corner of a certain number of folios – notably in the text *On the Properties of Circles* – make reading very difficult, sometimes impossible.

The collection is not written in a single hand; we can in fact recognise at least two. Nevertheless the numerous treatises by Ibn al-Haytham are all in a single hand, the script is *nasta'līq*, not very neat, the same script as in the copy of the treatise by the astronomer al-Khafrī, *Zubdat al-mabsūṭāt*, in the month of Rajab 1066, that is in May 1656. So the treatises by Ibn al-Haytham were copied at about this date, probably in some part of the Iranian world.

The text itself has been transcribed in black ink and the geometrical figures drawn in red ink. It has neither glosses nor additions in its margins and there seems to be nothing to indicate it has been compared with its original after the transcription was complete. Each folio measures 42.5 × 28 cm. We also observe that there are several series of numbering

---

[15] See vol. II, p. 522.

for the folios – evidence that the collection has been put together from several parts that have later been regrouped. In fact, in addition to the traces of a former numbering system, there are several others. Thus, the collection begins with a text by al-Kāshī in which we can recognise this old numbering, in Arabic numerals at the top of the page, a system which continues. A recent numbering in Indian numerals, at the bottom of the page, continues up to folio 493^v. When we examined the manuscript, we were able to make the following list, adopting the recent numbering:

1^v–10^v: al-Kāshī, *al-Risāla al-kamāliyya*

11^r: blank page

11^v–31^v: Muḥammad ibn Aḥmad al-Khafrī, *Zubdat al-mabsūṭāt*

32^r: blank page

32^v–270^v: al-Fārisī, *Tanqīḥ al-manāẓir*

271^r: title page

271^r–301^v: al-Fārisī, *Zayl tanqīḥ al-manāẓir*

301^v–307^v: al-Fārisī, *Taḥrīr maqāla fī ṣūrat al-kusūf*

308^r–309^v: *Fihrist muṣannafāt Ibn al-Haytham*

309^v–310^v: Ibn al-Haytham, *Fī ḥall shakk fī al-shakl 4 min al-maqāla 12 li-Uqlīdis*

310^v–311^r: Ibn al-Haytham, *Fī qismat al-miqdarayn al-mukhtalifayn*

311^v: blank page

312^r–326^v: Ibn al-Haytham, *Fī ḍaw' al-qamar*

326^v–339^v: Ibn al-Haytham, *Fī aḍwā' al-kawākib*

339^v–334^v: Ibn al-Haytham, *Fī kayfiyyat al-aẓlāl*

335^r–347^v: Ibn al-Haytham, *Fī al-ma'lūmāt*

348^r–368^r: Ibn al-Haytham, *Fī al-taḥlīl wa-al-tarkīb*

368^v–420^v: Ibn al-Haytham, *Fī hay'at ḥarakāt kull wāḥid min al-kawākib*

421^r–431^r: Ibn al-Haytham, *Fī khawāṣṣ al-dawā'ir*

431^v: blank page

432^r–432^v: Ibn al-Haytham, *Istikhrāj ḍil' al-muka''ab*

433^r–489^v: part of *Tafhīm* d'al-Bīrūnī, in another hand, with marginal glosses

490^r–491^v: part of *Fī khawāṣṣ al-dawā'ir*

492^r–493^v: treatise on algebra, anonymous and incomplete (late).

We note that the text *On the Properties of Circles* is made up of two parts (421^r–431^r and 490^r–491^v). But, whereas the first part (421^r–431^r) includes the title of the treatise as well as the colophon, the second part (490^r–491^v) is anonymous. It is for this reason that everyone who had

examined the manuscript had thought that the complete text was to be
found in the first part and had taken the second part as belonging to an
anonymous mathematical treatise that appears at the end of the collection.[16]
It is identifying this last part that has allowed me to assemble a complete
text of Ibn al-Haytham's work, to put it in order and finally establish it as a
text. The treatise appears in the following order:

$$421^r\text{--}421^v,\ 490^r\text{--}490^v,\ 422^r\text{--}428^v,\ 491^v,\ 491^r,\ 429^r\text{--}431^r.$$

We may also note that folio 491 is back to front. This kind of mistake
is not rare; it reappears several times in other treatises. These mistakes
have, in all probability, occurred when the collection was bound.

The bad state of the text *On the Properties of Circles* that we have
described above made it extremely difficult to reconstruct. Sometimes we
had no more than a few words to start from when reconstructing a whole
paragraph. In reconstructing we called upon every means at our disposal:
palaeography, philology and mathematics, as well as our familiarity with
the writings of Ibn al-Haytham. Further, our long and numerous interventions require us to indicate explicitly what has been done, not only for the
reader of Ibn al-Haytham's Arabic text, as one has to do when establishing
critical editions of texts, but also for the reader of the French and English
translations. In the English translation (as in the French one) each of our
interventions is isolated in the style <...>. Text between square brackets
[...] is an addition to the French text that is necessary for understanding the
English text.

---

[16] See Catalogue of the library of St Petersburg, no. 1588 and B. A. Rosenfeld in
Nauka, Moscow, 1974, p. 124, no. 16.

MATHEMATICAL COMMENTARY

Proposition 1. — *In a circle let there be an arbitrary chord* AC *cut in* E *by a chord* BD; *if* $\hat{BEC} = \hat{ABC}$, *then* $\overset{\frown}{BC} = \overset{\frown}{CD}$.

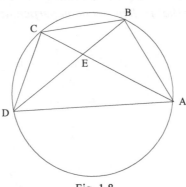

Fig. 1.8

The proof is immediate if we use the property of the interior angle. This property will be proved only later on – see Proposition 13. Before this, Ibn al-Haytham uses inscribed angles and triangles.

We have $\hat{BEC} = \hat{ABC}$, so the two triangles BEC and ABC are similar, hence

(1) $$\frac{EC}{BC} = \frac{BC}{AC} \implies BC^2 = EC \cdot AC.$$

Moreover, $\hat{BEC} = \hat{ABC}$, so $\hat{CED} = \hat{ADC}$; the two triangles EDC and ADC are similar; hence

(2) $$\frac{EC}{DC} = \frac{DC}{AC} \implies CD^2 = EC \cdot AC;$$

taken together, (1) and (2) give the result.

With the help of the property of the inscribed angle and that of the interior angle, we immediately have

$$\text{meas. } \hat{ABC} = \frac{1}{2}(\overset{\frown}{AD} + \overset{\frown}{DC}).$$

$$\text{meas. } \hat{BEC} = \frac{1}{2}(\overset{\frown}{BC} + \overset{\frown}{AD}).$$

The equality of the two angles leads to the result.

The first proposition belongs to a group that, as we shall see, includes Propositions 12 and 13. In the following four propositions – 2 to 6 – Ibn al-Haytham deals with parallel chords and similar ranges.

Proposition 2. — *The straight line joining the mid points of two parallel chords is a diameter that is their common perpendicular bisector.*

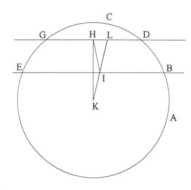

Fig. 1.9

Let there be two parallel chords BE and DG, and their respective mid points I and H; we have $\dfrac{DH}{DG} = \dfrac{BI}{BE} = k = \dfrac{1}{2} \Rightarrow HI$ is a diameter and the angle IHD is a right angle.

Ibn al-Haytham proves this proposition by *reductio ad absurdum*.

Let K be the centre of the circle. If HI did not pass through K, then KI would cut DG in L, $L \neq H$. But I is the mid point of BE, so the angle KIE is a right angle, hence the angle KLG is a right angle. In the same way, H is the mid point of DG, so the angle KHL is a right angle. So in the triangle KHL we should have two right angles, which is absurd.

In the following proposition Ibn al-Haytham considers a ratio $k \neq \dfrac{1}{2}$.

Proposition 3. — *Let there be two chords* EG *and* DB, *parallel and unequal, and such that* $\dfrac{HE}{EG} = \dfrac{IB}{BD} = k \neq \dfrac{1}{2}$; *then* HI *is not a diameter.*

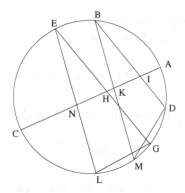

Fig. 1.10

Let us suppose that the two chords are unequal and that *HI* cuts the circle in *A* and *C*. Let us draw *EN* and *BK* perpendicular to *AC*, they cut the circle in *L* and *M* respectively. The triangles *EHN* and *BIK* are similar, we have

$$\frac{IB}{BK} = \frac{HE}{EN};$$

but by hypothesis

$$\frac{IB}{BD} = \frac{HE}{EG},$$

so we have

$$\frac{DB}{BK} = \frac{EG}{EN}.$$

If *AC* were a diameter we should have *EL* = 2*EN* and *BM* = 2*BK*, hence

$$\frac{DB}{BM} = \frac{EG}{EL},$$

the triangles *DBM* and *GEL* which have equal angles at *E* and *B* would thus be similar; we should have $E\hat{L}G = B\hat{M}D$, hence $\overset{\frown}{EAG} = \overset{\frown}{BAD}$, which is impossible. So *AB* is not a diameter.

Notes:

1) The reasoning assumes that the points *E* and *B* are on the same side of the straight line *HI*, that is that the half lines [*EG*) and [*BD*) are oriented in the same direction.

2) Ibn al-Haytham starts from $B\hat{D}M = E\hat{G}L$, which gives $\overparen{BAM} = \overparen{EAL}$, which would be impossible. In fact, we have

$$\overparen{BAM} = \overparen{BAD} + \overparen{DM} \text{ and } \overparen{EAL} = \overparen{EAG} + \overparen{GL} ;$$

now $\overparen{BAD} \neq \overparen{EAG}$ by hypothesis and $\overparen{DM} = \overparen{GL}$ because $\hat{E} = \hat{B}$; so

$$\overparen{BAM} \neq \overparen{EAL} .$$

3) If the chords DG and BE are *equal* and parallel, they are symmetrical with respect to the centre of the circle. If D and G have B and E as their respective images in the homothety $h(0; -1)$, then:

• if $\dfrac{\overline{DH}}{\overline{DG}} = \dfrac{\overline{BI}}{\overline{BE}} = k \neq \dfrac{1}{2}$, H has image I, and HI is a diameter.

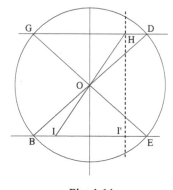

Fig. 1.11

• If $\dfrac{\overline{DH}}{\overline{DG}} = \dfrac{\overline{EI'}}{\overline{EB}} = k \neq \dfrac{1}{2}$; then $HI' \perp DG$ and $HI' \perp BE$, and HI' is not a diameter.

Proposition 4. — *Let there be a circle with centre* M, *two parallel chords* BD *and* EG *divided at points* I *and* K *respectively such that* $\dfrac{DB}{KB} = \dfrac{GE}{IE} = k \neq \dfrac{1}{2}$. *Let us suppose that* BE *and* KI *intersect in* H, *then* HM *is perpendicular to the two chords.*

By hypothesis we have

$$\frac{BK}{BD} = \frac{EI}{EG} .$$

The concurrent straight lines *BE*, *KI* and *HM* define two similar divisions:

$$\frac{BL}{BK} = \frac{EN}{EI}.$$

So we have

$$\frac{BL}{BD} = \frac{EN}{EG}.$$

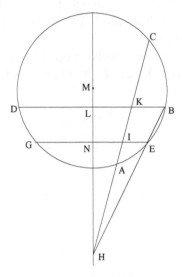

Fig. 1.12

So the diameter *HM* has cut the two parallel chords in the same ratio, so, by the second and third propositions, it is perpendicular to the two chords.

Notes:

1) The argument is applicable in both cases: *E* and *B* on the same side of *IK*, or *E* and *B* on opposite sides of *IK*.

2) In both cases we have a homothety with centre *H* in which *E* → *B* and *I* → *K*.

3) If *EG* = *DB*, then *EI* = *BK* and in consequence *EB* ∥ *IK*; the point *H* does not exist. The parallel to *EB* passing through *M* is the perpendicular bisector of *EG* and of *BD*, the ranges *B*, *K*, *L*, *D* and *E*, *I*, *N*, *G* are equal divisions, they correspond to one another in the translation defined by *BE*.

Proposition 5. — *Let there be two parallel chords* BH *and* DI, *cut orthogonally in* E *and* G *by a chord* AC *that is not a diameter. Let us suppose that* BE ≠ DG, *then*

$$\frac{BE}{EH} \neq \frac{DG}{GI}.$$

We may also state this proposition in an equivalent way: two unequal parallel chords are divided in unequal ratios by a chord that is perpendicular to them, if that chord is not a diameter. If the chord is a diameter, the ratios are equal.

Ibn al-Haytham proves this proposition by *reductio ad absurdum*: we draw the diameter *PN* parallel to *AC*, it cuts *BH* and *DI* in their mid points *N* and *P*.

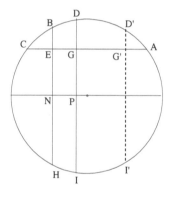

Fig. 1.13

If

$$\frac{BE}{EH} = \frac{DG}{GI},$$

we have

$$\frac{BE}{BH} = \frac{DG}{DI} \Rightarrow \frac{BE}{BN} = \frac{DG}{DP} \Rightarrow \frac{BE}{EN} = \frac{DG}{GP};$$

which is absurd, since $EN = GP$ and $BE \neq DG$.

Note: We have supposed $BE \neq DG$, which is the same as supposing that the parallel chords are unequal. We cannot have $BE = DG$ except if the chords BH and DI are symmetrical about the perpendicular bisector of AC, as BH and $D'I'$ are; in this case we have $BH = D'I'$ and $BE = D'G'$.

Proposition 6. — *In a circle with centre* L *let there be two parallel chords* BH *and* DI *divided in the same ratio, at* E *and* G *respectively, by the straight line* AC:

$$\frac{DG}{DI} = \frac{BE}{BH} = k.$$

Let there be a third chord OU parallel to the first two, such that it is cut at Q by the straight line AC; we have

$$\frac{UQ}{UO} \neq k.$$

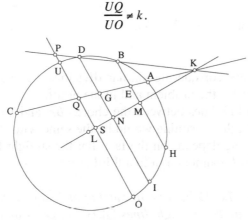

Fig. 1.14

The straight lines BD and AC intersect in K. From Proposition 4, the diameter KL cuts BH, DI and UO respectively in their mid points M, N and S respectively. We have

$$\frac{ND}{DI} = \frac{SU}{UO}.$$

So, if we had

$$\frac{DG}{DI} = \frac{UQ}{UO},$$

we should have

$$\frac{DN}{DG} = \frac{SU}{UQ},$$

hence

$$\frac{NG}{GD} = \frac{QS}{QU}.$$

But the straight line BD cuts OU in P outside the circle, and we have two similar divisions N, G, D and S, Q, P, hence

$$\frac{NG}{GD} = \frac{QS}{PQ};$$

thus, we have

$$\frac{QS}{PQ} = \frac{QS}{QU},$$

which is impossible since $PQ > UQ$.

Notes:

1) Thus, we are considering similar divisions on two chords, both in what we are given for the problem and in the proof.

2) If the chord OU lies between the chords BH and DI, we would have P inside the circle; the reasoning would be the same with $PQ < UQ$.

3) The reasoning depends on the fact that the straight line BD that cuts the circle in B and D cannot cut it in a third point.

Proposition 7. — *Let* D *be a point outside or inside a given circle. if we draw from this point two straight lines* DEB *and* DCA *that are secants and from the endpoint of one of the two chords cut off by these straight lines we draw a straight line parallel to the other chord, let it be* EG || AB, *then we have* GD · DC = DE².

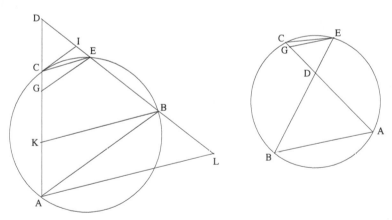

Fig. 1.15

We have

$$DA \cdot DC = DE \cdot DB \qquad \qquad \text{(the power of } D\text{)},$$

hence

$$\frac{DA}{DB} = \frac{DE}{DC}.$$

Moreover, $EG \parallel BA$, hence

$$\frac{DA}{DB} = \frac{DG}{DE}.$$

So we obtain

$$\frac{DE}{DC} = \frac{DG}{DE},$$

hence

$$DE^2 = DG \cdot DC.$$

In the same way, if $CI \parallel AB$, we have $DC^2 = DE \cdot DI$; if $BK \parallel EC$, we have $DB^2 = DA \cdot DK$ and if $AL \parallel EC$, we have $DA^2 = DB \cdot DL$.

The argument is the same in both cases of the figure: with D inside or D outside the circle; Ibn al-Haytham uses the power of the point D and homothetic triangles.

In the following three propositions – 8 to 10 – the data are the same.

Proposition 8. — *Let there be a circle with centre* D *and radius* R. *Let* A *be a point on this circle; if on the half line* DA *we take two points* E *and* H *such that* DE $\cdot$ DH $=$ R^2, *then for any point* B *of the circumference, other than* A *and* C, *we have*

$$E\hat{B}A = A\hat{B}H.$$

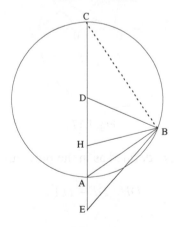

Fig. 1.16

By hypothesis we have $DE \cdot DH = DB^2$; hence

$$\frac{DE}{DB} = \frac{DB}{DH},$$

so the triangles BED and DBH are similar; hence

$$\frac{DB}{DH} = \frac{DE}{DB} = \frac{EB}{BH}.$$

But $DB = DA$, so

$$\frac{EB}{BH} = \frac{DA}{DH} = \frac{DE}{DA} = \frac{AE}{AH},$$

so A is the foot of the bisector of the angle EBH.

Note: The points E and H are harmonic conjugates with respect to A and C. The pencil $B\,(C, A, H, E)$ is a harmonic pencil. This proposition shows that if in a harmonic pencil two of the radiating lines are perpendicular, they are the bisectors of the angles between the other two.

Proposition 9. — Let us return to the figure for the previous proposition and let the second point of intersection of EB with the circle be called I, then $B\hat{D}I = B\hat{H}I$.

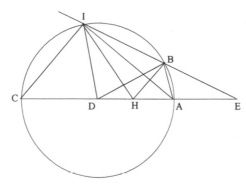

Fig. 1.17

The points E and H are defined as in the previous proposition:

$$DH \cdot DE = DA^2.$$

So we have

$$E\hat{B}A = A\hat{B}H = \frac{1}{2} E\hat{B}H ,$$

$$E\hat{I}A = A\hat{I}H = \frac{1}{2} E\hat{I}H .$$

Moreover,

$$B\hat{A}I = \frac{1}{2} B\hat{D}I , \ E\hat{B}H = E\hat{I}H + B\hat{H}I \text{ and } E\hat{B}A = E\hat{I}A + B\hat{A}I ,$$

hence, by doubling all the terms:

$$E\hat{I}H + B\hat{D}I = E\hat{B}H = E\hat{I}H + B\hat{H}I ;$$

so we have

$$B\hat{D}I = B\hat{H}I .$$

The argument in this proposition depends on that in the previous proposition and on the property that an angle inscribed in the circumference is equal to half the angle at the centre.

Note:

The proposition implies that the points B, I, D, H are concyclic. The circle passing through these four points is the transform of the straight line BE by an inversion with centre D that leaves the points of the circle ABI unchanged; E and H correspond to one another in this inversion while B and I are unchanged.

This proposition can be interpreted as follows, employing a style of expression different from that of Ibn al-Haytham: the inversion with centre D and power DA^2 transforms the chord BI of the circle with centre D and radius DA into the circle circumscribed about the triangle BID.

Proposition 10. — With the same data as in the two previous propositions, we have

$$(EB + BH) \cdot HI = CH \cdot HE.$$

On the extension of HB we mark off $BK = BE$.

Let us first note that in the manuscript text the end of a line is torn away and we have reconstructed it as follows: 'Q homologous to the point' (p. 96).

The circle circumscribed about the triangle *ECI* does not in general[17] pass through the point *K*, but through the point *Q* *symmetrical* to *K* with respect to *EC* as is established in the proof that follows. Let us now turn to the proof.

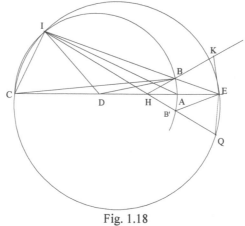

Fig. 1.18

We have

$$BE = BK \Rightarrow \hat{K} = \hat{E} = \frac{1}{2}\, H\hat{B}E = A\hat{B}E\,.$$

But the quadrilateral *ABIC* is inscribed in the given circle, so $I\hat{C}A = A\hat{B}E$ and consequently $H\hat{K}E = I\hat{C}A$.

In the quadrilateral *BHDI*, we have $I\hat{D}B = I\hat{H}B$ from the previous proposition, so $D\hat{H}I = D\hat{B}I = D\hat{I}B = B\hat{H}A$.

On the extension of *IH* we mark off $HQ = HK$; *K* and *Q* are thus symmetrical with respect to *ED*; so we have $I\hat{Q}E = H\hat{K}E = I\hat{C}E$.

Consequently, the circle circumscribed about *ICE* passes through *Q*, and the power of *H* with respect to this circle gives

$$HE \cdot HC = HI \cdot HQ = HI \cdot HK.$$

Note: As in Proposition 8, by hypothesis (*C*, *A*, *H*, *E*) is a harmonic division, so the pencil *B* (*C*, *A*, *H*, *E*) is a harmonic pencil in which two lines *BC* and *BA* are perpendicular; thus, these lines are the interior and exterior bisectors of the angle between the other two lines. The same applies to the pencil *I* (*C*, *A*, *H*, *E*).

---

[17] This circle does not pass through *K* unless *CE* is a diameter of it, that is if the angle *CIE* is a right angle.

Proposition 11. — *Let there be a circle with diameter* EA, *and let the two half lines* DG *and* DH *be symmetrical with respect to* AE. *Let the two points* B *and* C *be such that* $\overset{\frown}{BE} = \overset{\frown}{EC}$ *(so* BC $\perp$ AE *and* OB = OC*),* BC *cuts* DG *and* DH *in* P *and* N *respectively. From* B *let us draw* BI *and* BK *and from* C, CL *and* CM *which are respectively parallel to them. We have* BI · BK = CM · CL.

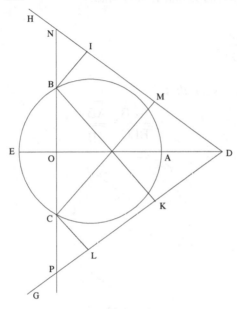

Fig. 1.19

We have, $ON = OP$, $BN = CP$, $CN = BP$; so

$$\frac{CN}{BN} = \frac{BP}{PC}.$$

But

$$BI \parallel CM \implies \frac{CN}{NB} = \frac{CM}{BI} \text{ (homothety with centre } N\text{)}$$

and

$$BK \parallel CL \implies \frac{BP}{PC} = \frac{BK}{CL} \text{ (homothety with centre } P\text{),}$$

so

$$\frac{CM}{BI} = \frac{BK}{CL},$$

hence the result.

Note: From the statement, *EA* is an axis of symmetry for the circle and the triangle *NDP*. The proof depends on this symmetry and on the triangles with vertex *N* and with vertex *P* being homothetic (*OND* and *OPD*; *CNM* and *BPK*; *INB* and *CPL*).

Proposition 12. — *Let there be two arcs of a circle cut off by a chord* AC *and divided by the points* B *and* D *such that*

(1)
$$\frac{\overset{\frown}{BA}}{\overset{\frown}{BC}} = \frac{\overset{\frown}{DC}}{\overset{\frown}{DA}},$$

then BD *cuts* AC *in* E *such that*

$$\frac{A\hat{E}B}{B\hat{E}C} = \frac{\overset{\frown}{AB}}{\overset{\frown}{CB}}.$$

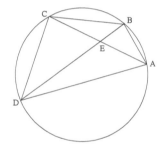

Fig. 1.20

We have

$$\frac{\overset{\frown}{AB}}{\overset{\frown}{BC}} = \frac{A\hat{C}B}{B\hat{A}C} \text{ and } \frac{\overset{\frown}{DC}}{\overset{\frown}{AD}} = \frac{D\hat{A}C}{D\hat{B}A} = \frac{D\hat{B}C}{D\hat{B}A}.$$

So we have

$$\frac{\overset{\frown}{AB}}{\overset{\frown}{BC}} = \frac{A\hat{C}B}{B\hat{A}C} = \frac{D\hat{B}C}{D\hat{B}A} = \frac{A\hat{C}B + D\hat{B}C}{B\hat{A}C + D\hat{B}A} = \frac{A\hat{E}B}{B\hat{E}C}.$$

In the proof Ibn al-Haytham uses the following two properties:

1) The ratio of two arcs is equal to the ratio of the inscribed angles that they subtend.

2) Proposition I.32 of Euclid's *Elements*: the sum of two angles of a triangle is equal to the non-adjacent exterior angle.

Proposition 13. — *If in a circle two chords intersect, then each of the angles in which they cut one another is equal to the angle subtended by the sum of the two arcs that lie between the two chords.*

The statement assumes that the point of intersection lies inside the circle. Ibn al-Haytham begins with this case. We draw *BH* parallel to *AC* (Fig. 1.21).

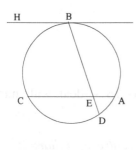 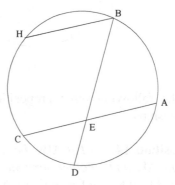

Fig. 1.21

There are two cases: *BH* is a tangent to the circle, or *BH* cuts the circle; in both cases, we have $H\hat{B}E = B\hat{E}A$.

If *BH* is a tangent to the circle and $BH \parallel AE$, we have $\overset{\frown}{BA} = \overset{\frown}{BC}$. The angle *HBD* is equal to the inscribed angle subtended by the arc *BCD*. We have

$$\overset{\frown}{BCD} = \overset{\frown}{BC} + \overset{\frown}{CD} = \overset{\frown}{AB} + \overset{\frown}{CD}.$$

If *BH* cuts the circle, then $\overset{\frown}{HC} = \overset{\frown}{BA}$. The angle *HBD* intercepts the arc *HCD*. We have

$$\overset{\frown}{HCD} = \overset{\frown}{HC} + \overset{\frown}{CD} = \overset{\frown}{AB} + \overset{\frown}{CD}.$$

So we conclude that the interior angle *AEB* is equal to an inscribed angle that cuts off an arc equal to the sum of the arcs *AB* and *CD*.

Similarly, the angle *BEC* is equal to an inscribed angle that intercepts the sum of the arcs *AD* and *BC*.

Ibn al-Haytham next considers the case in which the point of intersection lies outside the circle, and proves that the exterior angle *AEB* is equal to an inscribed angle that intercepts an arc equal to the difference of the two arcs *CB* and *AD* (Fig. 1.22).

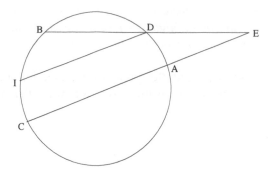

Fig. 1.22

The following three propositions – 14 to 16 – deal with metrical relationships.

Proposition 14. — *Let* ABC *be a circle with centre* E, *a diameter* AE, *a tangent* AL *and another tangent* BD *at* B, *another point of the circle, cutting* AL *in* K *and* EA *in* D; *we have*

$$BK \cdot DB = BL \cdot BE.$$

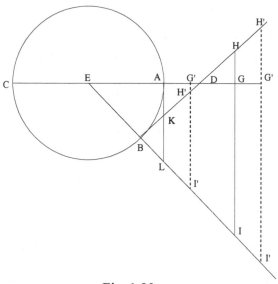

Fig. 1.23

The triangles AKD and BED are similar, so $\dfrac{AK}{EB} = \dfrac{AD}{BD}$; and we have $AK = BK$, hence

$$BK \cdot DB = AD \cdot BE.$$

The triangles AEL and BED are similar with $AE = EB$, so they are congruent and we have $LB = AD$, hence

$$DA \cdot BE = LB \cdot BE.$$

Consequently we have

(1) $\qquad KB \cdot BD = EB \cdot BL.$

Let us extend EA to G and let us draw HGI perpendicular to EA; we mark the point H on the tangent BD and the point I on the straight line EB. We have

$$\frac{HB}{BK} = \frac{BI}{BL} \qquad (Elements \text{ VI.2}),$$

hence

$$\frac{HB \cdot BD}{BK \cdot BD} = \frac{BI \cdot BE}{BL \cdot BE};$$

and from (1), we have

(2) $\qquad BH \cdot BD = BI \cdot BE.$

The result holds true for any line GHI perpendicular to the straight line AC, since Proposition 2 of Book VI of the *Elements* is applicable for any position of the point G.

We note that to establish (1), Ibn al-Haytham uses similar triangles and congruent triangles; and to establish (2), he uses (1) and homothetic triangles.

Ibn al-Haytham follows this proposition with two others that also deal with metrical properties of the circle. The proofs are immediate and seem to require no comment. We shall simply record the statements.

Proposition 15. — *Let there be a circle* ABCD, B *the mid point of the arc* AC *and* D *an arbitrary point on the arc; we have*

$$DA \cdot DC + DB^2 = AB^2.$$

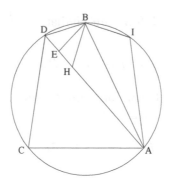

Fig. 1.24

Let us draw $BE \perp AD$. Let H be a point on EA such that $ED = EH$ and I a point on the circle such that $\overset{\frown}{BD} = \overset{\frown}{BI}$.

We have $BD = BH = BI$ and $B\hat{D}A = B\hat{H}D$, so $B\hat{I}A = B\hat{H}A$ and $B\hat{A}I = B\hat{A}H$. The triangles ABH and ABI are congruent, hence $AI = AH$. But $\overset{\frown}{CD} = \overset{\frown}{IA}$, so $AI = CD$ and $AH = CD$. Thus, we have

$$AE = AH + HE = CD + DE \quad \text{and} \quad AD = CD + 2ED;$$

from which it follows that

$$AD \cdot DC + DE^2 = CD^2 + 2ED \cdot CD + DE^2 = (CD + DE)^2 = AE^2$$

and

$$AD \cdot DC + DE^2 + EB^2 = AE^2 + EB^2,$$

hence

$$AD \cdot DC + DB^2 = AB^2.$$

Ibn al-Haytham thus establishes this metric relationship from equalities of arcs, from which he deduces equalities of chords and the congruence of two triangles.

Note: What he says gives no information about the arc AC that is under consideration. The argument is valid for any arc AC smaller than, equal to or greater than a semicircle.

Proposition 16. — *If in a circle we have two chords* AB *and* AC *such that* $\overset{\frown}{AB} < \overset{\frown}{AC}$ < *a semicircle, if* D *is the mid point of the arc* AB, *and* DE ⊥ AC, *then*

$$AC \cdot CB + BD^2 = CD^2.$$

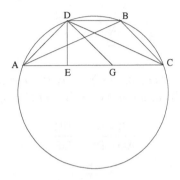

Fig. 1.25

$\overset{\frown}{AD} = \overset{\frown}{DB}$ implies $D\hat{C}B = D\hat{C}A$. Let us draw $EG = EA$; the triangle ADG is isosceles and we have $AD = DG = DB$.

The angles DBC and DGC have equal supplements $D\hat{A}G$ and $D\hat{G}A$, so $D\hat{B}C = D\hat{G}C$. The three angles of the triangles BDC and CDG are equal and we have $DB = DG$, so $CB = CG$. As E is the mid point of GA, we have

$$CA \cdot CG = CE^2 - EG^2$$

and

$$CA \cdot CG + GD^2 = CE^2 + DG^2 - EG^2 = CE^2 + ED^2 = CD^2;$$

from which we have

$$AC \cdot CB + BD^2 = CD^2.$$

Note: In Proposition 16, the point D is the mid point of the smaller of the two arcs and in Proposition 15, we consider the mid point of the greater arc. Propositions 15 and 16 are analogous to Proposition II.5 of the *Elements* when, instead of dividing a segment of a straight line, we divide an arc of a circle; the parts of the segment are then replaced by the corresponding chords of the circle.

Ibn al-Haytham introduces a new group of propositions, 17 to 23, apart from Proposition 21, which begins with a return to Proposition 94 of

Euclid's *Data* (Proposition 17). The propositions in this group are related as shown in the following diagram:

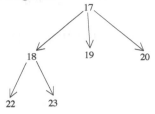

Proposition 17. — *Let there be a circle* ABC, *the arc* AC *with mid point* E, *the chords* ED *and* EB *that cut* AC *in* G *and* H *respectively, then*

$$\frac{AB + BC}{AD + DC} = \frac{BE}{DE}.$$

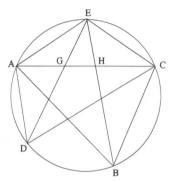

Fig. 1.26

We have $\overarc{AE} = \overarc{EC}$ which implies

$$E\hat{A}C = E\hat{C}A = E\hat{D}C = E\hat{C}G = A\hat{D}E.$$

The triangles *ECG* and *ECD* are similar, hence

$$\frac{ED}{EC} = \frac{EC}{EG} = \frac{DC}{CG}.$$

But *DG* is the bisector of *ADC*, so

$$\frac{DC}{CG} = \frac{DA}{AG} = \frac{DC + DA}{CG + AG} = \frac{DC + DA}{AC},$$

so

$$\frac{DC + DA}{DE} = \frac{AC}{EC}.$$

In the same way, we prove that

$$\frac{AB + BC}{BE} = \frac{AC}{CE};$$

hence the result.

Notes: In *The Knowns* (second part, Proposition 18), Ibn al-Haytham gives the same proof.

Moreover, in Euclid's *Data*, Proposition 94:[18] we find

1) $\frac{BA + BC}{BE} = \frac{AC}{CE}$, which implies the result.

2) $(BA + BC) \cdot EH = AC \cdot CE$ or $(AD + DC) \cdot EG = AC \cdot CE.$

Ibn al-Haytham's approach is no different from that of Euclid: we proceed by using the ratio of similarity of two triangles and by the property of the foot of the bisector, that is Proposition 3 of Book VI of the *Elements*.

Proposition 18. — *Let* ABC *be a circle,* AC *a diameter,* D *the mid point of a semicircle and* B *an arbitrary point on the other semicircle, then we have*

$$(AB + BC)^2 = 2\ BD^2.$$

As in the preceding proposition, we have

$$\frac{AB + BC}{BD} = \frac{AC}{CD};$$

so we have

$$\frac{(AB + BC)^2}{BD^2} = \frac{AC^2}{CD^2}.$$

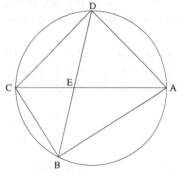

Fig. 1.27

[18] *Les Œuvres d'Euclide*, French trans. by F. Peyrard, Paris, 1819; repr. with an important additional Introduction by M. Jean Itard, Paris, 1966, p. 599.

But $AC^2 = 2CD^2$, hence the result. So we have a special case of the preceding proposition where AC is a diameter.

Proposition 19. — *Let there be a circle* ABCD *and the inscribed equilateral triangle* ADC; *let us draw the straight lines* DEB, AB *and* BC; *then we have*

$$AB + BC = BD.$$

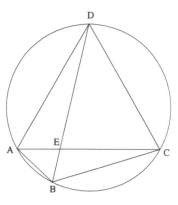

Fig. 1.28

From the preceding proposition, we have

$$\frac{AB + BC}{BD} = \frac{AC}{CD}.$$

But $AC = CD$; hence the result. We have a special case of Proposition 17 where AC is the side of the equilateral triangle.

Proposition 20. — *Let there be a circle* ABCD, AC *the side of the inscribed regular pentagon, the point* D *the mid point of the arc* ADC; *let us draw the straight line* DEB *cutting* AC *and let us join* AB, BC, BD, *then we have*

$$\frac{AB + BC + BD}{BD} = \frac{BD}{AB + BC}.$$

Ibn al-Haytham thus wishes to prove that BD is a mean proportional between the sum $(AB + BC + BD)$ and the sum $(AB + BC)$.

From Proposition 17, we have

$$\frac{AB + BC}{BD} = \frac{AC}{CD}.$$

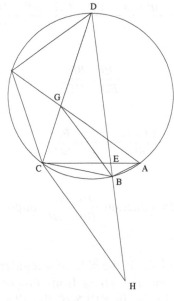

Fig. 1.29

But, from Proposition XIII.8 of the *Elements*, the straight line that joins *A* to the mid point of the arc *DC* divides the straight line *DC* at the point *G* such that

(1) $\qquad DC \cdot CG = DG^2$

and we have $DG = CA$, so

$$\frac{AB + BC}{BD} = \frac{DG}{CD};$$

but (1) implies

(2) $\qquad \dfrac{DG}{CD} = \dfrac{CG}{DG}.$

So if we extend *DB* to *H*, with $BH = AB + BC$, we have

$$\frac{HB}{BD} = \frac{AB + BC}{BD} = \frac{CG}{DG},$$

hence

$$\frac{HD}{BD} = \frac{AB + BC + BD}{BD} = \frac{CG + GD}{DG} = \frac{DC}{DG};$$

so, by (2) we have

$$\frac{HB}{BD} = \frac{BD}{HD},$$

that is

$$\frac{AB + BC}{BD} = \frac{BD}{AB + BC + BD};$$

hence the result.

We may note that the equation $\dfrac{DH}{DB} = \dfrac{DC}{DG}$ implies that CH and BG are parallel to one another.

Note: This time the chord AC is the side of a regular pentagon inscribed in the circle. Ibn al-Haytham thus starts from Proposition 17 and uses the property established by Euclid in XIII.8 of the *Elements* to prove that if a straight line is equal to the sum $BA + BC + BD$, it is divisible in extreme and mean ratio and the greater part is DB.

Proposition 21. — *Let there be a circle* ABC *with centre* D, *let* AC *be the side of the inscribed regular pentagon; the radius* DB *divides* AC *into two equal parts at* E; *and let* DG = BE, *then we have* EG = P_{10} *(*P_{10} *is the side of the regular decagon). Let* H *be the mid point of* DB; H *is also the mid point of* GE.

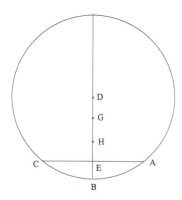

Fig. 1.30

From Hypsicles' Proposition I.1 (*Elements* XIV.1), we know that

$$DE = \frac{1}{2}(P_6 + P_{10}),$$

where P_6 is the side of the inscribed regular hexagon. But $P_6 = DB$ and $\frac{1}{2}P_6 = DH$, so $HE = \frac{1}{2}P_{10}$ and $GE = P_{10}$.

Note: It is true that here the chord AC is again by hypothesis the side of a regular pentagon, but the proposition is, nevertheless, independent of the preceding one. Here, Ibn al-Haytham starts from the result of Hypsicles' Proposition I.1. From this latter, if we consider the sides of the regular decagon and the hexagon inscribed in the same circle as the regular pentagon with side AC and apothem DE, we have

$$DE = \frac{1}{2}(P_6 + P_{10}).$$

Now $DB = P_6$, and Ibn al-Haytham's conclusion follows.

Proposition 20 deals with a special case of Proposition 17; it is concerned with the side of the regular pentagon. Proposition 21, which is out of place in this treatise, justifies its inclusion here only by providing a new property of the side of the pentagon.

Proposition 22. — Let us return to the figure for Proposition 18, and its hypotheses. We prove that the area $(ABCD) = \frac{1}{2}BD^2$.

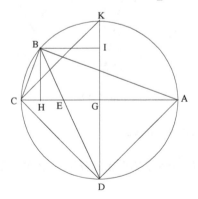

Fig. 1.31

We have $(AB + BC)^2 = 2BD^2$, hence $AB^2 + BC^2 + 2AB \cdot BC = 2BD^2$; but $AB^2 + BC^2 = AC^2 = 2AD^2$, so $AD^2 + AB \cdot BC = BD^2$. Moreover,

$$\text{area } (ABCD) = \text{area } (ABC) + \text{area } (ACD) = \frac{1}{2} AB \cdot BC + \frac{1}{2} AD^2,$$

so

$$\text{area } (ABCD) = \frac{1}{2} BD^2.$$

We can state this proposition as a corollary to Proposition 18: if AC is a diameter, then area $(ABCD) = \dfrac{1}{2} BD^2$.

Proposition 23. — *Let there be a circle ABC, let AC be a diameter, B the mid point of one of the semicircles, and D and E two points on the arc BC, then*

$$(DA + DC)^2 - (EA + EC)^2 = 2(EB^2 - DB^2).$$

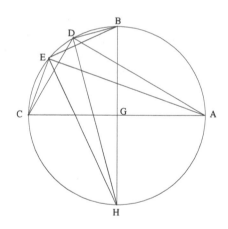

Fig. 1.32

From Proposition 18, we have

$$(DA + DC)^2 = 2DH^2 \text{ and } (EA + EC)^2 = 2EH^2;$$

but

$$HD > HE \Longrightarrow (DA + DC)^2 - (EA + EC)^2 = 2(DH^2 - EH^2);$$

now

$$DH^2 = HB^2 - DB^2 \text{ and } EH^2 = HB^2 - EB^2,$$

so

$$DH^2 - EH^2 = EB^2 - DB^2;$$

which gives us the conclusion.

We may note that the proof is carried out in the same way if the points D and E are on opposite sides of B.

The following two propositions – 24 and 25 – deal with calculations of areas of inscribed triangles.

Proposition 24. — *Let there be a circle* ABCD; *let* AC *and* BD *be two diameters that intersect in* E. *If* AC $\perp$ BE, G *is on* CB *and* GH $\perp$ DB, *then*

$$EB \cdot BH = \text{area (ABG)}.$$

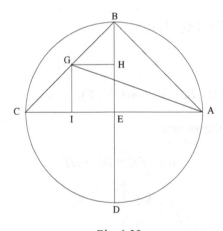

Fig. 1.33

We draw $GI \perp AC$. Now $GH \parallel AC$, hence $GI = HE$; so

$$AC \cdot GI = 2 \; \mathscr{A}(AGC),$$
$$AC \cdot HE = 2 \; \mathscr{A}(AGC),$$
$$AC \cdot BE = 2 \; \mathscr{A}(ABC);$$

hence, by subtracting,

$$AC \cdot BH = 2 \, [\mathscr{A}(ABC) - \mathscr{A}(AGC)] = 2 \; \mathscr{A}(ABG);$$

but $AC = 2\,EB$, hence

$$EB \cdot BH = \mathscr{A}(ABG).$$

Proposition 25. — *Let us return to the previous figure and let us mark* N *the centre of the circle,* E *and* G *two points on the arc* CB, GH $\perp$ BN *and* EI $\perp$ BN; EB *cuts* GH *in* K *and* GC *cuts* EI *in* M, *then*

$$BE \cdot EK = 2\ \text{area (AMG)}.$$

The triangles *BKH* and *BDE* are similar, so

$$\frac{EB}{BH} = \frac{ED}{HK} = \frac{BD}{BK},$$

hence

$$EB \cdot HK = ED \cdot BH$$

and

(1) $BE \cdot BK = BD \cdot BH;$

moreover,

(2) $EB^2 = BD \cdot BI$ (*Elements*, VI.8);

from (1) and (2) it follows that

$$BE \cdot EK = BD \cdot HI.$$

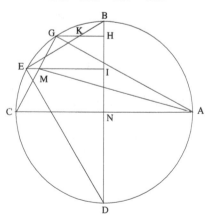

Fig. 1.34

But we have

(3) $$DB \cdot NH = AC \cdot NH = 2\ \mathscr{A}(AGC)$$

and

(4) $$DB \cdot NI = AC \cdot NI = 2\ \mathscr{A}(AMC);$$

from (3) and (4) it follows that

$$DB \cdot HI = 2\ [\mathscr{A}(AGC) - \mathscr{A}(AMC)] = 2\ \mathscr{A}(AGM).$$

So we have

$$BE \cdot EK = 2\ \mathscr{A}(AGM).$$

The following group is made up of six propositions – from 26 to 31 – which deal with concentric circles.

Proposition 26. — *Let there be two concentric circles with centre* G, AC = 2R *and* DE = 2r, *two diameters of the large and small circle respectively,* AHB *a tangent to the small circle at* H; *we have*

$$AB^2 + 4r^2 = 4\ R^2.$$

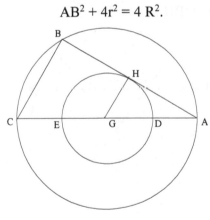

Fig. 1.35

AC is a diameter of the large circle, the angle *AHG* is a right angle, so *HG* ∥ *BC* and we have

$$\frac{AC}{AG} = \frac{AB}{AH} = \frac{BC}{GH};$$

now G is the mid point of AC, so H is the mid point of AB and $CB = 2GH =$ $2r$; hence the result.

Note: Ibn al-Haytham in fact wishes to prove that for any tangent to the circle (G, r), if we draw a diameter through one of its endpoints, the arc of the circle (G, R) which lies between the tangent and the diameter, an arc homologous to BC, is subtended by a chord of length $2r$.

Any chord of the circle (G, R) tangent to (G, r) thus has a length $2l$ such that

$$l^2 + r^2 = R^2.$$

We may note that 'H is the mid point of AB' is a result found in the *Collection* of Pappus (Proposition 77).[19]

We may also note that Ibn al-Haytham uses the fact that the two triangles AHG and ABC are homothetic.

Proposition 27. — *Let there be two concentric circles with centre* G, *a straight line cutting the two circles in* B *and* H *for* (G, R) *and in* E *and* D *for* (G, r) *and let there be a straight line* IEK *tangent to the latter circle in* E; *we have*

(1) $IK^2 + DE^2 = BH^2.$

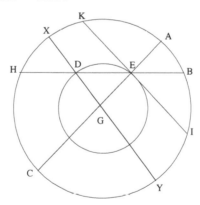

Fig. 1.36

The power of E with respect to the large circle gives

$$EI \cdot IK = EI^2 = EC \cdot EA = EB \cdot EH.$$

[19] French trans. Ver Eecke, p. 612.

If through D we draw the diameter XY, we have $DX = EA$, $DY = EC$, so

$$DX \cdot DY = DH \cdot DB = EA \cdot EC = EI^2,$$

so

$$DH \cdot DB = EB \cdot EH,$$

we have

$$HD = EB \text{ and } HE = DB.$$

Moreover, we have $KI = 2EI$, hence $4HE \cdot EB = IK^2$ and in consequence $4DB \cdot BE = IK^2$. But

$$DE = DB - BE, \text{ and } DE^2 = DB^2 + BE^2 - 2DB \cdot BE,$$

so

$$DE^2 + 4DB \cdot BE = (DB + BE)^2 = BH^2;$$

hence the conclusion.

Notes:

1) If as before we designate the length of a tangent as $2l$, the relation (1) may be rewritten as

$$(2) \quad 4l^2 + DE^2 = BH^2.$$

Equation (2) shows that the result proved in Proposition 26 for a diameter $ADEC$ can be extended to the case of a chord such as the chord $BEDH$.

2) The fact that EB and DH are equal derives from DE and HB having the same perpendicular bisector.

3) Ibn al-Haytham makes use of the equalities $EB = HD$ and $BD = HE$ proved in Pappus' *Collection* (Proposition 79).[20] On the other hand, these equalities follow immediately and accordingly do not permit us to infer that Ibn al-Haytham read Pappus.

Proposition 28. — *Let* BD *and* EI *be two concentric circles with centre* H, *let there be a straight line* BEID *that cuts them but does not pass through* H, *and* IG ⊥ BD; *we have*

---

[20] French trans. Ver Eecke, p. 613.

$BD^2 + GI^2 = 4R^2$ (*where* R *is the diameter of the large circle*).

We have

$$BD = EI + 2BE;$$

hence, after some calculation,

$$BD^2 = EI^2 + 4EB \cdot ED$$

and

$$BD^2 + GI^2 = 4EB \cdot ED + EG^2.$$

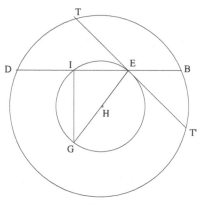

Fig. 1.37

Let us draw TT', the tangent at E, we have (power of E)

$$ET \cdot ET' = ET^2 = EB \cdot ED,$$

hence

$$TT'^2 = 4EB \cdot ED.$$

So we have

$$BD^2 + GI^2 = TT'^2 + EG^2;$$

but from Proposition 26: $TT'^2 + EG^2 = 4R^2$ (the square of the diameter of the large circle); hence the result.

Proposition 29. — With the data of the previous proposition, we have that the perpendicular erected at the point E of the small circle – EG – is equal to the perpendicular erected at the point B of the large circle – BC.

From Proposition 28 we have $AB^2 + EG^2 = 4R^2$; moreover, ABC is a triangle right-angled at B, so AC is the diameter and we have

$$AB^2 + BC^2 = 4R^2,$$

hence

$$EG = BC.$$

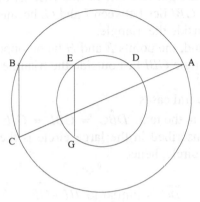

Fig. 1.38

Note: This is in fact a corollary of the preceding proposition.

Proposition 30. — *Let there be two concentric circles, and let* GI *be a diameter of the small circle, the straight lines* BI *and* BG *from a point* B *of the large circle cut the small circle in* E *and* H *and the large circle in* C *and* D; *we have*

$$\overset{\frown}{IH} + \overset{\frown}{GE} \ similar \ to \ \overset{\frown}{DC}.$$

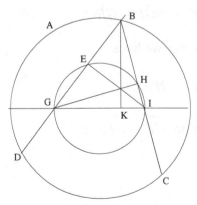

Fig. 1.39

The straight lines BK, GH and EI are the three heights of the triangle BGI. We have $I\hat{B}K = I\hat{G}H$ and $G\hat{I}E = G\hat{B}K$, so $I\hat{G}H + G\hat{I}E = I\hat{B}G$. The angles IGH, GIE, IBG intercept the arcs IH, GE respectively on the small circle and the arc CD on the large circle; hence the result.

Notes: The argument is constructed for the case in which E and H are on the same side of the straight line GI. In this case K, the foot of the height from B in the triangle GBI lies between I and G, because the orthocentre of the triangle BGI lies inside the triangle.

If, on the other hand, the points E and H lie on opposite sides of GI, the orthocentre of the triangle GBI lies outside the triangle and the point K lies on the extension of GI.

Thus, we have several cases:

a) The case given in the text: $D\hat{B}C = G\hat{B}I = G\hat{I}E + H\hat{G}I$.

The angle DBC inscribed in the large circle is the sum of two angles inscribed in the small circle, hence

$$\overset{\frown}{DC} \text{ is similar to } \overset{\frown}{IH} + \overset{\frown}{GE}.$$

b) Other cases: If E and B are on the same side of GI, and H on the other side, we have K beyond I

$$G\hat{B}I = G\hat{B}K - K\hat{B}I = G\hat{I}E - H\hat{G}I.$$

If H and B were on the same side of GI and E on the other side, K would lie beyond G and we should have

$$G\hat{B}I = K\hat{B}I - G\hat{B}K = H\hat{G}I - G\hat{I}E.$$

So for these two cases we have

$$D\hat{B}C = |G\hat{I}E - H\hat{G}I|,$$

hence $\overset{\frown}{DC}$ is similar to $\overset{\frown}{GE} - \overset{\frown}{IH}$, or to $\overset{\frown}{IH} - \overset{\frown}{GE}$.

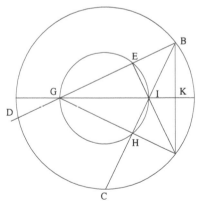

Fig. 1.40

Proposition 31. — *Let there be two concentric circles, with diameter* AC *and* DI *respectively;* A, D, I, C *are collinear. If* DI *is the diameter of the small circle and B a point on the large circle, we have*

$$BD^2 + BI^2 = AD^2 + DC^2.$$

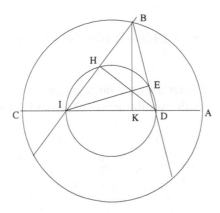

Fig. 1.41

The circle with diameter *IB* passes through *E* and *K*, so (power of *D*)

$$DB \cdot DE = DI \cdot DK.$$

The circle with diameter *BD* passes through *H* and *K*, so (power of *I*)

$$IB \cdot IH = ID \cdot IK.$$

Adding these two equalities term by term gives

(1) $DB \cdot DE + IB \cdot IH = ID^2.$

The points *B* and *A* are at the same distance from the centre, so their powers with respect to the small circle (small with respect to the large one) are equal:

$$BD \cdot BE = BH \cdot BI = AD \cdot AI,$$

hence

(2) $BD \cdot BE + BH \cdot BI = 2\,AD \cdot AI.$

If *DI* is the diameter of the small circle and *B* a point on the large circle, adding (1) and (2) term by term gives

$$BD^2 + BI^2 = ID^2 + 2\,AD \cdot AI;$$

but $AI = AD + DI$, so

$$ID^2 + 2AD \cdot AI = DI^2 + 2AD^2 + 2AD \cdot DI = AD^2 + AI^2 = AD^2 + CD^2;$$

hence the result.

If *DI* is now the diameter of the large circle and *B* a point on the small circle, we obtain the result by subtracting (2) from (1) term by term.

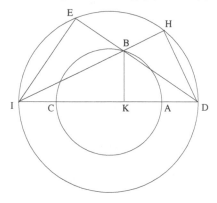

Fig. 1.42

Note: The method used to prove Proposition 31 is not applicable if the points *H* and *E* are on opposite sides of *AC*. However, the result is general and can be stated as follows without involving the circle (*DI*) and the points *H* and *E*:

If on a straight line we have two segments *AC* and *DI* with the same mid point, *O*, then for any point *B* of the circle (*O*, *OA*), we have

$$BD^2 + BI^2 = AD^2 + DC^2.$$

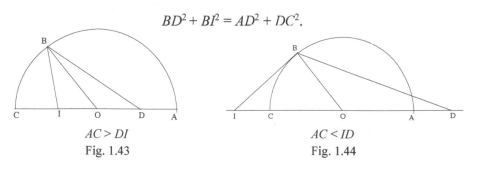

$AC > DI$	$AC < ID$
Fig. 1.43	Fig. 1.44

For any point B, we have

$$BI^2 + BD^2 = 2\,(BO^2 + OD^2) \text{ (median theorem)}$$

and for any point A, we have

$$AI^2 + AD^2 = 2\,(AO^2 + OD^2);$$

now by hypothesis $AI = DC$ and $OB = OA$, so

$$BD^2 + BI^2 = AD^2 + DC^2.$$

Note: The property established here was stated and proved in the same way by Ibn al-Haytham in his treatise *The Knowns*, Proposition 22.[21] Ibn al-Haytham's result is equivalent to the median theorem.

The last group of propositions in this book – from 32 to 43 – deals with tangent circles and homotheties. From Proposition 32 onwards, in fact, all the propositions involve two circles and, in general, except in Proposition 33, the argument calls upon one or another homothety that connects them.

Proposition 32. — *Let there be two circles* ABC *and* CDE *tangent at the point* C, *then the straight line* BCD *drawn in the two circles cuts off two similar arcs* CB *and* CD *and we have*

$$\frac{CB}{CD} = \frac{CA}{CE}.$$

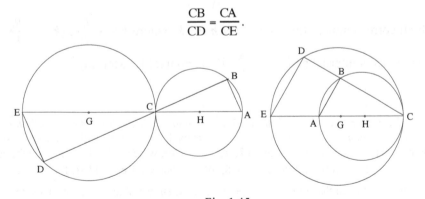

Fig. 1.45

[21] See pp. 327, 406–8.

From the fact that the homologous angles *BAC* and *DEC* are equal we deduce that the arcs *BC* and *CD* are similar. The right-angled triangles *CBA* and *CDE* are similar, hence

$$\frac{CB}{CD} = \frac{CA}{CE}.$$

Notes: The statement does not give any details about the circles; the contact can be internal or external and in this latter case, the circles can be equal or unequal.

The proposition may be rewritten: Let *AC* and *CE* be the diameters from the point of contact *C* and let *CBD* be a secant; we have

- the arcs *BC* and *CD* are similar;
- the arcs *AB* and *DE* are similar;
- $\dfrac{CB}{CD} = \dfrac{CA}{CE}.$

The argument depends upon the fact that *AB* and *ED* are parallel and the results in the statement follow from that immediately.

If we now call the radii of the two tangent circles R_H and R_G respectively, then to *any secant straight line* passing through *C* there correspond points *B* and *D* such that

$$\frac{\overline{CB}}{\overline{CD}} = \frac{\overline{CA}}{\overline{CE}} = \pm\frac{R_H}{R_G},$$

which corresponds to the homothety $h\,(C, k)$ where $k = \pm\dfrac{R_H}{R_G}$. ($k = +\dfrac{R_H}{R_G}$ if the contact is internal and $k = -\dfrac{R_H}{R_G}$ if the contact is external).[22]

---

[22] In his treatise *Fī taḥṣīl al-qawānīn al-handasiyya al-maḥdūda* (*On Obtaining Determinate Geometrical Theorems*), al-Sijzī states this same proposition and points out that he had proved it in his treatise on *Tangent Circles*, which today has yet to be found. Here is al-Sijzī's text (mss Paris, BN 2458, fol. 3ʳ; Istanbul, Reshit 1191, fol. 71ʳ⁻ᵛ):

'When two circles touch in a point and when we draw straight lines to the circumferences of the two circles, a ratio is also generated.

Let there be the two tangent circles in the two cases of the figure, *AB* and *AC*, the point *A* their <point of> contact. We have drawn the two straight lines *BAC* and *DAE*; then this generates the ratio of *AB* to *AC* <which is> equal to the ratio of *AD* to *AE*. We have proved this in the first proposition of our book on <*Tangent*> *circles*'.

Proposition 33. — *Let there be two circles* ABC *and* CDG *touching externally at* C *and let there be* BD *a common tangent, let us join* BC *and* DC *and* GD *and* AB, *then the angle* BCD *is a right angle and* GD *and* BA *intersect and are perpendicular.*

Ibn al-Haytham proves this proposition, for the cases when the circles are equal and when they are not. To prove that the angle *BCD* is a right angle, we consider the perpendicular to *EK* at *C*, it cuts the common tangent in *I*, we have that *IC* is a tangent to the two circles, so *IB* = *IC* = *ID*, so triangle *BCD* has a right angle at *C*.

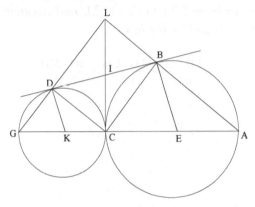

Fig. 1.46

To prove that *GD* and *AB* intersect and are perpendicular to one another, we deduce from the above that $B\hat{C}A + D\hat{C}G$ = one right angle, hence $D\hat{G}A + B\hat{A}G$ = one right angle, so the two straight lines *GD* and *AB* intersect at *L* and the angle *BLD* is a right angle.

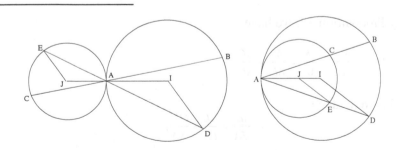

Al-Sijzī's result corresponds to the homothety $\left(A, \dfrac{AI}{AJ} \right)$. We may note that al-Sijzī, unlike Ibn al-Haytham, does not mention either the ratios of the chords *BD* and *CE*, or the fact that the arcs *BD* and *CE* are similar.

Ibn al-Haytham next proves that L lies on the tangent CI, using a method that is a little too long – which we shall examine later. It would have been enough to note that $CDLB$ is a rectangle, and that we know that $IB = IC = ID$: so I is the point of intersection of the diagonals CL and BD. If the circles were equal, $CDLB$ would be a square.

Proposition 34. — *Let $\mathscr{C}_1$ and $\mathscr{C}_2$, $\mathscr{C}_2 \subset \mathscr{C}_1$, be two circles that touch internally at the point A, let AC and AH be their respective diameters from the point A. Let L and E be two arbitrary points on $\mathscr{C}_2$, a secant passing through L cuts $\mathscr{C}_2$ in N and $\mathscr{C}_1$ in G and M, and another secant, passing through E, cuts $\mathscr{C}_1$ in B and D. We have*

$$\frac{EB \cdot ED}{EA^2} = \frac{LG \cdot LM}{LA^2} = \frac{NG \cdot NM}{NA^2}.$$

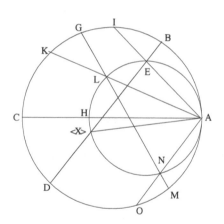

Fig. 1.47

By Proposition 32, we have

$$(1) \quad \frac{AC}{AH} = \frac{AK}{AL} = \frac{AI}{AE} = \frac{AO}{AN},$$

from this we deduce

$$\frac{IE}{EA} = \frac{KL}{LA} = \frac{ON}{NA},$$

hence

$$\frac{KL.\,LA}{LA^2} = \frac{EI.\,EA}{EA^2} = \frac{ON.\,NA}{NA^2},$$

but

$$KL \cdot LA = LG \cdot LM, \; EI \cdot EA = EB \cdot ED, \; ON \cdot NA = NG. \, NM;$$

hence the result.

Notes:

1) Ibn al-Haytham's statement can be rewritten as follows:

The ratio of the power of an arbitrary point of the small circle – small relative to the large circle – to the square of its distance from the point of contact is always the same.

In other words, let X be the second point in which EB meets the small circle, we shall again have $\dfrac{EB. \, ED}{EA^2} = \dfrac{XB. \, XD}{XA^2}$, so the common value of the two ratios associated with a secant does not depend on which secant we consider.

Let this ratio be k; if we put $AC = 2R$ and $AH = 2r$, we have

$$k = \frac{HC. \, HA}{HA^2} = \frac{HC}{HA} = \frac{R-r}{r}.$$

2) This relationship makes it clear that we have a homothety with centre A, $h\left(A, \dfrac{R}{r}\right)$.

3) We find the same property and the same proof in Ibn al-Haytham's treatise *The Knowns* (first part, Proposition 18).

Proposition 35. — *Let $\mathscr{C}_1$ (D, DC) and $\mathscr{C}_2$(H, HG) be two unequal circles touching one another at the point* C; *let there be a common tangent* BE *that cuts the extension of the diameter that passes through* D *and* H *in a point* I, *and let there be a secant passing through* I *that cuts $\mathscr{C}_1$ in* N *and* A *and $\mathscr{C}_2$ in* M *and* K; *we have that*

$$\overparen{ABN} \; and \; \overparen{KEM} \; are \; similar.$$

Now $D\hat{B}I = H\hat{E}I$ = a right angle, so $DB \parallel HE$, hence

(1) $\qquad \dfrac{DB}{HE} = \dfrac{ID}{IH},$

but

$$DB = DA = DN \text{ and } HE = HK = HM,$$

so

(2) $$\frac{DI}{IH} = \frac{DA}{HK} = \frac{DN}{HM},$$

hence $DA \parallel HK$ and $DN \parallel HM$. Accordingly we have $A\hat{D}N = K\hat{H}M$, so the arcs ABN and KEM are similar.

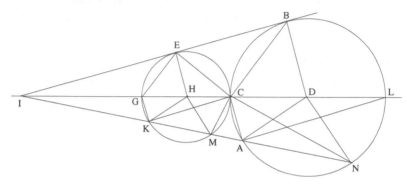

Fig. 1.48

Notes:

1) Let R_D and R_H be the radii of the circles. The point I exists if $R_H \neq R_D$. So Ibn al-Haytham intends to specify the position of the point I. He starts from the fact that DB and HE are parallel, which gives equation (1), which makes it clear that we have a homothety with centre I, $h\left(I, \dfrac{R_D}{R_H}\right)$ in which the points M, K, G, E, C of $\mathscr{C}_2$ have as their respective homologues the points N, A, C, B, L of $\mathscr{C}_1$. The straight lines HM, HK, HG, HE, HC are parallel to their homologues DN, DA, DC, DB, DL. Ibn al-Haytham deduces from this that the angles with vertex H are equal to their homologues with vertex D and that in consequence the arcs KM, MG, KE, EM, KC and the arcs AN, NC, AB, BN, AL that are homologous to them are similar.

Ibn al-Haytham did not, however, point out that the chords of the homologous arcs are parallel, something that can be deduced immediately from the fact that the homologous angles are equal.

The equality $A\hat{C}K$ = a right angle follows immediately.

We have $AL \parallel KC \Rightarrow A\hat{C}K = C\hat{A}L$ (alternate internal angles). Now the angle CAL is a right angle, hence the result. In the same way we have that angle NCM is a right angle.

2) This proposition is important and calls for several comments. So let us give a rigorous description of Ibn al-Haytham's reasoning. He begins with two unequal circles that touch one another externally. He next draws the external common tangent and constructs two homothetic triangles, which allows him to describe the position of the point of intersection of the tangent and the line of centres in terms of a ratio, and in this way to find the centre and the ratio of homothety, then to deduce that the homologous radii are parallel and the homologous arcs are similar. Thus, he has identified these concepts of the centre and ratio of a homothety, which he applies later.

Proposition 36. — *Let* $\mathscr{C}_1(I, ID)$ *and* $\mathscr{C}_2(H, HC)$ *be two circles one outside the other, equal or unequal. Let* K *be a point of the segment* CD *such that*

$$\frac{KC}{KD} = \frac{AC}{DG} = \frac{R_H}{R_I}.$$

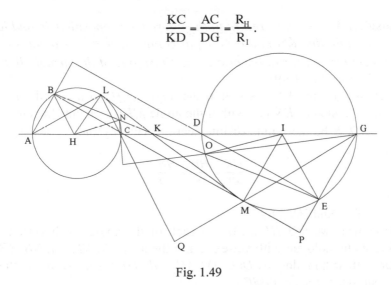

Fig. 1.49

If the straight line KL touches the circle $\mathscr{C}_2$, it also touches the circle $\mathscr{C}_1$.

We have $HL \perp KL$; the parallel to HL drawn through I cuts KL in M. We have

$$\frac{KC}{KD} = \frac{CH}{DI} = \frac{KH}{KI},$$

so

$$\frac{KH}{KI} = \frac{HL}{DI} \Rightarrow \frac{\overline{KH}}{\overline{KI}} = -\frac{R_H}{R_I}.$$

But $IM \parallel HL$, so

$$\frac{KH}{KI} = \frac{HL}{IM},$$

hence $IM = ID$ and M lies on $\mathscr{C}_1$.

Moreover, $HL \perp LM$, so $IM \perp LM$, the straight line LM is the tangent to the circle $\mathscr{C}_1$ at M.

Notes: In this proposition Ibn al-Haytham proves that the point M on $\mathscr{C}_1 (I, R_I)$ is homologous to L on $\mathscr{C}_2 (H, R_H)$ in the homothety $h\left(K, -\dfrac{R_I}{R_H} \right)$. If $\mathscr{C}_1 = \mathscr{C}_2$, the homothety becomes a central symmetry; a complete study of these two cases is carried out in *The Knowns* (second part, Proposition 24).

Proposition 37. — *Let us return to the figure for Proposition 36 and let us draw a straight line* KN *that cuts* $\mathscr{C}_2$ *in* B *and* N*; it will also cut* $\mathscr{C}_1$, *and the arcs cut off on each of the circles on either side of the straight line* KN *are similar* (see Fig. 1.49).

The straight line KN lies within the angle LKH, so it cuts the tangent LK. The extension of KN lies within the angle MKI, so it cuts the circle $\mathscr{C}_1$. Let B, N, O, E be the points of intersection with $\mathscr{C}_2$ and $\mathscr{C}_1$ respectively. We have

$$\frac{HK}{KI} = \frac{HB}{IE} = \frac{HN}{IO},$$

hence $HB \parallel IE$ and $HN \parallel IO$.

Moreover, we had $HL \parallel IM$. So each of the angles with vertex H is equal to its homologue with vertex I, and the arcs CN, NL, LB, NB, AB are similar to their homologues DO, OM, ME, OE, GE respectively, so the arc NLB is similar to the arc OME.

Note: The properties to which Ibn al-Haytham draws attention correspond to those of the homothety $h\left(K, -\dfrac{R_I}{R_H} \right)$: The points A, B, L, N, C of $\mathscr{C}_2$ have as their respective homologues the points G, E, M, O, D, so the arcs AB, BL, LN, NC, BN have as their homologues the arcs GE, EM, MO, OD, EO and are similar to them. We may note that two homologous arcs, for example, BN and EO, lie on opposite sides of the straight line that joins the endpoints of the arcs.

Proposition 38. — *Let us again return to the same figure; we have* CL ⊥ GM, CB ⊥ GE *and* AB ⊥ ED *(see Fig. 1.49).*

From Proposition 37, we have that $\widehat{AL}$ is similar to $\widehat{GM}$, hence $A\hat{C}L = M\hat{D}G$, so $L\hat{C}A + M\hat{G}D$ = a right angle and $CL \perp GM$.

We employ the same method in the other cases.

Note: In the homothety $h\left(K, -\dfrac{R_I}{R_H} \right)$, we have A, C, D, G on the line of centres where $G = h(A)$ and $D = h(C)$, then for any $X \in \mathscr{C}_2$, if $X' = h(X)$, we have $X' \in \mathscr{C}_1$, $AX \parallel GX'$ and $CX \parallel DX'$, hence $AX \perp DX'$ and $CX \perp GX'$.

This is precisely the procedure that Ibn al-Haytham presents, in different, but equivalent, terms.

Proposition 39. — *Let there be two circles $\mathscr{C}_1$ (H, HC) and $\mathscr{C}_2$ (I, IG) each exterior to the other, and* HI *meets them at the points* A, C, D, G, *in that order and with* AC > DG; *let there be a point* K *on the extension of* AG *such that* $\dfrac{KH}{KI} = \dfrac{AC}{DG}$ *and let* KE *be a tangent to* $\mathscr{C}_2$, *we have that* KE *is a tangent to* $\mathscr{C}_1$, *let it be the tangent at* L. *Conversely, if* KL *is a tangent to* $\mathscr{C}_1$; *then it is a tangent to* $\mathscr{C}_2$.

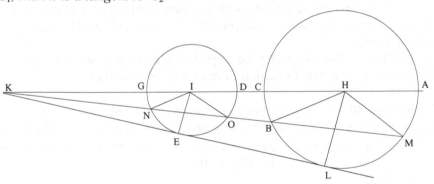

Fig. 1.50

We have $IE \perp KE$; through H we draw the parallel to IE, it cuts KE in L and we obtain $HL \perp KL$. We have $HL \parallel IE$, so

$$\frac{KH}{KI} = \frac{HL}{IE}.$$

But

$$\frac{HK}{KI} = \frac{AC}{DG} = \frac{\frac{1}{2}AC}{IE},$$

so $HL = \frac{1}{2}AC$ and L is a point on the circle $\mathscr{C}_1$, where $H\hat{L}K$ is a right angle, so KL is a tangent to the circle $\mathscr{C}_2$.

Similarly, if from K we draw a straight line that is a tangent to $\mathscr{C}_1$, we can prove that it is a tangent to $\mathscr{C}_2$.

Notes:

1) Here, as in Proposition 35, Ibn al-Haytham proves that the two properties of the point K, the centre of the positive homothety $h\left(K, \dfrac{R_H}{R_I}\right)$, are equivalent.

• K is the point of intersection of the line of centres and the external common tangent of two circles that touch externally.

• K lies on the straight line IH such that $\dfrac{\overline{KH}}{\overline{KI}} = \dfrac{R_H}{R_I}$.

2) As we have noted earlier, in Proposition 24 of the second part of *The Knowns*, Ibn al-Haytham considers the two cases of equal circles and unequal circles.

3) What we have here may be regarded as a study of the properties of the homothety as a transformation of one circle into another.

Proposition 40. — *Let us return to the figure for the previous proposition. If a straight line from* K *cuts* $\mathscr{C}_2$ *in N and O, then it cuts* $\mathscr{C}_1$ *in B and M, and the arcs cut off on the circles on the same side of the secant straight line are similar two by two* (see Fig. 1.50).

We have

$$\frac{HK}{KI} = \frac{HB}{IN} = \frac{HM}{IO},$$

so $HB \parallel IN$ and $HM \parallel IO$. We also have $HL \parallel IE$. So each of the angles with vertex I is equal to its homologue with vertex H and the arcs AB, BL, LM, MC are similar to their homologues DN, NE, EO, OG. So the arcs cut off by the straight line KO on the same side of the straight line are similar. The arc *NEO* is similar to the arc *BLM* and the arc *NGDO* is similar to the arc *BCAM*; which is the conclusion given in the statement.

Notes:

1) Ibn al-Haytham proves that the homologous arcs are similar. He does not prove that the chords that correspond to them are parallel: *AB* ∥ *DN*, *BL* ∥ *NE* and so on, a property that can be used to shorten the proof in certain cases (for example, Proposition 42 for unequal circles and Proposition 43).

2) The method Ibn al-Haytham adopts here is the one he had employed in Proposition 37. In the homothety $h\left(K, \dfrac{R_H}{R_I}\right)$, the points *N*, *E*, *O* of the circle C₂ have as their homologues the points *B*, *L*, *M* of the circle $\mathscr{C}_1$; the radii *IN*, *IE*, *IO* have as their respective homologues the radii *HB*, *HL*, *HM* which are parallel to them; the arcs *NE*, *EO*, *NO* have as their homologues the arcs *BL*, *LM*, *BM* which are similar to them; and finally two homologous arcs, for example, the arcs *NEO* and *BLM*, are on the same side of the straight line that joins their endpoints.

Proposition 41. — *Let there be two circles* $\mathscr{C}_1$ *and* $\mathscr{C}_2$, *equal or unequal, and let* AC *and* DG *be their diameters, where* A, C, D, G *are in that order, and let* BE *be an external common tangent,* B ∈ $\mathscr{C}_1$ *and* E ∈ $\mathscr{C}_2$. *We have*

DE ⊥ BC *and* GE ⊥ AB.

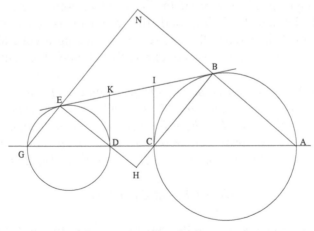

Fig. 1.51

First of all we have $BCA + EDG < \pi$, it is the same for the sum of two opposite angles; so BC and CD meet one another; let it be in H. We draw the tangents DK and CI, and we obtain $KD = KE$ and $IB = IC$, so

$$D\hat{K}I = 2\, D\hat{E}K \text{ and } C\hat{I}K = 2\, C\hat{B}I.$$

Now $D\hat{K}I + C\hat{I}K = \pi$, hence

$$D\hat{E}K + C\hat{B}I = \frac{\pi}{2},$$

hence the straight lines meet one another and $D\hat{H}C = \frac{\pi}{2}$.

Moreover, we have $B\hat{A}G + E\hat{G}A < \pi$, so the straight lines AB and GE meet one another; let it be in N. The angle N is the fourth angle of a quadrilateral $NBHE$ which has three right angles, at $\hat{H}, \hat{E}, \hat{A}$, so the fourth angle is a right angle.

Notes: If the point of intersection of BE and AG is called J, in the case where the *circles are unequal*, the homothety $h\left(J, \dfrac{AC}{DC}\right)$ gives $GE \parallel CB$, so $GE \perp AB$ and $ED \parallel BA$, so $ED \perp BC$; in the special case where the circles are equal, the two triangles ABC and DEG are right-angled and isosceles, the conclusion follows immediately and the correspondence between the two circles is a translation.

Ibn al-Haytham gives a proof that is valid in both cases, for unequal circles and for equal circles, and the problem is comparable with Proposition 33. It is perhaps for this reason – seeking a general proof for the two cases – that here he does not explicitly call upon the homothety.

Proposition 42. — *Let there be two circles each exterior to the other and unequal, with respective diameters* AC *and* DG, AC > DG *and the order of the points is* A, C, D, G. *Let there be a point* K *on the extension of* AG, *through which we can draw a common tangent to the two circles. If from* K *we draw a straight line to cut the two circles in* L, E, M, B *in that order, then*

$$DE \perp CB \text{ and } DL \perp CM.$$

The arcs AB and DE are similar, by Proposition 40, so $B\hat{C}A = E\hat{G}D$, hence $D\hat{C}H + C\hat{D}H = \frac{\pi}{2}$, hence the conclusion.

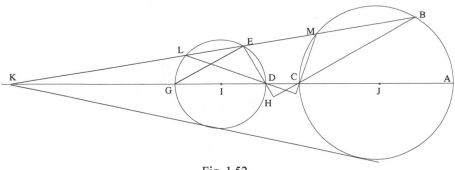

Fig. 1.52

In the same way, *DL* and *CM* meet one another and are perpendicular.

Notes:

1) In the homothety $h\left(K, \dfrac{R_I}{R_J}\right)$, we have *A, B, C, D* on the line of centres such that $G = h(C)$ and $D = h(A)$, then, for any point *M* of the circle $C(J, JC)$, if $M' = h(M)$, we have $M' \in C(I, IG)$ and $AM \parallel DM'$, $CM \parallel CM'$, hence $AM \perp GM'$ and $CM \perp DM'$. In this example, we have taken $M = B$ and $M' = E$.

2) Here Ibn al-Haytham defines the point called *K* in Propositions 39 and 40 as the point on the line of centres through which we can draw an external tangent to the two circles. It is, however, clear that he considers this point as defined by a ratio, because he refers to the property concerning similar arcs established in Proposition 40.

3) The property established here, starting from the point *K* as the centre of a positive homothety, corresponds to what is established in Proposition 38 with *K* as the centre of a negative homothety. It follows immediately from the fact that the homologous chords are parallel, which Ibn al-Haytham does not point out.

Proposition 43. — *Let there be two circles* $\mathscr{C}_1(I, IG)$ *and* $\mathscr{C}_2(H, HC)$, $\mathscr{C}_1 \subset \mathscr{C}_2$, AC *and* DG *their respective diameters* AC > DG; *the points are in the order* A, D, I, H, G, C. *Let* $K \in GD$, *be such that* $\dfrac{KD}{KG} = \dfrac{AD}{GC}$ *and let there be a straight line passing through* K *that cuts* $\mathscr{C}_1$ *in* E *and* M *and* $\mathscr{C}_2$ *in* B *and* L, *in the order* B, E, K, M, L. *We have that the arcs* CB, BA, AL *are similar to the arcs* GE, ED, DM.

$$\frac{DK}{KG} = \frac{AD}{GC} \Rightarrow \frac{AD}{DK} = \frac{GC}{KG} \Rightarrow \frac{AK}{DK} = \frac{KC}{KG} = \frac{AC}{DG} = \frac{CH}{GI} = \frac{HK}{IK}.$$

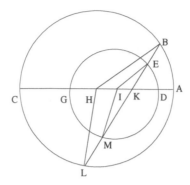

Fig. 1.53

So the point K divides the line segment HI externally in the ratio of the radii

$$\frac{\overline{KI}}{\overline{KH}} = \frac{R_I}{R_H}.$$

Moreover, we have

$$\frac{HK}{IK} = \frac{HB}{IE} \Rightarrow HB \parallel IE \text{ and } \frac{HK}{IK} = \frac{HL}{IM} \Rightarrow HL \parallel IM;$$

from which it follows that the angles at the centre with vertices H and I are equal, hence the arcs are similar.

Notes:

1) We could also deduce from this that the homologous chords are parallel: $AB \parallel DE$, $AL \parallel DM$; which Ibn al-Haytham does not point out.

2) The point K is the centre of a positive homothety. The method employed here is the same as the one found in Propositions 38 and 41.

In the homothety $h\left(K, \dfrac{R_I}{R_H}\right)$, the points A, B, C, L have as their respective homologues D, E, B, M; the radii HB, HL have homologues IE and IM, which are parallel to them; the arcs CB, BA, AL have as their homologues the arcs GE, ED, DM which are similar to them.

TRANSLATED TEXT

Al-Ḥasan ibn al-Ḥasan ibn al-Haytham

On the Properties of Circles
Fī khawāṣṣ al-dawā'ir

In the name of God, the Compassionate the Merciful

TREATISE BY AL-ḤASAN IBN AL-ḤASAN IBN AL-HAYTHAM

On the Properties of Circles

None of the geometrical figures has more properties; none invites more subtle investigation; none is more surprising in its complexity than the figure of the circle. The ancients and the moderns present several types of discussion on its properties, and on the effects of its subtlety. However, in regard to all the discussions we have found by mathematicians[1] on the properties of the circle, we have not seen that they exhausted all the properties that can occur in it. Since this was the case, we thought to investigate the properties of this figure, following step by step everything that can occur in it, giving proofs for everything that we have found that mathematicians have not [previously] described, as well as all the things that they have not proved in [those of] their books that have come down to us. We have examined this matter carefully and we have composed this treatise.

It is possible that our predecessors gave discussions of the properties of circles that have not come down to us; but this is not certain. Now we must not use the pretext of that possibility to justify our refraining from setting out what we have found on [matters] that have not come down to us. If anyone discovers, in a discussion by one of our predecessors, one of the things that we set out in this treatise, let him rest assured that it did not come down to us nor did we have sight of it; and let him not doubt that what we have set out on this subject is merely the effect of coincidence. It does indeed happen that people arrive at the same idea without intending to, and not deliberately. Further, we do not claim that what we have established regarding the properties of circles, combined with what all our predecessors established about them, exhausts all the properties of circles. Indeed, the properties of circles are numerous, and their number is almost infinite; which is to say that, whatever [number] of them has already been found and whatever [number] is found, it its always possible, later on, to

---

[1] Lit.: by people of this art.

find something more. Nevertheless, what this treatise contains is the end result of our investigation, [material] that we have not found in the books we inherited from our predecessors. It is God's help that we pray for in all things.

<1> If in a circle we draw an arbitrary chord and if in the circle we draw another chord that cuts this chord and makes with it an angle equal to the angle inscribed in the segment cut off by the first chord, then the second chord cuts off two equal arcs on either side of the first chord.

Example: Let there be a circle *ABC* in which we draw the chord *AC* and in which we also draw the chord *BED* in such a way that the angle *BEC* is equal to the angle inscribed in the segment *ABC*.

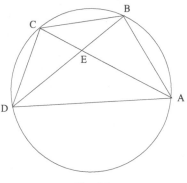

Fig. I.1

I say that the arc BC *is equal to the arc* CD.

Proof: We join *AB*, *BC*, *CD* and *DA*. Since the angle *BEC* is equal to the angle *ABC*, the product of *AC* and *CE* is equal to the square of *CB*. Since the angle *BEC* is equal to the angle *ABC*, the angle *CED* is equal to the angle *ADC*, and since the angle *CED* is equal to the angle *ADC*, the product of *AC* and *CE* is equal to the square of *CD*, so the square of *CB* is equal to the square of *CD*, so *BC* is equal to *CD* and the arc *BC* is equal to the arc *CD*. This is what we wanted to prove.

<2> If in a circle we draw two parallel chords, and if we divide each of them into two halves and if we join the two points of division by a straight line, then if we extend that straight line, it passes through the centre of the circle.

Example: Let there be a circle *ABC*, with centre *K*, in which we draw the two parallel chords *BE* and *DG*. We divide *BE* into two halves at the point *I*, we divide *DG* into two halves at the point *H* and we join *HI*.

I say that if we extend HI, *then it passes through the point* K.

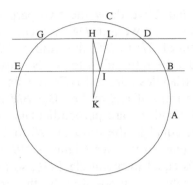

Fig. I.2

Proof: It cannot be otherwise; if that were possible, let it not pass through the centre. We join *KI*; it cuts the straight line *HI* because it is not in contact with it along a straight line, so let it cut it. We extend it; it will cut the straight line *DG*; let it cut it at the point *L*. Since *BE* is divided into two equal parts at the point *I*, the angle *KIE* is a right angle and the angle *KLG* is a right angle. We join *KH*; then the angle *KHL* is a right angle, so the two angles *L* and *H* of the triangle *KLH* are right angles; which is impossible. So the straight line *HI* passes through the centre of the circle. This is what we wanted to prove.

<3> If in a circle we draw two parallel chords, if we divide the two chords in the same ratio, other than the ratio of one half, if we join the two points of division with a straight line and if we extend that straight line, it does not pass through the centre of the circle.

Example: Let there be a circle *ABC* in which we draw the parallel chords *BD* and *EG*. We divide them at the points *I* and *H* in the same ratio other than the ratio of one half, and we join *IH*.

I say that IH *does not pass through the centre of the circle.*

Proof: This is not possible; if it were possible, let the line pass through the centre. We extend it on both sides to the points *A* and *C*; so *AC* would be a diameter of the circle and the two angles that are at the points *H* and *I* would not be right angles.

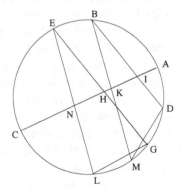

Fig. I.3

At the points B and E we draw the two perpendiculars BK, EN, we extend them to M and L and we join DM and GL. Since the ratio of DI to IB is equal to the ratio of GH to HE, the ratio of DB to BI is equal to the ratio of GE to EH. But the ratio of IB to BK is equal to the ratio of HE to EN, because the two triangles IBK and HEN are similar, so the ratio of DB to BK is equal to the ratio of GE to EN. But BM is twice BK and LE is twice EN because BM and EL are perpendiculars to the diameter. So the ratio of DB to BM is equal to the ratio of GE to EL; but the two angles DBM and GEL are equal, so the two triangles DBM and GEL are similar, and the angle BDM is equal to the angle EGL; so the two segments BAM and EAL are similar; which is impossible. So the straight line IH does not pass through the centre of the circle. This is what we wanted to prove.

So it is clear, from what we have proved, that if two <unequal> parallel chords cut a diameter of the circle and are not perpendicular to this diameter, then they will not be divided in the same ratio by the diameter.[2]

<4> If in a circle we draw two parallel chords, if we divide them in the same ratio and if we join the two points of division with a straight line such that the two angles that it makes with the two chords are not right angles, if we then extend the straight line that joined the two points of division, if we then join the two endpoints of the chords with a straight line that we extend to meet the straight line that passes through the two points of division, if from the point of intersection we then draw a straight line to the centre of the circle, then it will be perpendicular to the two chords.

Example: Let there be a circle ABC in which there are the two chords BD and EG that have been divided in the same <ratio>, other than the ratio of doubling, at the points I and K. We join KI and we extend it; <we join BE and we extend it>, it meets the straight line KI <at the> point H. We join the point H to the centre of the circle, let it be M, <with a straight line>; let it cut the two chords BD and EG at the points L and N.

I say that the straight line <HNL> *is perpendicular to the two chords* BD *and* EG.

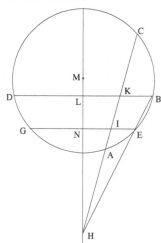

Fig. I.4

---
[2] See Note 3 in the special case where the chords are equal, p. 38.

Proof: The ratio of *DB* to <*BK* is equal to the ratio> of *GE* to *EI*, the ratio of *KB* to *BL* is equal to the ratio of *IE* to *EN*.[3] So the ratio of <*DB*> to *BL* is equal to the ratio of *GE* to *EN*. So if the straight lines *BD* and *EG* were not <perpendicular> to the diameter *MN*, then two parallel straight lines would be cut by this diameter in the same ratio <while not being> perpendicular to it; which is impossible. So the straight line *HM* is perpendicular to the two straight lines *BD* and *EG*. <This is what we wanted to prove>.

<5> If in a circle we draw an arbitrary chord that cuts off <two parts> on the circle <and if we draw> two unequal perpendiculars that end on the circumference of the circle on both sides, the chord <does not divide them> in the same ratio.

Example: Let there be a circle *ABC* in which there is <a chord *AC*. We draw> the two unequal perpendiculars *BE* and *DG*, and we then extend these two perpendiculars <to *H* and *I*>.

<*I say*> *that the straight line* AC *does not divide the two perpendiculars* BH *and* DI *in the <same> ratio.*

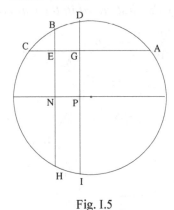

Fig. I.5

<*Proof*>: This is not possible; if it were possible, let the ratio of *BE* to *EH* <be equal to that ratio>. Let us consider the centre of the circle and let

---

[3] Ibn al-Haytham uses the following property: three concurrent straight lines define similar divisions on two parallel lines (an immediate consequence of the similarity of two triangles). A more general version of this result, for an arbitrary number of concurrent lines, is proved by Piero della Francesca (*c.* 1412–92) in his treatise on perspective (*De prospectiva pingendi*, Book I, section 8), using similar triangles and citing theorems from *Elements* VI (see J. V. Field, *Piero della Francesca: A Mathematician's Art*, London, 2005). In context, this becomes a proof of the convergence of perspective images of orthogonals (or images of any set of parallel lines).

us draw from it a diameter parallel to the chord *AC*; let it be <the diameter *PN*>. The diameter will be perpendicular to the two parallel straight lines. <So the ratio of *NB* to> *BE* is equal to the ratio of *PD* to *DG*, so the ratio of *NE* to *EB* is equal to the ratio of *PG* to *GD*. <But> *NE* is equal to *PG*, so the straight line *EB* is equal to the straight line *GD*; but by hypothesis they were unequal, which is impossible. So the two perpendiculars *BH* and *DI* are not divided in the same ratio by the chord *AC*. This is what we wanted to prove.

<6> If in a circle we draw an arbitrary chord and if we then draw in the circle two parallel chords that are divided in the same ratio by the first chord, then we cannot draw in the circle another chord that is parallel to the two chords and that is divided by the first chord in the ratio of the two chords that are parallel to it.

Example: Let there be a circle *ABC*; in it we draw an arbitrary chord, which is *AC*. We then draw in the circle two parallel chords, *BEH* and *DGI*, that are divided by the first chord in the same ratio.

I say that we cannot draw in the circle a third chord parallel to the first two and that is divided by the first chord in the ratio of the first two chords.

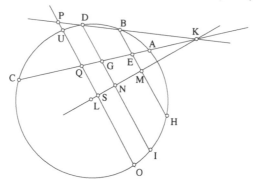

Fig. I.6

Proof: This is not possible; if it were possible, let us then draw the chord *UQO* which is divided at the point *Q* such that the ratio of *UQ* to *QO* is equal to the ratio of *DG* to *GI*. <We draw> *DB*, we extend it in the direction towards *B* and we draw *AC*; let the lines meet one another <at the point> *K*. <We mark> the centre of the circle, let it be *L*; we join *KL*; so the straight line *KL* is perpendicular to the parallel chords, as has been shown in Proposition 4. Let it cut the parallel chords at the points *M*, *N*, *S*; so it divides [each] of the chords into two equal parts. The ratio of *ND* to *DI* is equal to the ratio of *SU* to *UO*, and the ratio of *DI* to *DG* is equal to the ratio of *OU* to *UQ*, so the ratio of *ND* to *DG* is equal to the ratio of *SU* to *UQ*, and the ratio of *NG* to *GD* is equal to the ratio of *SQ* to *QU*. We extend the straight line *KBD* in the direction towards *D*, it thus meets the straight line *QU*; let it meet it at the point *P*; so the point *P* lies outside[4] the circle and the ratio of *NG* to *GD* is equal to the ratio of *SQ* to *QP*. But the ratio of *NG* to *GD* was equal to the ratio of <*SQ*> to *QU*, so the ratio of *SQ* to *QU* would be equal to the ratio of *SQ* <to *QP*>; which is impossible. So we cannot draw in the circle a third chord that is divided by a straight line <*AC* in the ratio of the two chords> *BH* and *DI*. This is what we wanted to prove.

<7> <Let there be a circle and an arbitrary point through which we draw> two straight lines that cut the circle; we draw the <two chords> that subtend the two arcs cut off by these two straight lines. Then, from the endpoint of one of the two chords, we draw a straight line parallel to the other chord. The parallel straight line then cuts off from the straight line on which it ends a straight line such that its product with the straight line that <passes> through the point and the endpoint of the arc is equal to the square of the straight line that runs between the point and the other end of the arc.

Example: Let there be a circle *ABC* and a given point *D*; from the point *D* we draw the two straight lines *DEB* and *DCA*, we join *AB* and *CE* and we draw *EG* parallel to *BA*.

I say that the product of GD *and* DC *is equal to the square of* DE.

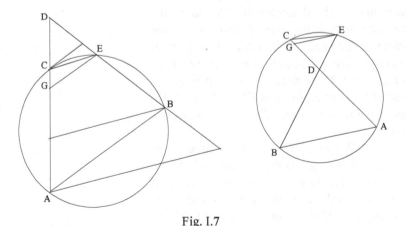

Fig. I.7

Proof: The product of *AD* and *DC* is equal to the product of *BD* and *DE*, so the ratio of *AD* to *DB* is equal to the ratio of *ED* to *DC*; <but the ratio of *AD* to *DB* is equal to the ratio of *DG* to *DE*>, so the ratio of *GD* to *DE* is equal to the ratio of *ED* to *DC*, and the product of *GD* and *DC* is equal to the square of *DE*.

In the same way, if we draw from the point *C* a straight line parallel to the straight line *BA*, we then show by an analogous proof that the product of *ED* and the straight line cut off from the straight line *ED* by the parallel is equal to the square of *DC*.

In the same way, if we draw from the point *B* a straight line parallel to the straight line *EC*, to cut the straight line *DA*; and in the same way if we draw from the point *A* a straight line parallel to the straight line *EC* <to cut the straight line *DB*>, the same result necessarily follows. This is what we wanted to prove.

<8> If in a circle we draw one of its diameters, which we then extend outside the circle, if we take an arbitrary point on it and if we then make the product of the straight line that lies between the exterior point and the centre of the circle and a part of that straight line equal to the square of the semidiameter, then if through the three points – the exterior point, the interior point and the endpoint of the diameter – <we cause to pass> three straight lines that meet one another in a point on the circumference of the circle, whatever that point may be, then the two angles formed by the three straight lines are equal.

Example: Let there be a circle *ABC* in which there is the diameter *AC*. We draw the diameter <*AC*> to the point *E* and we put the product of *ED* and <*DH*, which is a part of the straight line *DE*, equal to the square of the semidiameter. If we draw from the points> *E*, *A* and *H* three straight lines *EB*, *AB* and <*HB*, I say that the two angles *EBA* and *ABH*> are equal.

Proof: We join <*DB*, then the product of *ED* and *DH* is equal to the> square of *DB*, so the ratio of <*DE* to *DB* is equal to the ratio of *DB* to *DH*, so the triangle> *BED* is similar to the triangle <*DBH*>, and the ratio of *BD* to <*DH* is equal to the ratio of *EB*> to *BH*; it is

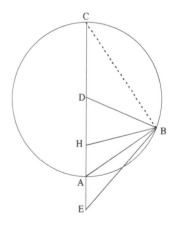

Fig. I.8

also equal to the ratio <of *DA*> to *DH* and it is equal to the ratio of *EA* <to *AH*. So the ratio of *AE* to *AH* is equal to the ratio> of *EB* to *BH* <and the angles *EBA* and *ABH* are equal.[5] This is what we wanted to prove>.

<9> If from an exterior point we draw a straight line that cuts the circle, and cuts off from it an arc smaller than a semicircle, if we then join the endpoints of the arc to an interior[6] point and if we also join the endpoints of the arc to the centre of the circle, then the two angles that are formed are equal.

<*Example*>: Let us draw from the point *E* the straight line *EBI* and let us join the straight lines *BH*, *IH*, *BD* and *ID*.

I say that the two angles BHI *and* BDI *are equal.*

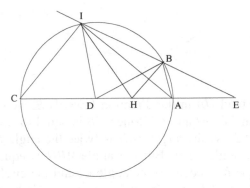

Fig. I.9

Proof: We join the straight lines *AB* and *AI*, then the two angles *EBA* and *HBA* are equal, and the angles *EIA* and *HIA* are equal. But the angle *EBH* exceeds the angle *EIH* by the angle *BHI* and the angle *EBA* exceeds the angle *EIA* by the angle *BAI*. But the excess of the half over the half is half the excess of the whole over the whole. So the angle *BAI* is half of the angle *BHI*, but the angle *BAI* and the angle *BDI* stand on the same arc which is *BI*, the angle *BDI* is at the centre and the angle *BAI* at the circumference; so the angle *BAI* is half the angle *BDI*. Thus, the two angles *BHI* and *BDI* are equal. This is what we wanted to prove.

<10> In the same way, let us return to the figure and let us draw a straight line from *C* to *I*.

---

[5] From Euclid, *Elements* VI.3.

[6] The interior point and the exterior point referred to in this statement are those of the previous proposition.

I say that the product of the sum of EB *and* BH *with* HI *is equal to the product of* CH *and* HE.

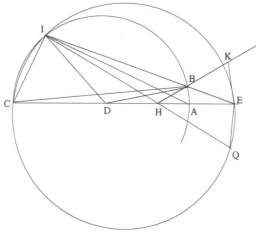

Fig. I.10

Proof: We extend *HB* in the direction towards *B*, we cut off *BK* equal to *BE* and we join *EK*, *EI* and *CI*. Since *EB* is equal to *BK*, the angle *E* is equal to the angle *K*, so the angle *EBH* is twice the angle *K*, and the angle *K* is equal to the angle *ABH*. But the angle *ABH* is equal to <the angle *ABE*>, so the angle *K* is equal to the angle *C*, and the circle circumscribed about the triangle *ECI* passes through the point <*Q* homologous to the point> *K*;[7] so the product of *KH* and *HI* is equal to the product of *CH* and *HE* <which is equal to the product of *QH* and *HI*>. But *KH* is equal to the sum of *EB* and *BH*, so the product of the sum of *EB* and *BH* with *HI* is equal to the product of *CH* and *HE*. This is what we wanted <to prove>.

<11> If in a circle we draw one of the diameters, which we extend outside <the circle, if we take> an arbitrary point on it and we draw two straight lines <that make> equal <angles> on either side of the diameter and <let them not meet the circumference of the circle, if we take> on the circumference of the circle on either side <of the endpoint of the diameter two points at equal distances from> this endpoint and <if we draw from one> of these two points two straight lines to the straight lines and from the

---

[7] In the manuscript, a part of the end of the line is effaced; we have replaced it by <*Q* homologous to the point>. The circle circumscribed about triangle *ECI* does not pass through the point *K*, it passes through the point *Q*, which is symmetrical with *K* with respect to *EC*.

other point two straight lines <parallel to these latter, then the product of these two straight lines> one with the other, is equal to the product of the two other straight lines, one with the other.

Example: Let there be a circle *ABC* in which we draw the diameter *AED*, on which we take the point *D* outside the circle. We draw from this point two straight lines *DG* and *DH* which do not meet the circle but are such that the two angles *GDE* and *HDE* are equal. On the circumference of the circle we take two points *B* and *C* whose distances from the point *E* are equal, we draw from the point *B* two straight lines *BI* and *BK* and we draw from the point *C* two straight lines *CM* and *CL* parallel to the straight lines *BI* and *BK*.

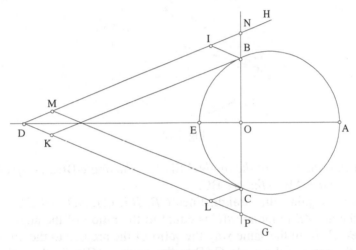

Fig. I.11

I say that the product of BI *and* BK *is equal to the product of* CL *and* CM.

Proof: We join *BOC*, it will be perpendicular to the diameter *EA*, because the arc *BE* is equal to the arc *EC*; we extend *BC* on either side to *N* and *P*: so *NO* is equal to *OP*, because the two angles *NDO* and *PDO* are equal; now *BO* is equal to *OC*, because the two arcs *BE* and *CE* are equal; finally *BN* is equal to *CP*, so the ratio of *CN* to *NB* is equal to the ratio of *BP* to *PC*. But the ratio of *CN* to *NB* is equal to the ratio of *CM* to *BI*, because the latter are parallel; now the ratio of *BP* to *PC* is equal to the ratio of *BK* to *CL*, so the ratio of *CM* to *BI* is equal to the ratio of *BK* to *CL*. So the product of *BI* and *BK* is equal to the product of *CL* and *CM*. This is what we wanted to prove.

<12> If in a circle we draw an arbitrary chord, if we divide the two arcs cut off by the chord in the same ratio by alternating and if we join the endpoints of the two arcs,[8] then the ratio of the two angles formed at the point of intersection, one to the other, is equal to the ratio of the two arcs <whose> chords we drew from one of the two points, the one to the other.

Example: Let there be a circle <ABC, and the two points B and D on either side> of the chord *AC* such that the ratio of the arc *AB* to the arc *BC* is equal to the ratio <of the arc *CD* to the arc> *DA*. We join *BED*.

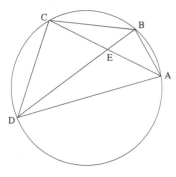

Fig. I.12

I say that the ratio of the angle AEB *to <the angle* BEC *is equal to the ratio of> the arc* AB *to the arc* BC.

Proof: We join <the straight lines *CB*, *BA*, *CA*, *AD* and *DC*>; so the ratio of the arc *AB* to the arc *BC* is equal to the ratio <of the angle *ACB* to the angle> *CAB*. In the same way the ratio of the arc *CD* to the arc <*AD* is equal to the ratio of the angle *CAD* to the angle> *ACD*. So the ratio of the angle *ACB* to the angle <*CAB* is equal to the ratio of the angle *DAC* to the angle> *ACD*. But the angle *DAC* is equal to the angle <*DBC* and the angle *ACD* is equal to the angle> *ABD*, so the ratio of the angle *ACB* <to the angle *CAB*> is equal to the ratio of the angle *CBE* to <the angle *DBA* and is equal to the ratio of the angle *AEB* to the angle> *BEC*, which is equal to the ratio of the whole to the whole. <So the ratio of the angle *AEB*> to the angle *BEC* <is equal to the ratio of the arc> *AB* to the arc *BC*. This is what <we wanted to prove>.

<13> <If we draw> in a circle <two chords that cut one another inside the circle, then each of the angles in which they cut one another> is equal

---

[8] That is, the points of division of the two arcs.

<to the angle that intercepts the sum[9] of the two arcs that lie between the two chords>.

<*Example*: Let there be a circle *ABC*> in which the two chords *AC* and *BD* cut one another at the point *E*.

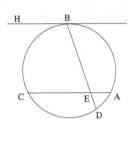 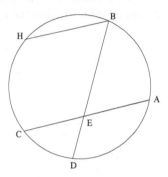

Fig. I.13.1

I say that the angle AEB *is equal to the angle that intercepts the sum of the two arcs* AB *and* CD.

Proof: From the point *B* we draw a straight line parallel to the straight line *AC*, let it be *BH*; the straight line *BH* is either a tangent to the circle or cuts it. If *BH* is a tangent to the circle, then the angle *HBE* is equal to the angle inscribed in the segment *BAD* that intercepts the arc *BCD*. If *BH* is a tangent to the circle, then the point *B* is the mid point of the arc *ABC*, so the arc *BC* is equal to the arc *AB*, the sum of the arcs *AB* and *CD* is equal to the arc intercepted by the angle *HBE* which is equal to the angle *AEB*. And the angle *BEC* itself is also equal to the angle that intercepts the sum of the remaining two arcs of the circle and which are the arcs *AD* and *CB*.

And if the straight line *BH* cuts the arc intercepted by the angle *BEC*, <then the angle *HBD* intercepts the arc *HCD*. But the arc *HCD* is equal to the sum of the arcs *HC* and *CD* and the arc *HC* is equal to the arc *AB*. Now the angle *HBD* is equal to the angle *AEB*, so the angle *AEB* is equal to the angle which intercepts the sum of the arcs *AB* and *CD*. And the angle *BEC* is itself also equal to the angle which intercepts the sum of the remaining two arcs of the circle and which are the arcs *AD* and *CB*>. This is what we wanted to prove.

[9] In such expressions, we have added the term 'sum' to conform with normal English usage.

We say in the same way that if in a circle we draw two chords that cut one another outside the circle, then the angle in which they cut one another is equal to the angle that intercepts the arc by which the greater of the two arcs that lie between the two straight lines exceeds the smaller one.

Let there be a circle *ABC* in which we draw the chords *AC* and *DB*; let them meet one another outside the circle at the point *E*.

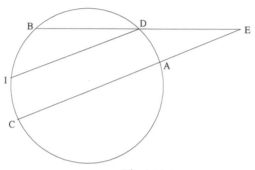

Fig. I.13.2

I say that the angle BEC *is equal to the angle <that> intercepts the arc by which the arc* BC *exceeds the arc* DA.

Proof: We draw the straight line *DI* parallel to the straight line *AC*, so the angle *BDI* is equal to the angle *BEC*. But the angle *BDI* is the angle that intercepts the arc *BI*, and the arc *BI* <is the arc by which the arc> *CB* <exceeds> the arc *DA*, because the arc *DA* is equal to the arc <*IC*. This is what we wanted to prove>.

<14> In a circle we draw one of the diameters, then we draw <from its endpoint, in one> of the two directions, <a tangent to the circle>, we next draw another tangent to the circle, <then we extend the diameter passing through the point of contact which meets the first tangent, we have that the product> of the two parts of the latter, the one with the other, is <equal> to the product <of what the first tangent cut off from the second tangent> and the straight line that is adjacent to it, which lies between <the point of contact and the point of intersection of the second tangent with the first diameter>.

<*Example*>: Let there be a circle *ABC*; in it we draw <the diameter *AEC*; we take the point *D* outside the circle. We draw a straight line that at its endpoint touches the circle>, such as *AL*. We draw the straight line *DB* <tangent to the circle in *B* and which meets *AL* in *K*, and we extend the diameter *EB*; so it meets *AL* in *L*.

I say that the product of> KB *and* BD *<is equal to the product of* EB *and* BL.

Proof: The straight line *DB* is a tangent to the circle at the point *B*, so the angle> *B* is a right angle. The triangle *AKD* is similar to the triangle *EBD*, so the ratio of *AK* to *EB* is equal to the ratio of *AD* to *DB*, so the product of *AK* and *DB* is equal to the product of *AD* and *EB*. Now *AK* is equal to *KB*, *EB* is equal to *EA* and the two triangles *AEL* and *EBD* are similar; but *AE* is equal to *EB*, so *LE* is equal to *ED* and *LB* is equal to *AD*, so the product of *DA* and *BE* is equal to the product of *EB* and *BL*, and the product of *KB* and *BD* is equal to the product of *EB* and *BL*. This is what we wanted to prove.

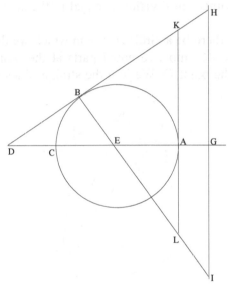

Fig. I.14

Let us return to the circle and the diameter. We extend the diameter in the direction of *A* as well. On it, we take a point *G*, we erect the perpendicular *HGI*, we draw the tangent *HB*, we draw *EB* and we extend it to *I*.

I say that the product of HB *and* BD *is equal to the product of* EB *and* BI.

Proof: We draw *AK*, a tangent to the circle, and we extend it to *L*, so the product of *KB* and *BD* is equal to the product of *EB* and *BL*; now the ratio of *HB* to *BK* is equal to the ratio of *IB* to *BL*, so the ratio of the product of *HB* and *BD* to the product of *KB* and *BD* is equal to the ratio of the product of *IB* and *BE* to the product of *LB* and *BE*. If we permute, the ratio of the product of *HB* and *BD* to the product of *IB* and *BE* is equal to the ratio of the product of *BK* and *BD* to the product of *LB* and *BE*. But the product of *BK* and *BD* is equal to the product of *LB* and *BE*, so the product of *HB* and *BD* is equal to the product of <*IB* and *BE*. This> is what we wanted to prove.

<15> In a circle we draw an arbitrary chord, <we then divide the arc subtended by this chord into two equal parts, then by means of an arbitrary point> into two unequal parts, and we draw the chords that subtend the arcs; then the product of the chord of the greater of the two parts <and the chord of the smaller plus the square of the> chord of the arc that lies between the two points <of division is equal to the square of the chord of half of the arc.>

<*Example*: Let there be a circle *ABC*> in which we draw the chord *AC*; we divide <the arc *AC* into two equal parts at the point *B* and into two unequal parts> at the point *D*. We join the straight lines <*AB, AD, DC* and *DB*.

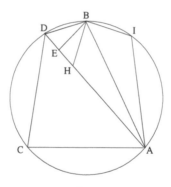

Fig. I.15

I say that the sum of the product of DA *and* DC> *and the square of* DB <*is equal to the square of* AB.

Proof: We draw from the point *B* the straight line *BE* perpendicular to the straight line *DA*, the straight line> *AB* is greater <than the straight line *AE*>. We cut off *EH* equal to *ED* and we join *BH*; it will be equal to *BD*. We cut off the arc *BI* equal to the arc *BD*; we join *AI* and *IB*; then the sum of the angle *I* and the angle *C* is equal to two right angles. But the angle *D* is equal to the angle *H*, so the angle *I* is equal to the angle *AHB*. But the angle *BAI* is equal to the angle *HAB*, and the straight line *AB* is common, so the triangle *AIB* is equal to the triangle *AHB*, and the straight line *AI* is equal to the straight line *AH*; but *AI* is equal to *DC*, because the arc is equal to the arc, so the straight line *AH* is equal to the straight line *DC*; but *HE* is equal to *ED*, so *AE* is equal to the sum of *ED* and *DC*, and the sum of the product of *AD* and *DC* and the square of *DE* is equal to the square of *AE*. We add the same square [EB^2] to both, then the sum of the product of *AD* and *DC* and the squares of *DE* and *EB* is equal to the sum of the squares of *AE* and *EB*, so the sum of the product of *AD* and *DC* and the square of *DB* is equal to the square of *AB*. This is what we wanted to prove.

<16> If in a circle we draw two chords, if we divide the small arc into two equal parts and if we join the chords, then the sum of the product of the chord of the large arc and the chord of the <arc> by which the large <arc> exceeds the small one and the square of the chord of half the small arc is equal to the square of the chord of the arc composed of half the small <arc> and the <arc> by which the greater <arc> exceeds the smaller one.

Example: Let there be a circle *ABC* in which we have drawn the two chords *AB* and *AC*. We divide the arc *AB* into two equal parts at the point *D* and we join the chords.

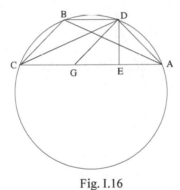

Fig. I.16

I say that the sum of the product of AC *and* CB *and the square of* BD *is equal to the square of* DC.

Proof: We draw the perpendicular *DE*, we cut off *EG* equal to *EA* and we join *DG*. So the angle *G* is equal to the angle *A* and the sum of the angle *A* and angle *DBC* is equal to two right angles; so the angle *DBC* is equal to the angle *DGC*; but the angle *BCD* is equal to the angle *DCG*, so the triangle *DBC* is equal to the <triangle *DGC*> and *BC* is thus equal to *CG*. The product of *AC* and *CB* is equal to the product of *AC* <and *CG*. But *GD*> is equal to *DA* and *DA* is equal to *DB*, so *GD* is equal to *DB*. And since the triangle *ADG* is isosceles, the product of *AC* <and> *CG* <is equal to the difference between the squares of *CE* and of *EG*, so the sum of the product of *AC* and *CG*> and the square of *GD* is equal to the square of *CD*. So the sum of the product of *AC* <and *BC*, which is equal to the product of

AC and *CG*>, and the square of *BD*, is equal to the square of *CD*. This is what we wanted to prove.

<**17**> If we draw an arbitrary chord <in a circle>, if we divide one of the two arcs <subtended> by that chord into two equal parts and if we draw from the point of <division two> arbitrary <straight lines> which cut the chord; <if we then join the endpoints of each of these straight lines to the endpoints of the chord> with straight lines, <then the ratio of the sum of the first> two straight lines to the sum of the last two straight lines <and the ratio of the straight lines that join the two endpoints to the point of division, are equal one to> the other.

Example: <Let there be a circle *ABC*; we divide the arc> *AC* into two equal parts at the point *E*. We draw from the point *E* two arbitrary straight lines *EHB* and *EGD*, we join the straight lines *AB*, *CB*, *AD*, *CD*.

I say that the ratio of the sum of AB *and* BC *to the sum of* AD *and* DC *is equal to the ratio of* BE *to* ED.

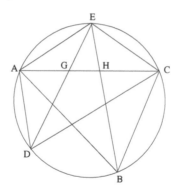

Fig. I.17

Proof: Let us join *AE* and *EC*. They are equal, so the two angles that are at the points *A* and *C* of the triangle *ABC* are equal and the angle *EAC* is equal to the angle *EDC*; so the angle *EDC* is equal to the angle *ECG*, and the triangle *EDC* is similar to the triangle *ECG*; so the ratio of *ED* to *EC* is equal to the ratio of *CE* to *EG* and is equal to the ratio of *DC* to *CG*. But the ratio of *DC* to *CG* is equal to the ratio of *DA* to *AG*, because the two angles at the point *D* are equal.[10] <So the ratio> of the sum of *AD* and *DC* to *AC* is equal to the ratio of *DE* to *EC*. So the ratio of the sum of *AD* and

---

[10] Euclid, *Elements* VI.3.

DC to *DE* is equal to the ratio of *AC* to *CE*.[11] By the same proof, we show that the ratio of the sum of *AB* and *BC* to *BE* is equal to the ratio of *AC* to *CE*. So the ratio of the sum of *AB* and *BC* to the sum of *AD* and *DC* is equal to the ratio of *BE* to *ED*. This is what we wanted to prove.

<18> If in a circle we draw one of the diameters and if we then divide one of the two semicircles into two equal parts; if we then draw from the point of division a straight line that cuts the diameter in an arbitrary point and if we join the two endpoints of the diameter to the endpoint of the straight line by two straight lines, then the square of the two straight lines, if they become a single straight line, is equal to twice the square of the straight line that cuts the diameter.

Example: In the circle *ABCD* we draw a diameter *AC*, <we divide the semicircle> *ADC* into two equal parts at the point *D*, we draw from this point a straight line *DEB* which cuts <the circle in *B* and we join> the two straight lines *AB* and *BC*.

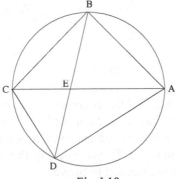

Fig. I.18

I say that the square of the sum of the two straight lines AB <*and* BC *if they become*> *a single* <*straight line*> *is twice the square of* BD.

Proof: We join <the two straight lines> *AD* and *DC*, so the ratio of the sum of *AB* and *BC* to <*BD*> is equal to the ratio of *AC* to *CD*, as we have proved earlier [in Proposition 17]. <So the ratio of the square of> *AB* and *BC*, if they become a single straight line, <to the square of *BD*, is equal to the ratio of the> square of *AC* to the square of *CD*. But the square of *AC* <is twice the square of *CD*, so the square of the sum of> *AB* and *BC* is twice <the square of *BD*. This is what we wanted to prove>.

---

[11] This property is found in the *Data*, Proposition 94. It uses the ratio of similarity of two triangles and the property of the foot of the bisector, *i.e.* of the point in which the bisector of an angle of a triangle meets the opposite side (Euclid, *Elements*, VI.3).

<19> <In a circle we draw the side of an> equilateral <triangle>, <then with a point we divide the arc subtended by this side of the triangle, we draw from the point of division a straight line that cuts the side> which subtends this <arc such that it reaches the circle; then we draw from the endpoint of the straight line> two straight lines <to the two endpoints> of the side, <these two straight lines combined are equal to this straight line>.

Example: We draw in the circle *ABCD* an <equilateral> triangle *ADC*, we draw from the point *D* the straight line *DEB* and we join *AB* and *BC*.

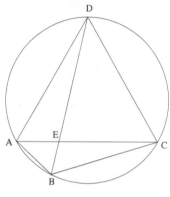

Fig. I.19

I say that the sum of AB *and* BC *is equal to the straight line* BD.

Proof: The ratio of the sum of *AB* and *BC* to *BD* is equal to the ratio of *AC* to *CD*, from what we have proved earlier.[12] Since the arc *AD* is equal to the arc *DC* and *AC* is equal to *CD*, accordingly the sum of *AB* and *BC* is equal to *BD*. This is what we wanted to prove.

<20> If in a circle we draw the side of a pentagon,[13] if we then divide the remainder of the circle into two equal parts and we draw from the point of division a straight line which cuts the side of the pentagon and ends on the circle, and then we join the two endpoints of the side of the pentagon to the endpoint of the straight line by two straight lines, then the sum of the two straight lines and the first straight line, if the three are aligned, is divisible in extreme and mean ratio,[14] and the greater part is the straight line <that is a> secant.

---

[12] Proposition 17.

[13] Throughout this text, the pentagon is assumed to be regular.

[14] The length of the first straight line is the mean proportional between the total length and the sum of the two other straight lines.

Example: Let there be a circle *ABCD* in which there is the side of the pentagon, which is *AC*. We divide the arc *ADC* into two equal parts at the point *D* and we draw the straight line *DEB* which cuts *AC*; we join *AB* and *BC*.

I say that if AB, BC, BD *are aligned, their sum is divided in extreme and mean ratio and <the greater part is* DB>.

Proof: We join *DC*, so the ratio of the sum of *AB* and *BC* to *BD* is equal to the ratio of *AC* to *CD*, from what we have proved earlier [Proposition 17]. But *DC* subtends the arcs of the circle,[15] so if we divide *DC* in extreme and mean ratio, the greater side will be equal to <*AC*>. We divide *DC* at *G*, so *DG* is equal to *AC*.[16] So the ratio of the sum of *AB* <and *BC* to> *BD* is equal to the ratio of *GD* to *CD*.[17]

We extend *DB* <to the point *H* such that *BH*> is equal to *AB* plus *BC*, so the ratio of *HB* to *DB* is equal to the ratio <of *CG* to *GD*>, so the ratio of *HD* to *BD* <is equal to the ratio of *CD* to *DG* which is equal to the ratio of *DB* to *BH*. <So the ratio of *HD* to *DB* is equal to the ratio of *DB* to *BH*.>

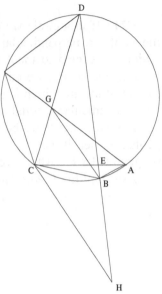

Fig. I.20[18]

[15] One of the arcs subtended by *DC* is two fifths of the circle, that is two of the arcs corresponding to the pentagon.

[16] To prove that *DG = CA*, it is sufficient to prove that *DG = DD'*, that is that the triangle *GDD'* is isosceles.

$$D\hat{G}D' = G\hat{D}D'',$$

$$D\hat{D}'G = D''\hat{A}G = G\hat{D}D'',$$

so

$$D\hat{G}D' = D\hat{D}'G$$

and the triangle *GDD'* is isosceles.

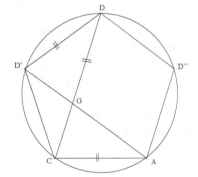

[17] From Euclid, *Elements* XIII.9, we have *GD/CD = CG/DG*. Perhaps the copyist has omitted something.

[18] The figure in the manuscript is incorrect.

<The ratio> of the sum of *AB*, *BC*, *BD* <to *BD* is thus equal to the ratio of *BD* to the sum of *AB* and> *BC*, so the straight lines *AB*, *BC*, <*BD*, if we imagine them aligned>, are divided in <extreme and mean ratio>. This is what we wanted to prove.

<21> If in a circle we draw the side of a pentagon and we draw from the centre a perpendicular to the side of the pentagon, a perpendicular which we extend to meet the circumference of the circle, if, on the side towards the centre, we cut off from the perpendicular a piece equal to the sagitta of the arc which is one fifth [*sc.* of the circle], then what remains of the perpendicular is equal to the side of the decagon.

Example: Let there be a circle *ABC* in which we draw the side of the pentagon; let it be *AC*. From the centre, which is *D*, we draw the perpendicular *DE* which we extend to *B*, and we cut off *GD* equal to *BE*.

I say that EG *is equal to the side of the decagon.*

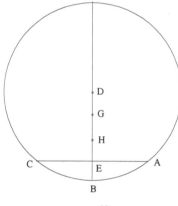

Fig. I.21[19]

Proof: Let us cut *DB* into two equal parts at the point *H*; so the straight line *GE* is divided into two equal parts at the point *H*. Now it has been proved in the last two books attached to Euclid's book[20] that the perpendicular *DE* is equal to half the side of the hexagon plus half the side of the decagon; now the straight line *DH* is half the side of the hexagon, so the straight line *EH* is half the side of the decagon, and the straight line *EG* is the side of the decagon. This is what we wanted to prove.

---

[19] This figure does not appear in the manuscript.
[20] Ibn al-Haytham is aware that the fourteenth and fifteenth books do not form part of Euclid's *Elements*. He knew their author as Hypsicles.

<22> If in a circle we draw one of the diameters, if we divide one of the semicircles into two equal parts and we draw from the point of division an arbitrary straight line that cuts the diameter and ends on the circumference, if we draw from its endpoint two straight lines to the two endpoints of the diameter and if from the point of division we also draw two straight lines to the two endpoints of the diameter, then the quadrilateral that is formed is half the square of the straight line drawn from the point of division.

Example: Let there be a circle *ABCD* in which we draw the diameter *AC*. We divide the arc *ADC* into two equal parts at the point *D*; we draw the arbitrary straight line *DEB* and we join the straight lines *AB*, *BC*, *CD* and *DA*.

I say that the quadrilateral <ABCD *is equal*> *to half the square of the straight line* BD.

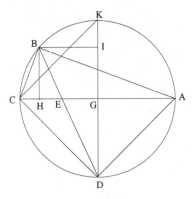

Fig. I.22

Proof: We mark the centre <*G*, we join> *DG* and we extend it to *K*; *DGK* will be perpendicular <to *AC*. We drop from the point> *B* the two perpendiculars *BH*, *BI*, then *AC* by <*BH*> is twice the triangle *ABC*, <and the product of *AC* and *DG*> is twice the triangle *ADC*, so the product <of *AC*> and *BH* <is equal to the product of *AC* and *GI* and the product of *AD* and *DC* is equal> to twice the square of *DG*. But *GC* is equal to *GD*, so the product <of *GC* and *DI* is equal to the quadrilateral> *ABCD*. But *AC* <is equal to *DK*> and the product of *KD* <and *DI* is equal to the square of *BD*>, so the quadrilateral <*ABCD* is equal to half the square of *BD*. This is what we wanted to prove>.

<23> In a circle we draw one of the diameters, we divide one of the semicircles into two equal parts, then on one of its quarters we take two points and we draw from each of them two straight lines to the endpoints of

the diameter and a straight line to the mid point, then the square of the greater sum of the two straight lines, if they become one single straight line, exceeds the square of the smaller sum of the two straight lines, if they become one single straight line, by twice the excess of the square of the greater of the two straight lines that join the two points to the mid point of the arc over the square of the smaller.

Example: Let there be a circle *ABC*; we draw the diameter *AC*, we divide the arc *ABC* into two equal parts at the point *B*, on the arc *BC* we take two points *D* and *E* and join the straight lines *AD*, *DC*, *AE*, *EC*, *EB*, *DB*.

I say that the excess of the square of the sum of AD *and* DC, *if they become one single straight line, over the square of the sum of* AE *and* EC, *if they become one single straight line, is twice the excess of the square of* EB *over the square of* DB.

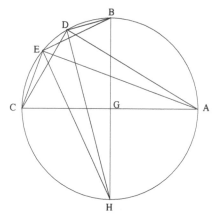

Fig. I.23

Proof: We mark the centre, let it be *G*; we join *BG* and we extend it to *H*, then the arc *AHC* is divided into two equal parts at the point *H*. We join *HD*; then the square of the sum of *AD* and *DC* – if they become a single straight line – is twice the square of *HD*; in the same way the square of the sum of *AE* and *EC* – if they become a single straight line – will be twice the square of *EH*, as we have proved in Proposition 18. So the excess of the square of the sum of *AD* and *DC* over the square of the sum of *AE* and *EC* is equal to the excess of twice the square of *HD* over twice the square of *HE*. But the excess of twice over twice is twice the excess of half over half. So the excess of the square of the sum of *AD* and *DC* over the square of the sum of *AE* and *EC* is twice <the excess of the square of *HD*> over the square of *HE*. But the excess of the square of *HD* over the square of *HE* is the excess of the <square of *EB* over the square of> *BD*.

The excess of the square of the sum of *AD* and *DC* over the square of the sum of *AE* and *EC* is thus <twice the excess of the square of *EB* over the square> of *DB*. This is what we wanted to prove.

<**24**> In a circle we draw <two perpendicular diameters and we join> the endpoints of one of these two diameters to one endpoint <of the other diameter. We draw from one endpoint> of the first diameter an <arbitrary> straight line to the chord that is opposite it; <from the point of intersection we drop a perpendicular to> the other diameter, then the product of the semidiameter <and the straight line cut off on it by this perpendicular is equal to the> triangle on the side towards the endpoint of the diameter <and which has as its base the straight line drawn from the endpoint of the first diameter.>

<*Example*: Let there be a circle *ABCD*>; the two diameters *AC* and *BD* cut one another <at the point *E*. We join *CB* and from the point *A* we draw a straight line that cuts *CB* in an> arbitrary <point>, let it be *G*; from <*G*> we drop <the perpendicular *GH* to *BD*.

I say that the product of EB *and* BH> *is equal to the triangle* ABG.

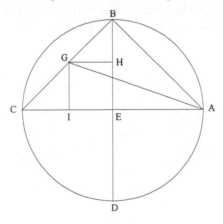

Fig. I.24

Proof: Let us draw the perpendicular *GI*, we have that the product of *AC* and *IG* is equal to twice the triangle *AGC*. But *GI* is equal to *HE*, so the product of *AC* and *EH* is twice the triangle *AGC*; but the product of *AC* and *EB* is twice the triangle *ABC*, so finally the product of *AC* and *BH* is equal to twice the triangle *ABG*; but *AC* is twice *EB*, so the product of *EB* and *BH* is equal to the triangle *ABG*. This is what we wanted to prove.

<25> Let us return to the circle and the two diameters; let the centre be <the point> *N* and on the arc *BC* let us take two points *G* and *E*. We join the straight lines *AG*, *GC*, *AE*, *EC*, *EB* and we draw the perpendiculars *GH* and *EI*; let the perpendicular *GH* cut the straight line *EB* at the point *K* and let the perpendicular *EI* cut the straight line *GC* at the point *M*; we join *AM*.

I say that the product of BE *and* EK *is twice triangle* AGM.

Proof: We join *DE*, so the product of *DB* and *BH* is equal to the product of *EB* and *BK*, because the two triangles *BED* and *BHK* are similar since each of the angles *E* and *H* is a right angle. But the product of *DB* and *BI* is equal to the square of *BE*; finally the product of *DB* and *HI* is equal to the product of *BE* and *EK*. In the same way, the product of *DB* and *NH* is equal to twice the triangle *AGC*, because *NH* is equal to the perpendicular dropped from the point *G* onto the straight line *AC* and *BD* is equal to *AC*, and the product of *BD* and *NI* is twice the triangle *AMC*, so the product of *DB* and *HI* is equal to the excess of twice the triangle *AGC* over twice the triangle *AMC*. So the product of *DB* and *HI* is twice the excess of the triangle *AGC* over the triangle *AMC*. But the product of *BE* and *EK* <is equal to the product of *DB* and *HI*>, so the product of *BE* and *EK* is equal to twice the excess of the triangle *AGC* over the triangle *AMC*. But the excess of the triangle *AGC* over the triangle *AMC* is equal to the triangle <*AMG*, because the straight lines *GH* and *MI*> are parallel, so the product of *BE* <and *EK* is equal to twice the triangle *AMG*. This is what we wanted to prove>.

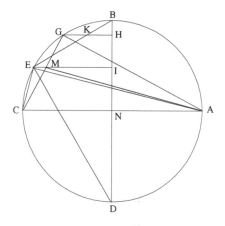

Fig. I.25[21]

---

[21] The figure in the manuscript is incorrect.

<26> If in two concentric circles <we draw a straight line that cuts the two circles and passes through their centre; and if through one of the points of intersection we draw the tangent to the small circle that meets the large circle when we extend it, then the chord of the arc cut off on> the large circle is equal <to the diameter of the small circle, the tangent is divided into two equal parts at the point of contact, and the square of the tangent plus the square of the diameter of the small circle is equal to the square of the diameter of the large circle>.

Example: Let there be circles <*ABC* and *DHE* with centre *G*; we draw from the point *A*> the straight line *AH* <a tangent to the small circle at *H* and we extend it to the point *B* on the large circle>.

<*I say that the chord of the arc* BC *is equal to the diameter of the circle* DHE, *that the straight line* AB *is divided into two equal parts at* H *and that the sum of the square of* AB> *and the square of the diameter of the circle* DHE *is equal to the square of the diameter of the circle* ABC.

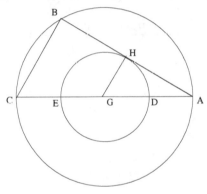

Fig. I.26

Proof: We join *AG* and we extend it to *C*; let it cut the circle *DHE* at the points *D* and *E*; so *AC* is the diameter of the large circle and *DE* the diameter of the small circle. We join *GH* and *CB*; then the angle *H* is a right angle and the angle *B* is a right angle, so the straight line *CB* is parallel to the straight line *GH*, and the ratio of *BA* to *AH* is equal to the ratio of *CA* to *AG*. But *CA* is twice *AG*, so *BA* is twice *AH*. So the tangent is divided into two equal parts by the point of contact. But since *CA* is twice *AG*, accordingly *CB* is twice *GH* and *GH* is the semidiameter of the circle *DHE*. The straight line *CB* is thus equal to the diameter of the circle *DHE*, and the sum of the square of *AB* and the square of *BC* is equal to the square of *AC*.

For any straight line that touches the circle *DHE*, if from one of its two endpoints we draw a diameter of the <large> circle and if we join the other endpoint to <the other> endpoint of the diameter, then the straight line that joins them is equal to the diameter of the small circle; thus, the arcs cut off by the tangent,[22] [which are] homologous to the arc *BC*, are always equal. Thus, the arcs cut off by the tangents are equal, so the tangents are equal. This is what we wanted to prove.

<27> If in two concentric circles we draw a straight line that cuts the two circles, then the two parts of the straight line that lie between the two circles are equal and the sum of the square of the part that is inside the small circle and the square of the tangent is equal to the square of the complete straight line.

Example: Let there be the two circles *ABC* and *DE* whose centre is *G*; we draw the straight line *BEDH* which cuts the two circles.

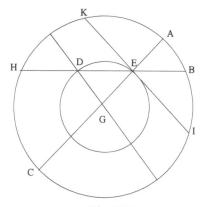

Fig. I.27

I say that <the sum of the square> of DE *<and the square> of the tangent is equal to the square of* BH.

Proof: <From the centre of the two circles>, let it be *G*, <we draw> the straight line *GEA*, we extend it <to the point *C* and we draw the tangent> to the small circle, let it be *EI*; <then the product of *CE* and *EA* is equal to the square of *EI*; but the product of *CE* and> *EA* is equal to the product of *HE* and *EB* <which is equal to the square of *EI*. We draw from the> point *G* a straight line to the point *D* and we extend it; <so the point *D* divides the diameter into two straight lines> equal to the two straight lines *AE* and *EC*; <the product of *HD* and> *DB* is equal to the product of <*HE* and *EB*; so the straight line *HD* is equal to the straight line *BE*. The product of *HE* and *EB* is equal to the square of *EI*>; in the same way the product of *CE* and *EA* is

---

[22] We are dealing with arcs that are homologous with the arc intercepted by the angle between the tangent and the diameter.

equal to the square of *EI*. But *EI* is equal to *EK*, so four times the product of *CE* and *EA* is equal to the square of *IK* and four times the product of *HE* and *EB* is equal to the square of *IK*. But *HD* is equal to *EB*, so *HE* is equal to *DB*, so four times the product of *DB* and *BE* is equal to the square of *IK*. But four times the product of *DB* and *BE*, plus the square of *ED*, is equal to the square of the sum of *DB* and *BE* which is *BH*; so the sum of the square of the tangent and the square of *ED* is equal to the square of *BH*. This is what we wanted to prove.

<28> If in two concentric circles we draw a straight line that cuts the two circles but does not pass through the centre and if, from one of the two points of intersection between the secant and the small circle, we draw a perpendicular to the secant, then the sum of the square of the secant and the square of the perpendicular is equal to the square of the diameter of the large circle.

Example: Let there be the two circles *ABC* and *EIG* with centre *H*. We draw the straight line *BEID* which cuts the two circles but does not pass through the centre. From the point *I* we draw the perpendicular *IG*.

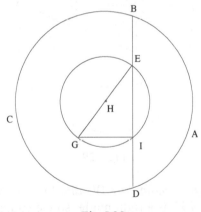

Fig. I.28

I say that the sum of the square of BD *and the square of* IG *is equal to the square of the diameter of the large circle.*

Proof: We join *EG*; it will be a diameter, because the angle *EIG* is a right angle; so the arc *EIG* is a semicircle and the straight line *EG* is a diameter of the circle *IG*. But the square of *BD* is four times the product of *DE* and *EB*, plus the square of *EI*, so the sum of the square of *BD* and the square of *IG* is four times the product of *DE* and *EB*, plus the square of *EI*, plus the square of *IG*; but the sum of the square of *EI* and the square of *IG* is equal to the square of *EG*. Now four times the product of *DE* and *EB* is the square of the tangent, so the sum of the square of *BD* and the square of *IG* is the square of the tangent, plus the square of *EG*, which is the diameter of the small circle. Now the sum of the square of the tangent and the square of the diameter of the small circle is the square of the diameter of the large circle, as has been proved in Proposition 26. This is what we wanted to prove.

<29> If in two concentric circles we draw a straight line that cuts the two circles and does not pass through the centre, and if, from its endpoint on the large circle and from its point of intersection with the small circle, we draw two perpendiculars which fall inside the two circles, then the two perpendiculars are equal.

Example: Let there be two circles *ABC*, *DEG* with the same centre; in them we draw a straight line *ADEB* that cuts the two circles, and from the points *B* and *E* we draw two perpendiculars *BC* and *EG*.

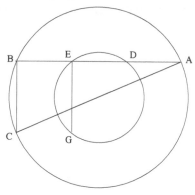

Fig. I.29

I say that the two perpendiculars BC *and* EG *are equal.*

Proof: The angle *ABC* is a right angle, so the straight line that joins the two points *A* and *C* is a diameter of the circle and the sum of the square of *AB* and the square of *BC* is equal to the square of the diameter of the large circle. But it has been proved in the previous proposition that the sum of the square of *AB* and the square of *EG* is equal to the square of the diameter of the large circle, so the sum of the square of *AB* and the square of *BC* is equal to the sum of the square of *AB* and the square of *EG* and the straight line *BC* is equal to the straight line *EG*. This is what we wanted to prove.

<30> If in two concentric circles we draw a diameter of the small circle, if from the endpoints of the diameter we draw two straight lines that

cut the small circle and meet one another on the circumference of the large circle and if we extend them until they meet <the circumference> of the large circle on the other side, then the arc cut off by these two straight lines on the large circle is similar to the sum of the two arcs cut off from the small circle.

Example: Let there be two circles *ABCD*, *IHEG* in which we draw the diameter *IG*; we draw from the points *I* and *G* two straight lines *IHB* and *GEB* and we extend them to *C* and *D*.

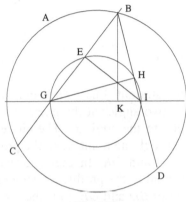

Fig. I.30

I say that the two arcs IH *and* GE *combined are similar to the arc* DC.

Proof: We join the straight lines *GH* and *IE* and we draw a perpendicular *BK*; so the triangle *BIK* will be similar to the triangle *GHI*; so the angle *IBK* is equal to the angle *IGH*, and the triangle *EGI* is similar to the triangle *BKG*; so the angle *GIE* is equal to the angle *KBG*, so the sum of the angles *GIE* and *IGH* is equal to the angle *CBD*, so the sum of the two arcs *IH* and *GE* is similar to the arc *DC*. This is what we wanted to prove.

<31> If in two concentric circles we draw one of the diameters, if from the endpoints of the diameter of one of them we then draw two straight lines which meet one another on the circumference of the other circle, then the sum of their squares is equal to the sum of the squares of the two parts of the diameter of the large circle.

Example: Let there be two circles *ABC* and *DEI*, whose centre is the same point; we draw the diameter *ADIC* and we draw from the points *D* and *I* the two straight lines *DE* and *IH* which meet one another at the point *B*.

I say that the sum of the squares of DB *and* BI *is equal to the sum of the squares of* AD *and* DC.

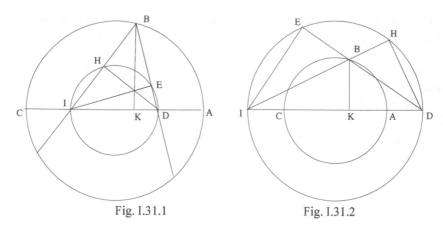

Fig. I.31.1   Fig. I.31.2

Proof: We join the straight lines *DH* and *IE* and we draw the perpendicular *BK*. So the two angles at the points *E* and *H* are right angles and each of them is equal to the angle *K*. The circle circumscribed about the triangle *IEB* passes through the point *K*, so the product of *BD* and *DE* is equal to the product of *ID* and *DK*. In the same way we prove that the product of *BI* and *IH* is equal to the product of *DI* and *IK*. So the square of *DI* is equal to the product of *BD* and *DE*, plus the product of *BI* and *IH*. So if they [*i. e. DE* and *IH*] meet on the circumference of the large circle, in the first case of the figure, we add the product of *DB* and *BE* to the product of *IB* and *BH*, and their sum is equal to twice the product of *IA* and *AD*, because if we extend *BD* until it ends on the large circle, then the part outside [the small circle] will be equal to *BE*; so the sum of the squares of *DB* and *BI* is equal to twice the product of *IA* and *AD*, plus the square of *DI*. But the sum of twice the product of *IA* and *AD* and the square of *DI* is the sum of the squares of *AD* and *AI*, that is to say *DC*. But if they meet on the circumference of the small circle, as in the second case for the figure, we subtract the product of *DB* and *BE* and that of *IB* and *BH* whose sum is twice the product of *DA* and *AI*, finally the [sum of the] squares of *DB* and *BI* is equal to the square of *DA* plus the square of *AI*, so the sum of the squares of *DB* and *BI* is equal to the sum of the squares of the two parts of the diameter. This is what we wanted to prove.

<**32**> If two circles touch one another and if from the point of contact we draw a straight line that cuts the two circles, then the two alternate of interior parts of the two circles from the inside of the contact or from the outside are similar, and the ratio of the two straight lines one to the other is equal to the ratio of one diameter to the other.

Example: Let there be the two circles *ABC* and *CDE* which touch at the point *C*. In the two circles we draw the straight line *BCD*.

<*I say that the two arcs* CB *and* CD *are> similar <and that the ratio of* CB *to* CD *is equal to the ratio of one diameter to the other>*.

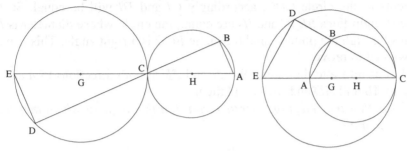

Fig. I.32

Proof: We mark the two centres, let them be *H* and *G*; we join *HG*; it passes through the point *C*. We extend *HG* to *A* and *E* and we join *AB* and *ED*; the two angles *B* and *D* are right angles, so the two straight lines *AB* and *ED* are parallel, the two angles *BAC* and *CED* are equal, the two arcs *BC* and *DC* are similar, the two arcs *AB* and *ED* that remain are also similar, and the ratio of *BC* to *CD* is equal to the ratio of *AC* to *CE*. This is what we wanted to prove.

<33> If two circles touch on their exteriors, if we draw a tangent to the two circles and if we join its endpoints to the point of contact, then the angle produced is a right angle.

Example: Let there be the circles *ABC* and *CDG* touching at the point *C*; let their centres be *E* and *K*. We draw a straight line *BD* tangent to the circles at the points *B* and *D* and we join <*BC*> and *DC*.

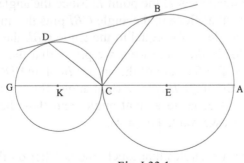

Fig. I.33.1

I say that the angle BCD *is a right angle.*

Proof: We join *EK*; it passes through the point *C*. We draw the straight line *CI* perpendicular to the straight line *EK*; so *CI* is a tangent to the two circles. Since the two straight lines *CI* and *BI* are tangents to the circle *ABC*, *BI* and *CI* will be equal; but since the straight lines *CI* and *DI* are tangents to the circle *CDG*, accordingly *CI* and *DI* will be equal. So the three straight lines *BI*, *IC* and *ID* are equal, the circle whose diameter is *BD* passes through the point *C*, and the angle *BCD* is a right angle. This is what we wanted to prove.

Let us return to the figure. We extend *EK* in both directions to *A* and *G*, we join *AB* and *GD* and we extend them.

I say that they meet one another and that the angle at which they meet is a right angle.

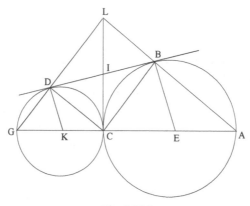

Fig. I.33.2

Proof: The two angles *BDG* and *DBA* are each greater than a right angle, so the sum of the two angles that are at the points *B* and *D* above the straight line *BD* is smaller than two right angles, so the two straight lines meet one another; let them meet at the point *L*. Since the angle *BCI* plus the angle *BCA* makes a right angle, and the angle *CBI* plus the angle *LBI* make a right angle and the angle *BCI* is equal to the angle *CBI*, the angle *BCA* is equal to the angle *LBI*. In the same way, we prove that the angle *DCG* is equal to the angle *LDI*. But the sum of the angles *BCA* and *DCG* is equal to a right angle, because the angle *BCD* is a right angle. So the sum of the two angles *DBL* and *LDB* is equal to a right angle, and thus the angle *L* is a right angle. This is what we wanted to prove.

<34> Let there be two circles that touch one another on the inside. We draw <two straight lines> that cut the two circles in arbitrary points and we join the point of contact and the points of intersection of the two straight

lines [with the circles], then the ratio of the product of two parts of one of the two straight lines, one with the other, to the square of the straight line that lies between the point of intersection and the point of contact, is equal to the ratio of the product of the two parts of the other straight line, one with the other, to the square of the straight line that lies between the point of intersection and the point of contact.

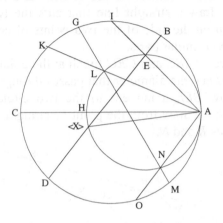

Fig. I.34[23]

Example: Let there be two circles *ABC* and *AEH* which touch at the point *A*. In these circles we draw the straight lines *BED* and *GLM* and we join *AE*, *AL*.

I say that the ratio of the product of BE *and* ED *to the square of* AE *is equal to the ratio of the product of* GL *and* LM *to the square of* LA *and that the ratio of the product of* GL *and* LM *to the square of* LA *is equal to the ratio of the product of* GN *and* NM *to the square of* NA.

Proof: We draw the straight lines *AE*, *AL* and *AN* to the points *I*, *K* and *O*, and we draw the common diameter; let it be *AHC*. So the ratio of *CA* to *AH* is equal to the ratio of *KA* to *AL*, which is equal to the ratio of *IA* to *AE* and equal to the ratio of *OA* to *AN*, as has been proved in Proposition 32. So the ratio of *IA* to *AE* is equal to the ratio of *KA* to *AL* and equal to the ratio of *OA* to *AN*; and the ratio of *IE* to *EA* is equal to the ratio of *KL* to *LA* and is equal to the ratio of *ON* to *NA*. So the ratio of the product of *KL* <and> *LA* to the square of *AL* is equal to the ratio of the product of *IE* and *EA* to the square of *EA* and is equal to the ratio of the product of *ON* and *NA* to the square of *NA*; so the ratio of the product of *BE* and *ED* to the

[23] In the figure in the manuscript, *GM* is parallel to *BD*.

square of *EA* is equal to the ratio of the product of *GL* and *LM* to the square of *LA* and is equal to the ratio of the product of *GN* and *NM* to the square of *NA*. This is what we wanted to prove.

<35> If we have two circles that touch on their exterior, if we draw a straight line which touches the two circles and which meets the diameter that passes through the two centres of the circles[24] and if from the point where they meet we draw a straight line that cuts the two circles in an arbitrary manner, then on the side of the two points of contact it cuts off from the two circles two similar parts.

Example: Two circles *ABC* and *CEG* touch at the point *C*, their centres are the points *D* and *H* and the diameter that passes through the two centres is *LDCHGI*. We draw *BEI*, a tangent to the two circles, it meets the diameter at the point *I*; we draw from the point *I* a straight line that cuts the two circles at the points *K* and *N*.

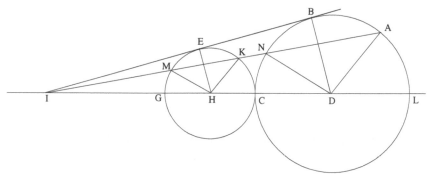

Fig. I.35.1

I say that the two parts [of circles] ABN and KEM are similar.

Proof: We join the straight lines *DA*, *DB*, *DN*, *HK*, *EH*, *HM*. Since the angles *DBI* and *HEI* are right angles, accordingly the two straight lines *DB* and *HE* are parallel, so the ratio of *DB* to *HE* is equal to the ratio of *DI* to *IH*. But *DB* is equal to *DA*, and *HE* is equal to *HK*. In the same way, *DN* is equal to *DB* and *HM* is equal to *HE*, so the ratio of *DI* to *IH* is equal to the ratio of *DA* to *HK* and is equal to the ratio of *DN* to *HM*, so the straight line *DA* is parallel to the straight line *HK*, and *DN* is parallel to the straight line *HM*,[25] so the angle *ADN* is equal to the angle *KHM* and the arc *ABN* is similar to the arc *KEM*. But since the angles that are at the point *D* are

---

[24] This assumes that the circles are unequal.
[25] That these lines are parallel follows from Euclid, *Elements*, VI.7.

equal to the angles that are at the point H,[26] the arc NC will also be similar to the arc MG, the arc AB will be similar to the arc KE, the arc BN will be similar to the arc <EM and the arc AL is similar to the arc> KC. This is what we wanted to prove.

Let us return to the figure and let us join the straight lines AC and KC.
I say that the angle ACK *is a right angle.*

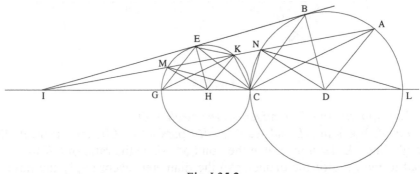

Fig. I.35.2

Proof: We join BC and CE, so the angle BCE is a right angle, from what has been proved in Proposition 33. Since the arc AB is similar to the arc <KE>, the angle ACB is equal to the angle EGK and the angle ECK is <their> common <supplement>, so the angle ACK is equal to the angle BCE; but the angle BCE is a right angle, so the angle ACK is a right angle.

In the same way, we prove that if we join the points N and M to the point C by two straight lines, then the angle formed at the point C is a right angle. This is what we wanted to prove.

<36> If we join the centres of two separate circles with a straight line and if we divide the part that lies between the two circles into two parts such that the ratio of one to the other is equal to the ratio of the diameter to the diameter, then the straight line drawn from the point of division and tangent to one of the circles is tangent to the other one, and the straight line drawn from the point of division and which is a secant for one of the two circles is a secant for the other circle and cuts off from the two circles two similar alternate parts.

<Example>: Let there be two separate circles ABC and DEG, their centres H and I; we draw the diameter that passes through their centres; let it be $AHCDIG$. Let us make the ratio of CK to KD equal to the ratio of AC

---

[26] The equality of the angles with vertex D and their homologues with vertex H follows from the straight lines DB, DC, DA, DN, DL being parallel to the straight lines HE, HG, HK, HM, HC that are their homologues in the homothety (I; IH/ID).

to *DG*, let us draw the straight line *KL* to be a tangent to the circle *ABC* and let us extend it in the direction towards *K*.

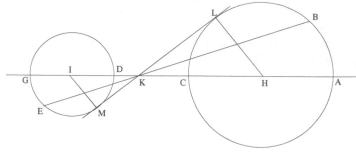

Fig. I.36

I say that the line is a tangent to the circle DEG.

Proof: We join *HL* and we draw *IM* parallel to *LH*, then it meets the straight line *LK*; let it meet it at the point *M*. Since the ratio of *CK* to *KD* is equal to the ratio of the diameter to the diameter, accordingly the ratio of *CK* to *KD* is equal to the ratio of *HC* to *DI* and is equal to the ratio of the whole to the whole, so the ratio of *HK* to *KI* is equal to the ratio of the semidiameter to the semidiameter, and the ratio of *HK* to *KI* is equal to the ratio of *HL* to the semidiameter <of the second circle>. But since *IM* is parallel to *HL*, the ratio of *HK* to *KI* is equal to the ratio of *HL* to *IM*, accordingly the straight line *IM* is the semidiameter of the circle *DEG*, and the point *M* lies on the circumference of the circle. But since *IM* is parallel to *HL*, and the angle *HLK* is a right angle, the angle *IMK* is a right angle, so the straight line *KM* is a tangent to the circle.

<37> In the same way, we draw from the point *K* a straight line that cuts the circle *ABC*, let the straight line be *KNB*; then it cuts the circle *DEG*, because it cuts the tangent. Let it cut the circle *DGE* at the points *O* and *E*.

I say that the part BLN *is similar to the part* OME.

Proof: We join the straight lines *HB*, *HN*, *IO* and *IE*; the ratio of *HK* to *KI* is equal to the ratio of *HB* to *IE* and is equal to the ratio of *HN* to *IO*, so the straight line *HB* is parallel to the straight line *IE*, and the straight line *HN* is parallel to the straight line *IO* and *HL* is parallel to *IM*; so the angles at the point *H* are equal to the angles at the point *I*, each of the angles is equal to its homologue, so the arcs intercepted by the equal angles are similar, the arc *CN* is similar to the arc *DO*, the arc *NL* is similar to the arc *OM*, the arc *LB* is similar to the arc *ME*, the arc *NB* is similar to the arc *OE* and the arc *AB* is similar to the arc *GE*; so the straight line *BNKOE* has divided the two circles into alternate similar arcs. This is what we wanted to prove.

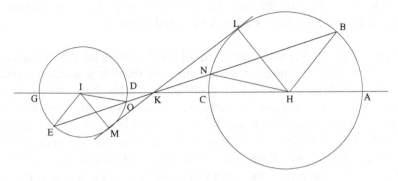

Fig. I.37

<38> Let us return to the figure. We join the straight lines *LC* and *GM*.

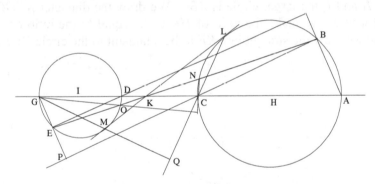

Fig. I.38

I say that they meet and enclose a right angle.

Proof: The arc *AL* is similar to the arc *GM*, so the angle *LCA*, plus the angle *MGD*, is a right angle; so if we draw the straight line *LC* in the direction towards *C*, then the angle which is below the straight line *GC*, plus the angle *MGD*, is a right angle. So the two straight lines meet; let them meet at the point *Q*, then the angle *Q* is a right angle.

In the same way, we join the straight lines *BC* and *GE*.

I say that they meet and enclose a right angle.

Proof: The arc *AB* is similar to the arc *GE*, so the angle *BCA*, plus the angle *EGD*, is a right angle; so if we draw the straight lines *BC* and *GE*, they meet one another; let them meet at the point *P*, so the angle *P* is a right angle.

In the same way, we prove that if we join the straight lines *AB* and *ED*, they meet one another and enclose a right angle. In the same way, we prove that if we join the straight lines *NC* and *OG*, they meet one another and enclose a right angle. This is what we wanted to prove.

<39> If in two separate and unequal circles we draw the diameter that passes through their centres, if we extend it on the side towards the smaller of the two circles and if we take a point on it outside the small circle, such that the ratio of the straight line between the centre of the large circle and this point to the straight line that lies between this point and the centre of the small circle is equal to the <ratio of the diameter to the diameter>, then the straight line drawn from this point and a tangent to one of the two circles will be a tangent to the other circle and any straight line drawn from this point that cuts one of the two circles, cuts the other circle and cuts off similar arcs from the two circles.

Example: Let there be two circles *ABC* and *DEG* whose centres are the points *H* and *I*; the larger circle is *ABC*. We draw the diameter *ACDG*, we extend it to *K*, we make the ratio of *HK* to *KI* equal to the ratio of *AC* to *DG* and we draw the straight line *KE* to be a tangent to the circle *DGE*.

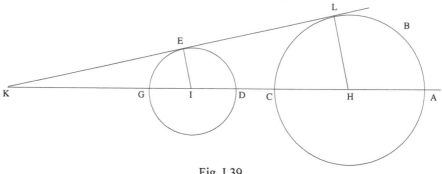

Fig. I.39

I say that if we extend it, it will be a tangent to the circle ABC.

Proof: We join *IE*; then the angle *E* is a right angle. We draw from the point *H* a straight line parallel to the straight line *IE*, then it meets the straight line *KE*; let it meet it at the point *L*. Since *HL* is parallel to *IE*, the ratio of *HK* to *KI* is equal to the ratio of *HL* to *IE*; but the ratio of *HK* to *KI* is equal to the ratio of the diameter *AC* to the diameter *DG*, so it is equal to the ratio of the semidiameter *AC* to the semidiameter *DG*, so the ratio of *HL* to *IE* is equal to the ratio of the semidiameter *AC* to the semidiameter *DG*. But *IE* is half of *DG*, so the straight line *HL* is half of *AC*, the point *L* lies on the circumference of the circle *ABC*; now the angle *HLK* is a right angle, so the straight line *KL* is a tangent to the circle *ABC*.

In the same way, we prove that if *KL* is a tangent to the circle *ABC*, then it is a tangent to the circle *DEG*.

<40> In the same way, we draw the straight line *KON* cutting the circle *DEG* and we extend it; it is clear that it also cuts the circle *ABC*, because it lies between the diameter and the tangent; let it cut it at the points *B* and *M*. We prove in the same way that if the straight line *KB* cuts the circle *ABC*, it is clear that it also cuts the circle *DEG*.

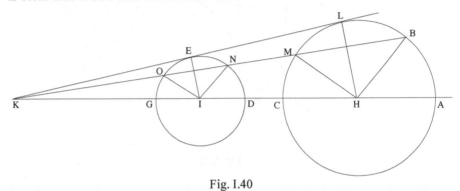

Fig. I.40

I say that this straight line cuts off similar arcs from the circles.
Proof: We join the straight lines *HB*, *HM*, *IN*, *IO*, then the ratio of *HK* to *KI* will be equal to the ratio of *HB* to *IN* and equal to the ratio of *HM* to *IO*, so <*HB*> is parallel to *IN* and *HM* is parallel to *IO*. But it has been proved that *HL* is parallel to *IE*, so the angles at the points *H* and *I* are equal – each is equal to its homologue; so the arcs *AB*, *BL*, *LM*, *MC* are similar to the arcs *DN*, *NE*, *EO*, *OG*. This is what we wanted to prove.

<41> If in two separate circles we draw a straight line that is a tangent to them, if we draw the diameter that passes through their centres and if we join the two points of contact and the points of intersection with two straight lines, then, if we extend them, they meet one another and enclose a right angle.

Example: Let there be the circles *ABC* and *DEG*, two separate circles; the straight line *BE* is a tangent to them and the straight line *ACDG* passes through their centres. We join the straight lines *BC* and *ED*.

I say that the straight lines BC *and* ED *meet one another and enclose a right angle.*

Proof: Each of the two angles *BCA* and *EDG* is less than a right angle, so their sum is less than two right angles; but the two opposite angles which are under the straight line *CD* are equal to them, so their sum is less than two right angles, so the two straight lines *BC* and *ED* meet one another below the straight line *CD*; let them meet at the point *H*. I say that the angle *BHE* is a right angle.

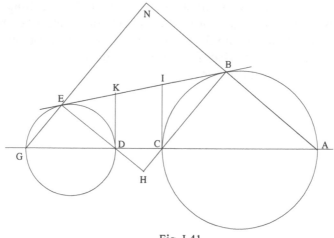

Fig. I.41

From the points *C* and *D* let us draw the two perpendiculars *CI* and *DK*, thus they are tangents to the two circles, the straight line *CI* is equal to the straight line *IB* and the straight line *DK* is equal to the straight line *KE*; so the angle *ICB* is equal to the angle *IBC*, and the angle *CIK* is twice the angle *CBI*. In the same way, we prove that the angle *DKI* is twice the angle *DEK*, so the sum of the two angles *CIK* <and *DKI*> combined is twice that of the two angles *CBI* and *DEK*. But the sum of the two angles *CIK* and *DKI* is equal to two right angles, so the sum of the two angles *CBI* and *DEK* is equal to a right angle; in conclusion the angle *CHD* is a right angle.

In the same way, we join the straight lines *AB* and *GE*.

I say that they meet one another and enclose a right angle.

Proof: The sum of the two angles *BAG* and *AGE* is less than two right angles, so the straight lines *AB* and *GE* meet one another; let them meet one another at the point *N*. Since *NBHE* is a quadrilateral, the sum of its four angles is equal to four right angles. But each of the angles at the points *B*, *H*, *E* is a right angle, so, finally, the angle *N* is a right angle. This is what we wanted to prove.

<42> If in two separate circles we draw the diameter that passes through their centres and we extend it, if we find the point from which one draws a tangent to the two circles, we draw from that point a straight line that cuts the two circles and we draw from each of the endpoints of the two similar parts cut off by this straight line, two straight lines to the endpoints

of the two diameters, and if we extend them, then they meet one another and enclose a right angle.

Example: Let there be two circles *ABC* and *DEG*. We draw the diameter *ACDG*; we extend it to *K*. Let the point *K* be the point from which one draws the tangent to the two circles. We draw the straight line *KLEMB* that cuts the two circles; it is clear that the arcs *AB* and *MC* are similar to the arcs *DE* and *LG*, as has been shown in Proposition 40. We join *BC* and *ED*.

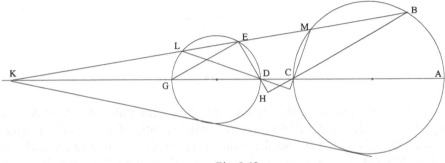

Fig. I.42

I say that the two straight lines BC *and* ED *meet one another and enclose a right angle.*

Proof: The arc *AB* is similar to the arc *DE*, so the sum of the angle *BCA* and the angle *EDG* is equal to a right angle. If we extend the two straight lines *BC* and *ED*, then the two angles formed below the straight line *CD* have a sum of a right <angle>, so the two straight lines meet one another below the straight line *CD*; let them meet one another at the point *H*; so the angle *CHD* is a right angle. In the same way, if we join the straight lines *DL* and *MC*, they meet one another and enclose a right angle. This is what we wanted to prove.

<43> Let there be two circles one of which encloses the other, and whose centres are different and let the circles not touch. We draw the common diameter and we divide the diameter of the small circle into two parts such that the ratio of one to the other is equal to the ratio of the two parts cut off on either side of the diameter of the small circle, one to the other. Any straight line drawn from the point of division and which cuts the two circles cuts off similar arcs from the two circles.

Example: Let there be the circles *ABC* and *DEG* with centres *H* and *I*. We draw the diameter *ADGC*, we divide *DG* into two parts at the point *K* and we make the ratio of *DK* to *KG* equal to the ratio of *AD* to *GC*. We draw the straight line *KEB* which we extend to *M* and *L*.

I say that the arcs CB, BA, AL *are similar to the arcs* GE, ED, DM.

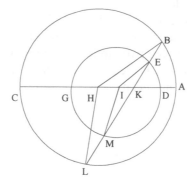

Fig. I.43

Proof: We join *IE*, *HB*, *IM* and *HL*. Since the ratio of *DK* to *KG* is equal to the ratio of *AD* to *GC*, accordingly the ratio of *AD* to *DK* is equal to the ratio of *CG* to *GK*. So the ratio of *AK* to *KD* is equal to the ratio of *CK* to *KG* and is equal to the ratio of the whole to the whole, that is to the ratio of *AC* to *DG*. And the ratio of *CK* to *KG* is equal to the ratio of the diameter *AC* to the diameter *DG*, so it is equal to the ratio of *CH* to *GI* and equal to the ratio of the remainder, which is *HK*, to the remainder, which is *IK*; so the ratio of *HK* to *KI* is equal to the ratio of the semidiameter to the semidiameter, the ratio of *HK* to *KI* is equal to the ratio of *BH* to *EI* and *BH* and *IE* are parallel; the angles *BHC* and *EIG* are equal, so the arc *BC* is similar to the arc *EG* and finally the arc *BA* is similar to <the arc> *ED*. But the ratio of *HK* to *KI* is also equal to the ratio of *HL* to *IM*, so the two straight lines *HL* and *IM* are parallel, the two angles *AHL* and *DIM* are equal and the two arcs *AL* and *DM* are similar; finally the arcs *LC* and *MG* are similar, so the arc *BAL* is similar to the arc *EDM* and the arc *BCL* is similar to the arc *EGM*. This is what we wanted to prove.

It is time to end this treatise. Thanks be given to God and to the benefit of His help; blessing and peace be upon Muḥammad and all that are his.

CHAPTER II

THE ANALYTICAL ART
IN THE TENTH TO ELEVENTH CENTURIES

INTRODUCTION

1. *The rebirth of a subject*

Among the many writings that mathematicians dedicated to 'analysis and synthesis' before the mid seventeenth century, that is among those that have come down to us, two undoubtedly stand out: a treatise by Ibrāhīm ibn Sinān (296/909–335/946), called *On the Method of Analysis and Synthesis in Geometrical Problems*, and a treatise by Ibn al-Haytham also called *Analysis and Synthesis*, whose text is presented in translation here. In form and in content, both these texts differ from all the other writings we know on the subject. Whereas the Greek philosophers, mathematicians and physicians who discussed the matter, from the fourth century BC onwards, did so only briefly, and have left us only some fragments, Ibrāhīm ibn Sinān and Ibn al-Haytham each composed a substantial work entirely devoted to analysis and synthesis. In fact, the Greek mathematicians who discussed the matter can be counted on the fingers of one hand: there are some lines in pseudo-Euclid,[1] a short fragment in Pappus[2] and another in Proclus.[3] Not that the terms 'analysis' and 'synthesis' were unknown to Greek mathematicians – Archimedes, Apollonius, Diophantus and others – but none of them felt the need to discuss the meaning of the

---

[1] This apocryphal paragraph was inserted after the fifth proposition of Book XIII of the *Elements*. See the French translation of F. Peyrard, *Les Œuvres d'Euclide*, Paris, 1966, p. 486.

[2] *Pappi Alexandrini Collectionis ... quae supersunt e libris manu scriptis edidit latina interpretatione et commentariis instruxit F. Hultsch*, Berlin, 1876–8; Pappus d'Alexandrie, *La Collection mathématique*, trans. P. Ver Eecke, Paris, 1982, vol. II, pp. 477–8. The text, which is no more than the beginning of the preface to Book VII, has appeared in a new edition, see A. Jones, *Book 7 of the Collection*, New York, 1986.

[3] Proclus, *In Primum Euclidis Elementorum librum Commentarii*, ed. G. Friedlein, Leipzig, 1873; reprod. Olms, 1967, p. 255, 8–26. See also the French translation by P. Ver Eecke, *Proclus: Les Commentaires sur le premier livre des* Éléments *d'Euclide*, Bruges, 1948, pp. 220–1.

terms. It is one thing, as we know, to employ a procedure, to adopt a certain approach; but it is another, quite different matter to set out the ideas that underlie a subject, as constituting a method or forming something as large as an area of research. One possibility is to follow the example set by Archimedes and do no more than identify the stages of the procedure; a second possibility is to explain briefly what underlies the procedure, and then indicate how it can be used and the conditions for its applicability: this is what Pappus and Proclus do for analysis and synthesis. And Pappus, in fact, in a short text, takes the trouble to describe the approach taken by Euclid, Aristaeus the Elder and Apollonius, to remind us of the direction of reasoning in analysis and synthesis, of their reversibility, and to distinguish between theoretical analysis and problematical analysis, and finally to refer to the conditions for their applicability. Pappus took no more than a page to deal with all these explanations. This immediately accounts for the conflicting interpretations that sprang up around the text of the Alexandrian mathematician.[4]

Was this fragment of Pappus and the fragment of Proclus translated into Arabic? Did later mathematicians have indirect knowledge of the Greek texts that dealt with this subject? We do not know. At the moment, the only text that we know is that of Galen, who picks up the definition of analysis and synthesis.[5] On the other hand, we know that mathematicians and philosophers who were also mathematicians rediscovered this subject when reflecting on one or other of the mathematical disciplines. Thus, the mathematician Thābit ibn Qurra (d. 901) composed a short paper called *On the Means of Arriving at Determining of the Construction of Geometrical Problems.*[6] Although he never once uses the terms 'analysis' or 'synthesis', Thābit ibn Qurra is certainly operating in their subject area, or at least in an area near it. In contrast, al-Fārābī is more explicit: although he touches on

---

[4] See, for example, J. Hintikka and U. Remes, *The Method of Analysis*, Dordrecht, 1974; M. Mahoney, 'Another look at geometrical analysis', *Archive for History of Exact Sciences*, vol. V, no. 3–4, 1968, pp. 318–48; R. Rashed, 'L'analyse et la synthèse selon Ibn al-Haytham', in R. Rashed (ed.), *Mathématiques et philosophie de l'antiquité à l'âge classique: Hommage à Jules Vuillemin*, Paris, 1991, pp. 131–62; reprinted in *Optique et mathématiques: recherches sur l'histoire de la pensée scientifique en arabe*, Variorum Reprints, Aldershot, 1992, XIV; and A. Behhoud, 'Greek geometrical analysis', *Centaurus*, 37, 1994, pp. 52–86.

[5] On Galen's text, see R. Rashed, 'La philosophie mathématique d'Ibn al-Haytham. II: *Les Connus*', *Mélanges de l'Institut Dominicain d'Etudes Orientales du Caire* (*MIDEO*), 21, 1993, pp. 87–275, Appendix: 'Un fragment de l'*Ars medica* de Galien sur l'analyse et la synthèse', pp. 272–5.

[6] *Kitāb Thābit ibn Qurra ilā Ibn Wahb fī al-taʾattī li-istikhrāj ʿamal al-masāʾil al-handasiyya* (see Appendix, Text 1).

the subject only in passing in his *Classification of the Sciences,*[7] he gives an explanation of it in his long book *On Music.*[8] It is in the course of the tenth century that research on analysis and synthesis sees renewed and more intensive activity. Apart from the brief texts concerned with this subject that one finds here and there, it is in the tenth century that we see the development of three additional types of research, clearly associated with three distinct points of view.

We find collections of selected problems, all treated by analysis and synthesis, or by only one of the two methods. Ibn Sinān,[9] Ibn Sahl,[10] al-Sijzī,[11] and others chose to adopt this style of composition, and have left us substantial research texts. In contrast, we find writings intended solely for teaching, in which the author seems to be showing beginners, by means of examples, how to proceed by analysis and synthesis. This was, apparently, the intention behind one of the lost writings of the philosopher mathematician Muḥammad ibn al-Haytham (who is not to be confused with al-Ḥasan ibn al-Haytham), in the *Book on Geometrical Analysis and Synthesis as an Example for Beginners, a Collection of Geometrical and Arithmetical Problems that I have Analysed and for which I have Carried out the Synthesis* (*Kitāb fī al-taḥlīl wa-al-tarkīb al-handasiyyin 'alā jihat al-tamthīl li-al-muta'allimīn, wa-huwa majmū' masā'il handasiyya wa-'adadiyya ḥallalathā wa-rakabathā*). Finally, we come across texts whose object of study is indeed the subject of analysis and synthesis. These texts are intended for mathematicians, young or old, carrying out research and are distinct from the two preceding types of text. The writings of Ibn Sinān and of Ibn al-Haytham belong to this third category. We might add to them

---

[7] Al-Fārābī, *Iḥṣā' al-'ulūm*, ed. 'Uthmān Amīn, 3rd ed., Cairo, 1968, pp. 99–100. Al-Fārābī notes that, in the *Elements*, Euclid proceeds only by synthesis, whereas other ancient mathematicians proceed by analysis and synthesis.

[8] Al-Fārābī, *Kitāb al-mūsīqā al-kabīr*, edited by Ghattās 'Abd al-Malik Khashaba, revised and introduced by Maḥmūd Aḥmad al-Hifnī, Cairo, n.d., pp. 185 to 187 and p. 205.

[9] Ibn Sinān, *al-Masā'il al-Mukhtāra* (*Anthology of Problems*), in R. Rashed and H. Bellosta, *Ibrāhīm ibn Sinān: Logique et géométrie au Xe siècle*, Leiden, 2000, Chap. V.

[10] See the *Book on the Synthesis of the Problems Analysed by Abū Sa'd al-'Alā' ibn Sahl*, in R. Rashed, *Géométrie et dioptrique au Xe siècle. Ibn Sahl, al-Qūhī et Ibn al-Haytham*, Paris, 1993; English trans. *Geometry and Dioptrics in Classical Islam*, London, 2005, pp. 444–85.

[11] Al-Sijzī, *Fī al-masā'il al-mukhtāra allatī jarrat baynahu wa-bayna muhandisī Shīrāz wa-Khurāsān wa-ta'līqātihā* (*Selected Problems Raised Between Him and the Geometers of Shīrāz and Khurāsān, and Commentaries on Them*), ms. Dublin, Chester Beatty 3652/7, fols 35–52.

the text by al-Sijzī, and another, later one by al-Samaw'al.[12] Let us underline it once more: a simple examination of these treatises is enough to convince one that they were not addressed only to pupils studying mathematics, but to fully fledged mathematicians concerned with the foundations of their discipline and interested in the theory of proof. As we shall see, the examples chosen by Ibn al-Haytham included problems arising from the most advanced research: for example, Apollonius' problem of the construction of a circle touching three given circles.

This diversity among the texts dedicated to analysis and synthesis in the tenth century seems to reflect a new situation, involving motivations and issues we can only guess at. Here we have a subject on the frontier of mathematics, logic and philosophy, investigated by mathematicians of this period, but starting from a practice that was already old – almost a millennium old; a subject given a title, ἀναλυόμενος (sc. τόπος), in the statements made by Pappus, one that was reactivated by the mathematicians of the ninth and tenth centuries, although at the time when they were beginning to move away from Hellenistic mathematics: tenth-century mathematics had benefited from new disciplines such as that of algebra, and had embarked upon new areas of geometrical research, for instance on projections and transformations. The reasons for the reactivation of this subject must no doubt be sought in this somewhat paradoxical aspect of the state of mathematics.

Let us begin with Ibn Sinān. He made a major contribution, and what he says provides information about the period that saw the revival of interest in analysis and synthesis. According to Ibn Sinān, we are concerned with the first third of the tenth century. It is at about this time that mathematicians resume discussion of the subject, and in particular of the question whether synthesis is, strictly, the converse of analysis. Ibn Sinān writes:

> We have now said enough about the method of analysis that is used by geometers, about the criticisms that can be made of it, about what may be false in these criticisms [...].[13]

In the introduction to his treatise, he lets us know that some people criticise the analysis employed by geometers and complain that it presents a synthesis that is not a converse. These might be the same mathematicians

---

[12] These are *Kitāb fī tashīl al-subul li-istikhrāj al-ashkāl al-handasiyya* (see Appendix, Text 2), and the lost book by al-Samaw'al that he refers to in his treatise on algebra, *al-Bāhir*, ed. S. Aḥmad and R. Rashed, Damascus, 1972, p. 74 of the Arabic text.

[13] *The Method of Analysis and of Synthesis*, in R. Rashed and H. Bellosta, *Ibrāhīm ibn Sinān: Logique et géométrie au Xᵉ siècle*, p. 224.

that Ibn Sinān mentions in his *Anthology of Problems*.[14] It remains that Ibn Sinān reminds us that this discussion had then only begun to take place and that the subject of analysis and synthesis had not often been addressed (*lam yakthur ... al-khawḍ fīhi*).[15]

In short, according to an eyewitness, from the beginning of the tenth century, mathematicians return to the subject of analysis and synthesis: the community of geometers takes possession of it as providing material for debate.

Besides providing this first-rate historical evidence, Ibn Sinān's book also suggests two epistemological points of reference. One is spelled out by the author in the definition of the purpose of his book: to give an answer to the objections certain people addressed to the geometers who make the analysis too short, and also to provide a correction to the procedure itself, so as to establish the canons to be respected to avoid beginners making mistakes. These beginners are not, however, mere pupils, but beginners in research who are already capable of reflecting on mathematical arguments. This, it seems to us, is the didactic element in Ibn Sinān's project. The other point of reference is an indirect one that is implicit in Ibn Sinān's book, as indeed it is in those of all his successors: he refers to the mathematical context of the tenth century.

Anyone who studies the development of mathematics between the ninth and eleventh centuries notices the striking diversity of the work, which is unprecedented in history. If we are not aware of this and do not grasp the reasons for it, we are condemned to a profound misunderstanding of the history of mathematics in this period and, in attempting to mitigate this incomprehension, to lose ourselves in reductionism.

The heirs to Hellenistic mathematics had, naturally enough, accumulated methods and results over more than two centuries of active research, and were led to conceive of disciplines unknown to the Greeks: algebra, integer Diophantine analysis, the algebraic theory of cubic equations, and so on. Moreover, drawn on by works on astronomy, on optics and on statics, mathematicians had, in a way, renewed Hellenistic geometry and had introduced new fields. Among the renewed disciplines we find – as we have already noted – infinitesimal geometry, spherical geometry and so on; the new fields are concerned with the geometry of position and form, and the study of geometrical transformations.

---

[14] This means mathematicians like Abū al-'Alā' ibn Karnīb and a certain Abū Yaḥyā, who are quoted, among others, by Ibrāhīm ibn Sinān himself in the *Anthology of Problems*.

[15] *The Method of Analysis and of Synthesis*, in R. Rashed and H. Bellosta, *Ibrāhīm ibn Sinān: Logique et géométrie au Xᵉ siècle*, p. 99, 12.

The mathematical language associated with the *quadrivium* obviously could not accommodate such diversity, and mathematicians were already similarly constrained by operating within the language of the theory of proportions. So they began to look towards other types of proof, sometimes algebraic ones, with the help of proportions, but also folding one plane into another (rabatment) in order to carry out projections. Finally, the overall picture – increasing diversity and frozen forms of expression – called, so to speak as a matter of necessity, for logical investigation and philosophical explanation. Philosophers such as al-Fārābī seem to have anticipated some of the difficulties to which this situation gave rise. For instance, al-Fārābī devised a new ontology for mathematical entities,[16] and, when he composed his mathematical encyclopaedia, a framework different from that of the *quadrivium*, and also different from that of knowledge as a whole.[17] But, for reasons that were both theoretical and practical, it was up to mathematicians alone to address these difficulties. And indeed, in their writings on analysis and synthesis, mathematicians were not slow to face up to these difficulties. Henceforth, the encyclopaedic aspect of analysis and synthesis leads to a live problem that in this context is also not set out explicitly: to give an account of the new disciplines and restore the unity of mathematics.

Now, at the end of the ninth century and the beginning of the tenth, the term 'mathematical', and even the term 'geometrical', described a set of scattered disciplines that could no longer be enclosed within the narrow framework of the *quadrivium*. It was, moreover, no longer possible to collect all these disciplines under a single name, for example, that of 'theory of magnitudes'. Under these conditions, how was it possible to conceive the unity of mathematics? The question was as necessary as it was difficult: there was not at the time, and for a long time had not been, any means of arriving at that unity. Algebra was still far from being the discipline of algebraic structures that it was destined to become, and it was in no way formalised. It could effect only a few partial unifications: for instance, of the geometry of conics and the theory of equations. Algebra as the science of structures was yet to be created; mathematicians had no choice but to look for another approach: it was a matter of finding a discipline that was logically prior to all the other mathematical disciplines – but on the historical level necessarily later than all of them – so that it would be capable of providing unifying principles. However, no specification of the nature of that discipline, of its methods, of its objects, was clear *a priori*. Analysis

---

[16] R. Rashed, 'Mathématiques et philosophie chez Avicenne', in *Études sur Avicenne*, directed by J. Jolivet and R. Rashed, Paris, 1984, pp. 29–39.

[17] See al-Fārābī, *Iḥṣā' al-'ulūm*.

and synthesis obviously played the part of a unifying discipline. Ibn Sinān does not concern himself with mathematics as a whole, but solely with geometry; nevertheless the unification that he brings to bear affects the family of procedures for analysis and synthesis, and the arguments that are deployed, so to say independently of the areas of geometry in which they are applied. The discipline that provides justification for the method, that is analysis and synthesis as a discipline, is a sort of programmatic logic, insofar as it allows us to associate an *ars inveniendi* with an *ars demonstrandi*.

Ibn Sinān's contribution has a particular importance: as far as we know, it is the first substantial text in the philosophical logic of mathematics. The author has reduced the fundamental problem of the unity of geometry to the logico-philosophical discipline of analysis and synthesis, thereby setting in train a long tradition that we can trace all through the tenth century and as far as the twelfth-century algebraist al-Samaw'al. It is, also, following on from Ibn Sinān – in opposition to him – that Ibn al-Haytham constructs his own system.

With Ibn Sinān, we have not yet reached the middle of the great century. Mathematical activity is at its zenith. The differentiation between disciplines is underway; the geometry of projections takes a great stride forward with such mathematicians as al-Qūhī and Ibn Sahl;[18] geometrical transformations become objects of study and of application among mathematicians; a field of geometrical constructions using conic sections comes into being and develops.[19] In geometrical proofs, more and more use is made of rabatments, images of points and asymptotic properties of conic sections to prove the existence of their points of intersection. In short, two types of demand have now come to the fore: we need to design frameworks for proofs relating to the new mathematical objects and, at the same time, to provide levels of existence for them. These two purposes are closely connected, and accomplishing them requires, in turn, that the methods have a solid basis in a discipline. This discipline must be sufficiently general (but without reducing matters to pure logic) so as to be able to supply levels of existence for the new geometrical objects; but the discipline must also be logically prior to all the mathematical disciplines, so as to supply a foundation for the various frameworks for proofs. This was the monumental task to which Ibn al-Haytham addressed himself, no doubt as a

---

[18] See *Géométrie et dioptrique au Xe siècle*.

[19] See *Les Mathématiques infinitésimales du IXe au XIe siècle*, vol. III: *Ibn al-Haytham. Théorie des coniques, constructions géométriques et géométrie pratique*, London, 2000. English translation: *Ibn al-Haytham's Theory of Conics, Geometrical Constructions and Practical Geometry. A History of Arabic Sciences and Mathematics*, vol. 3, Culture and Civilization in the Middle East, London, 2013.

matter of deliberate choice, but also of necessity. He was the one who took innovative research furthest in all branches of geometry, but also in arithmetic and in Euclidean number theory. These are in fact the very domains in which he did most work.

2. Analytical art: discipline and method

Ibn Sinān's treatise is entirely concerned with analysis and synthesis of 'geometrical problems', and only with them. The main body of the book corresponds perfectly with the title it is given. However, towards the end of the treatise, in a sentence whose meaning is less than clear, Ibn Sinān seems to suggest a possible extension. This is what he writes:

> If you consider carefully what their intention is there, you will find that it leads to the method of true analysis, which is used in the other sciences.[20]

Ibn Sinān gives no further information, and does not explain what he means by 'other sciences'. Perhaps he is simply referring to the other mathematical sciences, or perhaps he means to point to other disciplines. In any case, Ibn Sinān promises to compose an exhaustive treatise on this matter. Ibn Sinān, a mathematical genius who died at the age of thirty-eight, never wrote this treatise.

It seems that only Ibn al-Haytham fulfilled Ibn Sinān's purpose: his treatise is not limited to geometry, but considers the mathematical disciplines as a whole, though he does not include algebra. Thus, he examines analysis and synthesis in arithmetic, in geometry, in astronomy and in music, as if he were taking a literal view of the divisions in the *quadrivium*.[21] But that is an illusion that more careful examination does not take long to correct: we shall see that the essence is still geometry.

Just as the forms of the writings by Ibn Sinān and Ibn al-Haytham are different, so too their purposes are not the same: Ibn Sinān is entering a domain, Ibn al-Haytham wants to lay the foundations of a discipline. But this difference, obviously a crucial one, may not be apparent at first read-

---

[20] *The Method of Analysis and of Synthesis*, in R. Rashed and H. Bellosta, *Ibrāhīm ibn Sinān: Logique et géométrie au Xe siècle*, p. 154.

[21] Commenting on a similar act, J. Hintikka wrote: 'This meaning of the term "analysis" was naturally extended from the analysis of geometrical configurations to the "analysis" of physical or astronomical configurations. This is roughly the sense in which the first great modern scientists speak of analysis' (J. Hintikka, 'Kant and the tradition of analysis', in Paul Weingartner [ed.], *Deskription, Analytizität und Existenz*, Salzburg-Munich, 1966, p. 258).

ing. To grasp it, let us start by listening to Ibn Sinān talking about his own purpose.

Thus I have established in this book, in an exhaustive manner, a method intended for students, which contains all that is necessary for the solution of problems in geometry. In it I have set out in general terms the various classes of problems in geometry; next I have subdivided these classes and I have illustrated each of them by an example; then I have guided the student towards the path thanks to which he will be able to know into which of these classes he should place the problems that he will be set, by which he will know how to carry out the analysis of the problems – as well as the subdivisions and conditions necessary for that – and to carry out their synthesis – as well as the conditions necessary for that – then how he will know if the problem is one of those that have one solution or more than one, and in a general way, all that it is necessary to know in this matter.

I have pointed out into what kind of error geometers fall in analysis, from the fact that they commonly have a habit that has come upon them: to shorten in an excessive way. I have also indicated for what reason there can, for geometers, appear to be, in the propositions and the problems, a difference between analysis and synthesis, and I have shown that their analysis differs from the synthesis only because of the fact of the abbreviations, and that, if they had carried out the analysis in full as should be done, it would have been identical to the synthesis; doubt would then depart from the hearts of those who suspect them of producing in the synthesis things of which they had not made mention earlier in the analysis, these things, lines, surfaces and other things, that we see figuring in their synthesis, without mention having been made of them in the analysis; I have shown that and I have illustrated it by examples. I have presented a method thanks to which the analysis is such that it coincides with the synthesis; I have warned against the things that the geometers tolerate in analysis, and I have shown what kind of error follows if we tolerate them.[22]

Ibn Sinān's intention is clear, and his purpose is well articulated: to classify geometrical problems according to different criteria (the number of conditions, the number of solutions, and so on) in order to show how to proceed, in each category, by analysis and synthesis, and in order to point out where errors can occur so as to allow them to be avoided. So it is essentially a matter of devising a programmatic and pragmatic logic in which the problem of irreversibility takes on a very special importance. In that connection, Ibn Sinān would have made a useful source for recent writings on analysis and synthesis.

---

[22] *The Method of Analysis and of Synthesis*, pp. 96–8.

From Ibn Sinān onwards, and unlike him, mathematicians worked in succession on two other projects. The first was the project of al-Sijzī. Having read Thābit ibn Qurra and Ibn Sinān, he wrote a text – translated and analysed here – in which he considers the question of discovery in geometry. He accordingly examines the many ways of making a discovery, as it were grouping them around a principal way, which is that of analysis and synthesis. That is he arrives at a conception of an *ars inveniendi*, though without naming it as such. The second is the scheme of Ibn al-Haytham. He starts from the works of his predecessors, Ibn Sinān of course, and very probably Thābit ibn Qurra and al-Sijzī, but his aim is different: he wants to lay the foundations for a scientific art with its own rules and vocabulary. This time the word is used, it is indeed a matter of an *art* and, in fact, of an analytic art. Here again Ibn al-Haytham shows himself to be as he always was in each of the various areas of mathematics: he completes the tradition to which he belonged. In this case the tradition is one that begins with Ibn Qurra and includes many well-known scholars, among them, notably, Ibn Sinān and al-Sijzī.

Ibn al-Haytham begins by reminding us that mathematics is based on proofs. By proof, he means 'the syllogism that indicates, necessarily, the truth of its own conclusion'.[23] This syllogism is in turn made up 'of premises whose truth and validity is recognised by the understanding, without being troubled by any doubt in regard to them; and [the syllogism] has an order and arrangement of these premises such that they compel the listener to be convinced by their necessary consequences and to believe in the validity of what follows from their arrangement'.[24] The Art of analysis (*Ṣinā'at al-taḥlīl*) provides the method for obtaining these syllogisms, that is 'to hunt (*taṣayyud*) for their premises, to seek the devices for grasping them, and to find their arrangement'.[25] In this sense, the Art of analysis is an *ars demonstrandi*. It is also an *ars inveniendi*, insofar as it is thanks to this art that we are led 'to determine the unknowns in the mathematical sciences, and how to proceed in pursuing the search (*taṣayyud*) for the premises, which are the basis for the proofs that show the validity of what we determine regarding the unknowns in these sciences, and the method for arriving at the arrangement of the premises and the structure of their combination'.[26]

For Ibn al-Haytham, it is indeed an *ars* (τέχνη, *ṣinā'at*) *analytica* that he must design and construct. Now, as far as I know, no one before him

---

[23] See p. 221.
[24] *Ibid.*
[25] *Ibid.*
[26] *Ibid.*

had considered analysis and synthesis as an art, or, more precisely, as a two-fold art, of proof and of discovery. In the first, the analyst (*al-muḥallil*) must be familiar with the principles (*uṣūl*) of mathematics. This knowledge must be underpinned by 'ingenuity', and by 'intuition in this art (*ḥads ṣinā'ī*)' (p. 222). This intuition is indispensable for discovery and proves to be equally necessary when the synthesis is not a strict converse of the analysis but requires supplementary data and properties that have to be discovered. Knowledge of principles, ingenuity and intuition: these are the qualities that the analyst must possess in order to find mathematical unknowns. And finally he must know 'the laws' and 'the principles' of this analytic art. This necessary knowledge is the object of a discipline that is concerned with the foundations of mathematics, and which deals with the 'knowns'. This knowledge itself is to be constructed. This last characteristic is peculiar to Ibn al-Haytham, insofar as before him no one, not even Ibn Sinān, had imagined there could be an analytic art based on a mathematical discipline of its own. Ibn al-Haytham devotes a second treatise to this discipline, *The Knowns*, which he had promised in his treatise *Analysis and Synthesis*.[27] He presents this new discipline as the one that provides the analyst with 'the laws' of this art and 'the foundations' which will allow the discovery of properties to be completed and the premises to be understood; that is to say the discipline forms the very basis of mathematics and, as we have said, it is indeed necessary to be already familiar with this basis in order to construct the complete art of analysis: these are the notions called 'the knowns'.[28] We may note that, each time that he is considering a foundational problem, as in his treatise *On the Quadrature of the Circle*,[29] Ibn al-Haytham returns to these 'knowns'.

According to Ibn al-Haytham, a notion is called 'known' when it is always the same and does not admit of change, whether or not that notion is something thought by a subject who has understanding. The 'knowns' designate the invariant properties, independently of whether we know about them, and remain unchanged even when the other elements of the mathematical object vary. The analyst's aim, according to Ibn al-Haytham, is to arrive at these invariant properties. Once these fixed elements are found, his task is completed, and he can then set out upon the synthesis.

---

[27] See p. 230.

[28] *Ibid.*

[29] R. Rashed, *Les Mathématiques infinitésimales du IXe au XIe siècle*, vol. II: *Ibn al-Haytham*, London, 1993, p. 91–5; English trans. *Ibn al-Haytham and Analytical Mathematics. A History of Arabic Sciences and Mathematics*, vol. 2, Culture and Civilization in the Middle East, London, 2012, pp. 99–106.

The *ars inveniendi* is neither mechanical nor blind, rather it is by the exercise of ingenuity that it must lead to the 'knowns'.

Thus, to construct the analytic art demands the existence of a mathematical discipline that itself needs to be constructed. The latter includes the 'laws' and the 'principles' of the former. In this conception, the analytic art cannot be reduced to something merely logical, independent of mathematics, but its specifically logical part is contained in this mathematical discipline. It is for this reason that we must not allow ourselves to be misled by the fact that vocabulary is borrowed from Aristotle's *Organon*. From here on we see the limits on extending this art: it is to the limits of this discipline that we now need to turn our attention. We also see the differences between Ibn al-Haytham's project and those of Ibn Sinān and of al-Sijzī. The *ars inveniendi* must itself have a mathematical foundation.

3. *The analytical art and the new discipline: 'The Knowns'*

In his treatise *Analysis and Synthesis*, Ibn al-Haytham says that Euclid's *Data* 'includes many notions concerning these knowns, which are among the instruments of the art of analysis'. He goes on to say:

> the greater part of the art of analysis is based on these notions, but with the exception that there remain other notions among the knowns that are indispensible for the art of analysis, and which we need many times, that are deduced by analysis, that are not included in this book (the *Data*), and that we have not found in any book.[30]

Recognising the necessity of filling this gap in order to provide foundations for the art, Ibn al-Haytham promises to write an independent treatise, once this treatise on analysis and synthesis is completed, 'in which we shall show the essential characters of the known notions that we use in the mathematical sciences'.[31]

This is how, in his treatise *Analysis and Synthesis*, he introduces the known ideas he needs – as indeed he had done in his treatise *On the Quadrature of the Circle*[32] – before devoting a whole treatise to the study of known ideas in the mathematical sciences. This close relationship between the two texts – the *Analysis and Synthesis* and *The Knowns* – is so strongly emphasised by Ibn al-Haytham himself that it merits further attention.

---

[30] See p. 231.
[31] *Ibid.*
[32] *Les Mathématiques infinitésimales,* vol. II, chap. I.

Ibn al-Haytham wrote the treatise *The Knowns* in three parts: a long introduction – it occupies almost a third of the book – in which he sets out a complete treatment of 'known ideas', followed by a first part that deals with properties 'that none of the ancients has written about, and they have not written on this kind of thing',[33] and, finally, there is a second part that gives the properties of 'the same kind as those set out by Euclid in his book the *Data*, although nothing of this part is to be found in the book the *Data*'.[34] So although Ibn al-Haytham considers that in the *Data* Euclid contributed to this new discipline, 'the knowns', it is as a distant predecessor. In fact, simply reading through Ibn al-Haytham's treatise is enough to make one realise how strange it is and, if one may say so, how original. In the introduction, much effort is expended on defining the concept of a 'known', and in the two parts of the main body of the work we are not concerned either with geometry in general, or with one or other of the branches of it defined and recognised in the tradition. As we have said, everything is there to meet the needs of the analyst.

In the substantial body of text that serves as an introduction to his treatise, Ibn al-Haytham is at pains to give precise meaning to this concept of a 'known'. The word is not new, and it is found in the vocabulary of Arabic translations of Euclid. It is in fact the term by which Isḥāq ibn Ḥunayn translated the Greek δεδομένα; and thereafter the word was in constant use among mathematicians. Thus, Ibn al-Haytham refers, successively, to 'known in number', 'known in ratio', 'known in position', 'known in shape' and 'known in magnitude'. To go no further than the Euclidean sense, which is made explicit in the *Data*, is to understand nothing of the distance travelled by Ibn al-Haytham and his predecessors. To give just one example, let us consider the phrase 'known in position'. Euclid means by this no more than a single position, that can be determined completely. So, a segment known in position is a segment that is always in the same position, a position that we can determine. Ibn al-Haytham, on the other hand, defines position by the term *naṣba* (θέσις, 'situation'), whether this is in relation to a thing that is fixed or to one that moves. In short, Ibn al-Haytham explicitly introduces movement in order to speak of position, something that Euclid could not allow. We shall see later the significance of involving movement.

It is, so to speak, as a philosopher that Ibn al-Haytham is at pains to determine the sense of the word 'known'. He begins by returning to what characterises apodictic knowledge, that is its being invariable both on the ontological level and as conceived by the mind. According to Ibn al-

[33] See *On the Knowns*, p. 385.
[34] See *On the Knowns*, p. 410.

Haytham, there are no objects of such knowledge except invariable concepts to which the subject who has the knowledge accords a credibility that
is itself invariable, having the additional knowledge that this is the case.
But, for our mathematician, this invariability of the concepts, of the ideas
of phenomena, implies the other two: the invariability of credibility, and
the knowledge the subject has of it. In other words, the invariability of concepts is ontologically and logically prior to the invariability of credibility,
and the knowledge that the subject has of this credibility. It is just this that
leads Ibn al-Haytham to adopt an unequivocal realism in mathematics
when he maintains that 'the known is in truth any notion that does not
admit of change, whether or not it is believed by someone who believes'.[35]

Ibn al-Haytham then sets out some conditions that this apodictic
knowledge satisfies: its necessity – it is not relative either in regard to a
place or a time; the nature of the credibility accorded to it by the subject –
we are concerned with conscious credibility. Thus, it is not enough to know
that a concept is invariable in order to know that we know it, but that
credibility must be invariable, and we must know that is so. This awareness
of the invariability of the credibility of the concept is acquired either by the
evidence of its necessity – as for the assertion 'the whole is greater than the
part'; or as the result of a demonstrative syllogism, when we are dealing
with a mathematical proposition. 'The known' belongs to this last species,
and only to that one: on the ontological level, it is an invariable concept,
independent of any subject that knows it; on the level of what is knowable,
it is characterised by an invariable credibility, which is either the result of
evidence of its necessity, or the conclusion of a proof.

To this theory, with its air of Platonism, Ibn al-Haytham adds a distinction which, for its part, looks Aristotelian: distinguishing something known
in actuality (in actu) from something known in potentiality (in potentia).
There is, however, no ontological difference, between these two varieties
of 'knowns', but simply one of understanding: something known in
potentiality is a known that is just as real as something known in actuality,
it is simply awaiting a subject to know it.

Any historian who is interested in more than mere mathematical results
cannot help but be disconcerted by this philosophical digression, which the
author presents as an integral part of his mathematical exposition. We may
ask why Ibn al-Haytham felt the need to work out this philosophical theory, which is in fact rather sketchy, in order to discuss 'knowns'. Perhaps
he did so in an attempt to give a philosophical answer to a mathematical
question that could not yet be given a mathematical answer. Everything
seems to suggest this was so, particularly since this kind of response is not

---

[35] See On the Knowns, p. 364.

an exceptional occurrence in the history of mathematics and of the sciences. Exactly what is happening? Ibn al-Haytham's problem, inherited from his predecessors, at least from the time of the Banū Mūsā, then developed and widened by Ibn al-Haytham himself, is to provide justification for permanence or change in the properties of a geometrical entity that has been subjected to transformation or motion. He needs to know what happens to its extension, its position, its form and its magnitude. While geometry was a subject in which the concepts of motion and transformation did not occur, this question did not present itself as an urgent one. But the situation is completely altered once one introduces motion and geometrical transformations, as had been done by Ibn al-Haytham's predecessors, and above all by Ibn al-Haytham himself. He is thoroughly aware of this when, in his treatise *Analysis and Synthesis*, he describes, in connection with knowns, what separates him from Euclid:

> [...] all the knowns mentioned by Euclid, in his book called the *Data*, are included within the sum total of the parts that we have mentioned; and in what we have mentioned, there are some things that Euclid did not mention: these are movable things known in position.[36]

In other words, while Euclid's knowns define position, form, and magnitude, as properties inherent to figures, in a geometry that is concerned only with figures, Ibn al-Haytham's *Knowns* defines the same properties, but for figures and places that move with a continuous motion or are subjected to transformations. This difference brings many others with it, and they are not insignificant: they affect the conception of a geometrical object and that of space. In Euclid, geometrical research is concerned with the properties of invariant figures; in Ibn al-Haytham and his immediate predecessors, we begin to be interested in the relations between figures in a space – this is indeed why Ibn al-Haytham felt obliged to write his treatise *On Place*.[37] So the difficult part is to provide justification for this new concept of a 'known', to be able to speak of the invariant properties of a figure, of a place, of a geometrical object, in motion or subjected to a transformation. Neither Ibn al-Haytham, nor his successors in the following eight centuries, were capable of providing a mathematical answer to that mathematical problem. It is not uncommon that, in situations like this one, a mathematician offers a philosophical answer. Once in possession of this concept of a 'known', Ibn al-Haytham paints a picture of the different knowns in the mathematical sciences. But, when writing the

---

[36] Emphasis added, see p. 234.
[37] See Chapter III.

Knowns, his position changes slightly, and this modification tells us a great deal about the ultimate basis of his theory.

In his treatise *Analysis and Synthesis*, which, according to his own account, was written only a little before the treatise *The Knowns*, once the 'known' has been defined in general, Ibn al-Haytham introduces successively: things known in number, known in magnitude, known in ratio (numerical and non-numerical), known in position and known in form. He then proceeds to make several classifications of problems: theoretical and practical, by the number of solutions for practical problems, and so on; a classification that is repeated for each of the four mathematical disciplines. Although the matter is quickly settled in regard to astronomy and music, since analysis and synthesis in these subjects reduces to the practices in geometry and in arithmetic respectively, the disciplines are nevertheless treated individually. In the second part of the same treatise, Ibn al-Haytham proposes problems on the level of research, or, as he puts it, 'problems of analysis that involve some difficulties' – six in all, three in arithmetic and three in geometry. On these two points, the treatise *The Knowns* differs from *Analysis and Synthesis*. The *quadrivium* has vanished from the introduction as well as from the two parts of the book. Moreover, in the latter, the essential matter of the new discipline – the 'knowns' – deals with geometry. Let us look more closely at this point, which we think is a very important one.

In the long introduction to his book *The Knowns*, Ibn al-Haytham abandons the language of the *quadrivium* in favour of that of the 'categories'. Thus, he begins by referring to the Aristotelian subdivision of quantity, because his account is limited to only those 'knowns' that relate to quantity. He then refers to the elements of discrete quantity: the phonemes of language and the numbers. In the first case, the 'knowns' deal with the essence of the phoneme, the number and the combination of phonemes. For the numbers, the 'knowns' are: essence, quantity, the properties of their nature (*tabī'a*) (perfect, deficient, square, and so on), and their association (commensurability, ratio, addition, subtraction, factors, and so on). Once these subdivisions of discrete quantity have been set out, Ibn al-Haytham does not return to them, and does not investigate even the simplest example of them in the other parts of the book. He then sets out the subdivisions of continuous quantity: segments of a straight line, surfaces, solids, weights and time. In fact, only the first three feature in the course of the development.

This classification is certainly thoroughly traditional. Its content, however, is much less so. In fact, from the first, one cannot fail to be aware of the care, so to say both for the whole and for the relation of parts, that runs through all of Ibn al-Haytham's account: when he is dealing with one of the elements of a figure, he does not consider it only as a magnitude, but also as a specimen of the family to which it belongs. Knowledge of that element will thus, and this is the significant part, take account of its magnitude, its position, its form, as well as of its relations with other things: in short, of the properties of the space. A considerable step has been taken, and Ibn al-Haytham devotes a large part of his introduction to explaining these concepts. By way of an example, let us take the central concept of position.

To define position Ibn al-Haytham employs three concepts: motion, order and relation. Thus, the position of a point – considered as the endpoint of a line – is known when its distance (or distances) from another point (or other points) remains invariable. There are several cases to consider: the point P is fixed and the other points are also fixed; the point P moves in circular motion about the fixed point, but without the distance between them changing; the point P and the other points all share the same motion, which leaves the distances between P and each of the other points unchanged.

In the same way, the position of a line is defined in relation to fixed points; in this case, the line moves with no motion, except increase and diminution, and the distances between its points, and two points, or more, do not vary. This line will be said to be *of absolutely known position*. The position of the line could also be located by its relation to a single fixed point, and in this case the known concepts would be the unchanging distances between any point of the line and this fixed point, irrespective of whether the line itself is fixed or in motion. We can also locate the position of the line by its relation to another line, irrespective of whether the latter is fixed or in motion. Again, we can locate the position of the line by its relation to a moving point or a set of moving points, and in this case the known concepts will be the unchanging distances between each point of the line and each of the moving points; the line must then move with the same motion, and in the same direction, as the motion of the points concerned. Finally, we can locate the position of the line by its relation to a fixed line, and in this case the known concept is that of the angle formed by the intersection of these two lines or of their extensions, irrespective of whether the line whose position we seek to know is fixed or in motion, provided that the angle formed remains invariable. If the line, or its extension, does not cut the line by its relation to which it would be known in position, it will be known in any case if the two lines are cut by a straight line that forms a known angle with each of them.

Ibn al-Haytham takes his list further still and locates the position of the line by its relation to a moving line, then by its relation to a fixed surface, and finally by its relation to a moving surface. He returns to the analogous task of defining the position of a surface and the position of a solid body, and examining other concepts: of known form, of known magnitude and in a known ratio.

We see from the first, when we examine the long introduction to his book, that Ibn al-Haytham has included motion as a basic concept of geometry, a concept that is necessary for defining the position and form of any geometrical magnitude, and as guarantee of its continuity. An examination of this introduction shows, further, that as an heir to Archimedes and also to Apollonius, Ibn al-Haytham distinguishes explicitly between positional properties and metrical ones. Even if a positional property can be presented using measures of distances and angles, that is in a metrical form, Ibn al-Haytham nevertheless prefers to describe what derives from position in positional terms. The essential matter at this stage is to locate the position – say of a point – without involving any system of coordinates, but only in relation to points and lines, fixed or moving; thus, we are, so to speak, dealing with a geometry that is descriptive in the proper sense of the term. The objective that Ibn al-Haytham sets himself in *The Knowns* is clear: to identify the invariant relationships that allow one to describe position, form, magnitude and ratio. Each group of relationships will make up a chapter in the geometry that is to follow, or in the subject he has named 'The Knowns'.

The two chapters that follow this introduction are brimful of powerful and penetrating insights. In the first chapter, the author is chiefly concerned with properties of position and form. He deals with certain sets of points and some point-to-point transformations: homothety, similarity, translation, as well as other birational transformations of order 2. But, whereas he describes the nature of the first three, he simply uses the remainder without description. The first propositions of the chapter are concerned with this description, before we turn to an examination of some properties, such as the homothetic image of a circle, the transform of a circle under a translation, and so on. In the second chapter, Ibn al-Haytham tries to find the simplest geometrical methods for defining the positions of points as well as the relations between them, starting from elements that are known. In short, throughout these two chapters, Ibn al-Haytham studies loci, straight and circular, as well as their transforms.

The research work in these two chapters represents, at most, only a partial realisation of the project set out in the introduction, a sketch of what the new discipline promised. However the rewards are sufficiently substantial to indicate the direction for this research and to shed light on its significance. The results give the analyst some invariant properties of position and form for a number of geometrical objects, obtained by motion, by transformation and by taking plane sections. All these are necessary elements for laying the foundations for the art of analysis.

But this achievement, an important piece of geometry, and one that, moreover, points the way for future developments, nevertheless cannot disguise the gulf between the project and its realisation. The project is concerned with the disciplines of the *quadrivium*, according to the treatise *Analysis and Synthesis*; discrete quantity as well as continuous, according to the introduction to the treatise *The Knowns*. The realisation deals only with geometry. The gulf seems not to have escaped the attention of Ibn al-Haytham: indeed he seems to have provided a justification for it in advance. It was by considering the example of geometry that he conceived knowns peculiar to each discipline of the *quadrivium*, or to each subdivision of quantity. These special knowns are – let us remind ourselves – for number: the essence of number, its quantity, its nature (perfect, square and so on), the associations of numbers (ratio, addition, subtraction, commensurability and so on). In the treatise *The Knowns*, that is where he is concerned with the new discipline, once he has referred to these properties, Ibn al-Haytham forgets about them and also about arithmetic itself. These knowns, after being drawn to our attention in the introduction, are then not mentioned again. The same happens for all the other special knowns, except for those of geometry. Thus, there is every indication that they appear here purely and simply for the sake of completeness in regard to the other disciplines that deal with quantity. Their presence, which one might describe as allusive, is, however, indispensable for providing the generality required by the method of analysis and synthesis, which according to Ibn al-Haytham was based on 'knowns'. This method, as Ibn al-Haytham explained well in his treatise *Analysis and Synthesis*, must be applicable to all the disciplines of the *quadrivium*. But Ibn al-Haytham is too deep a thinker to be satisfied by the juxtaposition of 'knowns' of diverse and heterogeneous origins as guarantors of the required generality. And it is here that the philosophical theory of knowns he has worked out comes into play: it is this theory that gives the discussion of the 'knowns' its unity and thus its generality. Thus, this philosophical theory comes into play twice: to justify the inclusion of motion and transformation as basic concepts for geometry; to ensure that there is unity in the discussion of 'knowns' in the disciplines concerning discrete quantity as well as continuous quantity. We can see that this theory is not something derived from elsewhere that has been incorporated into Ibn al-Haytham's theory. Many centuries later it was to be replaced by another, *Analysis Situs*; but that is another story.

In his treatise *Analysis and Synthesis*, Ibn al-Haytham returns to the method, that of analysis and synthesis, to examine its application in each of the disciplines. That is to say he 'actualises' or adapts the method for each of them. Thus, he begins with a classification of the mathematical concepts and propositions into two classes: theoretical (*'ilmī*) and practical (*'amalī*). While the theoretical class is the same as what is found in his predecessors, insofar as it deals with properties that are specific and thus in their essence necessary to the object under consideration, the practical class is synonymous with 'action', and is thus different from one discipline to another as can easily be verified. That is, the pair theoretical/practical is not identical with the well-known pair 'theoretical'/'problematic' that we meet in Pappus' *Collection*, III. Thus, for Ibn al-Haytham, to find a perfect number or to find two squares whose sum is equal to a given square (Diophantus, II.8) is as practical, in arithmetic, as the construction of an equilateral triangle on a given segment of a straight line. As has been the case since Thābit ibn Qurra, practical analysis includes both the determination of an unknown magnitude and the construction of a geometrical figure. Theoretical analysis is of the same kind for each of the complete sets of disciplines. According to Ibn al-Haytham, the same is true of practical analysis, though with the difference that the latter is divided into three species according to whether or not there exists a diorism, and whether, in the latter case, we have a single solution or many solutions.

Ibn al-Haytham then explains what analysis means in each case and gives examples to illustrate the application of the method. So it remains to examine all the mathematical and logical problems raised by this research done by Ibn al-Haytham. The mathematical problems, some of which form part of the advanced research of the time, are systematically noted and commented upon here. As for the logical questions, they are of two types: the philosophico-logical ones that Ibn al-Haytham raises, and the questions a logician of our own time might recognise as underlying his text. The first type will also be noted and commented upon; the second type will be the subject of a separate text.[38]

---

[38] We intend, in fact, to write a book on analysis and synthesis in ancient and classical Arabic mathematics.

4. History of the texts

On Analysis and Synthesis[39]

The authenticity of the treatise *Analysis and Synthesis* and its attribution to al-Ḥasan ibn al-Haytham are not in doubt. The manuscript tradition establishes both without the slightest ambiguity. Al-Qifṭī, Ibn Abī Uṣaybiʿa and the copyist of the Lahore manuscript are at one: they all mention this title in their lists of the works of Ibn al-Haytham.[40] Finally, in this treatise Ibn al-Haytham himself cites two other of his writings: *The Knowns* and the *Commentary on Euclid's Postulates*.

This treatise has come down to us in four manuscripts:

1) Dublin, Chester Beatty Library, no. 3652/12, fols 69ᵛ–86ʳ, in the numbering in Arabic numerals. This manuscript, here called B, is a copy made at Baghdad and completed on Saturday 23 Jumādā al-ūlā 612 of the Hegira, that is Saturday morning 19 September 1215, as is indicated in the colophon. The writing is elegant *naskhī*, and the figures are drawn by the copyist.

2) Istanbul, Reshit collection, no. 1191/1, fols 1ᵛ–30ᵛ. This manuscript, here called R, belongs to the collection of the famous copyist and scholar, Muṣṭafā Ṣidqī,[41] and has thus been copied before the mid eighteenth century. The writing is *nastaʿlīq*, and the figures are drawn. As for the date of his copy, we cannot be very clear, but it seems to us to be earlier than that of manuscript Q, which we shall discuss below.

3) Cairo, Dār al-Kutub, Taymūr collection, Riyāḍa, no. 323. This manuscript has 68 numbered pages, and here we shall call it Q. It belongs to the collection of Muṣṭafā Ṣidqī. The writing is good *nastaʿlīq*, and the figures have not been drawn.

4) Kuibychev, Lenin Library V.I – a collection now transferred to St Petersburg – fols 348ʳ–368ʳ (former numbering fols 316ʳ–336ʳ), and here we shall call it S.[42] We may note that one leaf, pages 351ʳ and 351ᵛ, is inverted.

Comparing these four manuscripts two by two enables us to show without the shadow of a doubt that the Istanbul manuscript, Reshit 1191/1,

---

[39] For the original Arabic, see *Les Mathématiques infinitésimales du IXᵉ au XIᵉ siècle*, vol. IV: *Méthodes géométriques, transformations ponctuelles et philosophie des mathématiques*, London, 2002, pp. 230–391.

[40] See *Les Mathématiques infinitésimales*, vol. II, pp. 532–3; English trans. pp. 420–1.

[41] R. Rashed, *Géométrie et dioptrique au Xᵉ siècle*, p. CXXXVI.

[42] For a description of this manuscript, see the history of the text *On the Properties of Circles*, Chapter I, pp. 32–4.

is a copy made from the one in Dublin, Chester Beatty 3652/12 and from that alone. We do not need to work through all the details of the comparison here, but simply to refer to some facts:

1) In relation to B, R is missing 8 phrases of more than two words and 29 individual words. On the other hand, in relation to R, B is missing only a single word, which is in no way significant; that is it could have been added by the copyist of R: the word مثل. The copyist of B has repeated a long paragraph, 82ʳ, 38–82ᵛ, 16 (19 lines each with about fifteen words, which might correspond to a page in his model). He noticed his mistake and has written above the line, at the start of the repetition, the word خطأ (error). The copyist of R followed him blindly and has included the word in the form خطأ while repeating the paragraph in its entirety. This repetition that, on its own, constitutes irrefutable proof, is not unique; we have another example. The copyist of B repeats a sentence fol. 70ᵛ, 11–12; the copyist of R follows him and repeats the same sentence (fol. 3ʳ, 18–19).

2) We can find at least 35 mistakes in Arabic in B repeated in R.

3) Where letters in B have been effaced the copyist of R has left empty spaces.

4) All the mathematical mistakes that occur in B are preserved in R.

5) All the words and phrases lacking in B are missing in R.

In regard to a wider context, the copyist of R transcribed the other treatises of B.[43]

If we look at the texts of al-Sijzī that are in R but not in B, such as *The Asymptotes to an Equilateral Hyperbola*, we can easily show, by a careful reading of B, that these treatises were included in it before being removed from B. So the copyist of R transcribed B before the loss of these texts.

Thanks to a similar comparison, we can show that no. 323 of the Taymūr collection in Dār al-Kutub, called Q, is also a copy of B, and only of B, as can easily be verified.

Finally, a comparison of R with Q shows that in relation to R, Q has 37 omissions of a word and 34 omissions of a phrase of more than two words, whereas in relation to Q, R has 33 omissions of a word and nine omissions of a phrase.

Comparing B and S shows that, in relation to S, B shows 68 omissions of a word, 41 omissions of a phrase (of more than two words, on one occasion of 32 words), so that starting from B alone we cannot obtain a secure

---

[43] See for example P. Crozet, 'À propos des figures dans les manuscrits arabes de géométrie: l'exemple de Siǧzī', in Y. Ibish (ed.), *Editing Islamic Manuscripts on Science*, Proceedings of the Fourth Conference of al-Furqān Islamic Heritage Foundation, 29th–30th November 1997, London, 1999, pp. 131–63.

text. On the other hand, in S, by comparison with B we can note 69 omissions of a word, but only 15 omissions of a phrase.

Systematic comparison of these manuscripts with the help of omissions, additions and different types of error allow us to establish the following *stemma*:

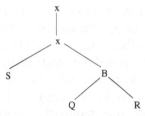

Note that the introduction, that is the most philosophical part of the treatise, has been printed as an appendix to our study of Ibn al-Haytham's text.[44] Similarly, the French translation of his theorem on perfect numbers, as well as a discussion of the history of this theorem, have been the subject of an earlier publication.[45] The only critical edition of this text, as well as the only translation of it, are those we published more than twenty years ago.[46] It is that *editio princeps* and the French translation to which we are returning here, with the necessary emendations, as a basis for the present English version.

The Knowns[47]

The authenticity of this text and its attribution to al-Ḥasan ibn al-Haytham are not in doubt. The treatise is promised in *Analysis and Synthesis*, and *The Knowns* includes a reference to another text by Ibn al-Haytham: *On Measurement*.[48] Moreover, it appears in the list of the works of al-Ḥasan ibn al-Haytham copied by Ibn Abī Uṣaybiʻa,[49] and it is also

---

[44] 'L'analyse et la synthèse selon Ibn al-Haytham', in *Mathématiques et philosophie de l'Antiquité à l'âge classique. Études en hommage à Jules Vuillemin*, éditées par R. Rashed, Paris, Éditions du CNRS, 1991, pp. 131–62; reprod. in *Optique et mathématiques: recherches sur l'histoire de la pensée scientifique en arabe*, Variorum Reprints CS388, Aldershot, 1992, XIV.

[45] 'Ibn al-Haytham et les nombres parfaits', *Historia Mathematica*, 16, 1989, pp. 343–52; repr. in *Optique et mathématiques*, XI.

[46] 'La philosophie mathématique d'Ibn al-Haytham. I: L'analyse et la synthèse', *MIDEO*, 20, 1991, pp. 31–231.

[47] For the original Arabic, see *Les Mathématiques infinitésimales*, vol. IV, pp. 445–583.

[48] See *Les Mathématiques infinitésimales*, vol. II, pp. 524–5; English trans. *Ibn al-Haytham and Analytical Mathematics*, pp. 422–3 and below, p. 377.

[49] Ibn Abī Uṣaybiʻa, *'Uyūn al-anbā' fī ṭabaqāt al-aṭibbā'*, ed. N. Riḍā, Beirut, 1965; ed. A. Müller, p. 98.

noted by the copyist of the Lahore manuscript.[50] The text itself has come down to us in two manuscripts, which have served to establish the text:

1) Paris, Bibliothèque Nationale, no. 2458, fols 11^v–26^r, here called B. This is a copy produced in the region of Khusrū Kerd, close to Nishapur, and finished on Sunday the ninth of Dhū al-Ḥijja 539, that is Sunday 3 June 1145,[51] as is indicated by the colophon. The manuscript was copied in *naskhī* by Ibn Asʿad al-Bayhaqī, who also drew the figures. It forms part of a collection that contains other important mathematical treatises, such as the *Algebra* of al-Khayyām and three treatises by al-Sijzī. The collection belonged to Melchissedech Thévenot, who died in 1692.

The number of omissions in the copy of Ibn al-Haytham's text is extremely small: five words, one geometrical symbol and two connecting letters. The copyist did in fact revise his copy by checking it against his original, as we can see from the number of words and phrases that he has added in the margin, indicating where they belong in the text. He has noted some glosses in the margin, in which there are references to propositions in Euclid. We note, however, that the order of two folios has been reversed, an error that was certainly made after the copying was completed. The treatise thus appears in the following order: 11^v, 13^r–14^v, 12^r, 12^v, 15^r–26^r.

2) The second manuscript belongs to the collection in the library of Kuibychev that we mentioned earlier,[52] fols 335^r–347^v (formerly numbered fols 303^v–315^v), here called S. We note an omission of several pages in our edition (*Les Mathématiques infinitésimales*, vol. II, pp. 481–517), as well as nine omissions, each time of a phrase, and 44 omissions of a word. On the other hand, this manuscript includes six words that are absent in B. This shows that what we have is a manuscript tradition different from that of B.

---

[50] A. Heinen, 'Ibn al-Haiṯams Autobiographie in einer Handschrift aus dem Jahr 556 H / 1161 A.D.', *Die islamische Welt zwischen Mittelalter und Neuzeit, Festschrift für Hans Robert zum 65*, Beirut, 1979, pp. 254–79.

[51] According to concordance tables, this date corresponds to 2 June 1145, which was, however, a Saturday not a Sunday. The tables have this month beginning on 25 May 1145, which assumes that the lunar crescent was visible on 24 May in the evening. In the place where the manuscript was copied, it is perfectly possible that the lunar crescent was not visible, locally, until the evening of 25 May, so we can take the date of Sunday 3 June 1145 as the date of completion of the copy.

[52] See p. 151, n. 42.

We return here to our *editio princeps* of the treatise *The Knowns*, as well as its first translation[53] making improvements where necessary.

In establishing these texts, we followed the very rigorous rules that we have set out and explained more than once. The French translation also followed the method that we decided upon earlier: to translate literally, but without conflicting with the stylistic rules of French, thus in language that respected the sense and, as far as is possible, the letter of the Arabic text. The English translation, whose initial draft was made from the French, follows the same principles and has, throughout, been compared with the Arabic original.

---

[53] 'La philosophie mathématique d'Ibn al-Haytham. II: Les Connus', *MIDEO*, 21, 1993, pp. 87–275. The only work concerned with Ibn al-Haytham's treatise is L. A. Sédillot, 'Du *Traité* des Connus géométriques de Hassan ben Haithem', *Journal asiatique*, 13, 1834, pp. 435–58.

MATHEMATICAL COMMENTARY

1. The double classification of analysis and synthesis

Preliminary propositions

The first chapter of the treatise is devoted, in its entirety, to illustrating the classification of the different kinds of analysis that was set up in the introduction, and the classification of the forms they take in the mathematical sciences: arithmetic, geometry, astronomy and music. The intention is obviously as much logical and methodological as it is didactic; so we shall not encounter any new mathematical research in this chapter. But, before setting out on this work, Ibn al-Haytham first states three propositions relevant to analyse as a whole, both theoretical and practical. These propositions, all of which are taken from his book *The Knowns*, again ensure continuity with that work. So it is likely that they tell us something about the direction in which the mathematician's thoughts were turning. It is no surprise that they are all concerned with geometrical transformations. Let us look at them one by one.

Proposition 1. — *Given two fixed points* A *and* B *and two segments* G *and* E, *to show that the point* C *defined by the relation* $\dfrac{CA}{CB} = \dfrac{G}{E} = k$ *lies on a circle whose centre and radius are known.*

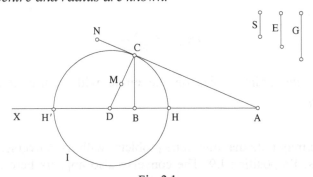

Fig. 2.1

We assume $k > 1$. If C satisfies the conditions, we have $CA > CB$. Let us extend AB and construct the straight half line CM such that $A\hat{C}M = C\hat{B}X$; since $C\hat{B}X > B\hat{C}A$, the straight half line CM lies outside $A\hat{C}B$. We have $A\hat{C}M + C\hat{A}B < 2$ right angles; so CM cuts AB at the point D beyond B.

The triangles ACD and BCD have a common angle D and $A\hat{C}D = C\hat{B}D$; so they are similar, hence

$$\frac{AD}{DC} = \frac{CD}{DB} = \frac{CA}{CB} = k,$$

so

$$\frac{AD}{DC} \cdot \frac{DC}{DB} = \frac{AD}{DB} = k^2.$$

From which it follows that

$$\frac{AB}{BD} = k^2 - 1,$$

so

$$DB = \frac{AB}{k^2 - 1}.$$

So the point D is determinate, and

$$DA = AB \cdot \frac{k^2}{k^2 - 1}.$$

Moreover,

$$DA \cdot DB = DC^2,$$

hence

$$DC = AB \cdot \frac{k}{k^2 - 1}.$$

That is, the point C lies on the circle with centre D and radius $R = \frac{k}{k^2 - 1} \cdot AB$.

First we may note that this same problem, with its converse, appears in *The Knowns*, Proposition I.9. The converse also appears here in Problem 20.

Next we may note that, for the time being, Ibn al-Haytham has only carried out the analysis: if C satisfies $\dfrac{CA}{CB} = k$, then C lies on the circle with centre D and radius $R = \dfrac{k}{k^2 - 1} \cdot AB$. The converse: any point C of the circle (D, R) satisfies $\dfrac{CA}{CB} = k$ will be dealt with later, as we said.

This converse shows that the points H and H', the points of intersection of the circle and the straight line AB, divide the segment AB in the ratio k; so (A, B, H, H') is a harmonic range.

We may note that, in a different form, this problem was studied by Ibn Sinān.[1] In the statement of the problem Ibn Sinān assumes *a priori* that the locus of the points is a circle. He describes Apollonius' analysis, as well as a synthesis by his own grandfather Thābit ibn Qurra.[2] As it is presented, Ibn al-Haytham's analysis seems to be a more solid version of that of Apollonius.

Proposition 2. — *Given a fixed circle with centre* E *and radius* R, *and a fixed point* C, *if with any point* A *of the circle we associate the point* D *on* CA *produced such that* $\dfrac{CA}{AD} = k$, *then* D *lies on a circle whose centre and radius are known.*

Let A be an arbitrary point on the circle (E, R) and D the point on CA produced such that $\dfrac{CA}{AD} = k$, then $\dfrac{CD}{CA} = \dfrac{k+1}{k} = k_1$ is known.

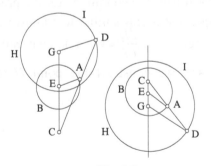

Fig. 2.2

[1] R. Rashed and H. Bellosta, *Ibrāhīm ibn Sinān: Logique et géométrie au Xe siècle*, Leiden, 2000, pp. 627–35.

[2] *Ibid.*, p. 633.

Let G be a point on CE such that $DG \parallel EA$, and the triangles CEA and CGD are homothetic, so $\dfrac{GD}{EA} = \dfrac{DC}{CA} = \dfrac{GC}{CE} = k_1$, so $CG = k_1 CE$ and $GD = k_1 EA = k_1 R$. The point D lies on the circle with centre G and radius $k_1 R$, that is the circle homothetic with the given circle in the homothety $\left(C, \dfrac{k+1}{k} R \right)$.

Here Ibn al-Haytham does not investigate the converse: any point D lying on the circle $\left(C, \dfrac{k+1}{k} R \right)$ satisfies $\dfrac{CA}{AD} = k$; whereas the problem, with the converse, appears in *The Knowns*, Proposition I.3.

Proposition 3. — *Given a fixed point* C *and a fixed straight line* AB, C $\notin$ AB *and* D *is an arbitrary point on the straight line* AB; *the point* E *defined by* $C\hat{D}E = \alpha$ *(a given angle) and* $\dfrac{CD}{DE} = k$, *a given ratio, lies on a fixed straight line.*

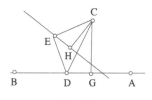

Fig. 2.3

Let E be a point that fulfills the conditions for the problem; the triangle CDE has a known shape, that is it is similar to a known triangle; so $G\hat{C}H$ $D\hat{C}E = \beta$, a known angle, and $\dfrac{CD}{CE} = k_1$, a known ratio. We draw CG perpendicular to AB, and we construct the point H such that $G\hat{C}H = D\hat{C}E = \beta$ and

(1) $\qquad \dfrac{GC}{CH} = \dfrac{CD}{CE} = k_1.$

We have

$$CH = \dfrac{1}{k_1} GC,$$

so H is a known point.

From (1) we obtain

$$\dfrac{GC}{CD} = \dfrac{CH}{CE},$$

thus the triangles CHE and GCD are similar, and consequently angle CHE is a right angle. So the point E lies on the straight line Δ which is the perpendicular to the straight line CH at H.

We may note that in the similarity with centre C, with angle β and ratio $\dfrac{1}{k_1}$, the point G has image H, because $CH = \dfrac{1}{k_1} GC$ and $G\hat{C}H = \beta$. The given straight line AB, perpendicular to CG at G, has as its image the straight line Δ, perpendicular to CH at H, and any point D of AB has as its image a point E lying on Δ.

Thus, the analysis has led Ibn al-Haytham to give an account of the properties of a similarity. He considers the same problem, and its converse, in his treatise *The Knowns*, Proposition I.4.

We have just seen that at the start of his first chapter Ibn al-Haytham presents three propositions to which he will return in his treatise *The Knowns*, and that he presents them as having a bearing on analysis as a whole. Looking at the matter more closely, we may in fact observe that they have some common characteristics. All three are of the following form: if the position of a point is defined by means of known elements and has a property P, then the point lies on a known line L, a circle or a straight line. However, in the first proposition this line or straight line is obtained as a locus of points, whereas in the following two propositions it is the result of a transformation of a figure by a similarity. In the first proposition, in fact, we prove that the set of points C such that $\dfrac{CA}{CB} = k$ is a circle whose centre lies on AB; the endpoints of the diameter are harmonic conjugates of A and B in the same ratio k. This circle, associated with the harmonic range, will reappear in Problem 20 and in the last two transformations – a homothety and a similarity – that Ibn al-Haytham uses in Problem 21. So the motivation behind the first proposition is clear, and is connected with the introduction of these transformations. It is, moreover, specifically these two that are treated in Propositions 2 and 3. These first three propositions serve as an introduction for all the others, and stand apart from them as providing methods that will be employed later on. They give an early indication of what *The Knowns* have also told us, namely that geometrical transformations played an important part in Ibn al-Haytham's reflections on analysis and synthesis.

Analysis and synthesis in arithmetic
  1. *The theoretical section of the arithmetical problems*
     1.1. *Synthesis as the converse of analysis*

Proposition 4. — *Let* $(a_n)_{n \geq 1}$ *be a sequence of positive integers; if*

$$\frac{a_1}{a_2} = \frac{a_2}{a_3} = \ldots = \frac{a_{n-1}}{a_n},$$

then

$$\frac{a_2 - a_1}{a_1} = \frac{a_n - a_1}{\sum\limits_{i=1}^{n-1} a_i}$$

(P) $\Rightarrow$ (Q).

Analysis: With the help of *Elements* VII.11 and 12, Ibn al-Haytham proves that

(P) $\Rightarrow$ $\dfrac{a_2 - a_1}{a_1} = \dfrac{\sum\limits_{i=1}^{n-1}(a_{i+1} - a_i)}{\sum\limits_{i=1}^{n-1}(a_i)}$

(P) $\Rightarrow$ (T).

For $(P) \Rightarrow (Q)$ to be true, it is necessary that

$$\sum_{i=1}^{n-1}(a_{i+1} - a_i) = a_n - a_1.$$

Ibn al-Haytham proves that this equation holds (moreover, it holds whether or not the integers are proportional).

Synthesis: We know, as we have seen in the course of the analysis, that

(1) $a_1 < a_2 < \ldots < a_n \Rightarrow \sum\limits_{i=1}^{n-1}(a_{i+1} - a_i) = a_n - a_1$

and that

(2) $(P) \Rightarrow (T)$;

from (1) and (2) we obtain $(P) \Rightarrow (Q)$.

Accordingly the condition is necessary and sufficient, and the synthesis is indeed the converse of the analysis. The only difference between analysis and synthesis is the order in which the premises are arranged; the synthesis is derived from implication being transitive.

1.2. *Analysis leading to an impossibility: reductio ad absurdum*

The preceding analysis led to a condition that was satisfied by what was given. This time we have an analysis that ends in an impossibility. The analysis is itself a proof if it is taken as a demonstration by *reductio ad absurdum*.

Proposition 5. — *If* $\dfrac{a_1}{a_2} = \dfrac{a_2}{a_3} = \dots = \dfrac{a_{n-1}}{a_n}$, *then* $\dfrac{a_2 - a_1}{a_1} = \dfrac{a_n}{\sum\limits_{i=1}^{n-1} a_i}$ *is impossible.*

From the preceding proposition, if

$$\frac{a_1}{a_2} = \frac{a_2}{a_3} = \dots = \frac{a_{n-1}}{a_n},$$

then

$$\frac{a_2 - a_1}{a_1} = \frac{a_3 - a_2}{a_2} = \dots = \frac{a_n - a_{n-1}}{a_{n-1}} = \frac{a_n - a_1}{\sum\limits_{i=1}^{n-1} a_i}.$$

It requires that $a_n = a_n - a_1$, which is impossible since $a_1 \neq 0$.

2. *The practical section of the arithmetical problems*

2.1. *Practical section with discussion: synthesis as the converse of analysis*

Proposition 6. — *To divide two given numbers according to two given ratios.*

$$x_1 + x_2 = a,$$
(1)
$$y_1 + y_2 = b,$$
$$\frac{x_1}{y_1} = k_1, \quad \frac{x_2}{y_2} = k_2, \text{ with } k_1 > k_2.$$

The first equation can be rewritten as

$$k_1 y_1 + k_2 y_2 = a$$

and the hypothesis $k_1 > k_2$ implies $k_2 b < a < k_1 b$, or again

(2) $$k_2 < \frac{a}{b} < k_1,$$

a necessary condition for the system of equations in (1) to allow of a solution. We have

$$k_1 y_1 + k_2 (b - y_1) = a,$$

hence

$$(k_1 - k_2) y_1 = a - k_2 b$$

and

$$y_1 = \frac{a - k_2 b}{k_1 - k_2}, \quad y_2 = \frac{k_1 b - a}{k_1 - k_2};$$

from which we find x_1 and x_2.

So if condition (2) is satisfied, the four numbers x_1, x_2, y_1, y_2 are positive and rational and give a unique solution. As this condition is possible, that is it is a condition that does not lead to a contradiction, the analysis can have a converse and its converse is the synthesis.

2.2. *Analysis leading to an impossibility: reductio ad absurdum*
Here Ibn al-Haytham proves that, if condition (2) is not satisfied, then the analysis results in an impossibility, and in this case it can be regarded as a proof by *reductio ad absurdum*.

2.3. *Practical section, without discussion, problems with a unique solution: synthesis as the converse of analysis*

Proposition 7. — *Given an arbitrary number* AB, *to partition this number into two parts* AC *and* CB *where* AC < CB, *then into two other parts* AD *and* DB *where* AD > DB, *such that* CB = 2 DB *and* AD = 3 AC.

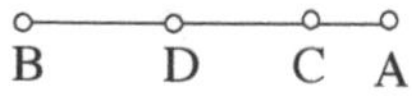

B   D   C   A

Fig. 2.4

This problem can be rewritten as a set of four first-degree equations in four unknowns. Let the given number be n; n is positive and rational:

$$x_1 + x_2 = n,$$
$$y_1 + y_2 = n,$$
$$x_1 = py_2,$$
$$y_1 = qx_2,$$

where p, q are integers, $p > 1, q > 1$.
We have a solution

$$x_1 = \frac{p(q-1)}{pq-1}n, \quad x_2 = \frac{p-1}{pq-1}n, \quad y_1 = \frac{q(p-1)}{pq-1}n, \quad y_2 = \frac{q-1}{pq-1}n;$$

for given n, p and q, the values of x_1, x_2, y_1, y_2 can be integers or fractions, but in any case the solution is unique.

Here analysis always leads to a rational solution, without requiring conditions or discussion. Moreover, as Ibn al-Haytham writes: 'If we invert this analysis, that allows us to complete the procedure and we establish the proof that this proposition is true.'[3]

2.4 *Practical section, without discussion, when there is an infinite number of solutions*

Proposition 8. — *To find two square numbers whose sum is a square.*
The solution corresponds to a modified statement: given an arbitrary square number, to find another square number such that the sum of the two is a square number. This problem has an infinite number of solutions.
So let the problem be to find positive rational numbers that satisfy

$$x^2 + a^2 = z^2, \text{ where } a \text{ is a given positive rational number.}$$

It is necessary that $z > x$; we put

$$x = t$$
$$z = t + u;$$

we have

$$\left(x = \frac{a^2 - u^2}{2u}, \ y = a, \ z = \frac{a^2 + u^2}{2u} \right),$$

[3] See p. 249.

a solution that depends on a parameter u.

So in this case analysis gives us an algorithm that Ibn al-Haytham summarises as follows:

> The analysis has arrived at supposing [we have] a square, an arbitrary square $[a^2]$, from which we then cut off a square $[u^2]$, an arbitrary square, subject to the condition that it is smaller than the first $[a^2 > u^2]$; then we divide the remainder $[a^2 - u^2]$ into two equal parts $\left[\dfrac{a^2 - u^2}{2}\right]$, next we divide the half by the side of the square that was removed $\left[\dfrac{a^2 - u^2}{2u}\right]$, we multiply the result of the division by itself $\left[\left(\dfrac{a^2 - u^2}{2u}\right)^2\right]$, then we add the result of the product to the first square $\left[\left(\dfrac{a^2 - u^2}{2u}\right)^2 + a^2\right]$.[4]

Here too the synthesis is the converse of the analysis.

Under the expression 'practical analysis with discussion or without discussion' (p. 225), Ibn al-Haytham silently brings together the two branches of algebra: determinate analysis and indeterminate analysis. In fact, in his work determinate analysis is represented by the practical section without a discussion with a unique solution and indeterminate analysis by the practical section without discussion with an infinity of solutions.

Analysis and synthesis in geometry

1. *Theoretical section on geometrical problems*
    1.1. *The multiplicity of analysis and auxiliary constructions*
    The example Ibn al-Haytham takes here is the famous proposition that is *Elements* I.20.

Proposition 9. — *The sum of any two sides of a triangle is greater than the third side.*

Ibn al-Haytham gives two of the various possible analyses for this inequality in a triangle, in each case with the required auxiliary construction. He emphasises that it is possible to give several analyses different from the two he has presented here.

---

[4] See p. 250.

1.2 Analysis that leads to an impossibility: reductio ad absurdum

Proposition 10. — *The sum of any two sides of a triangle is equal to the remaining side.*

2. Practical section on geometrical problems
    2.1. Practical section with discussion

Proposition 11. — *To divide a given segment* AB *into two segments that enclose a rectangle of given area* C.

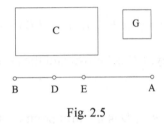

Fig. 2.5

This problem is that of constructing a focus of an ellipse – Apollonius, *Conics* III.45. Indeed Apollonius uses it several times in *Cutting off a Ratio*. The same problem had been considered in a similar way by Ibn Sinān.[5]

Analysis: Take a point D on AB such that $AD \cdot DB = C$. If $AD = DB$, then

$$AD \cdot DB = \left(\frac{1}{2}AB\right)^2,$$

hence

$$C = \left(\frac{1}{2}AB\right)^2.$$

If $AD \neq DB$, then

$$AD \cdot DB < \left(\frac{1}{2}AB\right)^2,$$

hence

---

[5] Apollonius, *Les Coniques*, Tome 2.1: *Livres II et III*, commentaire historique et mathématique, édition et traduction du texte arabe par R. Rashed, Berlin, 2010; Apollonius de Perge, *La section des droites selon des rapports*, commentaire historique et mathématique, édition et traduction du texte arabe par Roshdi Rashed et Hélène Bellosta, Berlin, 2009; R. Rashed and H. Bellosta, *Ibrāhīm ibn Sinān: Logique et géométrie au Xe siècle*, pp. 131–3.

$$C < \left(\frac{1}{2}AB\right)^2 ,$$

because if D exists it necessarily follows that

$$C \leq \left(\frac{1}{2}AB\right)^2 .$$

If $C < \left(\frac{1}{2}AB\right)^2$ and E is the mid point of AB, then $C < EB^2$. Let us put $EB^2 - C = G$, so G is known. We have

$$C = AD \cdot DB = (AE + ED)(AE - ED) = AE^2 - ED^2,$$

so $G = ED^2$. Consequently, ED is known and the point D is also known.

Synthesis: If $C = \left(\frac{1}{2}AB\right)^2$, D is the mid point of AB; we then have

$$AD \cdot DB = \left(\frac{1}{2}AB\right)^2 = C.$$

If $C < \left(\frac{1}{2}AB\right)^2$, and E is the mid point of AB, we put $EB^2 - C = G = DE^2$, hence we have DE, and in consequence D. We then have

$$AD \cdot DB = (AE + ED)(BE - ED) = BE^2 - ED^2 = EB^2 - G = C.$$

If $C > \left(\frac{1}{2}AB\right)^2$, the problem is impossible. Ibn al-Haytham proves this by *reductio ad absurdum*.

This appears a supplementary result since it has been proved that the condition $C \leq \left(\frac{1}{2}AB\right)^2$ is necessary.

We may note that this problem is the same as that of finding two numbers x and y when we know their sum and their product. In Propositions VI.27 and 28 of Euclid's *Elements*, it appears in the form of an application of areas with a defect (ἔλλειψις).

Proposition 12. — *From a given point* A *to draw a perpendicular to a given straight line* BC, *where* A *does not lie on* BC (Fig. II.1.18, p. 257).

This problem is the same as *Elements* I.12. Moreover, it belongs in the following section – containing practical geometrical problems, without discussion and with a single solution. It should have appeared after Problem 13. Problems 12 and 13 are, moreover, the two cases for the problem: from a point A to draw the perpendicular to a given straight line BC, $A \notin BC$ (Problem 12), $A \in BC$ (Problem 13).

The reversal of the order of these two problems must be the result of an accident to the text that occurred some time ago, since it is reproduced in all the manuscripts.

2.2. *Practical section on geometrical problems without discussion and with a single solution*

Proposition 13. — *From a given point* A *to draw a straight line perpendicular to a given straight line* BC, *when the point* A *lies on* BC (Fig. II.1.20 of the text, p. 258).

2.3. *Practical section on geometrical problems without discussion having an infinite number of solutions*

Proposition 14. — To *construct a circle tangent to a given straight line* CD *and to a given circle* AB, *where the straight line lies outside the circle.*

Ibn al-Haytham considers only the case in which the two circles touch on the outside.[6]

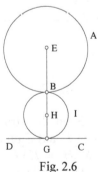

Fig. 2.6

---

[6] Al-Qūhī considers the same problem in his text *The Book of the Centres of Tangent Circles that Lie on Lines by the Method of Analysis*, ms. Paris, BN 2457, fols 19ʳ–21ʳ. He considers two cases: circles that make exterior contact and circles that make interior contact. See P. Abgrall, 'Les cercles tangents d'al-Qūhī', *Arabic Sciences and Philosophy*, 5.2, 1995, pp. 263–95.

Let E be the centre of the given circle, R its radius, H the centre of the required circle, G its point of contact with the straight line and B the point of contact of the two circles.

From *Elements* III.12, the points E, B, H are collinear, and $HG \perp CD$.

Analysis: 1) Let us suppose that E, H, G are collinear. Then $EG \perp CD$, so G is known; EG cuts the given circle in B, and H is the mid point of BG. Hence the synthesis for this case:

Synthesis: From E we drop a perpendicular to CD, let it be EG; it cuts the circle in B. Let H be the mid point of BG; the circle (H, HB) is a solution to the problem, it touches the straight line in G and the circle in B.

Analysis: 2) Let us suppose E, H, G are not collinear; we have $HG = HB < HE$. We extend HG by a length $GK = BE = R$; we have $HK = HE = HB + R$. If G is known, K is known, and the equality $HK = HE$ implies $H\hat{K}E = K\hat{E}H$; we can construct the straight line EH that cuts KG in H.

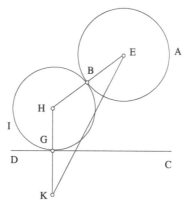

Fig. 2.7

Synthesis: Given any point G on the straight line CD, we construct $GK \perp CD$ such that $GK = R$; given K and E on either side of CD, the angle HKE is acute; we construct $K\hat{E}H = H\hat{K}E$. We have $HK = HE$ and $BE = GK = R$, hence $HB = HG$. The circle with centre H and radius HB touches the circle (E, EB) because H, B, E are collinear, and touches the straight line CD because $HG \perp CD$.

Thus, with every point G of the straight line CD there is associated a circle that touches both the given straight line and the given circle. So the problem admits of an infinite number of solutions.

Notes:

1) With each point $G \in CD$ there is associated a point K lying on a straight line $D \parallel CD$ at a distance R. The point H, the centre of the required circle, is equidistant from the point E and the straight line D; the point H lies on the parabola $\mathscr{P}$, with focus E and directrix D. Every point of the parabola $\mathscr{P}$ provides a solution to the problem.

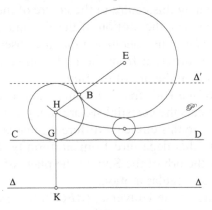

Fig. 2.8

2) If CD is a segment, a 'straight line with endpoints' in ancient geometry, the set of points H is an arc of a parabola.

3) If we consider the straight line Δ', symmetrical to Δ with respect to DC, any point of the parabola $\mathscr{P}'$ with focus E and directrix Δ' is the centre of a circle that touches CD and the circle E; the circles thus touch one another internally ($HK = HE = HB - R$).

Finally we may note that, in the course of all his explanations about analysis and synthesis in geometry, Ibn al-Haytham has avoided raising the question of whether one can construct a converse.

Analysis and synthesis in astronomy

This is the same as in geometry and in arithmetic because Ibn al-Haytham writes:

> As for problems that refer to astronomy, most of them reduce to numerical problems or to geometrical problems; their examples are the examples we gave earlier.[7]

---

[7] See p. 264.

But, among these problems, we can identify a particular group, problems 'that refer to explanations of the motions of the heavenly bodies'.[8] In this particular group that relates to celestial kinematics, Ibn al-Haytham illustrates analysis by considering the example of the motion of the Sun.

The problem is an old one. The ancients showed that the angles with their vertex at the centre of the instrument, the angles swept out in equal times by the radius joining this centre to the centre of the Sun, are unequal. Now, for these astronomers, the motion of the Sun must be regular, that is circular and uniform, hence the conclusion that the observed motion, that is the apparent motion, is different from the real motion, and that this effect results from the position of the orb of the Sun.

Now, the shape of the Universe is a sphere, and the centre of the Sun moves in a plane that cuts the celestial sphere in a great circle. The motion of the Sun with respect to this circle is 'different', that is, it is not a uniform circular motion. From this departure from uniformity, the ancients determined the position of the orb of the Sun, in the plane of the great circle, an orb that then described a regular motion.

Let E be the centre of the Universe, ($ABCD$) the great circle in which the plane cuts the Universe; the orb of the Sun being a circle in this plane, its centre lies in this plane; let G be this centre and ($HIMN$) the circle, the centre of the Sun describes the circle ($HIMN$) in a regular motion.

If the points G and E were identical, the arcs traversed on the two circles in the same time would be similar, which is impossible because the motion on the circle ($HIMN$) is regular and that on ($ABCD$) is 'different'; so $G \neq E$.

If the Sun is in H, it is seen in A; if it is in K, it is seen in L. It has traversed the arc HK on its orb and the arc AL on the circle G. We have HGK $H\hat{G}K > A\hat{E}L$, so the motion of the Sun on the circle (E) in the neighbourhood of A is slower than its motion on the orb. Let $BD \perp AC$; the arc AD is a quarter of a circle, the arc HN is greater than a quarter of a circle, the arc DC is a quarter of a circle and the arc NM is smaller than a quarter of a circle. The arcs BAD and BCD are semicircles. The arc IHN is greater than a semicircle, the arc IMN is smaller and the motion on ($HIMN$) is regular. So the apparent motion of the Sun on BAD is faster than that on BCD, and this is indeed what is seen.

---

[8] See p. 264.

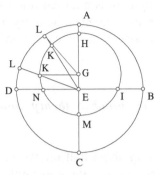

Fig. 2.9a

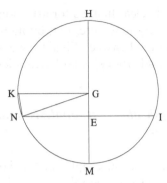

Fig. 2.9b

Let us determine the ratio $\dfrac{EG}{GH}$.

Since the motion is regular, the arcs that are traversed are proportional to the time taken to traverse them. If t_1 and t_2 are the times, in hours, that the Sun takes to traverse the arcs IHN and IMN, and if we draw GK parallel to EN, we have

$$\frac{\widehat{IHN} - \widehat{IMN}}{t_1 - t_2} = \frac{2\widehat{KN}}{t_1 - t_2} = \frac{360°}{24},$$

and the arc KN is thus known

$$\frac{GE}{GH} = \frac{GE}{GN} = \sin \widehat{KN}.$$

So the argument allows us to calculate the ratio $\dfrac{GE}{GH}$ but not to find the distance EG.

Thus, proceeding by analysis, Ibn al-Haytham has proved that if the Universe is a sphere with centre E, and the motion of the Sun is circular and uniform on a circle with centre G and radius GH, then 1) $G \neq E$; 2) the ratio $\dfrac{GE}{GH}$ is known.

Analysis in music

Here Ibn al-Haytham is still more laconic than in astronomy. He does no more than remind us that such analysis reduces to analysis of numerical problems and considers an example: the interval of an octave is composed of the interval of a fourth and the interval of a fifth.

It is clear that, as for astronomy, Ibn al-Haytham is not presenting anything very new here, but, by mentioning these two subjects, he is trying to be comprehensive. Finally, we should emphasise that an exception has been made for celestial kinematics.

2. *Applications of analysis and synthesis in number theory and in geometry*

The second chapter of the Treatise makes up about half the work, and comprises six examples in all, divided into two groups, three examples on number theory and three on geometry. Are arithmetic and geometry not the mathematical disciplines to which all the others can be reduced? On this point, at first glance, Ibn al-Haytham is in line with tradition. But, here as elsewhere, we must not let ourselves be deceived: though the bottles are the same, the wine is different. The terms 'arithmetic' and 'geometry' have already experienced serious changes of meaning.

Finally, we need to ask about the author's intention in writing this work and about his choice of examples. For this it is best to look to Ibn al-Haytham's text. And, indeed, no one could explain matters better than he does himself in presenting this second chapter in the following terms:

> It remains for us to set out problems of analysis involving some difficulties, so that analysis becomes a tool to be used by anyone who works through this treatise and a guide for anyone who is trying to acquire the art of analysis; so that this analysis may be directed by the propositions that are used in it and by the complementary results that are added to its objects so that he can exercise the art of analysis.[9]

So Ibn al-Haytham's intention seems to be transparent: to provide his readers with *difficult* examples, to give them practice in the art of analysis and to make them familiar with exercising it, and in particular to lead them to seek out the auxiliary constructions that are so necessary in the application of analysis: his purpose is obviously methodological and didactic. Although the term 'methodology' can often be misleading, what we have here is the presentation of several 'models' or 'model problems' for carrying out analysis and synthesis: six models in all, which correspond to six research situations from which the reader can draw inspiration, or, at the least, which he can use as a basis for imitation. What we mean by 'model' is much more than a mere illustration. The proof of this is that some of these models, as Ibn al-Haytham himself admits, refer to difficult

[9] See p. 268.

questions. This deliberate choice shows that his intention is not purely didactic. Ibn al-Haytham chooses research problems of the time: a theorem on perfect numbers, the construction of a circle to touch three given circles, and so on. Thus, we are dealing with problems that were subjects of debate at the time, and specifically so among mathematicians in the tradition to which Ibn al-Haytham himself belonged. So everything points to his wanting to tell his reader, so to speak *in vivo*, how to progress, step by step, along the path of analysis, and to find the 'complementary results' necessary for the purpose.

But, although this reasoning helps us understand his choices, it is not sufficient to explain the area from which these items are taken, notably in geometry. For this, we need to bear in mind the 'double' of this treatise, namely the treatise *The Knowns*. As we shall see, in the course of his geometrical analysis it is precisely the properties of position and form that are, above all, of interest to Ibn al-Haytham.

So let us turn our attention to these 'model problems'.

Number theory

Perfect numbers

Ibn al-Haytham seeks to prove the theorem about even perfect numbers that can be rewritten as follows:

Theorem. — *Let* n *be an even number,* $\sigma_0(n)$ *the sum of the proper divisors of* n; *the following conditions are equivalents*:

(a) *if* $n = 2^p(2^{p+1} - 1)$, *where* $(2^{p+1} - 1)$ *is prime, then* $\sigma_0(n) = n$.

(b) *If* $\sigma_0(n) = n$, *then* $n = 2^p(2^{p+1} - 1)$, *where* $(2^{p+1} - 1)$ *is prime.*

Condition (a) is none other than Proposition IX.35 of Euclid's *Elements*, while condition (b) was to be given a definitive proof only by Euler. But, as far as I know, the first attempt to prove it was that by Ibn al-Haytham. In any case, he was the one who stated the condition and tried to prove it.

To understand the choice of the example of perfect numbers, we need only remember that research into the properties of these numbers had been revived by Thābit ibn Qurra,[10] and that al-Khāzin had also been interested

---

[10] See F. Woepcke, 'Notice sur une théorie ajoutée par Thābit Ben Qorrah à l'arithmétique spéculative des grecs', *Journal Asiatique*, IV, 2, 1852, pp. 420–9; R. Rashed, 'Nombres amiables, parties aliquotes et nombres figurés aux XIIIe et XIVe

in them. [11] Nearer to Ibn al-Haytham's time, we meet al-Anṭākī, [12] and, among his contemporaries, al-Baghdādī. [13] These mark stages in the long journey of research before Ibn al-Haytham and in his time.

Employing the analytical approach, we suppose that the number is n, and let its proper divisors also have been found, and we suppose their sum is equal to the number. So this number has divisors and these have properties: we need to find them. Because of this Ibn al-Haytham first proves

(1) $$\sigma_0\left(2^p\right) = 1 + 2 + \ldots + 2^{p-1} = 2^p - 1,$$

hence, when n is an even perfect number,

(2) $$n = \sigma_0(n) \neq 2^p.$$

So a perfect number cannot be of the form 2^p. Ibn al-Haytham establishes this result by *reductio ad absurdum*:

If $n = 2^p$, then $n - 1 = 1 + 2 + \ldots + 2^{p-1}$; and, from (1), we obtain $n = n - 1$. Thus, if a number of the form 2^k has as its proper divisors each of the terms that precede it – in the same way as the perfect number – it is nevertheless not equal to their sum. On the other hand, the definition of a perfect number is that it is equal to the sum of its divisors.

We consider an even number n and a sequence D_1 of its proper divisors that forms a geometric progression with common ratio 2 that ends with $n/2$: $2^{p-1}g$, $2^{p-2}g$, ..., $2g$, g where $2^p g = n$. We suppose that the other divisors also form a geometric progression D_2 with common ratio 2 : 1, 2, ..., 2^{q-1}, 2^q and that $g = 2 \cdot 2^q - 1$. The two sequences of divisors, excluding 1, appear as pairs of divisors; so we have $p = q$, which Ibn al-Haytham also establishes by *reductio ad absurdum*. The sum of the divisors of D_1 is

siècles', *Archive for History of Exact Sciences*, 28, 1983, p. 107–47; repr. in *Entre arithmétique et algèbre: Recherches sur l'histoire des mathématiques arabes*, Paris, 1984, pp. 259–99.

[11] A. Anbouba, 'Un traité d'Abū Ja'far al-Khāzin sur les triangles rectangles numériques', *Journal for the History of Arabic Science*, 3.1, 1979, pp. 134–78, esp. p. 157.

[12] R. Rashed, 'Ibn al-Haytham et le théorème de Wilson', *Archive for History of Exact Sciences*, 22.4, 1980, pp. 305–21; repr. in *Entre arithmétique et algèbre*, pp. 227–43 and 'Ibn al-Haytham et les nombres parfaits', *Historia Mathematica*, 16, 1989, pp. 343–52; repr. in *Optique et mathématiques: Recherches sur l'histoire de la pensée scientifique en arabe*, Variorum CS388, Aldershot, 1992, XI.

[13] R. Rashed, 'Nombres amiables, parties aliquotes et nombres figurés'.

$(2^p - 1) g = n - g$ and the sum of the divisors of D_2 is $2^{q+1} - 1 = g$, so the complete sum is $n - g + g = n$ and n is perfect.

Finally, Ibn al-Haytham proves that g is prime.

Let us suppose that g is not prime; there exists $d|g$, $d \neq 1$. But $d|n$, so $d \in D_1 \cup D_2$. Now $d < g$, so $d \notin D_1$; and on the other hand $d \neq 2^k$, so $d \notin D_2$, because the terms of D_2 are the divisors of $2^{q+1} = g + 1$. It follows that $d = 1$. So it is clear that Ibn al-Haytham is in fact only giving a partial converse of Euclid's theorem. He does not prove that, among *all* even numbers, only Euclid's are perfect; he merely proves that, among even numbers of the form $2^p(2^{q+1} - 1)$, only Euclid's are perfect.

Ibn al-Haytham then performs the synthesis. He takes a number $n = 2^p g$, where g is a number such that

$$g = 2^{p+1} - 1 = \sum_{k=0}^{p} 2^k.$$

We have

$$n = g \sum_{k=0}^{P-1} 2^k + \sum_{k=0}^{p} 2^k.$$

Each number in D_1 or D_2 (for $p = q$) is indeed a divisor of n. Let us suppose that d is a divisor of n, then there exists e, a divisor of n, such that $d \cdot e = n = 2^p g$; we have

$$\frac{e}{g} = \frac{2^p}{d}.$$

If g is a factor of e, then d is a factor of 2^p and $d \in D_2$. If g is not a factor of e, $(g, e) = 1$ because g is prime; accordingly e is a factor of 2^p, and $e = 2^k$ $(1 \leq k \leq p)$; so $d = g2^{p-k}$, $d \in D_1$. Any divisor of n appears in D_1 or in D_2. We conclude that n is equal to the sum of his divisors; so n is perfect.

This half-failure should not obscure the whole: we have a serious attempt to find a property that characterises all even perfect numbers. This 'model problem' cannot be a mere illustration of analysis and synthesis in arithmetic designed for beginners; it is a piece of ongoing research in which Ibn al-Haytham is applying this method in number theory.

In considering the theorem on perfect numbers, Ibn al-Haytham is dealing with an important example in number theory, as the subject is defined by Euclid. In the course of his analysis Ibn al-Haytham raises the problem of the existence of these numbers, of their form, of the 'cause' ('*illa*) by which they have this form and finally the problem of their identification as a class of numbers; that is he proposes a criterion for distinguishing them as a class of numbers. Moreover, this is the reason that led him to prove the converse of Euclid's proposition. It is precisely this research into existence and form that justifies employing the method of analysis in arithmetic, even by way of analogy. In the two following examples, Ibn al-Haytham turns back towards the other tradition in number theory in the tenth century, the tradition of rational Diophantine analysis.

Two indeterminate systems of equations of the first degree
On this occasion also, Ibn al-Haytham is not concerned merely with providing solutions to the systems in rational numbers, but rather with establishing the existence, form and number of the solutions. So in each case the analysis must lead us to express these elements as clearly as possible, as is apparent from the text. Here we shall provide only the statements of the problems.

The first system may be written as

$$\frac{1}{2}x + \frac{2}{3}y = s,$$

$$\frac{1}{3}y + \frac{3}{4}z = s,$$

$$\frac{1}{4}z + \frac{1}{2}x = s.$$

Ibn al-Haytham begins by proving that

$$y = \frac{3}{8}z, \ x = \frac{10}{8}z \text{ et } x = \frac{10}{3}y;$$

which tells him that the required numbers have known ratios one to another: so they exist and are positive rational numbers. As for their form, Ibn al-Haytham proves in his synthesis that to any integer $n \equiv 0 \pmod 8$ there corresponds a solution $x = 10\frac{n}{8}, \ y = 3\frac{n}{8}, \ z = n$, in $\mathbf{Q}^+$.

The second problem is stated as follows: if $k_1, \ k_2, \ k_3$ are given ratios, and a and b two given numbers, to partition a and b so that

$$a = x_1 + x_2 + x_3$$

(*)

$$b = y_1 + y_2 + y_3$$

where $\dfrac{x_1}{y_1} = k_1$, $\dfrac{x_2}{y_2} = k_2$, $\dfrac{x_3}{y_3} = k_3$ $\quad (k_1 > k_2 > k_3 > 0)$.

Here again Ibn al-Haytham makes a point of establishing the existence, form and number of the solutions. While employing a different style of expression, let us follow Ibn al-Haytham's reasoning. The first equation of (*) can be rewritten

$$a = k_1 y_1 + k_2 y_2 + k_3 y_3.$$

But, $k_1 > k_2 > k_3 \Rightarrow k_1 b > a > k_3 b$, hence the necessary condition

$$k_1 > \frac{a}{b} > k_3.$$

Let us put

$$y_1 + y_3 = t,$$

then

$$y_2 = b - t \qquad\qquad (t < b)$$

and

$$a = k_1 y_1 + k_2(b - t) + k_3(t - y_1),$$

hence

$$y_1(k_1 - k_3) = a - k_2(b - t) - k_3 t,$$

hence

$$y_1 = \frac{a - k_2 b + t(k_2 - k_3)}{k_1 - k_3},$$

$$y_2 = b - t,$$

$$y_3 = \frac{k_2 b - a + t(k_1 - k_2)}{k_1 - k_3}.$$

Discussion: First of all we need to know whether the conditions $k_1 > \dfrac{a}{b} > k_3$ and $0 < t < b$ are sufficient to make y_1, y_2, y_3 positive.

- If $\dfrac{a}{b} = k_2$, the three numbers y_1, y_2, y_3 are positive for $0 < t < b$,

$$y_1 = \frac{k_2 - k_3}{k_1 - k_3}t, \quad y_2 = b - t, \quad y_3 = \frac{k_1 - k_2}{k_1 - k_3}t.$$

- If $k_3 < \dfrac{a}{b} < k_2$, $a < k_2 b$, we have $y_2 > 0$, $y_3 > 0$; but

$$y_1 > 0 \Leftrightarrow (k_2 - k_3)t - (k_2 b - a) > 0 \Leftrightarrow t > \frac{k_2 b - a}{k_2 - k_3};$$

the three numbers are positive if

$$b > t > \frac{k_2 b - a}{k_2 - k_3}.$$

- If $k_2 < \dfrac{a}{b} < k_1$, $k_2 b < a$, we then have $y_1 > 0$, $y_2 > 0$; but $y_3 > 0$ requires that

$$b > t > \frac{a - bk_2}{k_1 - k_2}.$$

Notes:

1) In the course of his synthesis, Ibn al-Haytham distinguishes three cases:

- $\dfrac{a}{b} = k_2$; he takes as parameter $BM = y_2 = b - t$.

- If $\dfrac{a}{b} \neq k_2$, Ibn al-Haytham takes as parameter $k = \dfrac{x_1 + x_3}{y_1 + y_3}$; we have

$$k = \frac{a - x_2}{b - y_2} = \frac{a - k_2(b - t)}{t} = k_2 + \frac{a - bk_2}{t}.$$

- If $\dfrac{a}{b} < k_2$, we have $b > t > \dfrac{k_2 b - a}{k_2 - k_3}$; from which we deduce

$$\frac{k_2 b - a}{b} < \frac{k_2 b - a}{t} < k_2 - k_3$$

and in consequence

$$k_3 < k < \frac{a}{b},$$

which is the condition Ibn al-Haytham imposes on the ratio $k = \frac{U}{F}$.

- If $\frac{a}{b} > k_2$, we have $b > t > \frac{a - bk_2}{k_1 - k_2}$, hence

$$\frac{a - bk_2}{b} < \frac{a - bk_2}{t} < k_1 - k_2,$$

and in consequence

$$\frac{a}{b} < k < k_1,$$

without again encountering the condition Ibn al-Haytham imposed on the ratio $k = \frac{S}{O}$.

2) The method Ibn al-Haytham uses has the aim of reducing this problem to Problem 6. He accordingly takes as auxiliary unknowns $X = x_1 + x_3$, $Y = y_1 + y_3$ and a parameter $k = \frac{x_1 + x_2}{y_1 + y_3}$; so we need $k_1 > k > k_3$.

The initial system can be rewritten

$$X + x_2 = a,$$

$$Y + y_2 = b,$$

$$\frac{X}{Y} = k, \quad \frac{x_2}{y_2} = k_2,$$

which corresponds to Problem 6.

Here the ratios $\frac{a}{b}$ and k_2 are given. From the investigation of Problem 6, if $\frac{a}{b} < k_2$, we need to choose k in the interval $\left]k_3, \frac{a}{b}\right[$ in order to have $k < \frac{a}{b} < k_2$; and if $\frac{a}{b} > k_2$, we need to choose k in the interval $\left]\frac{a}{b}, k_1\right[$ in order to then find

$$kY + k_2(b - Y) = a,$$

hence

$$Y = \frac{a - k_2 b}{k - k_2}, \quad y_2 = \frac{bk - a}{k - k_2};$$

from which we deduce X and x_2.

It remains to solve the system

$$x_1 + x_3 = X,$$

$$y_1 + y_3 = Y,$$

$$\frac{x_1}{y_1} = k_1, \quad \frac{x_3}{y_3} = k_3.$$

We know that $k_1 > \dfrac{X}{Y} > k_3$, from the choice of the parameter k; so this system has a unique solution if X and Y are taken to be known. Ibn al-Haytham is thus led to address an additional problem: to find a ratio lying between two given ratios.

Geometrical problems

Ibn al-Haytham chooses three problems, of which the first, and simplest, is a problem in plane geometry; the second deals with geometrical transformations and the third is concerned with a geometrical construction. This sequence of choices does not seem to arise purely as a matter of chance, but relates to three areas of geometry, in which there had been perceptible development.

Problem in plane geometry

The first problem is the simplest. It can be written:
Given three points A, C, B *in that order, and a straight line* DG, *to find a point* E *of that straight line such that* EC *bisects the angle* AEB.

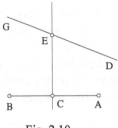

Fig. 2.10

Analysis: If EC bisects the angle AEB, we have $\dfrac{CA}{CB} = \dfrac{EA}{EB}$.

1) If C is the mid point of AB, we have $CA = CB$, hence $EA = EB$, and E lies on the perpendicular bisector of AB.

2) If $AC \neq CB$, $\dfrac{EA}{EB} \neq 1$ is a known ratio; so the point E lies on a known circle, let CI be its diameter (see Problem 1). So E is both on this circle and on the straight line DG.

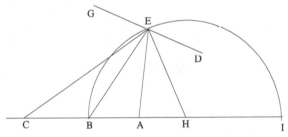

Fig. 2.11

Synthesis:
1) We draw Δ to be the perpendicular bisector of AB. If DG is not perpendicular to AB, Δ cuts DG at the point E, and we have $EA = EB$. The triangle EAB is isosceles, and the height EC bisects the angle at E; the problem has one solution.

If $DG \perp AB$ and $DG \neq \Delta$, the point E does not exist. If $DG = \Delta$, any point of DG provides a solution (Fig. 2.10).

2) Ibn al-Haytham takes $CA > CB$, and defines H by

(1) $$\frac{CH}{HB} = \frac{CA}{CB} > 1.$$

There exist two points H that satisfy the equation. Ibn al-Haytham finds the one between C and B, and the other one beyond B. He chooses the latter without giving details, perhaps because of the analogy with Problem 1 (in that problem the point D, which corresponds to the point H here, was defined by a different procedure, and lay on the extension of AB).

In any case, as the point H lies beyond B, we have

$$\frac{CH}{HB} = \frac{CA}{CB} = \frac{AC+CH}{CB+BH} = \frac{AH}{CH},$$

hence

$$CH^2 = HA \cdot HB.$$

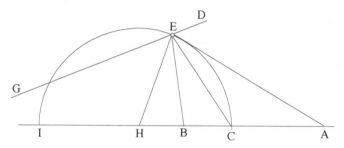

Fig. 2.12.1

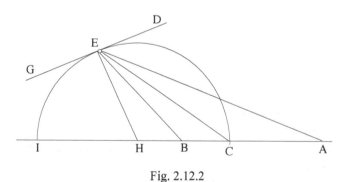

Fig. 2.12.2

Ibn al-Haytham then proves that the circle (H, HC) with diameter CI is the known circle of the analysis. Indeed if this circle cuts DG in E, we have

$$HE = HC \text{ and } \frac{AH}{HE} = \frac{AH}{HC} = \frac{AC}{CB} = \frac{CH}{HB} = \frac{HE}{HB}.$$

So the triangles AHE and BHE are similar, hence

$$\frac{AH}{HE} = \frac{AE}{EB} \quad \text{and} \quad \frac{AE}{EB} = \frac{CA}{CB}.$$

We may note that this proof for a point E of the circle (H, HC) is a proof of the converse that was not supplied in Problem 1; it establishes that any point E of the circle (C, CH) satisfies

$$\frac{EA}{EB} = \frac{CA}{CB}.$$

Discussion: The existence of the point E depends on the distance h from the point H to the straight line DG. Let R be the radius of the circle:

$h > R$	the problem has no solution
$h = R$	the problem has one solution
$h < R$	the problem has two solutions.

We may note that in this problem Ibn al-Haytham deals with the set of points E such that $\dfrac{EA}{EB} = k$.

If $k = 1$, the set is the straight line Δ the perpendicular bisector of AB;

If $k \neq 1$, the set is the circle with diameter CI, where C is given, and I is the harmonic conjugate of C with respect to A, B.

Problem solved with the help of transformations

The second problem in this group is more than merely more complicated, and in it Ibn al-Haytham proceeds by means of geometrical transformations. Its statement is:

Given a fixed point A, *a circle centre* G *and a straight line* BC, *to find a point* D *on the circle* (G) *and a point* H *on* BC *such that the angle ADH is equal to a given angle and* $\dfrac{DA}{DH}$ *is equal to a given ratio.*

The data for the problem show that the given point A and the required points D and H define a triangle of 'known shape', that is similar to a given triangle. So we can put the angle $DAH = \alpha$, a known angle, and $\dfrac{AH}{AD} = k$, a known ratio.

1) The point H can be found from D by one of the *similarities*:

$$S_1(A, \alpha, k) \text{ or } S_2(A, -\alpha, k).$$

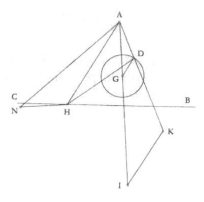

Fig. 2.13

To establish the existence of the point H is to prove that it lies at the intersection of the given straight line BC and one or other of the circles $\mathcal{C}_1 = S_1(G)$ or $\mathcal{C}_2 = S_2(G)$.

2) Similarly we can say that D is found from H by one of the *similarities*

$$S_1' = S_1^{-1} \quad \text{or} \quad S_2' = S_2^{-1}.$$

So if the point D exists, it lies at the intersection of the circle G and one or other of the two straight lines

$$D_1 = S_1'(BC) \quad \text{or} \quad D_2 = S_2'(BC).$$

In both cases, the synthesis compels us to discuss the intersection of a straight line and a circle.

Ibn al-Haytham proposes two analyses of this problem. In the first, he begins by using a *homothety* with centre A in which the circle (G, GD) has as its image the circle (I, IK); then he uses a *similarity* with centre A, in which the circle (I, IK) has as its image the circle (N, NH). The combination of the homothety and the similarity gives one of the similarities mentioned in 1).

In his second analysis, Ibn al-Haytham proves that D lies on the straight line found from BC by one of the similarities mentioned in 2). He then gives two syntheses, which contain nothing of particular interest. For each of these two syntheses, he points out that it compels us to discuss the intersection of a straight line and a circle. This discussion also gives him the number of solutions.

Construction of a circle to touch three given circles

The third geometrical problem – the last in the second chapter and thus the last in the Treatise – is the most important of all, equally for the place it occupies and for its history and the challenge it posed. This last 'model problem' occupies about a fifth of the Treatise as a whole. On the other hand, it is the problem that had already been posed by Apollonius, one to which Pappus and many others had returned. Finally, this same problem had been an object of controversies among Ibn al-Haytham's predecessors. So we have a problem with a noble lineage, but still open and thus a matter for ongoing research. In regard to the nature of the problem itself, historians of geometry have emphasised the profundity of the problem and the difficulty it presented at that time. J. L. Coolidge sees it as marking exactly the limit of what Greek mathematicians could do.[14]

The history of this problem is too well known to be worth further discussion here. Ver Eecke has already provided two accounts of it.[15] Let us simply note that Apollonius poses the problem in his now-lost book *The Tangencies*. It is probably this book that was translated into Arabic under the title *The Tangent Circles* (*al-Dawā'ir al-mumāssa*), which the tenth-century biobibliographer al-Nadīm mentions as among the books by Apollonius known in Arabic.[16] This Arabic translation has now itself been lost. The most direct testimony is still that of Pappus, who reports that this book included some propositions that 'they seem to have been numerous, but for them too we shall take only one'. This, it seems, sets out the question raised by Apollonius:

> If any three elements are given successively in position, such as points, or straight lines and circles, to describe a circle that, passing through each of the given points (in the case where points are given), or tangent to each of the given lines.[17]

A simple exercise in combinatorics gives the problems to be solved, which Pappus lists: 1) three points, 2) three straight lines, 3) two points and one straight line, 4) two straight lines and one point, 5) two points and a circle, 6) two circles and one point, 7) two straight lines and one circle, 8)

---

[14] J. L. Coolidge, *A History of Geometrical Methods*, Oxford, 1940; repr. Dover, 1963, pp. 51–2.

[15] See, for example, P. Ver Eecke's introduction to his translation of Apollonius' *Conics*, Paris, 1959, pp. XXV–XXX.

[16] Al-Nadīm, *Kitāb al-fihrist*, ed. R. Tajaddud, Teheran, 1971, p. 326.

[17] Pappus d'Alexandrie, *La Collection mathématique*, French trans. Paul Ver Eecke, Paris/Bruges, 1933, II.2, p. 483.

two circles and one straight line, 9) one point, one straight line and one circle, 10) three circles.

It is the last problem that is of interest here. However, we do not know what Apollonius' solution was, or even whether he proposed one. We are no better informed about the terms of Pappus' solution, since that seems to have been lost in Pappus' own time. But all this had already become a matter of legend that, as Ver Eecke writes, 'excited the curiosity of the greatest mathematians of recent centuries'.[18] Among these we should mention Viète, Descartes, Newton and, later, L. Carnot, T. Simpson, L. Euler, N. Fus, J. Lambert and Gergonne, among others. But this 'curiosity of the greatest mathematians' made its appearance well before the seventeenth century, since its effects can be seen in the mid ninth century and in the first half of the tenth century.[19] To understand this renewed interest, we need to look at the renewal of research in geometry, notably in the area of the theory of conics and that of geometrical constructions. In the case of the present problem at least, this renewal of interest is linked with names and books that antedate Ibn al-Haytham. One figure who plays a central part is Ibn Sinān, who is also important because Ibn al-Haytham picks up on his work. Ibn Sinān, the representative of one of the great scientific dynasties of the time, the dynasty of the descendants of Thābit ibn Qurra, sheds some light on the investigations into this problem carried out in the course of the first half of the tenth century. Thus, we know, thanks to him, that a representative of another scientific dynasty – the Banū Karnīb – Abū al-'Alā', took an interest in this construction. A third mathematician, no negligible figure, Abū Yaḥyā, one of the teachers of the famous Abū al-Wafā' al-Būzjānī, takes up this same problem on his own account. Ibn Sinān reports and criticises the solutions put forward by his two predecessors. He himself not only takes an interest in this problem, but also, in all probability, in Apollonius' book *The Tangencies*, to the point of composing a book that has the same title as the Arabic translation of Apollonius' work: *The Tangent Circles (al-Dawā'ir al-mumāssa)*. In his autobiography, Ibn Sinān tells us that in this book he deals with 'in which ways circles and straight lines are tangent to one another and pass through [particular] points, and about other things'.[20] This book, which has thirteen chapters, is, according

---

[18] *Les Coniques d'Apollonius de Perge*, trans. Paul Ver Eecke, p. XXVI.

[19] We may note that al-Nadīm attributes to the astronomer and mathematician Ḥabash al-Ḥāsib (still alive in 859) a book with the title *Book on Three Tangent Circles and How the Contact is Made (Kitāb al-dawā'ir al-thalāth al-mutamāssa wa-kayfiyya al-ittiṣāl)*, p. 334.

[20] R. Rashed and H. Bellosta, *Ibrāhīm ibn Sinān: Logique et géométrie au Xe siècle*, p. 12.

to its author himself, closely connected with questions of analysis and synthesis. It has not come down to us. Ibn Sinān wrote a supplement (*tatimma*) to this work, a collection of forty-one problems, 'difficult problems on circles, straight lines, triangles, tangent circles and other things, in which I have used only the method of analysis'.[21] Ibn Sinān in fact supplies, among other things, the analysis for the problem with which we are concerned here.

It is eminently reasonable to suppose that Ibn al-Haytham knew one or the other of Ibn Sinān's books – if not all of them. We have shown that in his research on hour lines for sundials he started from Ibn Sinān's work, but also in opposition to him. The same is true in *Analysis and Synthesis*. That is to say that, together with al-Khāzin, Ibn Sahl and al-Qūhī, Ibn Sinān is one of the leading figures of the tradition that Ibn al-Haytham is trying to take forward as far as possible. So the real question is to find out why he picked up on this problem, and what differentiates his treatment of it from that of Ibn Sinān.[22]

In investigating the problem of constructing a circle that touches three given circles, Ibn Sinān makes the same hypotheses as those made later by Ibn al-Haytham: the circles lie outside one another, their centres are not collinear, and the required circle touches them externally. Ibn Sinān then distinguishes three cases for the given circles $\mathscr{C}_1(K, R_1)$, $\mathscr{C}_2(H, R_2)$ and $\mathscr{C}_3(I, R_3)$ (see Fig. 2.14 below).

The first case is that of equal circles $R_1 = R_2 = R_3$.

The solution is immediate. The point L, the centre of the required circle, is the centre of the circle circumscribed about the triangle KHI, and its radius is $r = LK - R_1$. As we shall see, this case is ignored by Ibn al-Haytham.

The second case is that of two equal circles, $R_1 = R_2$.

Ibn Sinān considers the circle $\mathscr{C}(I, R_3 + R_1)$ if $R_3 < R_1$, or $\mathscr{C}(I, R_3 - R_1)$ if $R_3 > R_1$. The problem reduces to that of finding a circle that touches this circle and passes through the points K and H, a problem he has solved in his *Anthology of Problems*; but he provides only one solution, the second is obvious.

The third case is that in which the three circles are different from one another. Let R_3 be the smallest; Ibn Sinān reduces the problem to finding a circle that passes through the point I and touches the two circles $(K, R_1 - R_3)$ and $(H, R_2 - R_3)$. The solution has an error in the reasoning of the analysis,

[21] *Ibid.*, p. 16.
[22] *Ibid.*, Chap. V.

which leads Ibn Sinān to believe that a ratio is known from the data, which is not true.[23] It is this particular ratio that Ibn Sinān uses in the synthesis, in two different methods, to reduce this problem to another one that he has solved in *The Tangent Circles*: to construct a circle passing through a given point A and, at A, tangent to a given straight line, and tangent to a given circle.

Thus, Ibn Sinān's analysis consists of proving that, in the three cases he considers, constructing the circle touching three given circles can be reduced to a problem that has already been solved. But we are left with the fact that the third case, that is the general case, raises the difficulty we have mentioned.

So we have a problem that Abū al-'Alā' ibn Karnīb and Abū Yaḥyā worked on, and one the solutions to which were criticised by Ibn Sinān; who, although an eminent and esteemed mathematician, seems himself also to have been unable to provide a solution. So there is a challenge that Ibn al-Haytham will take up. This case is, moreover, far from unique. In addition, this construction problem is connected with analysis and synthesis. So, with this background, it is not hard to understand the reasons that spurred Ibn al-Haytham into attacking this problem. He engages with it in a manner different from that of Ibn Sinān: he is interested only in the general case $R_1 < R_2 < R_3$. His analysis, as we shall see in some detail, is different from that of Ibn Sinān: if the required circle $\mathscr{C}(L, r)$ exists, then the circle $\mathscr{C}(L, r + R_1)$ passes through the point K, the centre of $\mathscr{C}_1$, and cuts the straight lines KH and KI in two points S and O. The required point L is thus the centre of the circle circumscribed about the triangle KSO. So the analysis leads to the problem of determining the two points S and O from what is given; which led Ibn al-Haytham to an auxiliary construction for which analysis leads him to distinguish two cases, each time with a discussion.

Thus, Ibn al-Haytham's analysis is different from that of Ibn Sinān. However, it is from the latter's analysis that Ibn al-Haytham has started his own. In fact, his analysis brings in the circle KSO: this circle is none other than the circle to which Ibn Sinān appeals in the general case. Now it is precisely in the course of his investigation of the construction of this circle that Ibn Sinān makes the mistake in his reasoning to which we have already referred.

Thus, matters proceed as if Ibn al-Haytham has noticed this mistake and had picked up the problem starting from the same auxiliary circle as

[23] See R. Rashed and H. Bellosta, *Ibrāhīm ibn Sinān: Logique et géométrie au Xe siècle*, Chap. V.

Ibn Sinān. If that is true, then we may risk advancing hypotheses: when following the steps of Ibn Sinān's construction, Ibn al-Haytham, who had done more than anyone to develop construction methods using conic sections, does not use the differences $LH - LK = R_2 - R_1$ and $LI - LK = R_3 - R_1$ (see geometrical commentary below), which are, moreover, obvious from the figure, and which would have allowed him to construct L as the point of intersection of two branches of a hyperbola. However, in *On the Completion of the Conics*, Ibn al-Haytham has no hesitation in using intersections of conics, even for constructing solutions to plane problems like this one. Perhaps he wanted to follow the tradition that problems in the plane are to be solved by means of straightedge and compasses. Perhaps also, while having noticed this new possibility, he nevertheless wanted to continue along the route laid out by his predecessor, and to correct his work. Whatever we choose to believe, here as in other cases Ibn al-Haytham has conceived his construction in the terms laid down by Ibn Sinān and also against him. So let us examine Ibn al-Haytham's solution.

Let $\mathscr{C}_1$ (K, R_1), $\mathscr{C}_2$ (H, R_2), $\mathscr{C}_3$ (I, R_3) *be three given circles, each one outside each of the others;* K, H, I, *which are not collinear, being their centres;* R_1, R_2, R_3 *the radii such that* $R_1 < R_2 < R_3$.

Let us put $HI = d_1$, $KI = d_2$, $KH = d_3$, *and* $H\hat{K}I = \hat{\alpha} < 2$ right angles. Thus, by hypothesis $d_3 > R_1 + R_2$, $d_2 > R_1 + R_3$ and $d_1 > R_2 + R_3$.

To construct a circle $\mathscr{C}(L, r)$ touching the three circles (Fig. 2.14).

If such a circle $\mathscr{C}(L, r)$ exists, then the circle $\mathscr{C}(L, r + R_1)$ passes through K and cuts the straight lines HK and IK in S and O respectively. The aim of Ibn al-Haytham's analysis is to prove that the points S and O are 'known', that is determinate from the data for the problem, and that the required point L is consequently the centre of the circle circumscribed about the triangle KSO, a known triangle.

Ibn al-Haytham supposes that L lies inside the salient angle HKI, and in this case at least one of the angles LKH and LKI is acute; hence the three cases of the figure that are considered (Figs 2.15, 2.16, 2.17).

But it is possible for L to lie outside the salient angle – in the position L_1 of Figure 2.14, and in this case at least one of the angles LKH and LKI is obtuse. Ibn al-Haytham does not consider this possibility. Moreover, he does not address the problem of the number of possible solutions.

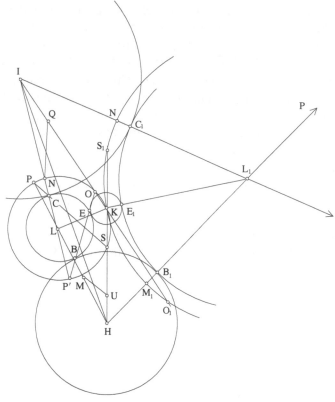

Fig. 2.14

In all cases of the figure, the circle $\mathscr{C}(L, r + R_1)$ cuts $[HL)$ in M and P and $[IL)$ in N and P', here $HM < HK < HP$ and $IN < IK < IP'$; we obtain

$$HM = R_2 - R_1, \ IN = R_3 - R_1, \ PM = NP' = 2KL = 2(r + R_1).$$

On the straight half lines $[HK)$ and $[IK)$, the positions of the points S and O with respect to the point K depend on the case of the figure. In the three cases investigated by Ibn al-Haytham, we have $IO < IK$, where $HS < HK$ (Figs 2.14 and 2.15), $HS = HK$ (Fig. 2.16) and $HS > HK$ (Fig. 2.17). But in Figure 2.14, we have instead $IO_1 > IK$ and $HS_1 > HK$.

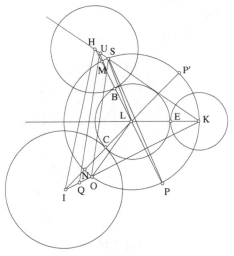

Fig. 2.15

In all the cases, we can write

$$HM \cdot HP = HS \cdot HK,$$

hence

$$\frac{HP}{HS} = \frac{HK}{HM} = \frac{d_3}{R_2 - R_1} = \lambda_1 \quad (\text{where } \lambda_1 > 1)$$

and

$$IN \cdot IP' = IO \cdot IK$$

hence

$$\frac{IP'}{IO} = \frac{IK}{IN} = \frac{d_2}{R_3 - R_1} = \lambda_2 \quad (\text{where } \lambda_2 > 1).$$

Ibn al-Haytham next defines U on $[HK)$ and Q on $[IK)$ by

$$\frac{HM}{HU} = \lambda_1 \text{ and } \frac{IN}{IQ} = \lambda_2,$$

which implies $MU \parallel PS$ and $NQ \parallel P'O$, and consequently that U lies between H and S, and Q between I and O.

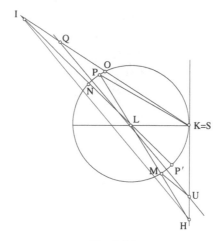

Fig. 2.16

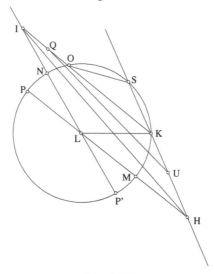

Fig. 2.17

We also have

$$HM^2 = HK \cdot HU \text{ and } IN^2 = IK \cdot IQ,$$

hence

$$HU = \frac{HM^2}{HK} = \frac{(R_2 - R_1)^2}{d_3} < d_3 = HK \text{ (because } R_2 - R_1 < R_2 + R_1 < d_3),$$

and

$$IQ = \frac{IN^2}{IK} = \frac{(R_3 - R_1)^2}{d_2} < d_2 = KI.$$

So the points U and Q are known points, U on the segment KH, Q on the segment KI and UKQ is thus a known triangle.

We have

$$\lambda_1 = \frac{HP}{HS} = \frac{HM}{HU} = \frac{HP - HM}{HS - HU} = \frac{MP}{US}$$

and

$$\lambda_2 = \frac{IP'}{IO} = \frac{IN}{IQ} = \frac{IP' - IN}{IO - IQ} = \frac{NP'}{OQ}.$$

On the other hand, if $K \neq O$ and $K \neq S$, in the triangle OKS we have $O\hat{K}S = \alpha$ or $O\hat{K}S = \pi - \alpha$, and consequently

$$OS = 2LK \sin \alpha = MP \sin \alpha = NP' \sin \alpha.$$

From which we deduce

$$\frac{OS}{US} = \frac{OS}{MP} \cdot \frac{MP}{US} = \lambda_1 \sin \alpha$$

and

$$\frac{OS}{OQ} = \frac{OS}{NP'} \cdot \frac{NP'}{OQ} = \lambda_2 \sin \alpha.$$

But, if $K = S$ (Fig. 2.16), we have $OS = OK = 2LK \sin \alpha$, and the preceding result remains true. If $K = O$, the preceding result is still true.

Thus, for all cases of the figure, analysis leads to a known triangle UKQ and two points S and O on the straight half lines $[UK)$ and $[QK)$, defined by

$$\frac{US}{OS} = k \text{ and } \frac{OQ}{OS} = k',$$

where

$$k = \frac{1}{\lambda_1 \sin \alpha} \text{ and } k' = \frac{1}{\lambda_2 \sin \alpha}.$$

In order to establish, from these last equations, the conclusion that S and O are two known points, Ibn al-Haytham considers an auxiliary problem:

> Given a triangle KUQ and two ratios k and k', to find a pair of points (S, O), where S is on [UK) and O on [QK) such that

$$\frac{US}{OS} = k \ \text{ and } \ \frac{OQ}{OS} = k'.^{24}$$

The data for the auxiliary problem – the angle $UKQ = \alpha$, KU, KQ, k and k' – are all expressed in terms of those of the initial problem:

$$KU = KH - HU = d_3 - \frac{(R_2 - R_1)^2}{d_3},$$

$$KQ = KI - IQ = d_2 - \frac{(R_3 - R_1)^2}{d_2},$$

$$k = \frac{US}{OS} = \frac{R_2 - R_1}{d_3 \sin \alpha},$$

$$k' = \frac{OQ}{OS} = \frac{R_3 - R_1}{d_2 \sin \alpha};$$

from which we obtain

$$\frac{US}{OQ} = \frac{d_2(R_2 - R_1)}{d_3(R_3 - R_1)},$$

$$\frac{KU}{KQ} = \left[\frac{d_3^2 - (R_2 - R_1)^2}{d_2^2 - (R_3 - R_1)^2} \right] \cdot \frac{d_2}{d_3}.$$

In his analysis of this problem, Ibn al-Haytham distinguishes two cases:

1. $\dfrac{k}{k'} = \dfrac{US}{OQ} = \dfrac{KU}{KQ},$ which corresponds to $SO \parallel UQ$;

2. $\dfrac{k}{k'} = \dfrac{US}{OQ} \neq \dfrac{KU}{KQ}.$

^{24} See p. 199.

In both cases, a discussion is necessary, but it does not appear either in the analysis or in the synthesis. In fact, for the required point S to give a solution to the problem of constructing the circle with centre L, we need S to lie on the segment UK or beyond K. In case 1, there can be either one or two solutions, and in case 2, there can be no solution, or one or two solutions (see auxiliary problem).

Synthesis: Let us return to the three given circles. The circle $\mathscr{C}_1(K, R_1)$ cuts HK in E and IK in G.

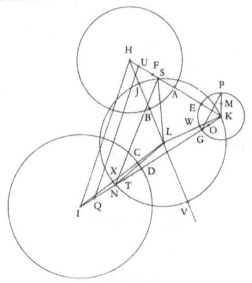

Fig. 2.18

Ibn al-Haytham calls upon $R_2 - R_1 = HF$, where F lies on $[HK]$, $R_3 - R_1 = IT$, where T lies on $[IK]$, and he defines U on $[HK]$ and Q on $[IK]$ by $HK \cdot HU = HF^2$ and $IK \cdot IQ = IT^2$; these are the points U and Q of the analysis (see Fig. 2.18 above and Figs II.1.45–46, pp. 297–8).

The points S and O of the analysis here become S and N – the letters of the figures change in the synthesis.

To derive the ratios $\dfrac{SN}{US}$ et $\dfrac{SN}{QN}$ in terms of the data, Ibn al-Haytham uses an auxiliary construction on the circle $\mathscr{C}_1(K, R_1)$: if $H\hat{K}I = \hat{\alpha}$, we have arc $GE = \hat{\alpha}$; we construct P such that arc $GP = 2\hat{\alpha}$, for $\hat{\alpha} < \dfrac{\pi}{2}$, $\hat{\alpha} = \dfrac{\pi}{2}$, $\hat{\alpha} > \dfrac{\pi}{2}$; so we have $GP = 2 R_1 \sin \alpha$, an equality that holds in all three cases of the figure.

The points M on $[PK]$ and O on $[KG]$ are defined by

$$\frac{2R_1}{PM} = \frac{d_3}{R_2 - R_1} \quad \text{and} \quad \frac{2R_1}{GO} = \frac{d_2}{R_3 - R_1}.$$

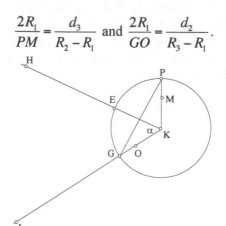

Fig. 2.19

Finally, let us put

$$\frac{SN}{US} = \frac{GP}{PM},$$

hence

$$\frac{SN}{US} = \frac{d_3}{R_2 - R_1}\sin\alpha$$

and

$$\frac{SN}{QN} = \frac{GP}{GO},$$

hence

$$\frac{SN}{QN} = \frac{d_2}{R_3 - R_1}\sin\alpha.$$

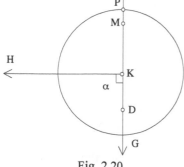

Fig. 2.20

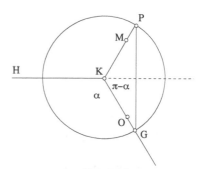

Fig. 2.21

Thus, Ibn al-Haytham has set out the expressions for the ratios investigated in his analysis, where he had merely asserted that they are known. He has obtained the points S and N. If these two points are different from K, then SKN is a triangle. To prove that L, the centre of its circumscribed circle, is the centre we require, he argues by *reductio ad absurdum*. In the synthesis, he does not consider the case in which one of the points S and N is identified with the point K. This possibility had, however, appeared in the second case in his analysis. If, for example, we suppose that $S = K$, then the point L lies at the intersection of the perpendicular bisector of $[KN]$ and the perpendicular to the straight line KH at K.

We may offer a brief conclusion. Ibn al-Haytham's analysis for the three cases he investigates is correct; it is correct also for the fourth case, which he seems not to have noticed, though providing that the investigation of the auxiliary problem must itself be correct. As we shall see shortly, this last is, however, incomplete. It should have included discussions of which there is no trace in the treatise. Indeed, Ibn al-Haytham thinks that the auxiliary problem has a single unique solution in all the cases, whereas it can have two or none. We may wonder why these discussions do not appear. To look into this, we must first return to the auxiliary problem. For the synthesis, we shall merely note the auxiliary constructions that distinguish it.

Auxiliary problem

Let us put $KU = b$, $KQ = c$, $UQ = a$, $U\hat{K}Q = \alpha$, $U\hat{Q}K = \beta$. Let us also put $US = y$, $OQ = z$, $SO = x$, where $x > 0$, $y > 0$, $z > 0$.

Let us follow Ibn al-Haytham and distinguish the two cases of his analysis.

I) $$\frac{y}{z} = \frac{b}{c} \Rightarrow \frac{b}{c} = \frac{k}{k'}.$$

In this case, we have $OS \parallel UQ$, and consequently

$$\frac{OS}{SK} = \frac{a}{b};$$

but by hypothesis

$$\frac{US}{OS} = k$$

hence

$$\frac{SU}{SK} = k\frac{a}{b}.$$

If $k = \frac{b}{a}$, $k\frac{a}{b} = 1$; one point S provides a solution to the problem, that is the mid point of $[UK]$.

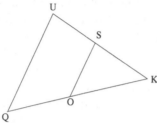

Fig. 2.22

If $k\frac{a}{b} \neq 1$, two points S on the straight line $[UK)$ satisfy $\frac{SU}{SK} = k\frac{a}{b}$.

If $k\frac{a}{b} > 1, k > \frac{b}{a}$, the two points provide solutions to the problem, S_1 on $[UK]$ and S_2 beyond K.

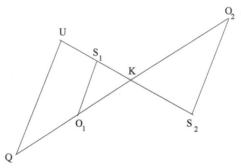

Fig. 2.23

If $k\frac{a}{b} < 1, k < \frac{b}{a}$, the point S_1 on $[UK]$ provides a solution to the problem; the second point, S_2, lies beyond U, it does not lie on the straight half line $[UK)$.

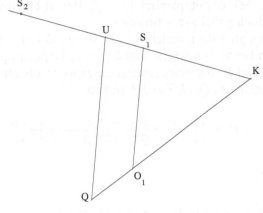

Fig. 2.24

Having found the point S, the point O is found from it, because $SO \parallel UQ$. Ibn al-Haytham considers only the point S_1.[25]

II)
$$\frac{y}{z} \neq \frac{b}{c} \Rightarrow \frac{b}{c} \neq \frac{k}{k'} \text{ or } k'b \neq kc.$$

[25] Let us take up this discussion again using a different method:
$$\frac{OS}{SK} = \frac{a}{b} \Leftrightarrow \frac{x}{|b-y|} = \frac{a}{b}.$$

• If $y < b$, $\dfrac{x}{b-y} = \dfrac{a}{b} \Leftrightarrow \dfrac{y}{b-y} = k\dfrac{a}{b} \Leftrightarrow y = \dfrac{kab}{b+ka}$, which gives $0 < y < b$, hence the solution:

$$x = \frac{ab}{b+ka} = \frac{ac}{c+k'a}, \ y = \frac{kab}{b+ka}, \ z = \frac{k'ac}{c+k'a} < c.$$

That solution gives S_1 on $[UK]$ and O_1 on $[QK]$; it exists for any value of k and $k' = k\dfrac{c}{b}$.

• If $y > b$, $\dfrac{x}{y-b} = \dfrac{a}{b} \Leftrightarrow \dfrac{y}{y-b} = k\dfrac{a}{b} \Leftrightarrow y = \dfrac{kab}{ak-b}$, we need $0 < y$, hence the condition $k > \dfrac{b}{a}$, hence $k' > \dfrac{c}{a}$ because by hypothesis $\dfrac{k}{b} = \dfrac{k'}{c}$.

If this condition is satisfied, we have the solution:
$$x = \frac{ab}{ka-b} = \frac{ac}{k'a-c}, \ y = \frac{kab}{ka-b}, \ z = \frac{k'ac}{k'a-c} > c.$$

This solution gives S_2 on the straight half line $[UK)$ beyond K, and O_2 on the straight half line $[QK)$ also beyond K. This solution exists only if $k > \dfrac{b}{a}$.

In this case, *SO* is not parallel to *UQ*. Ibn al-Haytham sets out a procedure for reducing this to the first case.

He draws through *S* the parallel to *UQ* – it cuts *KQ* in *T*; and through *U* he draws the parallel to *SO*, which cuts *KQ* in *J*. Depending on the values given for b, c, k and k', there are several cases to be distinguished for the positions of the points *K*, *Q*, *O*, *T* and *J*. In fact,

$$ST \parallel UQ \Rightarrow \frac{US}{QT} = \frac{b}{c} \Rightarrow \frac{OQ}{QT} = \frac{OQ}{US} \cdot \frac{US}{QT} = \frac{z}{y} \cdot \frac{b}{c} = \frac{k'b}{kc}.$$

We can have

1) $\dfrac{y}{z} > \dfrac{b}{c} \Leftrightarrow \dfrac{k}{k'} > \dfrac{b}{c} \Leftrightarrow kc > k'b$, hence $QT > QO$.

In this case, $J \in [Qz')$ and $J\hat{U}K = Q\hat{U}K + J\hat{U}Q$.

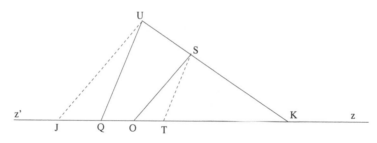

Fig. 2.25

2) $\dfrac{y}{z} < \dfrac{b}{c} \Leftrightarrow \dfrac{k}{k'} < \dfrac{b}{c} \Leftrightarrow kc < k'b$, hence $QT < QO$.

In this case $J \in [Qz)$ and $J\hat{U}K = Q\hat{U}K - J\hat{U}Q$.

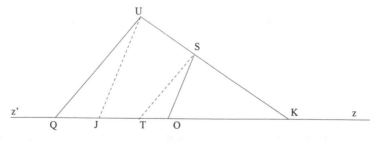

Fig. 2.26

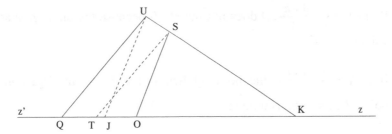

Fig. 2.27

In both cases, we have $OT = |OQ - QT|$, and we infer

$$\frac{OQ}{OT} = \frac{k'b}{|k'b - kc|} \quad \text{and} \quad \frac{OS}{OT} = \frac{b}{|k'b - kc|}, \quad \text{because } \frac{OQ}{OS} = k'.$$

We have drawn $UJ \parallel SO$; because triangles UQJ and STO are similar we may write

$$m = \frac{UJ}{JQ} = \frac{SO}{OT} = \frac{b}{|k'b - kc|}.$$

Moreover, we have $U\hat{Q}K = \beta$, which is given. But the point J can lie on the straight half line $[Qz)$ or on its extension. So there are two cases:

1) $kc > k'b$, $m = \dfrac{UJ}{JQ} = \dfrac{b}{kc - k'b}$ and $U\hat{Q}J = \pi - \beta$ (Fig. 2.25).

2) $kc < k'b$, $m = \dfrac{UJ}{JQ} = \dfrac{b}{k'b - kc}$ and $U\hat{Q}J = \beta$ (Figs 2.26 and 2.27).

In both cases, the ratio $\dfrac{UJ}{JQ} = m$ and the angle UQJ are known. Ibn al-Haytham concludes from this that 'the triangle UJQ is of known shape'. But, since the points U and Q are known, if $m = 1$, then J lies on Δ, the perpendicular bisector of $[UQ]$; and if $m \neq 1$, J lies on a circle Γ (the circle that is the locus of points M such that $\dfrac{UM}{MQ} = m$).

1. $\dfrac{c}{b} > \dfrac{k'}{k}$; J is on the straight half line $[Qz')$.

If $m = 1 \Leftrightarrow \dfrac{c}{b} = \dfrac{1+k'}{k}$, Δ does not cut $[Qz')$ because the angle β is acute; J does not exist.

If $m > 1 \Leftrightarrow \dfrac{c}{b} < \dfrac{1+k'}{k}$, the point Q lies inside Γ, Γ cuts $[Qz')$ in one point; J exists and is unique.

If $m < 1 \Leftrightarrow \dfrac{c}{b} > \dfrac{1+k'}{k}$, U lies inside Γ and Q outside it; Γ does not cut $[Qz')$, J does not exist.

2. $\dfrac{c}{b} < \dfrac{k'}{k}$; J is on the straight half line $[Qz)$.

If $m = 1 \Leftrightarrow \dfrac{c}{b} = \dfrac{k'-1}{k}$, Δ cuts $[Qz)$ because β is acute. J exists and is unique.

If $m > 1 \Leftrightarrow \dfrac{c}{b} > \dfrac{k'-1}{k}$, Q lies inside Γ, then Γ cuts $[Qz)$ in one point; J exists and is unique.

If $m < 1 \Leftrightarrow \dfrac{c}{b} < \dfrac{k'-1}{k}$, the circle Γ can cut $[Qz)$, touch it, or not cut it.

In this case, let us suppose MN is a diameter of Γ, P its centre and R its radius.

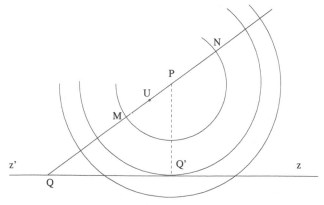

Fig. 2.28

We have

$$\frac{MU}{MQ} = \frac{NU}{NQ} = m,$$

hence

$$\frac{MU + MQ}{MQ} = m + 1, \quad MQ = \frac{a}{m+1};$$

in the same way we have

$$\frac{NQ - NU}{NQ} = 1 - m,$$

hence

$$NQ = \frac{a}{1-m}.$$

From which it follows that

$$MN = NQ - MQ = \frac{2am}{1-m^2},$$

hence

$$R = \frac{am}{1-m^2}.$$

Moreover,

$$PQ = \frac{NQ + MQ}{2} = \frac{a}{1-m^2}.$$

If $PQ' \perp [Qz)$, we have

$$PQ' = PQ \sin \beta = \frac{a \sin \beta}{1-m^2};$$

the circle will cut $[Qz)$ in two points if $PQ' < R$, that is if

$$\frac{a \sin \beta}{1-m^2} < \frac{am}{1-m^2},$$

that is if $m > \sin \beta$.

We have

$$m > \sin \beta \Leftrightarrow \frac{b}{k'b - kc} > \sin \beta \Leftrightarrow \frac{c}{b} > \frac{k' \sin \beta - 1}{k \sin \beta} \Leftrightarrow \frac{c}{b} > \frac{k'}{k} - \frac{1}{k \sin \beta}.$$

So we obtain

- if $\dfrac{c}{b} < \dfrac{k'}{k} - \dfrac{1}{k \sin \beta}$, the point J does not exist;

- if $\dfrac{c}{b} = \dfrac{k'}{k} - \dfrac{1}{k \sin \beta}$, there exists a unique point J on $[Qz)$;

- if $\dfrac{k'-1}{k} > \dfrac{c}{b} > \dfrac{k'}{k} - \dfrac{1}{k \sin b}$, there exist two points J_1 and J_2 on $[Qz)$.

Below is a summary of the complete discussion when it is assumed that β is acute, and we suppose that the two numbers $\dfrac{k'}{k} - \dfrac{1}{k \sin \beta}$ and $\dfrac{k'}{k} - \dfrac{1}{k}$ are positive.

$$\dfrac{c}{b} \in \left] 0, \dfrac{k'}{k} - \dfrac{1}{k \sin \beta} \right[\qquad \text{no point } J$$

$$\dfrac{c}{b} = \dfrac{k'}{k} - \dfrac{1}{k \sin \beta} \qquad 1 \text{ point } J$$

$$\dfrac{c}{b} \in \left] \dfrac{k'}{k} - \dfrac{1}{k \sin \beta}, \dfrac{k'-1}{k} \right[\qquad 2 \text{ points } J_1 \text{ and } J_2 \left. \right\} \text{ case } 2, J \text{ on } [Qz)$$

$$\dfrac{c}{b} = \dfrac{k'-1}{k} \qquad 1 \text{ point } J$$

$$\dfrac{c}{b} \in \left] \dfrac{k'-1}{k}, \dfrac{k'}{k} \right[\qquad 1 \text{ point } J$$

$$\dfrac{c}{b} \in \left] \dfrac{k'}{k}, \dfrac{1+k'}{k} \right[\qquad 1 \text{ point } J$$

$$\dfrac{c}{b} = \dfrac{1+k'}{k} \qquad \text{no point } J \left. \right\} \text{ case } 1, J \text{ on } [Qz')$$

$$\dfrac{c}{b} \in \left] \dfrac{1+k'}{k}, +\infty \right[\qquad \text{no point } J$$

Once we have obtained the point J, the triangle KUJ is known, and we find SO as in case 1, since $SO \parallel UJ$. To investigate the position of O would demand a supplementary discussion when J is on $[Qz')$, a discussion we shall not give here.

We may also note that the method Ibn al-Haytham uses in case II assumes $S \neq K$. If $S = K$, the parallel to SO drawn through U is parallel to QK; the point J accordingly recedes to infinity.

We have just seen that the investigation of the auxiliary problem, although correct, is not complete. Ibn al-Haytham seems to think that the problem has a unique solution in every case. But we have seen that there can be two solutions, or none.

If the conclusions we have drawn from our investigation are correct, we should ask ourselves about how this discussion came to escape Ibn al-Haytham. We can see only two lapses. The first arises from the fact that he took it that a point of a given straight line, defined by the ratio of its distances from two known points, is unique; he takes the one that lies on the segment joining the two points and neglects the one on the extension of the segment. The second lapse arises from the assertion that a triangle is defined, up to a similarity, if we are given an angle and the ratio of one side adjacent to that angle to the side that is opposite it.

It remains that Ibn al-Haytham has reduced the problem of the construction of the circle touching three given circles to the problem of the existence of the two points required in the auxiliary problem. So it is on the discussion of this latter problem that the number of solutions to the initial problem depends. Now this discussion, a very complicated one, as we have seen, was one Ibn al-Haytham did not engage upon.

Geometrical commentary on the problem
Let us return to the data for the problem. If the circle $\mathscr{C}(L, r)$ exists, we have $LK = r + R_1$, $LH = r + R_2$ and $LI = r + R_3$, hence

(1) $LH - LK = R_2 - R_1$

and

(2) $LI - LK = R_3 - R_1.$

From (1) $L \in \mathscr{H}_1$, the branch around the focus K of a hyperbola whose second focus is H; and from (2) $L \in \mathscr{H}_2$, the branch around the focus K of a hyperbola whose second focus is I. So the problem of constructing the tangent circle reduces to one of the problems that Ibn al-Haytham proposed in a particular area of geometry: geometrical construction using conic sections. The whole problem is now to find out whether $\mathscr{H}_1$ and $\mathscr{H}_2$ cut one another.

If we examine the special case in which the centres of $\mathscr{C}_1$, $\mathscr{C}_2$, $\mathscr{C}_3$, that is K, H, I, are collinear, the two branches $\mathscr{H}_1$ and $\mathscr{H}_2$ then have the same axis, and it is clear that $\mathscr{H}_1$ and $\mathscr{H}_2$ can have 0, 1 or 2 common points; consequently, the problem itself admits of 0, 1 or 2 solutions. If $H\hat{K}I = \alpha = 0$, we have the following figures:

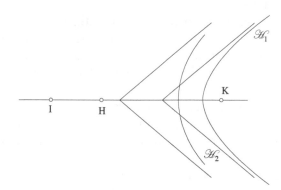

Fig. 2.29

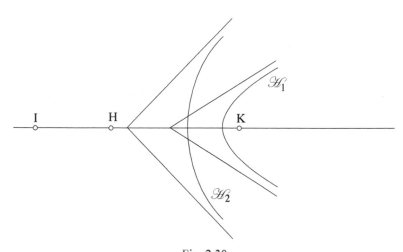

Fig. 2.30

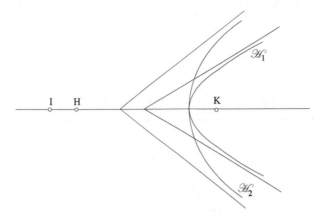

Fig. 2.31

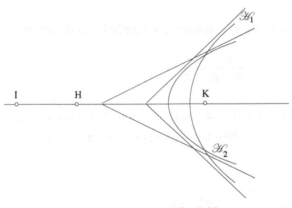

Fig. 2.32

We may note that $\mathcal{H}_1$ and $\mathcal{H}_2$ have the same vertex if and only if

$$KH - (R_2 - R_1) = KI - (R_3 - R_1) \Leftrightarrow KI - KH = R_3 - R_2 \Leftrightarrow d_3 - d_2 = R_3 - R_2.$$

If $H\hat{K}I = \alpha = \pi$, $\mathcal{H}_1$ and $\mathcal{H}_2$ cut one another in two points that are symmetrical with respect to the straight line HK.

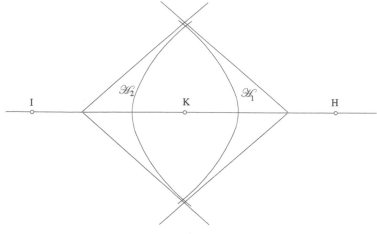

Fig. 2.33

We may also note that the asymptotes are parallel if and only if

$$\frac{KH}{R_2 - R_1} = \frac{KI}{R_3 - R_1}.$$

If in addition, $H\hat{K}I = \alpha = 0$, $\mathcal{H}_1$ and $\mathcal{H}_2$ have their common point at infinity, the curves $\mathcal{C}_1$, $\mathcal{C}_2$, $\mathcal{C}_3$ then have two common tangents.

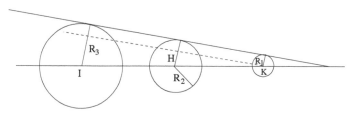

Fig. 2.34

As we have seen, this special case with collinear centres was not considered by Ibn al-Haytham, who supposes that KHI is a true triangle. The investigation of the intersection of $\mathcal{H}_1$ and $\mathcal{H}_2$ then becomes complicated. We can, however, reduce this investigation to that of the intersection of a straight line Δ and a branch of a hyperbola, using an algebraic method.

Let there be an orthonormal coordinate system (Kx, Ky); let us put $0 < H\hat{K}I < \pi$. We have $K(0, 0)$, $H(d_3, 0)$, $I(d_2 \cos \alpha, d_2 \sin \alpha)$, $L(x, y)$.

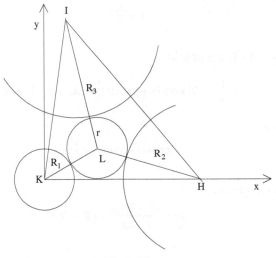

Fig. 2.35

Thus, the data satisfy the inequalities

$$R_1 < R_2 < R_3, \quad d_2 > R_3 + R_1, \quad d_3 > R_2 + R_1,$$

$$d_2^2 + d_3^2 - 2d_2 d_3 \cos \alpha > (R_2 + R_3)^2.$$

The circle $\mathscr{C}(L, r)$ is a solution to the problem if and only if

$$LK = r + R_1, \ LH = r + R_2 \text{ and } LI = r + R_3,$$

hence

(1) $$x^2 + y^2 = (R_1 + r)^2;$$

(2) $$(d_3 - x)^2 + y^2 = (R_2 + r)^2;$$

(3) $$(d_2 \cos \alpha - x)^2 + (d_2 \sin \alpha - y)^2 = (R_3 + r)^2.$$

From (1) and (2) we obtain

(4) $$d_3(d_3 - 2x) = (R_2 - R_1)(R_2 + R_1 + 2r),$$

and from (4) we have

$$x < \frac{d_3}{2}.$$

From (1) and (3) we obtain

(5) $d_2\left[d_2 - 2(x\cos\alpha + y\sin\alpha)\right] = (R_3 - R_1)(R_3 + R_1 + 2r),$

and from (5) we have

$$x\cos\alpha + y\sin\alpha < \frac{d_2}{2}.$$

From (4) we obtain

$$2r = \frac{d_3^2 - 2d_3 x}{R_2 - R_1} - (R_2 + R_1),$$

hence

(6) $2(r + R_1) = \dfrac{d_3^2 - 2d_3 x}{R_2 - R_1} - (R_2 - R_1).$

From (1) and (6) we obtain

$$x^2 + y^2 = \frac{1}{4(R_2 - R_1)^2}\left[d_3^2 - (R_2 - R_1)^2 - 2d_3 x\right]^2,$$

which can be rewritten

$$4(x^2 - d_3 x)\left[(R_2 - R_1)^2 - d_3^2\right] + 4y^2(R_2 - R_1)^2 = \left[d_3^2 - (R_2 - R_1)^2\right]^2;$$

hence

(7) $\dfrac{4\left(x - \dfrac{d_3}{2}\right)^2}{(R_2 - R_1)^2} - \dfrac{4y^2}{d_3^2 - (R_2 - R_1)^2} = 1,$

the equation of the hyperbola $\mathcal{H}_1$ with centre $\left(\frac{d_3}{2}, 0\right)$, foci K and H and transverse axis $(R_2 - R_1)$.

From the condition $x < \dfrac{d_3}{2}$, the point L lies on the branch of $\mathcal{H}_1$ that passes round the focus K.

From (4) and (5) we obtain

(8) $$R_3 - R_2 = \frac{d_2^2 - 2d_2(x\cos\alpha + y\sin\alpha)}{R_3 - R_1} - \frac{d_3^2 - 2d_3 x}{R_2 - R_1}$$

which is the equation of the straight line Δ.

We may note that by eliminating r between (1) and (5), we find the equation of $\mathcal{H}_2$, with axis KI and foci K and I, which can be written

$$x^2 + y^2 = \frac{1}{4(R_3 - R_1)^2}\left[d_2^2 - (R_3 - R_1)^2 - 2d_2(x\cos\alpha + y\sin\alpha)\right]^2,$$

where $(x\cos\alpha + y\sin\alpha) < \dfrac{d_2}{2}$; or again

$$\frac{4\left(x\cos\alpha + y\sin\alpha - \dfrac{d_2}{2}\right)^2}{(R_3 - R_1)^2} - \frac{4(-x\sin\alpha + y\cos\alpha)^2}{d_2^2 - (R_3 - R_1)^2} = 1.$$

Eliminating $x^2 + y^2$ between the equations of $\mathcal{H}_1$ and $\mathcal{H}_2$ give us back the equation of the straight line Δ.

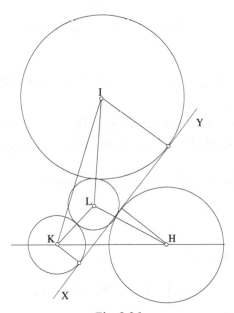

Fig. 2.36

Now, the investigation of Δ and $\mathcal{H}_1$ brings in six parameters: $R_1, R_2, R_3,$ d_2, d_3 and α. We shall not present it here. Let us note, however, that if Δ is parallel to an asymptote, Δ and $\mathcal{H}_1$ have one common point at infinity; to this point there corresponds a straight line tangent to the three given circles. This is, for example, the case in Figure 2.36, in which the circle L, on the one hand, and the straight line XY, on the other, are tangents to the three given circles. This straight line corresponds to the case in which, in Figure 2.14, the point L_1 recedes to infinity.

Algebraic commentary on the auxiliary problem
The algebraic reading of the auxiliary problem is not from Ibn al-Haytham. It allows us to look at his text in a different way, which might shed light on its development.

Let us set out with the same data as before. We are presented with several cases for the figure, according to the positions of S and O with respect to K on the straight half line $[UK)$ and the straight half line $[QK)$, respectively. In all the cases for the figure, we can write

$$SO^2 = KS^2 + KO^2 - 2KS \cdot KO \cos S\hat{K}O.$$

Let us put $SO = x$, $US = y$, $OQ = z$, (where x, y, z are positive unknowns), hence

$$x^2 = (b-y)^2 + (c-z)^2 - 2|b-y| \cdot |c-z| \cos S\hat{K}O.$$

If $b - y$ and $c - z$ have the same sign, $S\hat{K}O = \alpha$, and if $b - y$ and $c - z$ have opposite signs, $S\hat{K}O = \pi - \alpha$, so, in every case, x, y, z satisfy

(1) $$\begin{cases} x^2 = (b-y)^2 + (c-z)^2 - 2(b-y)(c-z)\cos\alpha, \\ \dfrac{y}{x} = k \quad \text{and} \quad \dfrac{z}{x} = k'. \end{cases}$$

Eliminating y and z gives

(2) $$(b - kx)^2 + (c - k'x)^2 - 2(b - kx)(c - k'x)\cos\alpha - x^2 = 0,$$

and if we make use of $a^2 = b^2 + c^2 - 2bc\cos\alpha$, (2) can be rewritten

(3) $x^2(k^2 + k'^2 - 2kk' \cos\alpha - 1) - 2x[bk + ck' - (kc + k'b)\cos\alpha] + a^2 = 0$,

an equation whose discriminant is

$$\Delta = [bk + ck' - (kc + k'b)\cos\alpha]^2 - a^2(k^2 + k'^2 - 2kk'\cos\alpha - 1)$$

which, after manipulation and simplification, can be rewritten

$$\Delta = a^2 - \sin^2\alpha(kc - k'b)^2 ;$$

hence

$$\Delta \geq 0 \Leftrightarrow |kc - k'b| \leq \frac{a}{\sin\alpha} \Leftrightarrow |kc - k'b| \leq \frac{b}{\sin\beta}.$$

1.　　　$kc = k'b$; then we have $\dfrac{c}{b} = \dfrac{k'}{k}$ and $\Delta = a^2$.

Let us put $\dfrac{k}{b} = \dfrac{k'}{c} = \lambda$, hence $k = \lambda b$ and $k' = \lambda c$; from (2) we have

$$(1 - \lambda x)^2(b^2 + c^2 - 2bc\,\cos\alpha) - x^2 = 0 \Leftrightarrow x^2 = a^2(1 - \lambda x)^2$$

$$\Leftrightarrow x = a|1 - \lambda x| \Leftrightarrow x(1 + a\lambda) = a \text{ or } x(a\lambda - 1) = a$$

$$\Leftrightarrow x = \frac{ab}{b + ak} = \frac{ac}{c + ak'} \text{ or } x = \frac{ab}{ak - b} = \frac{ac}{ak' - c}.$$

The first root gives a solution for any value of k; the second gives a solution if $\dfrac{k}{b} = \dfrac{k'}{c} > \dfrac{1}{a}$. So the problem has at least one solution; it will have a second one if $\dfrac{k}{b} > \dfrac{1}{a}$.

We may note that $kc = k'b \Leftrightarrow \dfrac{k}{k'} = \dfrac{b}{c}$, now $\dfrac{k}{k'} = \dfrac{y}{z}$, hence $\dfrac{y}{z} = \dfrac{b}{c}$ which implies $SO \parallel UQ$; so this case corresponds to Ibn al-Haytham's case I.

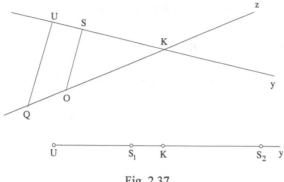

Fig. 2.37

2. $kc > k'b$; we thus have $\dfrac{c}{b} > \dfrac{k'}{k}$.

$$\Delta \geq 0 \Leftrightarrow \frac{b}{\sin\beta} \geq kc - k'b \Leftrightarrow kc \leq b\left(k' + \frac{1}{\sin\beta}\right) \Leftrightarrow \frac{c}{b} \leq \frac{k'}{k} + \frac{1}{k\sin\beta}.$$

3. $kc < k'b$; we thus have $\dfrac{c}{b} < \dfrac{k'}{k}$

$$\Delta \geq 0 \Leftrightarrow \frac{b}{\sin\beta} \geq k'b - kc \Leftrightarrow kc \geq b\left(k' - \frac{1}{\sin\beta}\right) \Leftrightarrow \frac{c}{b} \geq \frac{k'}{k} - \frac{1}{k\sin\beta}.$$

From 1, 2 and 3, it follows that

$$\Delta \geq 0 \Leftrightarrow \frac{k'}{k} - \frac{1}{k\sin\beta} \leq \frac{c}{b} \leq \frac{k'}{k} + \frac{1}{k\sin\beta}.$$

If this double condition is satisfied, the equation has two roots or one double root, but these roots do not provide solutions of the initial problem unless they are positive, which introduces discussions that we shall not enter into here.

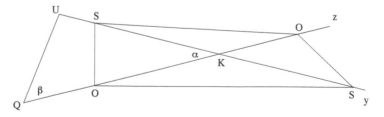

Fig. 2.38

Let us, however, examine the special cases in which one of the points, S or O, is identified with the point K. The equations (1), (2) or (3) still apply; from which we obtain:

- $S = K \Leftrightarrow y = b$, $x = \dfrac{b}{k}$ and $z = \dfrac{k'b}{k}$.

From (2) we shall have $S = K$ if and only if

$$\left(c - \frac{k'b}{k}\right)^2 = \left(\frac{b}{k}\right)^2 \Leftrightarrow kc - k'b = \pm b \Leftrightarrow \frac{c}{b} = \frac{k' \pm 1}{k};$$

$$\frac{c}{b} = \frac{k'+1}{k} \Rightarrow z = \frac{k'c}{1+k'} < c,$$

so O lies on $[QK)$;

$$\frac{c}{b} = \frac{k'-1}{k}$$

then does not give a solution unless $k' > 1$; we then have

$$z = \frac{k'c}{k'-1} > c,$$

the point O lies on $[K, z)$.

- $O - K \Leftrightarrow z = c$, $x = \dfrac{c}{k'}$ and $y = \dfrac{kc}{k'}$.

From (2) we shall have $O = K$ if and only if

$$\left(b - \frac{kc}{k'}\right)^2 = \left(\frac{c}{k'}\right)^2 \Leftrightarrow k'b - kc = \pm c \Leftrightarrow \frac{c}{b} = \frac{k'}{k \pm 1};$$

$$\frac{c}{b} = \frac{k'}{k+1} \Rightarrow y = \frac{kb}{k+1} < b,$$

so S lies on $[UK)$;

$$\frac{c}{b} = \frac{k'}{k-1}$$

does not give a solution unless $k > 1$; then we have

$$y = \frac{kb}{k-1} > b,$$

the point S lies on $[Ky)$.

We have just seen that equation (3) has the root $x = \frac{b}{k}$ when $\frac{c}{b} = \frac{k' \pm 1}{k}$ and has the root $x = \frac{c}{k'}$ when $\frac{c}{b} = \frac{k'}{k \pm 1}$. So in each of these four eventualities it has a second root that is obtained, for example, by using the product of the roots $P = \frac{a^2}{k^2 + k'^2 - 2kk' \cos\alpha - 1}$; the second root does not lead to a solution to the problem unless we have

$$k^2 + k'^2 - 2kk' \cos\alpha - 1 > 0.$$

To sum up, we find both cases Ibn al-Haytham investigated and special cases to which he did not draw attention.

TRANSLATED TEXT

Al-Ḥasan ibn al-Ḥasan ibn al-Haytham

On Analysis and Synthesis
Fī al-taḥlīl wa-al-tarkīb

In the name of God, the Compassionate the Merciful

TREATISE BY AL-ḤASAN IBN AL-ḤASAN IBN AL-HAYTHAM

On Analysis and Synthesis

Every science and every study has a purpose, and that purpose is the peak to which one ascends, and to which aspire the spirits of those who seek after it zealously, with the aim of reaching it and mastering it. The mathematical sciences are founded on proofs; the ends they seek to attain are determination of unknowns, in the parts of the mathematical sciences, and to establish proofs that demonstrate the truth of their results.[1] The peak to which the spirit aspires in those who investigate these sciences with zeal is to obtain proofs by which one deduces the unknowns in these sciences. A proof is a syllogism that indicates, necessarily, the truth of its own conclusion. This syllogism is made up of premises whose truth and validity are recognised by the understanding, without being troubled by any doubt in regard to them; and [the syllogism] has an order and arrangement of these premises such that they compel the listener to be convinced by their necessary consequences and to believe in the validity of what follows from their arrangement.

The method used to obtain these syllogisms is to hunt for their premises, to seek the devices for grasping them and to find their arrangement. The art by which we pursue the search for these premises, and arrive at the arrangement that leads to those of their results that we are working on, is called the art of analysis. Everything that has come to light in the mathematical sciences is owing to this art alone.

We explain in this treatise how to proceed by the art of analysis, which leads to determine the unknowns in the mathematical sciences, and how to proceed in pursuing the search for the premises, which are the basis for the proofs that show the validity of what we determine regarding the unknowns in these sciences, and the method for arriving at the arrangement of the

---

[1] The Arabic term used here is *ma'nā*. The term is used in a wide variety of contexts, to mean notion or concept, and sometimes proposition or result and so on. We have taken the liberty of translating the word according to the context.

premises and the structure of their combination. We also show the nature of these premises and the inverse of their arrangement, which is the demonstrative syllogism, and it is that which is known as synthesis; it has in fact been called synthesis because it is the combining of the premises deduced by analysis, and it is syllogistic synthesis. Furthermore, we divide this art into its subdivisions, we describe their rules and laws, and we display the details of its parts. We also provide our help in regard to all the principles used in this art and that are required by it, and thus we begin by saying:

We say that the mode of proceeding in analysis is to suppose the *quaesitum* completely accomplished and finished, then we examine the properties of this object, necessary consequences of this object and of its genus, then what follows from these necessary consequences, then the necessary consequences of these last, until we end up with a thing that is given in the *quaesitum*, which is not impossible for it. Here is how we generally proceed in analysis. When this examination ends up with the thing that is given, we break off the examination of this thing, and the person carrying out the examination stops there; what is given is the thing that we cannot reject, and that nothing can prevent [our accepting].

The mode of proceeding in synthesis consists of supposing what is given, which is what analysis arrived at and where the person examining it stopped, then add to it the property we found [in the analysis], then to that we add the property we found before that last; in this order we follow the inverse of the order followed in the analysis; in fact if we follow this route, the progression arrives at the thing we seek, because it was the first object in the analysis; if we invert the order, the first then becomes the last; and if the inverse order ends with the first *quaesitum*, then that order will be a demonstrative syllogism, and the first supposed *quaesitum* will be its conclusion; then the *quaesitum* will exist, and moreover, its validity will be certain, because it is the conclusion of a demonstrative syllogism that shows, as a necessity, the validity of its conclusion.

The art of analysis requires prior knowledge of the principles of mathematics and of their application so that the analyst bears these principles in mind during his practice of analysis, and furthermore has recourse to intuition in this art; for no art is in fact complete, for someone who practises it, except through an intuition of the method that leads to the *quaesitum*. We have recourse to intuition in the art of analysis when the analyst does not find in the subject of the problem given properties that, when combined, lead to the *quaesitum*; in that case, the analyst needs intuition. What he needs to grasp by intuition is an addition, this he adds to the object, so that, once this is added, it produces properties of the object that lead, with this addition, to properties which, once combined, have as their result the thing we are looking for.

In what follows in this account, we shall give examples of all that we have just mentioned, examples which display all the concepts that we have determined: the mode of obtaining them will appear, those among them that were hidden will be unveiled, moreover, the validity of what we have determined and arranged will be verified and will become certain, once that we have given a detailed account of this art, that we have set it out in order, and that we have acknowledged all its species and all its parts.

This art is divided up in accordance with the division of its objects, because the method for analysing each of the species of its objects is unlike the method in the analysis of the remaining species. The objects of this art are unknowns in the parts of the mathematical sciences; these unknowns in the parts of the mathematical sciences are themselves divided according to the subdivisions of all the parts of these sciences. Now, the parts of these sciences are divided first of all into two subdivisions: theoretical and practical; every part of the mathematical sciences is in fact either theoretical or practical. That part among them which is theoretical is that in which we are seeking to know the truth of a property necessary to that part on account of its essence and its form. That part which is practical is that we seek to carry out and bring things about in existence by action. Both for the theoretical and for the practical we give examples from the parts of each species of the mathematical sciences, so as to show that what we have set out is true.

The concepts proper to the theoretical part of the science of numbers[2] follow the example in our statement: for two square numbers, the ratio of one to the other is equal to the ratio of the side to the side multiplied by itself; and the example in our statement: given successive numbers in proportion, and that are the smallest numbers in their ratio, each of the <numbers> at the two ends is prime to the other;[3] and the example in our statement: for two numbers, one of which measures the other, the one that is measured has a part that has the same name as the one that measures it. This is the style of all the theoretical concepts in the science of numbers.

The concepts proper to the practical part of the science of numbers follow the example in our statement: to find two square numbers such that their sum is a square; and the example in our statement: to find successive numbers in the same ratio, as many of them as we wish; and the example in

---

[2] Although this expression is in the singular in Arabic it is translated by the plural in French. A literal translation of the Arabic term would be 'theory of number'.

[3] Compare Euclid, *Elements*, IX.15.

our statement: to find a perfect number.[4] This is the style of all the practical results in the science of numbers.

The theoretical concepts of geometry follow the example in our statement: <the sum of> two sides of a triangle is greater than the remaining side; and the example in our statement: the sum of the three angles of a triangle is equal to two right angles; and the example in our statement: the sides and the opposite angles in plane figures with parallel sides are equal two by two.

The practical concepts of geometry follow the example in our statement: to construct an equilateral triangle on a given straight line; and the example in our statement: to construct on a given straight line an angle equal to a given angle; and the example in our statement: to construct a square equal to a given figure.

The theoretical concepts of astronomy follow the example in our statement: the centre of the sphere of the Sun lies away from the centre of the Universe; and follow the example in our statement: the motion of Gemini is opposite to the order of succession of the signs of the Zodiac;[5] and follow the example in our statement: the sphere of the fixed stars is higher than the spheres of the wandering stars.

The practical concepts of astronomy do not form part of astronomy itself, but to its proofs, as for example: take away a ratio from a ratio, or add a ratio to a ratio, or draw from a point a perpendicular to one of the lines that are imagined in astronomy, or to construct a triangle on one of the lines in astronomy. All these propositions reduce to ones in the science of numbers or in geometry. We can mention in this area the construction of the instruments with which we observe heavenly bodies, and this forms no part of the body of theoretical mathematical sciences.

The theoretical concepts of music follow the example in our statement: the interval of an octave is composed of the interval of a fourth and the interval of a fifth; and follow the example in our statement: the interval of an octave is made twice by fifteen [single-tone] steps in interval;[6] and follow the example in our statement: the interval of a fourth divides into more than two tones.

---

[4] Lit.: the perfect number.

[5] The constellation of Gemini participates in the diurnal motion, which takes place in a retrograde sense about the axis of the world. The apparent annual progression of the Sun through the sign of the Zodiac takes place in the direct sense.

[6] Ibn al-Haytham's expression seems to be designed to avoid referring to intervals less than a single 'tone'. For the Greek music theory on which he is drawing, see Andrew Barker, *Greek Musical Writings*, 2 vols, Cambridge, 1984, 1989 [vol. 1, *The Musician and his Art*, vol. 2, *Harmonic and Acoustic Theory*].

The practical concepts of music are the composition of the degrees [of intervals of a single tone]; they reduce to the science of numbers, because they reduce to the composition of numerical ratios.

As for the practice of music, that is to say manual practice which consists of striking strings and instruments and of combining sounds, it does not enter into this examination.

No concept, in one or other of the mathematical sciences, can be anything but theoretical or practical. The practical part then divides into two subdivisions, with discussion or without discussion. The [examples in the] part with discussion follow the example in our statement in the parts of the science of numbers: to divide two known numbers in two known ratios; if we do not impose the condition that one of the two ratios is to be greater than the ratio of one of the divided numbers to the other, and that the other ratio is to be smaller than the ratio of the two numbers divided one by the other, it will not be possible for these two numbers to be divided in these two ratios[7] – this condition is called discussion; and follow the example in our statement: to find the greatest number that measures two known numbers; if we do not impose on the two numbers the condition that they must be commensurable, there cannot exist a number that measures them [both] – this condition is the discussion; and follow the example in our statement: to find a third number in proportion with two known numbers;[8] if we do not impose on the two numbers the condition that they must be commensurable, there cannot exist a third number in proportion with these two numbers.

Anything that has a discussion in the parts of geometry follows the example in our statement: from three given straight lines to construct a triangle; if we do not impose on the straight lines the condition that the sum of any two of them must be greater than the third, we shall not be able to construct a triangle from these three straight lines; and following the example in our statement: to draw in a known circle a chord equal to a known straight line; if we do not impose on the straight line the condition that it must not be greater than the diameter of the circle, we shall not be able to draw the chord in the circle; and following the example in our statement: from a known point to draw to a known straight line a straight line that is perpendicular to it; if we do not impose on the straight line the

---

[7] If a and b are the given numbers, k_1 and k_2 the given ratios, we require a_1 and a_2, b_1 and b_2 such that $a_1 + a_2 = a$, $b_1 + b_2 = b$, $a_1/b_1 = k_1$, $a_2/b_2 = k_2$. This problem is the sixth one in the text; we have proved that if $k_1 < k_2$, it is necessary that $k_1 < a/b < k_2$.

[8] Given a and b, to find x such that $a/b = b/x$, $x = b^2/a$.

condition that it must not be finite, then perhaps this will not be possible. These three conditions are the discussion for these three propositions.

As for astronomy and music, they do not involve discussion, because they do not contain practical concepts except in their proofs and in their syllogisms; and everything that is in these procedures is numerical or geometrical, and discussion of them belongs with discussion in the science of numbers and in geometry.

The part without discussion divides into two subdivisions, indeterminate and non-indeterminate. An indeterminate <problem> is one that has several solutions, and the non-indeterminate is that which has a single solution, that is to say that it cannot be completed except in a single way.

In the parts of the science of numbers, the indeterminate follows the example in our statement: to find two square numbers such that their sum is a square. [The problem in] this statement can have several solutions, that is to say there can exist many squares, an infinite number, such that each pair among them has a square sum;[9] and follows the example in our statement: to find a number that has given parts;[10] many numbers can be found, an infinite number, each of which has these same parts.

And following the example in our statement in the parts of geometry: to construct a circle that touches two given known circles; [the solution to] this proposition can be constructed in many ways, since the circle that is constructed can touch the two circles with its convex edge on the convex edges of the two circles; it can touch one of the two circles with its convex edge on the convex edge of the latter, and touch the other with its concave edge on the convex edge of it; it can touch each of the two circles with its concave edge on one of the convex edges of the two circles – the construction of this circle is thus carried out in three ways;[11] and follows the example in our statement: from a given point to draw a straight line that is a tangent to a given circle;[12] this construction is carried out in two ways,

---

[9] The word may seem ambiguous. The author means to say that there exist pairs of square numbers, an infinite number of them, such that the sum of the two terms of each pair is a square.

[10] This is a problem of factorisation of integers. Ibn al-Haytham's vocabulary is derived from Euclid. That is 'parts' means what would now be called 'factors' (see Euclid, *Elements*, VII, Definition 3).

[11] Here we have three types of problems that each have an infinity of solutions. The required circle can have 1) exterior contact with each of the given circles A and B; 2) interior contact with A and B; 3) interior contact with A (or B) and exterior contact with B (or A).

[12] It is taken for granted that the point lies outside the circle, otherwise the problem requires a discussion.

because if we join the point to the centre of the circle with a straight line, we can draw from the point two straight lines [one] on either side of that straight line, each of them being a tangent to the circle. In the science of numbers and geometry, the examples of these propositions are many; in the problems without discussion there can be indeterminate problems, and the examples that we have mentioned are sufficient [to stand] for all of them.

In astronomy there cannot be practical parts, except in its proofs, which reduce to the science of numbers or to geometry. It remains that among the motions of the heavenly bodies there can be some that can be produced in two ways, like the motion of the Sun, which can be according to two orbs: one of which has its centre at the centre of the Universe,[13] and the other [orb] is its epicycle and has its centre on the circumference of the first one; the motion of the Sun can be according to a single orb[14] whose centre lies away from the centre of the Universe. However, this proposition cannot be said to be practical, because it is as such, in one manner, and cannot be in the other one.

In the parts of music there can be indeterminate practical parts; however, their construction reduces to the science of numbers, as in [the proposition introduced by] our statement: to divide the interval of an octave into the two intervals of a fifth and a fourth; the division of this interval in fact takes place under two [different] assumptions; we can do things so that the interval of a fourth precedes the interval of a fifth and we can do things so that the interval of a fifth precedes the interval of a fourth; and in the example in our statement: to divide the interval of a fourth into three intervals; this interval, that is a fourth, divides into two tones and a remainder; this remainder can be at the beginning, or can be in the middle, or can be at the end; so the division can be done in three ways. Nevertheless, these subdivisions reduce to the science of numbers, because they divide according to the division of the numerical ratios through which the intervals are found according to ratios.[15]

It is clear from all that we have proved about the division of the parts of the mathematical sciences that they first divide into two subdivisions. One of the subdivisions then divides into three [further] subdivisions. It follows that analysis in the parts of these sciences must be divided in

---

[13] This is the circle called the *deferent*; its centre is *U*, the centre of the Universe; the Sun *S* moves on the epicycle whose centre *P* moves round the deferent.

[14] The circle usually called an *eccentric*, whose centre *E* is different from *U*.

[15] The three sections of the practical part are:
- with discussion (*maḥdūd*)
- without discussion (*ghayr maḥdūd*): - determinate (*ghayr sayyāl*)
                                        - indeterminate (*sayyāl*)

accordance with these subdivisions. Analysis in the theoretical part is of a single genus. Analysis in the practical part is also of a single genus, but it is divided into three species. Let us now show how analysis works in these subdivisions.

Analysis in the theoretical part is of a single genus. It is true that one and the same theoretical part can be analysed in several manners, but these manners nevertheless derive from a single genus. This is in effect because, if the *quaesitum* is theoretical, the analysis must be carried out by following through the properties only of the object of the notion we are seeking. So if it is analysed in several ways, that is if in analysing it several methods are used, then the analysis, for each of these methods, is carried out by seeking only its properties, once we suppose that the *quaesitum* is completely and perfectly given. If we do not, in some way, find properties for the *quaesitum* that lead to a property that exists in it, and such that, when we combine it with others, it produces this *quaesitum*, then the analyst must add to this object some additions that do not make it less like what it really is; he must then examine the properties of the object with the addition – it must in fact happen that other properties come to the object as a result of the addition; if, with this addition, the analysis is completed, then it is that one [*i.e.* that analysis] which, if we reverse it, produces the *quaesitum*; otherwise, we add to this addition another addition, and so on until there come, with these additions, properties which, once reversed and combined, produce what is required. These additions cannot be carried out except by using mathematical intuition, by which one grasps the premises. This intuition is the one we mentioned earlier, and the law of this intuition is to seek an addition such that, if one attaches it to the first object, there results from their combination a property or several properties that did not exist before this addition [was made]. If the analyst continues with this method, he cannot but arrive at a given property, or at a false property. If this method leads to a given property, then the proposition under investigation is valid and has a reality; but if the method derives a false property, the proposition under investigation is false and has no reality. We shall then show by means of examples how to add these additions, how to seek their properties, how to reverse them and how to reconstruct them.

If the analysis then leads to a given property and one that has reality, then, if we reconstruct this analysis, we show from that by true [*i.e.* deductive] proof that the proposition we are seeking to establish is true and beyond doubt. But, if the analysis leads to data that are impossible, this indicates that the proposition we are seeking to establish is impossible, and the analysis itself will be a proof of the falsity of the statement, if the analysis is presented as a proof by *reductio ad absurdum*; proof by *reductio ad absurdum* in fact reduces to supposing the assertion as true in with what it conveyed, and then to examine its necessary consequences. But, in the analysis that leads to something impossible, we have supposed the assertion as true in what it conveyed, and then examined its necessary consequences; these consequences have thus led to something impossible. Analysis that leads to something impossible is thus a proof by *reductio ad absurdum* of the falsity of the proposition under investigation. This is how analysis is carried out in theoretical parts of the mathematical propositions, as in their synthesis.

Analysis for the practical part belongs to the family of ingenious procedures. We are in fact seeking to do something that belongs to these subtle practices, now all the subtle practices belong to the family of ingenious procedures. The first thing that the analyst must do in connection with analysis in these practical parts, is, after having supposed that the *quaesitum* is entirely completed and perfect, to examine its necessary properties, if it [the *quaesitum*] exists and has the qualities required in practice, and to examine what necessarily follows from these properties, and what necessarily follows from these last consequences, until he arrives at something that is given, as we have shown in analysis in the theoretical part. If the analyst does not see properties that lead to the *quaesitum*, he adds to the object additions, from which properties arise, as we have shown in the example in the theoretical part, and he examines the properties of what is produced until he ends up with something given; if he thus ends up with something given, then for each of these properties he examines how one can find this property, and how to think up a stratagem for finding it, so that it exists, and to realize it *in actu* according to the quality that derives necessarily from the form of the proposition whose reality we are seeking to establish. In reflecting on the way to discover each of these properties, and in imagining a stratagem for bringing this property into existence, it becomes apparent to him that this property requires a condition and a discussion, or does not require them. If it is one of those properties that require a condition, then it is apparent to him that this property may not exist, that its existence may not occur, or that it may exist; it is when he leans in this direction that it becomes apparent to him that the *quaesitum* needs a discussion. It is then that he must assume the existence of that property or that proposition whose existence he has assumed, and ask himself when it is possible for that existence to be achieved, and when it is not possible for it to be achieved. If the quality by which the existent of this property or this *quaesitum* becomes a reality is then fixed for the analyst, the analysis is complete, and the discovery of what is required is complete. If in the course of his reflexion and his careful consideration of a means of finding the properties and the concepts by which the *quaesitum* is obtained, he does not encounter anything that makes it impossible to find them [or anything] that obstructs a part of them, then the *quaesitum* does not need either a condition or a discussion. In this case, he works to bring into action those properties that have appeared; but in bringing these properties and these concepts into action, it becomes apparent to him that these properties, or one of these properties, come about in several ways, or come about in only a single way. If each of these properties comes about only in a single way, then the *quaesitum* is not indeterminate. And if these properties, or one of them, come about in several ways, then the *quaesitum* can be obtained in several ways. So if, in this part, analysis also ends with something impossible, then this *quaesitum* cannot be obtained. All these subdivisions that make up the structure of analysis in the practical part belong to the same genus, and the method for analysis in them is similar to analysis in the theoretical part, except for the difference between analysis in the theoretical part and analysis in the practical part, that analysis in the theoretical part is a search for a property that pertains to the concept we seek and that exists in it, whereas analysis in the practical part consists of conceiving of the stratagem for finding the required concept and for bringing it into action, and that the method for finding it and bringing it into action is to bring into action each of the properties that appear in the analysis.

What we have just described is the whole set of the parts of analysis, and the mode of proceeding in each of these parts; when we mention examples, each of these parts will be illuminated and unveiled, as well as the art of analysis and its existence *in actu* will appear.

As for the laws that govern this art and its foundations, from which we complete our discovery of the properties and our grasp of the premises, and which are the basis of mathematics, of which we have said earlier that knowing them before we start is indeed necessary for completing the art of analysis, these are the notions called 'the knowns'. The knowns are divided into five groups, which are: known in number, known in magnitude, known in ratio, known in position, known in shape.[16] Euclid's book translated as the *Data* includes many notions concerning these knowns, which are among the instruments of the art of analysis; and the greater part of the art of analysis is based on these notions, but with the exception that there remain other notions among the knowns that are indispensible for the art of

---

[16] Ibn al-Haytham uses the term *ṣūra* (form). We have translated it as 'shape' so as to distinguish it from other uses of 'form' and because Ibn al-Haytham himself refers to Euclid's *Data*.

analysis, and which we need many times, that are deduced by analysis, that are not included in this book, and that we have not found in any book. We shall show in this book which knowns we use in the examples of analysis in this treatise, the ones that exist in the books, and also those that are not mentioned. We summarise each of these known notions, and we reveal what it really is; and, once this [present] treatise is finished, we shall straightaway set about [writing] an independent treatise on the knowns, in which we shall show the essential characters of the known notions that we use in the mathematical sciences; we shall deal exhaustively with all its parts, and we shall mention everything that concerns them.

Here we say: a known, in general terms, is that which does not change, because anything that changes and has change in its nature, has no determinate or assignable reality. If it does not have a determinate and assignable reality that is its essential character, it is not correct [to say] that it is known, because it is possible that all that we know about it may change from what it was; a thing will not be known unless it is fixed, in a single state, which is its essential character, which is proper to it. If this is the case, a known is that which does not change. Now that the essential character of the known is established, let us then explain each of the known notions that we have mentioned earlier and which are the basic elements of the art of analysis.

We say: a thing* known in number is one whose number does not change, and the number is a unit or a sum made up of units; a thing known in number is one whose units do not change, that is they neither increase nor decrease. A thing known in magnitude is one whose magnitude does not change, because a known is a thing that does not change. What is known of a thing of known magnitude is its magnitude. A thing known in magnitude is one whose magnitude does not change. Magnitudes can be divided into two groups: natural and imaginary. Natural magnitudes are sensible bodies, their surfaces, their dimensions, which are their length, their width and their depth. Imaginary magnitudes are the magnitudes derived from sensible magnitudes by the imagination; these are the dimensions that are a straight line, a surface, and, in the mathematical sense, a solid. We have set out these notions in our book the *Commentary on Euclid's Postulates*; and apart from that, these notions are well known to anyone who has studied geometry: their fame is such that we do not need to define them here. A thing known in magnitude is one whose magnitude does not change; but the magnitude is the dimension or the dimensions, so a thing known in magnitude is one whose dimension or dimensions do not

---

* We add 'thing'

change, that is to say that its dimension or dimensions do not increase or decrease.

A thing known in ratio is one whose ratio does not change. But a ratio is the measure of the quantity of what is compared to the quantity of that to which it is compared. But a ratio cannot exist except between two magnitudes of the same species, and which are jointly described by the same definition. Ratio comes in two species, which are numbers[17] and magnitudes. As for a ratio that is in numbers which are greater than one, it reduces entirely to a single base, which is: one of the two numbers is [some] parts of the other number, if we find the ratio of the smaller to the greater, and if we find the ratio of the greater to the smaller; and if we find the ratio of equality one to the other, each of them will be [some] parts of the other even when equal to it, because each of the units that are in the number is a part of the other number, and every number greater than one is a collection of units; and every number is [some] parts of every number; thus for two numbers, one is [some] parts of the other; a thing known in ratio, among numbers, [means that] there are two numbers such that the parts of one in relation to [those of] the other do not change, that is to say that the units of each of them do not increase or decrease. As for a ratio that is in magnitudes, it can be divided into two types: [it can be] a numerical ratio or a non-numerical ratio. We have demonstrated the distinction between each of these two ratios in our book, the *Commentary on <Euclid's> Postulates*, and in that book we have proved that each of these two ratios among magnitudes exists. Here we shall shed light on each of these two ratios by a brief discussion, which will make clear what they mean. The numerical ratio existing between two magnitudes is the one by which the ratio of one of its two magnitudes to the other is equal to the ratio of a number to a number. And the non-numerical ratio is the one by which the ratio of its two magnitudes is not the ratio of a number to a number; but the ratio by which one of its two magnitudes to the other is the ratio of a number to a number, is one by which one of its two magnitudes is a part or parts of the other, that is to say that we can divide up each of them [*sc.* the magnitudes] into equal parts in such a way that each of the parts of one of them is equal to each of the parts of the other, or that one is measured by the other. The non-numerical ratio is that for which this is not possible.

A known ratio, which is between two magnitudes, divides into two types. One of the types is when the ratio of one of the two magnitudes to

---

[17] Lit.: 'number' – singular which indicates a plural.

the other is equal to the ratio of a known number to a known number; and the other type is when the ratio of one of the two magnitudes to the other is equal to the ratio of a known magnitude, that we can find and take note of, to a known magnitude, that we can find and take note of. It is possible to unite the two types under this type; we then say: a known ratio that is between two magnitudes is such that the ratio of one of its two magnitudes to the other is equal to a ratio of a known magnitude that we can find and take note of, to a known magnitude that we can find and take note of; because, for two magnitudes for which the ratio of the one to the other is equal to the ratio of a known number to a known number, it is possible to find two magnitudes that are in their ratio. Thus, the known ratio between two magnitudes is such that we can find two known magnitudes that are in the ratio of its two magnitudes. If we find two known magnitudes in the ratio of two magnitudes, then the ratio between these two magnitudes does not change, because the two known magnitudes do not change, it being given that they are known.

A thing known in position is one whose position does not change. As for what position is, that is place (naṣba),[18] and place is established with respect to a thing whose location is defined. Position is for a body, for a surface, for a line and for a point. Position for a body divides into two types: either it can be relative to a fixed thing, or it can be relative to a thing that moves. Something that is relative to a fixed thing is a thing that does not undergo displacement or move with any kind of motion; a body known in position relative to a fixed thing is one in which the distance from each of its points to the fixed points that are in the fixed thing is [always] the same distance and does not change; this type is the one called known in position absolutely. As for a body known in position relative to a thing that moves, it is one in which the distance from each of its points to every point of this movable thing is [always] the same distance, which does not change. It follows that for this body, which is of known position and which has such a property, when the body to which it is related moves, [the body in question] moves with a motion equal to its motion, so that the distances between each of its points and any point of the thing to which it is related are the same distances as those which there were between them, as [if it were] a determinate part among the parts of the movable body, and as [if it were] the determinate organ among the organs of a human being. The distances from each point of the determinate part among the parts of the body to each point of the remaining parts of the body do not change;

---

[18] This is our translation for the term naṣba usually used to translate the Greek term θέσις.

however, if this body moves, the part moves with its motion and the distances from each point of the part to each point of the remainder of the body are the same distances and do not change. The part is said to be of known position with respect to this or that, and we cannot refer to this thing known in position without, while referring to it, referring to the other thing with respect to which it is known in position. In the same way, surfaces of known position are also divided into two types and their state as regards position is like the state for bodies, without any difference: either their position is relative to surfaces, or to lines, or to fixed points; or their position is relative to surfaces, or lines, or points that are movable; and these surfaces will be in motion through the motion of the things to which their position is related.

In the same way, the position of lines divides into two parts according to the same division as for surfaces; and in the same way for points, if we say: a point of a position known absolutely is one whose position is relative to a fixed point or points, and is one that does not suffer displacement or move. If we say that the point is of known position with respect to a movable thing, it will be one whose distance from any point of that movable thing is [always] the same distance, which does not change. And if the thing moves, the point moves through its motion, as the centre of a circle, in which the distance to each point of the circumference of the circle is the same distance, which does not change; and nevertheless, if the circle moves, its centre moves with it, as for the centre of the sphere, and as for the vertex of a cone; and for this there are numerous examples. Thus, [the manner in which] a thing is known in position is divided into two subdivisions in each of the magnitudes, which are a line, a surface and a body: it is the same for points.

As for known in form, it exists only for figures; thus a figure of known form is one whose angles are known, and for which the ratios of the sides one to another are known. Figures exist in surfaces and in bodies; plane figures can include figures of known form, and solid figures can include figures of known form.

What we have mentioned are all the parts of knowns, they are all used in the art of analysis; all the knowns mentioned by Euclid, in his book called the *Data*, are included within the sum total of the parts that we have mentioned; and in what we have mentioned, there are some things that Euclid did not mention: these are movable things known in position. There also remains among these parts a notion that none of the ancients mentioned, and that we have not found in any book; this is one of the notions that we need in the art of analysis, and one whose usefulness for solving problems is increasing; in this place we mention certain of its parts so as to use them in examples of analysis and to show how to use these knowns, and how the need for this notion makes itself felt; and we shall show where it proves successful in the art of analysis, and the inadequacy of the knowns that are in the books in dealing with all the parts of known ideas; we shall then deal with all the parts of the knowns, and we shall give an exhaustive account of them in this treatise, to which we shall now turn.

<FIRST CHAPTER>

– 1 – One <of the results> that we shall set out immediately is that for any pair of points of known position, from which are drawn two straight lines that meet one another in a single point and are such that the ratio of one of the straight lines to the other is a known ratio, then this point [of intersection] lies on the circumference of a circle of known position.

Example: Let there be two points *A* and *B* known in position, from which are drawn the two straight lines *AC* and *BC* such that the ratio of *AC* to *CB* is equal to a known ratio which is the ratio of *G* to *E*, the ratio of the greater to the smaller.

We say that the point C *lies on the circumference of a circle known in position.*

Proof: We join *AB* and we extend it in the direction of *B* to [the point] *D*. At the point *C* on the straight line *AC*, we construct an angle equal to the angle *CBD*; let the resulting straight line be drawn in the direction of *B* and let the angle be *ACM*. The straight line *CM* will then lie outside the triangle *ACB*, because the angle *ACM* is equal to the angle *CBD* which is greater than the angle *ACB*.

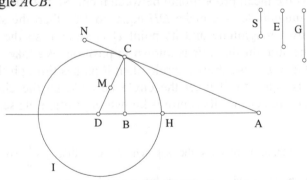

Fig. II.1.1

I say first that the straight line CM *meets the straight line* BD.

We extend the straight line *AC* to the <point> *N*. The angle *NCM* is then equal to the angle *CBA* and the angle *CBA* is greater than the angle *CAB*, because *AC* is greater than *CB* and this [is so] because the ratio of *AC* to *CB* is the ratio of the greater to the smaller; so the angle *NCM* is greater than the angle *CAB*, so the sum of the two angles *MCA* and *CAB* is smaller than two right angles and the straight lines *CM* and *AB* meet one another in the direction towards *B*; let them meet one another at the point *D*. Thus, the two triangles *ACD* and *BCD* are similar because the angle *ACD* is equal to the angle *CBD* and the angle *ADC* is common to the two triangles; so it remains that the angle *BCD* is equal to the angle *CAD*. So the ratio of *AD* to *DC* is equal to the ratio of *CD* to *DB* and is equal to the ratio of *AC* to *CB*. But the ratio of *AC* to *CB* is equal to the ratio of *G* to *E* which is known, so the ratio of *AD* to *DC* is equal to the ratio of *G* to *E*. We put the ratio of *E* to *S*[19] equal to the ratio of *G* to *E*, so the ratio of *E* to *S* is equal to the ratio of *CD* to *DB*. So the ratio of *AD* to *DB* is equal to the ratio of *G* to *S*. And the ratio of *G* to *E* is known, so the ratio of *E* to *S* is known and the ratio of *G* to *S* is known, as has been proved in the eighth proposition of the *Data*. So the ratio of *AD* to *DB* is a known ratio, so the ratio of *AB* to *BD* is known, as has been proved in the fifth proposition and eighth proposition of the *Data*. But *AB* is known in magnitude and in position, so the straight line *BD* is known in magnitude as has been proved in the second proposition of the *Data*. So the point *D* is known, the straight line *AD* is of known magnitude, the straight line *DB* is of known magnitude and the surface enclosed by the two straight lines *AD* and *DB* is of known magnitude as has been proved in the fiftieth proposition of the *Data*.[20] But the surface enclosed by the two straight lines *AD* and *DB* is equal to the square of *DC*, because *DC* is the mean proportional between them. So the straight line *DC* is of known magnitude. We make *DH* equal to *DC*, then the straight line *DH* is of known magnitude and its point *D* is known, so the point *H* is known and the straight line *DH* is known in position. We take *D* as centre and with distance *DH* we draw a circle, then it passes through the point *C*, because *DC* is equal to *DH*; let the circle be *HCI*; so the circle *HCI* is known in position, because its centre is known in position, its semidiameter

---

[19] Ibn al-Haytham introduces the segment *S* such that $\frac{G}{E} = \frac{E}{S}$ so as to apply

Proposition 8 of the *Data* and draw the conclusion that $\frac{G}{S}$ is known.

$$\frac{G}{E} = \frac{E}{S} = k \Rightarrow \frac{G}{S} = \left(\frac{G}{E}\right)^2 = k^2.$$

[20] This conclusion can be seen to follow from the converse of Proposition 55 (ed. Heiberg), Proposition 56 (in the recension by al-Ṭūsī).

is of known magnitude and it passes through the point C; so the point C lies on the circumference of a circle known in position which is the circle HCI. This is what we wanted to prove.

– **2** – We also say: if there exists a circle known in magnitude and in position and a point known in position, if from the point we draw a straight line as far as the circumference of the circle and extend it until the ratio of the first straight line to the second straight line is a known ratio, then the point that is the endpoint of the second straight line lies on the circumference of a circle known in magnitude and in position.

Example: The circle AB is known in magnitude and in position and the point C is known. From the point C we draw the straight line CA that we extend to D, so that the ratio of CA to AD is known.

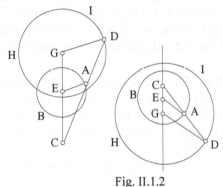

Fig. II.1.2

I say that the point D *lies on the circumference of a circle known in magnitude and in position.*

Proof: We mark off the centre of the circle, let it be E; we join CE, we extend it in the direction towards E; we join EA and we imagine <a straight line> DG parallel to the straight line AE. So the ratio of GD to EA is equal to the ratio of DC to CA and is equal to the ratio of GC to CE. But the ratio of DC to CA is known, because the ratio of DA to AC is known, as has been proved in the sixth proposition of the *Data*. So the ratio of GD to EA is known, the ratio of GC to CE is known, EA is of known magnitude and EC is of known magnitude. So the straight line GD is of known magnitude; but the straight line GC is of known magnitude, as has been proved in the second proposition of the *Data*. But since the two points C and E are known in position, accordingly the straight line CE is known in position, as has been proved in the twenty-fifth proposition of the *Data*.[21] So the

---

[21] This is Proposition 26 in Heiberg's edition and in the recension by al-Ṭūsī.

straight line CG is known in magnitude and in position and its point C is known, so its point G is known as has been proved in the twenty-sixth proposition of the *Data*.[22] We take the point G as centre and, with distance GD of known magnitude, we draw a circle, let it be the circle DHI; then the circle DHI is known in magnitude and in position, because its centre is known in position and its semidiameter is of known magnitude. Now the point D lies on the circumference of this circle, so the point D lies on the circumference of a circle known in magnitude and in position. This is what we wanted to prove.

– 3 – We also say: if we have a straight line known in position and a given point C that does not lie on the straight line, if from that point we draw a straight line to the straight line known in position and which is inclined to it, making a known angle, so that the ratio of the two straight lines that were generated, one to the other, is a known ratio, then the point that is the endpoint of the second straight line lies on a straight line whose position is known.

Example: The straight line AB is known in position and the point C is known; we draw the straight line CD to the point D on the known straight line AB. <The straight line> CD is inclined to the straight line DE, and encloses a known angle with DE, which is the angle CDE, so that the ratio of CD to DE is a known ratio.

I say that the point E *lies on a straight line known in position.*

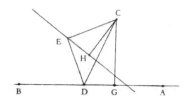

Fig. II.1.3

Proof: We join CE, so the triangle CDE is known in form, as has been proved in the thirty-ninth proposition of the *Data*;[23] so the angle DCE is known and the angle CED is known. From the point C we draw a perpendicular to the straight line AB, let it be CG; CG is thus known in position as has been proved in the twenty-ninth proposition of the *Data*.[24] The straight line AB is known in position and cuts the straight line CG. So the point G is

[22] This is Proposition 27 in Heiberg's edition and in the recension by al-Ṭūsī.

[23] This is Proposition 41 in Heiberg's edition and in the recension by al-Ṭūsī.

[24] This is Proposition 30 in Heiberg's edition and in the recension by al-Ṭūsī.

known, as has been proved in Proposition 24 of the *Data*.[25] The straight line *CG* has known endpoints, so it is known in magnitude and in position. On the straight line *CG* we construct the angle *GCH* equal to the known angle *DCE*; so the straight line *CH* is known in position, as has been proved in the twenty-eighth proposition of the *Data*.[26] We put the ratio of *GC* to *CH* equal to the ratio of *DC* to *CE*, which is known. So *CH* is known in magnitude, as has been proved in the first proposition of the *Data*. We join *EH*. Since the angle *GCH* is equal to the angle *DCE*, the angle *GCD* is equal to the angle *HCE*. But since the ratio of *GC* to *CH* is equal to the ratio of *DC* to *CE*, the ratio of *GC* to *CD* is equal to the ratio of *HC* to *CE*. So the triangle *HCE* is similar to the triangle *GCD*, so the angle *CHE* is equal to the angle *CGD*. But the angle *CGD* is a right angle, so the angle *CHE* is a right angle. So from the known point *H* we have drawn the straight line *HE* which encloses a known angle with *HC*, which is known in position. The straight line *HE* is known in position, so the point *E* lies on a straight line known in position. This is what we wanted to prove.

These notions from *The Knowns* that we have set out are sufficient for what we use and prove in this treatise on the way of proceeding by analysis.

– 4 – Let us now show, by examples, the way to proceed by analysis and let us mention, for each of the parts into which we have divided up all the concepts found by analysis, an example thanks to which we reveal the way of proceeding for solving the problems that belong to that part and the way of proceeding by analysis for resolving them.

We say that the example in the theoretical part of numerical problems is like our statement: if we have successive numbers in proportion and if we subtract from each, from the second and the last, [a number] equal to the first, then the ratio of what remains from the second to the first is equal to the ratio of what remains from the last to the sum of all the numbers that precede it.

The way of proceeding by analysis to solve this problem is to suppose the statement to be perfectly correct and to examine the properties of the numbers to which this statement applies, then <to examine> what necessarily follows from these properties and what necessarily follows from what is necessary, until we end up with a property that was given, as we have set out earlier.

---

[25] This is Proposition 25 in Heiberg's edition and in the recension by al-Ṭūsī.

[26] This is Proposition 29 in Heiberg's edition and in the recension by al-Ṭūsī.

Let the successive numbers in proportion be the numbers *A*, *BC*, *DE*, *GH*. We subtract from *BC* – the second – <a number> *CM* equal to *A* and we subtract from *GH* – the last – <a number> *LH* equal to *A*.

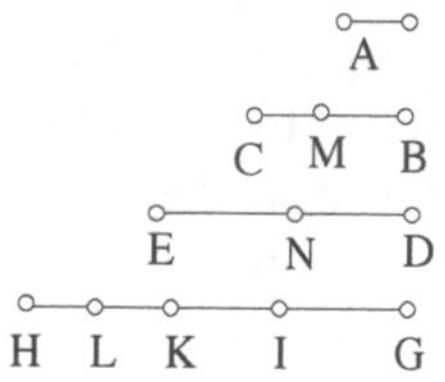

$$\overset{\circ\!\!-\!\!\circ}{A}$$

$$\underset{C \quad M \quad B}{\circ\!\!-\!\!\circ\!\!-\!\!\circ}$$

$$\underset{E \quad\quad N \quad\quad D}{\circ\!\!-\!\!\circ\!\!-\!\!\circ}$$

$$\underset{H \;\; L \;\; K \quad I \quad\quad G}{\circ\!\!-\!\!\circ\!\!-\!\!\circ\!\!-\!\!\circ\!\!-\!\!\circ}$$

Fig. II.1.4

I say that the ratio of BM *to* A *is equal to the ratio of* GL *to the sum of* DE, BC *and* A.

We suppose this is so and we examine the properties of these numbers that are the subject of the stated notion that must be investigated to decide whether it is true or false. If we examine the properties of this proposition, the first thing that is apparent is that the second [number] is greater than the first, because we cannot subtract from the second something that is equal to the first unless the second is greater than the first. And if the second is greater than the first, then each of the remaining numbers is greater than its predecessor. But since these numbers are proportional, we must investigate the properties of proportional numbers. But since we have taken away <certain quantities> from some of these numbers, we must investigate the properties of proportional numbers from which we have taken away <certain quantities>. It has been proved in the twelfth proposition of the seventh book of Euclid's work[27] that if from two numbers we subtract two numbers such that the ratio of the whole to the whole is equal to the ratio of the [part] subtracted to the [part] subtracted, then the ratio of the remainder to the remainder is equal to the ratio of the whole to the whole. It follows that if, from each of these numbers, we subtract the number that precedes it, then the ratio of the remainders one to another is equal to the ratio of the numbers subtracted one to another. If we permute the ratio, then the ratio of the remainder of one of the numbers to what has been subtracted from it is equal to the ratio of the remainder of each of these numbers to what has been subtracted from it. This investigation shows the intuition <formed> by the art that made us to add a supplement to the object; this supplement is the subtraction of each number from the number that follows it. So from the number *DE*, the third, we take away *NE* which is equal to *BC* and from *GH*, the fourth, [we take away] *IH* which is equal to *DE*; so the ratio of *GI*

---

[27] This is Proposition 11 in Heiberg's edition.

to *DN* is equal to the ratio of *GH* to *DE*, which is equal to the ratio of *IH* to *EN*, so the ratio of *GI* to *IH* is equal to the ratio of *DN* to *NE*. In the same way, the ratio of *DN* to *NE* is equal to the ratio of *BM* to *MC*, so the ratio of *GI* to *IH* is equal to the ratio of *DN* to *NE* and to the ratio of *BM* to *MC*.

If we examine the properties of the proportional numbers in a second way, then there exists a ratio of one of the earlier numbers to its homologue among the later ones that is equal to the ratio of all the preceding numbers to all the succeeding ones, because that has been proved in Proposition 13 of the seventh book of Euclid's work.[28] So the ratio of the sum of *GI*, *DN* and *BM* to the sum of *IH*, *NE* and *MC* is equal to the ratio of *BM* to *MC*, and *MC* is equal to *A*. So the ratio of the sum of *GI*, *DN* and *BM* to the sum of *IH*, *NE* and *MC* is equal to the ratio of *BM* to *A*. But *IH*, *NE* and *MC* are the numbers *DE*, *BC* and *A*, so the ratio of the sum of *GI*, *DN* and *BM* to the sum of *DE*, *BC* and *A* is equal to the ratio of *BM* to *A*. But the statement was that the ratio of *BM* to *A* is equal to the ratio of *GL* to the sum of *DE*, *BC* and *A*. The sum of the remainders, which are *GI*, *DN* and *BM*, is thus equal to the number *GL*.

Let us now investigate: if these remainders – which are *GI*, *DN* and *BM* – have a sum equal to the remainder that is *GL*, in this case the statement is true and it is a valid proposition; if the sum of these remainders was not equal to the remainder which is *GL*, then the statement would be false and would not have any validity. But we have taken away *IH* which is equal to *DE* and *DE* is greater than *BC*, so *IH* is greater than *BC*. So we take away from *IH* something equal to *BC*; let it be *KH*. But *BC* is greater than *A*, so *KH* is greater than *A* and *LH* is equal to *A*, so *KH* is greater than *LH*. But since *LH* is equal to *A* and *MC* is equal to *A*, *LH* is equal to *MC*; since *KH* is equal to *BC* and *LH* is equal to *MC*, *KL* is equal to *BM*, and since *IH* is equal to *DE* and *KH* is equal to *NE*, *IK* is equal to *DN*. The remainders, which are *GI*, *IK* and *KL*, are equal to the remainders, which are *GI*, *DN* and *BM*. But <the sum> of the remainders *GI*, *IK* and *KL* is the remainder *GL* taken as a whole. And if the remainders have a sum equal to *GL*, then the statement is true and is not subject to any doubt. What we have set out is the analysis of this problem, from which we have shown how to proceed by analysis for this problem and for any problem [that is] numerical, theoretical and true.[29]

The synthesis for this problem is: we assume we have successive numbers in proportion, let them be *A*, *BC*, *DE* and *GH*; from the second

---

[28] This is Proposition 12 in Heiberg's edition.

[29] The 'problem' concerned is in fact a proposition in number theory. So it can be either true or false.

and from the last we take away something that is equal to the first, that is to say *MC* and *LH*. We then take away from *GH* something that is equal to *DE*, let it be *HI*, and we take away from *HI* something that is equal to *BC*, let it be *HK*. It is clear then that the remainders – which are *GI*, *IK* and *KL*, whose sum is *GL* – are equal to the amounts by which the magnitudes *GH*, *DE*, *BC* and *A* exceed one another. But this premise is the one arrived at by the analysis.

In the same way, the ratio <of the sum> of these remainders to <the sum> of the magnitudes *DE*, *BC* and *A* is equal to the ratio of *BM* to *A*. It is obvious that the remainders *GI*, *IK* and *KL* are equal to the amounts by which the magnitudes *GH*, *DE*, *BC* and *A* exceed one another, because we have subtracted the magnitudes *HI*, *HK* and *HL* which are equal to the magnitudes *DE*, *BC* and *A*. That the ratios of these remainders to the magnitudes *DE*, *BC* and *A* are equal to the ratio of *BM* to *A* is proved like this: the ratio of *GH* to *HI* is equal to the ratio of *HI* which is subtracted, to *HK* which is subtracted, and is equal to the ratio of *GI* which is the remainder to *IK* which is the remainder. In the same way, the ratio of *IH* to *HK* is equal to the ratio of *HK* to *HL* and is equal to the ratio of the remainder, that is *IK*, to the remainder, that is *KL*. But the ratio of *IH* to *HK* is equal to the ratio of *GH* to *HI*, which is equal to the ratio of *GI* to *IK*, so the ratio of *GI* to *IK* is equal to the ratio of *IK* to *KL*. If we permute, the ratio of *GI* to *IH* is equal to the ratio of *IK* to *KH*.

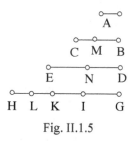

Fig. II.1.5

In the same way, we prove that the ratio of *IK* to *KH* is equal to the ratio of *KL* to *LH*, so the ratio of *GI* to *IH* is equal to the ratio of *IK* to *KH* and is equal to the ratio of *KL* to *LH*. But the ratio of one of the preceding [numbers] to one of the succeeding ones is equal to the ratio of all the preceding ones[30] to all the succeeding ones. So the ratio of *KL* to *LH* is equal to the ratio of *GL* to the sum of *IH*, *KH* and *LH*; but *IH* is equal to *DE*, *KH* is equal to *BC*, *LH* is equal to *A* and *KL* is equal to *BM*, so the ratio of

---

[30] That is the sum of all the preceding numbers.

BM to *A* is equal to the ratio of *GL* to the sum of *DE*, *BC* and *A*. This is what we wanted to prove.

This proof is the converse of the preceding analysis, that is to say that the premises used in this proof are the premises which appeared in the analysis, but their order is inverted compared with their order in the analysis.

– **5** – For the example of what leads to something impossible, it is as when we say in this same proposition: if we have successive number proportion and if we subtract from the second a <number> equal to the first, then the ratio of the number that remains from the second one to the first is equal to the ratio of the last number to the sum of all the numbers that precede it.

If we carry out an analysis of this proposition, then the method of analysis is the one we have set out: we examine the properties of the successive proportional numbers and the properties of what we subtract from them. The analysis ends up with subtracting from each number the number that precedes it and we have remainders such that the ratio of the sum of the remainders <to the sum> of the numbers that have been subtracted is equal to the ratio of the remainder of the second <number> to the first number. But the numbers from which we have subtracted the numbers that precede them are numbers measured by the first, and the numbers subtracted are all the numbers that precede the last one; so the ratio of the sum of the remainders to the sum of the numbers that precede the last one is the ratio of the remainder of the second to the first number. Now, the statement is that the ratio of the remainder of the second to the first number is equal to the ratio of the last number to the sum of the numbers that precede it. It follows from the analysis that the sum of the remainders is equal to the last number. If we subtract from the last number[31] the number that precedes it, then <the sum> of the remainders will be less than the last number by a quantity equal to the first number. Because we have proved in the first analysis that the sum of the remainders is equal to the number *GL* and that *LH* is equal to the first. This second analysis thus leads to the sum of the remainders being the number *GL*; now it was required that the sum of the remainders should be equal to the whole of *GH*. It is thus necessary from this analysis that *GL* is equal to *GH*, which is impossible. The analysis in which we supposed the statement to be <true>, that is that the ratio of the remainder of the second to the first is

---

[31] Lit.: we subtract from the last number. It is obvious that by 'last', he means a number that has a predecessor.

equal to the ratio of the whole last number to the sum of all the numbers that precede it, led to this impossibility.

If analysis has led to a false proposition, the proposition under investigation is false and has no truth, because the impossibility comes from having supposed the proposition under investigation could hold. This analysis is itself a proof that the proposition under investigation is impossible, if we make this analysis into a proof by *reductio ad absurdum* as we have shown earlier. According to this example we shall carry out the analysis of numerical and theoretical propositions, if they are false.

<6> For the example in the practical part with discussion for numerical problems, this is our statement: to divide two given numbers in two given ratios.

Let the two numbers be AB and CD and the two ratios the ratio of H to I and the ratio of K to L.

The analysis of this problem will be carried out in the following way: we suppose that the two numbers have been divided at the two points E and G, that the ratio of AE to CG is equal to the ratio of H to I, that the ratio of EB to GD is equal to the ratio of K to L and that the ratio of H to I is not equal the ratio of K to L. So the ratio of AE to CG is not equal to the ratio of EB to GD. It is necessary for the analyst to examine the properties of the different ratios. If he examines the properties of the different ratios, it will become clear to him that one of the two ratios is greater than the other. It follows that one of the two ratios, AE to CG and EB to GD, is greater than the other. That is all that can be brought to light here. So if the analyst does not add something to this object that can bring to light a supplementary property, the investigation of this proposition will not be completed. Now this addition requires intuition in order that the addition shall give rise to a supplementary property. The addition which gives rise to a supplementary property is to increase the smaller of the two ratios so that it becomes like the greater one or to decrease the greater ratio so that it becomes like the smaller one. Let the ratio of EB to GD be smaller than the ratio of AE to CG. We put the ratio of EB to GM equal to the ratio of AE to CG, so GM will be smaller than GD and the ratio of AB to CM will be equal to the ratio of AE to CG. But the ratio of AE to CG is equal to the ratio of H to I. So the ratio of AB to CM is equal to the ratio of H to I. But the ratio of AB to CM is greater than the ratio of AB to CD, so the ratio of AB to CD is smaller than the ratio of H to I. In the same way, since the ratio of EB to GD is smaller than the ratio of AE to CG, the ratio of EB to GD is equal to the ratio of AE to a number greater than GC; let this be the whole of the number GP. So the ratio of AE to PG is equal to the ratio of EB to GD and is equal to the ratio of the whole of AB to the whole of PD. So the ratio of AB to PD is equal to the ratio of EB to GD. But the ratio of EB to GD is equal to the ratio of K to L, so the ratio of AB to PD is equal to the ratio of K to L. But the ratio of AB to PD is smaller than the ratio of AB to CD, so the ratio of AB to CD is greater than the ratio of K to L. So the ratio of AB to CD is greater than one of the two ratios we assumed and smaller than the other ratio.

So analysis has led to showing that the ratio of one of the two numbers we took to the other one is greater than one of the two ratios we assumed and smaller than the other ratio, and that one of the two given ratios is the ratio of one of the two numbers to a part of the other, and that the other ratio is the ratio of this number to a number that is greater than the other one. At this stage, let the analyst examine the ratio of the two given numbers; if it is greater than one of the two ratios and smaller than the other, then the *quaesitum* is possible, and if it is neither greater than one of the two ratios nor smaller than the other one, then the *quaesitum* is not possible.

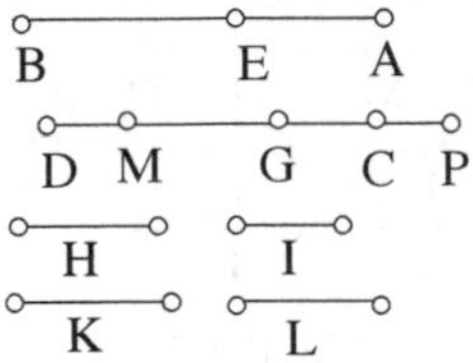

Fig. II.1.6

Analysis has also arrived at [the result] that the ratio of AE to PG is equal to the ratio of EB to GD, so the ratio of AE to EB is equal to the ratio of PG to GD. We also find that the ratio of AB to CM is equal to the ratio of AE to CG, so the ratio of AE to CG is equal to the ratio of EB to GM. So the ratio of AE to EB is equal to the ratio of CG to GM. But the ratio of AE to EB is equal to the ratio of PG to GD, so the ratio of CG to GM is equal to the ratio of PG to GD and is equal to the ratio of the remainder, which is PC, to the remainder, which is MD.

So analysis has arrived at [the result] that the ratio of two parts of CM, one to the other, is equal to the ratio of CP – which is the amount by which PD exceeds DC – to MC, which is the difference between the whole of CM and CD. This proposition is possible and is not difficult, that is to say that it is possible to divide CM into two parts such that the ratio of one to the other is equal to the ratio of PC, which is the excess, to MD, which is the difference.

If the analysis arrives at a possible proposition, if this analysis is then inverted and composed,[32] it produces what we are looking for; the properties that have been brought to light by the analysis will be premises from which we build up a demonstrative syllogism that produces the *quaesitum*.

The synthesis of this problem is carried out as we shall describe: we suppose <we are given> the two magnitudes and the two ratios; let the ratio of one of the two magnitudes to the other be greater than one of the two ratios and smaller than the other ratio. We put the ratio of *AB* to *CM* equal to the ratio of *H* to *I* which is the greater of the two ratios, so *CM* will be smaller than *CD*. We put the ratio of *AB* to *DP* equal to the ratio of *K* to *L* which is the smaller of the two ratios, thus *DP* will be greater than *CD*. We put the ratio of *CG* to *GM* equal to the ratio of *PC* to *MD* and we put the ratio of *AE* to *EB* equal to the ratio of *CG* to *GM*.

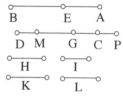

Fig. II.1.7

I say that the ratio of AE *to* GC *is equal to the ratio of* H *to* I *and that the ratio of* EB *to* GD *is equal to the ratio of* K *to* L.

Proof: The ratio of *CG* to *GM* is equal to the ratio of *CP* to *MD*, so the ratio of *CG* to *GM* is equal to the ratio of *PG* to *GD*. But the ratio of *CG* to *GM* is equal to the ratio of *AE* to *EB*, so the ratio of *AE* to *EB* is equal to the ratio of *PG* to *GD*. So if we permute, the ratio of *AE* to *PG* will be equal to the ratio of *EB* to *GD* and equal to the ratio of the whole of *AB* to the whole of *PD*. But the ratio of *AB* to *PD* is equal to the ratio of *K* to *L*, so the ratio of *EB* to *GD* is equal to the ratio of *K* to *L*.

In the same way, since the ratio of *AE* to *EB* is equal to the ratio of *CG* to *GM*, the ratio of *AE* to *CG* is equal to the ratio of *EB* to *GM* and is equal to the ratio of the whole of *AB* to the whole of *CM*; so the ratio of *AE* to *CG* is equal to the ratio of *AB* to *CM*. But the ratio of *AB* to *CM* is equal to the ratio of *H* to *I*, so the ratio of *AE* to *CG* is equal to the ratio of *H* to *I*. So we have divided up each of the two numbers *AB* and *CD* into two parts, so that the ratio of one of the two parts of *AB* to one of the two parts of *CD* is

---

[32] That is to say, it becomes a synthesis.

equal to the ratio of H to I and the ratio of the other part of AB to the other part of CD is equal to the ratio of K to L. This is what we wanted to prove.

The way to carry out the synthesis of this problem is as follows. All the premises that we have used in the division and in the proof that the division is correct are the properties that have been brought to light in the course of the analysis. They have been laid out in detail thanks to the additions and pushing forward the search. But we obtain this construction by supposing that the ratio of one of the two magnitudes to the other is greater than one of the two ratios and smaller than the other ratio. Now this proposition is the discussion of this problem, because we have completed it only after having taken this proposition as a condition.

So it remains for us to prove that if the ratio of the two numbers is not greater than one of the two ratios and is not smaller than the other, then the two numbers cannot be divided up in the two ratios.[33]

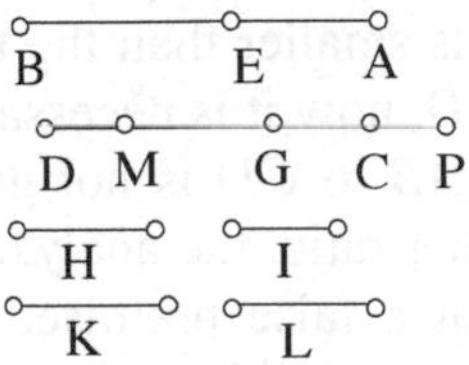

Fig. II.1.8

Let us return to the two numbers and the two ratios. Let the ratio of AB to CD not be greater than one of the two ratios nor smaller than the other, then the ratio of AB to CD will either be equal to one of the two ratios or be greater or smaller than the two of them.

Let the ratio of AB to CD first be equal to one of the two ratios, which [say] is the ratio of H to I. We suppose that the two numbers have been divided in the two ratios as we did before, and that the ratio of AE to CG is equal to the ratio of H to I. So the ratio of EB to GD is equal to the ratio of K to L. Since the ratio of AB to CD is equal to the ratio of H to I and the ratio of AE to CG is equal to the ratio of H to I, the ratio of AE to CG is equal to the ratio of AB to CD. So the ratio of EB to GD is equal to the ratio of AB to CD and is equal to the ratio of H to I. But the ratio of EB to GD was equal to the ratio of K to L, so the ratio of H to I is equal to the ratio of K to L. But by hypothesis these two ratios were different, which is impossible.

So the analysis has arrived at a premise which is not given, so we cannot carry out the synthesis corresponding to this analysis, because the

---

[33] The condition is indeed a necessary one.

last premise at which the analysis arrived is not given. If it is not possible to carry out the synthesis corresponding to the analysis, we cannot complete required division, nor establish the proof that it is correct.

If the ratio of AB to CD is greater than [both of] the two ratios, let us suppose <we are given> what we are required to find, that is to say that the ratio of AE to CG is equal to the ratio of H to I and that the ratio of EB to GD is equal to the ratio of K to L. So the ratio of AB to CD will be greater than the ratio of AE to CG and greater than the ratio of EB to GD. We put the ratio of AE to PG equal to the ratio of AB to CD; PG will thus be smaller than CG. We put the ratio of EB to GM equal to the ratio of AB to CD; GM will thus be smaller than GD. PM will thus be smaller than the whole of CD. But the ratio of AE to PG is equal to the ratio of EB to GM, so the ratio of AE to PG is equal to the ratio of AB to PM. But the ratio of AE to PG is equal to the ratio of AB to CD, so the ratio of AB to PM is equal to the ratio of AB to CD, so CD is equal to PM, which is impossible.

If the ratio of AB to CD is smaller than the two ratios, then the sum of PG and GM is greater than CD; now it is necessary that it is equal to it.

Thus, when the ratio of AB to CD is not greater than one of the two ratios nor smaller than the other ratio, the analysis arrives at a false premise. And if the analysis arrives at a false premise, this analysis is proof that what we are looking for is not possible and cannot exist. If we make this analysis into a proof by *reductio ad absurdum*, as we have done in this analysis, what we have shown is a proof of the discussion.

<7> For the example in the practical part without discussion for numerical problems that appear in a single way, this is our statement: to divide a known number, twice, into two parts, so that the greater part in the first division is twice the smaller part in the second division and the greater part in the second division is three times the smaller part in the first division.

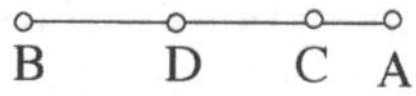

Fig. II.1.9

Let the given number be AB; we wish to divide AB, twice, into two parts in the way that we have described. Let us suppose that the number AB has been divided into two parts, twice, at the two points C and D: the first division at the point C, the greater part being CB, and the second division at the point D, the greater part being AD. CB will be twice BD, so CD is equal to DB. But AD is three times AC, so DC is twice AC; now CD is equal to DB, so BC is four times CA and AB is five times AC. But AB is known, so AC is known and each of the <numbers> AC and CB is known; DB is half of *BC*, so *BD* is known. The two parts *AC* and *CB* are known and the two parts *AD* and *DB* are known.

Analysis has arrived at parts that are known, and whose ratio to the number as a whole is known. But any number can be divided into parts whose ratio to the number as a whole is known. And if there are fractions in the parts, then if we multiply the number by the homonymous numbers of the fractions, the numbers will all be integers.

The analysis has arrived at a possible concept: the division of a number into known parts. If we invert this analysis, that allows us to complete the procedure and we establish the proof that this proposition is true. This analysis is one of those that does not need an addition to the object.

We shall carry out the synthesis for this problem by cutting off[34] a fifth from the number *AB*; that is the premise arrived at by the analysis, let it [the part removed] be *AC*; we divide *CB* into two equal parts at the point *D*.

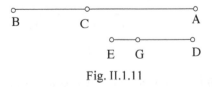

Fig. II.1.10

We say: we have divided AB *in the two ratios that were required.*

Proof: AB is five times *AC*, so *BC* is four times *CA*. *BD* is half *BC*, so *CB* is twice *BD*, which is one of the two <numbers> we seek. But since *CB* is four times *CA* and *CD* is half *CB*, *CD* is twice *CA*. So *DA* is three times *AC* and it is the second <number> that we seek. So *AB* has been divided twice in the manner required. This is what we wanted to do.

This part, [one] among the practical parts, can be completed only in a single way, since a single number has only a single firth [part] and its four fifths can be divided into two halves only by a single [form of] division; so a number can be divided in the two ratios we mentioned only in a single way.

<8> For the example in the practical part without discussion for indeterminate numerical problems, this is our statement: to find two square numbers whose sum is a square.

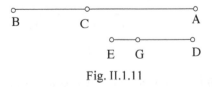

Fig. II.1.11

---

[34] Lit.: by dividing.

Let us suppose an example has been found and that these two numbers are *AC* and *CB*, so *AB* is a square. Let the number *DE* be the side of the square *AB* and the number *DG* the side of the square *AC*, so the square of *DE* is the number *AB* and the square of *DG* is the number *AC*. The amount by which the square of *DE* exceeds the square of *DG* is thus the number *CB* and the amount by which the square of *DE* exceeds the square of *DG* is the square of *EG* plus twice the product of *DG* and *GE*. The sum of the square of *EG* plus twice the product of *DG* and *GE* is a square number, because it is equal to *CB* which is a square. If from the square *CB* we cut off the square of *EG*, the remainder is twice the product of *DG* and *GE*; so half of it is the product of *DG* and *GE*. But if we divide the product of *DG* and *GE*, by *EG*, from the division we obtain *GD*. So if we cut off from the square *CB* the square of *EG* and if we take half of the remainder that we [then] divide by *EG*, from the division we obtain *GD*; if we next multiply *GD* by itself, we have *AC*, and *AC* plus *CB* is *AB* which is the square of *DE*.

The analysis has arrived at supposing [we have] a square, an arbitrary square, from which we then cut off a square, an arbitrary square, subject to the condition that it is smaller than the first; then we divide the remainder into two equal parts, next we divide the half by the side of the square that was removed, we multiply the result of the division by itself, then we add the result of the product to the first square.

This problem is possible and is not difficult; thus since this problem is possible, if we carry out the synthesis that corresponds to this analysis, the synthesis arrives at the existence of the *quaesitum* and, in addition, proves the truth of the *quaesitum*.

The synthesis of this problem is carried out in the following way: we suppose we have an arbitrary square number, let it be *AC*; we cut off from it an arbitrary square, let it be the square whose side is *DG*; we divide up what remains of *AC* into two equal parts and we divide the half by the number *DG*; let *GE* be what we obtain from the division. We multiply *GE* by itself, let *CB* be <the product>; so *CB* will be a square and *CA* a square.

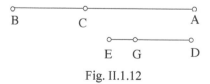

Fig. II.1.12

I say that AB *which is the sum of the two squares is a square.*

Proof: *AC* is the square of *DG* plus twice the product of *DG* and *GE*, and *CB* is the square of *GE*, so the whole of *AB* is the square of *DG* plus the square of *GE* plus twice the product of *DG* and *GE*. But the square of *DG* plus the square of *GE* plus twice the product of *DG* and *GE* is the square of *DE*, so the number *AB* is the square of *DE*, so *AB* is a square, it is the sum of *AC* and *CB*, which are squares. So we have found two square numbers whose sum is a square, they are the two numbers *AC* and *CB*. This is what we wanted to do.

This problem is indeterminate, that is to say that it can have many solutions. In fact, if we suppose that instead of the square *AC* we have another square, different from *AC*, and if we do what we did for *AC*, we obtain two squares whose sum is a square. This is proved as for the two squares *AC* and *CB*. If from the square of *AB* we cut off a square, different from the square *AC*, that is a square whose side is different from *DG*, and if we do what we did for *DG*, we obtain a square different from the square *CB*, and the sum of this square and the square *AC* will be a square.

This example provides a pattern for practical numerical problems, that are indeterminate and without discussion.

We have completed the parts of analysis for numerical problems.

<9> As regards geometrical problems, the example in the theoretical part of geometrical problems is as in our statement: the sum of two sides of a triangle is greater than the remaining side.

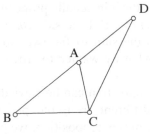

Fig. II.1.13

The analysis for this problem consists of supposing the statement true in its given form. So the sum of the two sides *AB* and *AC* is greater than *BC*. We examine the properties of the triangle to bring to light a property that leads to this. If we examine the properties of the triangle as it is, we do not find a property that leads to this statement being true. So the analyst must exercise intuition in regard to an addition that he adds to this proposition to generate a property or properties that are not found in this triangle as it is.

One of these additions[35] that we can add to generate a supplementary property consists of putting the two sides into a single line; we extend *BA* and we cut off from it <a straight line> equal to *AC*, let it be *AD*. So we have *BD* greater than *BC*. Let us join *CD*, we have the triangle *BDC* whose side *DB* is greater than the side *BC*. Now it has been proved in the eighteenth proposition of the first book of Euclid's work <the *Elements*> that the greatest side of any triangle subtends the greatest angle; so the angle *BCD* is greater than the angle *BDC*. But the angle *BDC* is equal to the angle *ACD* since *AD* is equal to *AC*. So the angle *BCD* is greater than the angle *ACD*, this is indeed how matters stand.

The analysis has arrived at a given proposition about which there is no doubt, that is that the angle *BCD* is greater than the angle *ACD*.

The synthesis corresponding to this analysis is carried out as we shall describe it: let us extend *BA*, as we did in the analysis; let us cut off *AD* equal to *AC* and let us join *DC*. So the angle *BCD* is greater than the angle *ACD*, this is the premise at which the analysis arrived, and it is the one that we start from at the beginning of the proof. But the angle *ACD* is equal to the angle *ADC*, because *AC* is equal to *AD*; now this premise was proved before the last premise, so the angle *BCD* of the triangle *BCD* is greater than the angle *BDC*. Thus, the side *BD* is greater than the side *BC* from what has been proved in the nineteenth proposition of the first book of Euclid's work <the *Elements*>. But the side *BD* is equal to the sum of the two sides *BA* and *AC*, so the sum of the two sides *BA* and *AC* is greater than the side *BC*. This is what we wanted to prove.

The analysis of this proposition can be carried out in another way, that is we add an addition different from the preceding one.[36] Among the additions that are possible in this proposition, we put *BD* equal to *AB*, since if *BC* were not greater than *BA*, the sum of *BA* and *AC* would be greater than *BC* and we dispense with the proof of this.

If *BC* is greater than *BA*, there remains *AC* greater than *CD*, so the angle *ADC* is greater than the angle *CAD*. But this is so because it is obtuse; in fact, the angle *BDA* is equal to the angle *BAD*, because the side *BA* is equal to the side *BD* and the sum of two angles of a triangle is smaller than two right angles; so the angle *BDA* is smaller than a right

---

[35] Here the addition is an auxiliary construction, the construction of the sum of two line segments.

[36] Lit.: an addition different from the addition that was added in that way [of carrying out the analysis].

angle, so the angle *ADC* is greater than a right angle and it is thus greater than the angle *DAC*. So the analysis has arrived at a premise that has already been proved: the angle *ADC* is greater than the angle *DAC* and the side *BA* is equal to the side *BD*.

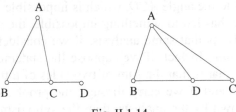

Fig. II.1.14

The synthesis corresponding to this analysis is carried out in the following way: we suppose we have the triangle, we cut off *BD* equal to *BA*, and we join *AD*; so the angle *BAD* is equal to the angle *BDA* and their sum is smaller than two right angles. So the angle *BDA* is smaller than a right angle, so the angle *ADC* is greater than a right angle. But the sum of the two angles *ADC* and *DAC* is smaller than two right angles, thus the angle *ADC* is greater than the angle *DAC*, the side *AC* is thus greater than the side *CD*. But the side *AB* is equal to the side *BD*, so the sum of the two sides *BA* and *AC* is greater than the side *BC*. This is what we wanted to prove.

We can carry out an analysis of this proposition in ways different from these two, but these two ways are sufficient for the purpose we set ourselves, that is to show by these two ways, specimens of those geometrical propositions that can be treated by analysis in several ways.

<10> Analysis that leads to something impossible in the theoretical part of geometrical problems is like the example in our statement of this proposition: the sum of two sides of a triangle is equal to the third side.

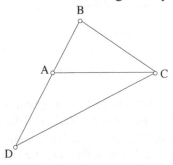

Fig. II.1.15

The analysis for this case will be carried out like the previous analysis, that is we extend *BA* and we cut off *AD* equal to *AC*; so *BD* will be equal to *BC* and the angle *BCD* will thus be equal to the angle *BDC*. But the angle *BDC* is equal to the angle *ACD*, because *AD* is equal to *AC*, so the angle *BCD* will be equal to the angle *ACD*, which is impossible.

As the analysis has led to something impossible, the statement is false. The proof it is false is that same analysis, if we consider it as a proof by *reductio ad absurdum*. In fact, if we suppose the statement to hold in the form it was given, that is that the sum of two sides of a triangle is equal to the remaining side and if we carry through the proof using the premises that have been shown by the analysis, then the syllogism will be demonstrative and it leads necessarily to the impossibility that followed in the analysis.

This example provides a pattern for carrying out the analysis of theoretical geometrical problems that lead to an impossibility; it is following the pattern of this proof, which is by *reductio ad absurdum* and generated by the analysis, that we shall carry out the proof that the statement is false.

<11> For the example in the practical part of geometrical problems with discussion, this is our statement: to divide a given straight line into two parts such that the area enclosed by the two parts is equal to a given area.

Let the straight line be *AB* and the area *C*; let us suppose that the straight line has been divided at the point *D* and that the area enclosed by the two straight lines *AD* and *DB* is equal to the area *C*.

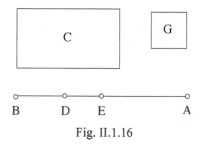

Fig. II.1.16

If we examine the properties of this proposition, we find that the two straight lines *AD* and *DB* are either equal or different. If they are equal, then the area *C* is equal to the square of half of the straight line *AB*. If the two straight lines *AD* and *DB* are different, then the area enclosed by the two straight lines *AD* and *DB* is smaller than the square of half the straight line $<AB>$.[37] So the area C will be smaller than the square of half the straight line which is AB. But it is not possible to divide the straight line AB into two parts such that the area they enclose is greater than the square of half of the straight line. So if the area C is equal to the square of half the straight line AB, the analysis arrives at the fact that the straight line AB has been divided into two equal parts and this is possible. If the area C is smaller than the square of half of the straight line, let the amount by which the square of half the straight line exceeds the area C be the area G. Let us divide AB into two equal parts at the point E. The area G will be equal to the square of DE because the square of EB is equal to the product of AD and DB plus the square of DE, the square of EB is equal to the sum of the two areas C and G and the product of AD and DB is equal to the area C. So the square of DE is equal to the area G and the square of EB is known, so the sum of the two areas C and G is known and the area C is known, so the area G is known, because if from a known magnitude we take away a known magnitude, the remainder is $<a>$ known $<magnitude>$ as has been proved in the fourth proposition of the *Data*. But the area G is equal to the square of ED, so the square of ED is known, so the straight line ED is known; but the straight line EB is known and the point E is known, so the point D is known.

The analysis has arrived at the straight line AB being divided in a known point which is the point D and that ED is known; now if ED is known, we can find it.

In addition, we have shown in the analysis that the area C is not greater than the square of half the straight line AB.

The synthesis of this problem will be carried out in the following way: if the area C is equal to the square of half the straight line AB, we divide the straight line AB into two equal parts; the area enclosed by the two halves is thus equal to the area C. If the area C is smaller than the square of half the straight line AB, we divide the straight line AB into two equal parts at the point E and from the square of EB we take away the area C; there remains the area G. We put the square of ED equal to the area G, the area enclosed by the two straight lines AD and DB will be equal to the area C, because the area enclosed by the two straight lines AD and DB plus the square of DE is equal to the square of EB, so the square of DE is the amount by

---

[37] The product of two numbers whose sum is constant is a maximum when the two numbers are equal. Indeed, $4xy = (x + y)^2 - (x - y)^2$, hence $xy = \left(\frac{x+y}{2}\right)^2 - \left(\frac{x-y}{2}\right)^2$. So the maximum value of xy is the square of half the sum.

which the square of *EB* exceeds the area enclosed by the two straight lines *AD* and *DB*. But the area *G* is the amount by which the square of *EB* exceeds the area *C*, so the area enclosed by the two straight lines *AD* and *DB* is equal to the area *C*. So we have divided the straight line *AB* into two equal parts at the point *E* so that the area enclosed by the two straight lines *AD* and *DB* is equal to the area *C*. This is what we wanted to prove.

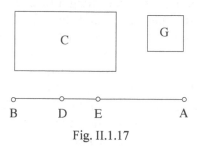

Fig. II.1.17

So it remains to prove that if the area *C* is greater than the square of half the straight line *AB*, then it is not possible to divide the straight line *AB* into two parts such that the area enclosed by these two parts is equal to the area *C*, which is a proof of the discussion. Indeed, if the straight line *AB* is divided into two parts, then either the division is in the mid point of the straight line, or the two parts are different. If the division is at the mid point of the straight line, then the area enclosed by the two parts is equal to the square of half the straight line. If the two parts are different, then the area enclosed by the two parts is smaller than the square of half the straight line. For any division by which the straight line *AB* is divided into two parts, the area enclosed by the two parts is accordingly not greater than the square of half the straight line. So if the area *C* is greater than the square of half the straight line, then the straight line is not divided into two parts that enclose an area equal to the area *C*.

<12> This is our statement: to draw, from a known point to a known unlimited straight line, a straight line that is perpendicular to it.

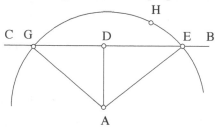

Fig. II.1.18

Let *A* be the point and *BC* the straight line; we wish to draw from the point *A* to *BC* a straight line that is perpendicular to it. We suppose that this has been done and that the perpendicular is *AD*. If the analyst examines the property of this straight line, he notices that any straight line drawn from the point *A* to the straight line *BC*, apart from the straight line *AD*, will be greater than the straight line *AD*, because if from the point *A* we draw another straight line to the straight line *BC*, we generate a triangle such that one of its angles is a right angle; each of the two remaining angles is thus acute. We draw the arbitrary straight line *AE*, so *AE* will be greater than *AD*, because the angle *ADE* is greater than the angle *AED*. It also follows that if we make the straight line *DG* equal to the straight line *DE* and if we join *AG*, then *AG* will be equal to *AE* and *GE* will be divided into two equal parts by *AD*. It follows that if from the point *A* to the straight line *BC* we draw two equal straight lines, if we divide the straight line that lies between them into two equal parts, and if we join the point of division and the point *A* with a straight line, then this straight line that joins them is perpendicular to the straight line *BC*. But if *AG* and *AE* are equal, then the circle with centre at the point *A* and with semidiameter the straight line *AE* cuts the straight line *BC* at the two points *G* and *E* such that a part of this circle lies beyond the straight line *BC*.

So the analysis has arrived at something possible, that is: to draw with centre *A* a circle that is cut by the straight line *BC*.

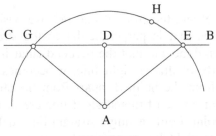

Fig. II.1.19

We carry out the synthesis of this problem by supposing that beyond the straight line *BC* we have a point, such as the point *H*; with centre *A* and distance *AH*, we draw a circle, it cuts the straight line *BC* in two points; let it [the straight line] cut it at the points *E* and *G*; we join *AE* and *AG*, we divide *EG* into two equal parts at the point *D* and we join *AD*. The two straight lines *ED* and *DA* are equal to the two straight lines *GD* and *DA* and the base *AE* is equal to the base *AG*, so the angle *ADE* is equal to the angle

ADG; so they are right angles and the straight line *AD* is perpendicular to the straight line *BC*. This is what we wanted to prove.

It is clear that it is possible to draw from the point *A* to the straight line *BC* only a single perpendicular, because if we draw two perpendiculars from the point *A* to the straight line *BC*, a triangle is formed in which two angles are right angles, which is impossible.

<13> For the example in the practical part without discussion which takes place in [only] one way, this is our statement: to draw from a given point on a known straight line a straight line that is perpendicular to it.

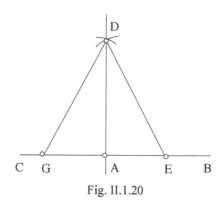

Fig. II.1.20

Let the point be *A* and the straight line *BC*; we wish to draw from the point *A* a straight line that is perpendicular to the straight line *BC*. We suppose this has been done and that the perpendicular is *AD*. If the analyst examines the property of this straight line, it becomes clear that for any straight line drawn from the point *A*, apart from the straight line *AD*, the angles that lie on either side of this straight line are different, and that we can draw from the point *A* only a single straight line such that the angles to either side <of this straight line> are equal. It next becomes clear that if from the point *D* we draw two straight lines to two points of the straight line *BC* on either side of the point *A*, such that their distances from the point *A* are equal, then they will be equal. Thus, the triangle that is formed is isosceles and the point *A* is the mid point of its base. Let it be the triangle *DEG*, such that *EA* is equal to *AG*. So the analysis has arrived at something possible, that is: to construct on a segment of the straight line *BC* an isosceles triangle such that the point *A* divides its base into two equal parts; and this is something possible.

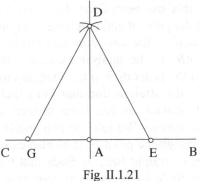

Fig. II.1.21

Synthesis of this problem: on either side of the point A let us cut off from the straight line BC two equal straight lines, such as the two straight lines AE and AG. On the straight line EG we construct an equilateral triangle, let the triangle be EDG; so this triangle will be isosceles. We join AD, the two triangles that lie on either side of AD have equal angles, so the angle EAD is equal to the angle GAD and the straight line AD will be perpendicular to the straight line BC. This is what we wanted to prove.

<14> For the example in the practical part of geometrical problems without discussion and indeterminate, this is our statement: given a circle, and given an unlimited straight line lying outside this circle, to construct a circle that touches the given circle and at the same time touches the straight line.

Let the circle be AB and the straight line CD; we wish to draw a circle that touches the circle AB and touches the straight line CD.

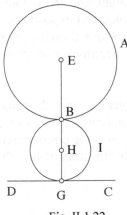

Fig. II.1.22

We suppose that this has been done. Let there be a circle *BIG*, let it touch the circle *AB* at the point *B* and touch the straight line *CD* at the point *G*; let there be the point *H*, the centre of this circle, and the point *E*, the centre of the circle *AB*. If the analyst examines the properties of this proposition, as well as the properties of the tangent circle, he finds that, for two circles that touch, the straight line that joins their two centres passes through the point of contact, as has been proved in the third book of Euclid's work <the *Elements*>. We join the two points *E* and *H*; the straight line *EH* thus passes through the point *B*. If he also examines the property of the circle that touches a straight line, he finds that the straight line drawn from the centre of the circle to the point of contact is perpendicular to the straight line tangent. We join the line *HG* perpendicular to the straight line *CD*. But the straight line *HG* either is a continuation of the straight line *EH* or it is not a continuation of it. If the two straight lines *EH* and *HG* are continuations of one another, as in the first case of the figure, then the line *EG* is straight and it is perpendicular to the straight line *CD*. But the point *E* is known, because it is the centre of the known circle, and the straight line *CD* is known in position, by hypothesis, since it has been given. Now, from the known point *E* we have drawn to the straight line *CD*, known in position, the straight line *EG*, which makes a known angle with it. So the straight line *EG* is known in position, as has been proved in Proposition 29 of the *Data*.[38] But the straight line *CD* is known in position, so the point *G* is known, as has been proved in Proposition 24 of the *Data*.[39] So the two points *E* and *G* are known, the straight line *EG* is then known in magnitude and in position and the circle *AB* is known in position, so the point *B* is known. The straight line *BG* is accordingly known in magnitude and it is divided into two equal parts at the point *H* since *BHG* is a straight line, so the point *H* is known, the straight line *HB* is known in magnitude and the circle *BIG* is known in magnitude and in position.

The analysis has arrived at our drawing, from a known point of the straight line *EG* known in position, and drawing a circle of known magnitude; this is possible.

The synthesis of this problem is carried out as we shall describe: let us draw from the point *E* a perpendicular to the straight line *CD*, let it be *EG*. This perpendicular necessarily cuts the circumference of the circle *AB*; let it cut it at the point *B*; we divide the straight line *BG* into two equal parts at

---

[38] This is Proposition 30 in Heiberg's edition and in the recension of al-Ṭūsī.

[39] This is Proposition 25 in Heiberg's edition and in the recension of al-Ṭūsī.

the point *H* and we take *H* as centre; using the point *H* and distance *HB* we draw the circle *BIG*.

I say that the circle BIG *touches the circle* AB *and touches the straight line* CD.

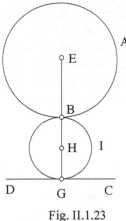

Fig. II.1.23

Proof: The straight line *HE* is a diameter of the two circles *AB* and *BIG*,[40] so the perpendicular drawn from the point *B* to the straight line *EH* is a tangent to the two circles, so the two circles touch [one another]. But since *CD* is perpendicular to the diameter *BHG*, accordingly the circle *BIG* touches the straight line *CD*. So we have drawn a circle that touches the circle *AB* and touches the straight line *CD*, which is the circle *BIG*. This is what we wanted to do.

If the two straight lines *EH* and *HG* are not continuations of one another as in the second case for the figure, then if the analyst examines the properties of this figure, he finds that the straight line *BH* is equal to the straight line *HG*; so we find that the straight line *EH* exceeds the straight line *HG* by a magnitude equal to the straight line *BE*. But *BE* is of known magnitude since it is the semidiameter of the circle *AB*, which is known in magnitude and in position, because it was given. So if we add to the straight line *HG* a straight line equal to the straight line *EB*, it will be equal to the straight line *EH*. We extend the straight line *HG* in the direction of *G* and we cut off from it *GK* equal to the semidiameter of the circle *AB*. So *KH* will be equal to *HE*; we join *EK*. So the triangle *EHK* will be isosceles. But if the point *G* is known in position, <the straight line> *HGK* will be known in position as has been proved in the twenty-eighth proposition of

---

[40] He means that this segment *HE* lies on the diameter of the two circles.

the *Data*,[41] the straight line *GK* will be known in magnitude and in position and the point *K* will thus be known. But the point *E* is known, by hypothesis, so the straight line *EK* has known endpoints, so it is known in magnitude and in position, as has been proved in the twenty-fifth proposition of the *Data*.[42] The angle *EKH* is known because its two sides are known in position and the angle *KEH* is known, because it is equal to the angle *EKH*. So the straight line *EH* is known in position, the triangle *EKH* has known angles and the straight line *KH* is known in position, so the two straight lines *KH* and *EH* are known in position; now they cut one another at the point *H*, so the point *H* is known.

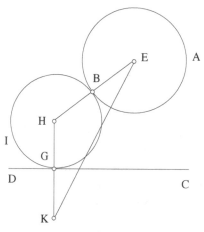

Fig. II.1.24

The analysis has arrived at [the conclusion] that, when the point *G* is known, the straight line *GH*, which is the semidiameter of the tangent circle, is known in position and thus the point *H*, which is the centre of the circle, is known. This is something possible and is not difficult, that is to say that we consider a point on the straight line *CD* from which is drawn a perpendicular to the straight line *CD* and we cut off from it a straight line equal to the semidiameter of the circle *AB*.

The synthesis of this problem is carried out in the following way: given the circle and the straight line, we consider an arbitrary point *G* on the straight line *CD*. From this point we draw the perpendicular *GH*, we extend it in the direction of *G*, we cut off from it *GK* equal to the semidiameter of

---

[41] This is Proposition 29 in Heiberg's edition and in the recension of al-Ṭūsī.

[42] This is Proposition 26 in Heiberg's edition and in the recension of al-Ṭūsī.

the circle *AB* and we join the straight line *EK*. So the angle *EKG* is acute, because the angle *CGK* is a right angle. At the point *E* on the straight line *EK* we construct an angle equal to the angle *EKH*, let it be the angle *KEB*. So the straight line *EB* meets the straight line *KH*, let it meet it at the point *H*. Since the angle *KEH* is equal to the angle *EKH*, the straight line *EH* is equal to the straight line *KH*, but the straight line *EB* is equal to the straight line *KG*; there remains the straight line *BH* equal to the straight line *GH*. We take the point *H* as centre and with distance *HG* we draw a circle, it accordingly passes through the point *B*, since *HB* is equal to *HG*; let the circle be *BIG*. Since the straight line *HE* is a common diameter of the two circles *AB* and *IG* and since the point *B* is common to the two circles, the circle *IG* touches the circle *AB*. But since *CD* is perpendicular to the straight line *GH*, the circle *GI* touches the straight line *CD*. So we have drawn a circle that touches the circle *AB* and touches the straight line *CD*, that is to say the circle *BIG*. This is what we wanted to do.

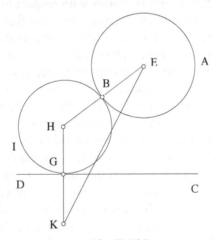

Fig. II.1.25

It becomes clear from our hypothesis concerning the point *G* that it is possible to construct many circles, an infinite number, each of them touching the circle *AB* and the straight line *CD*. So this problem is indeterminate, since through any point we take on the straight line *CD* we can draw a perpendicular to the straight line *CD* and follow the same procedure for this as we did for the perpendicular *GH*. On this perpendicular we choose a point such that if we take it as the centre of a circle, this circle will touch the circle *AB* and the straight line *CD*. But, as it is possible to draw from the point *E* a perpendicular to the straight line *CD*, it is also possible to construct a circle which touches the circle AB and touches the straight line CD, from the first property we mentioned.

We have dealt exhaustively with the examples of all the subdivisions of geometrical problems.

As for problems that refer to astronomy, most of them reduce to numerical problems or to geometrical problems; their examples are the examples we gave earlier; among these problems there are those that refer to explanations of the motions of the heavenly bodies.

<15> We present an example from which we show the analysis that leads to concepts deduced from astronomy; one of these examples is the motion of the Sun.

When the ancients observed the motion of the Sun and they measured it with respect to the centres of instruments, by means of which they observed the Sun, and which were taken as being at the centre of the Universe, they found that its motion varies with respect to the centres of the instruments, that is they found that in equal times the Sun traverses unequal angles with respect to the centres of the instruments. But they were convinced that the motions of the celestial bodies cannot be anything but uniform, similar, simple and not compound, because the substance of celestial bodies is a simple substance, not compound, and there is no variation in it.[43] So, when they found that their motions were variable, while at the same time assuming that their motions were uniform, they believed that the position of their orb meant that their apparent motions[44] were different from their real motions and they determined the position of the orb by analysis. But they had found that the centre of the Sun moves in a single fixed plane that cuts the Universe. For them, it was established that the form of the Universe is a spherical figure, it followed that the plane in which the centre of the Sun moves cuts the sphere of the Universe and it followed that a circle is formed in the surface of the sphere of the Universe, whose centre is the centre of the Universe. They then determined the position of this circle, they considered the motion of the Sun in relation to the circumference of this circle and they found it variable.[45] Starting from this variation,[46] they determined the position of the orb of the Sun, which

---

[43] That is, the substance is homogeneous.

[44] Lit.: what of their motions appears to the sight.

[45] 'Variable' means the motion is not both circular and uniform, *i.e.* that it is 'anomalistic'. See p. 172.

[46] *I.e.* anomaly.

moves the Sun with regular motion. Their determination of the position of this orb was carried out by analysis as we shall describe.

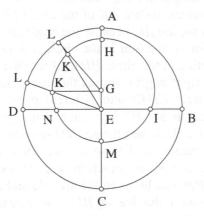

Fig. II.1.26

Let the circle, whose centre is the centre of the Universe in the plane in which the centre of the Sun moves, be the circle *ABCD* with centre *E*. Since the centre of the Sun always moves in the plane of this circle, it is necessary that the centre of the regular motion by which the Sun moves is also in the plane of this circle; let *G* be the centre of the regular motion and let the centre of the Sun move with a regular motion on the circumference of the circle *HIMN*. If the centre of the circle *HIMN* were the centre of the circle *ABCD*, the Sun would have traversed two similar arcs of the two circles in the same time, and the motion of the Sun along the circumference of the circle *HIMN* would also be variable.[47] Now the motion of the Sun along the circumference of the circle *HIMN* is uniform, by hypothesis, so the centre of the circle *HIMN* is not the centre of the circle *ABCD*, so the point *G* is not the point *E* and the point *G* is distinct from the point *E*. The analysis has arrived at the centre of the regular motion being different from the centre of the variable motion,[48] which is the centre of the Universe.

They carried out the synthesis corresponding to this analysis by joining the point *E* and the point *G* with a straight line that they extended on both sides to the two points *A* and *C*. They drew from the point *E* the straight line *EKL*, which cuts the two circles, and they joined *GK*. If the Sun is seen as lying on the straight line *EL*, then on the circle *ABC* it has traversed the arc *AL* and it has traversed the arc *HK* of the circle *HIMN*, because in its

---

[47] *I.e.* anomalistic.

[48] *I.e.* anomalistic motion.

revolution the Sun must pass through the point H and in relation to the circle $ABCD$ it is seen at the point A. If it comes to the point K, then it is seen as lying on the straight line EKL and thus it traverses the arc AL of the circle $ABCD$ and traverses the arc HK of the circle $HIMN$. But the arc HK is a little greater than the arc AL, because the angle HGK is greater than the angle AEL. The motion of the Sun in the circle $ABCD$ in this position will be slower than its motion in the circle $HIMN$. From the point E we draw the straight line BED at right angles, it cuts the circle $ABCD$ into four equal parts and cuts the circle $HIMN$ into different parts; so the arc AD is a quarter of a circle and the arc HN is greater than a quarter of a circle; the arc DC is a quarter of a circle and the arc NM is smaller than a quarter of a circle. We draw GK parallel to EN, so the arc HK is a quarter of a circle and the arc KN is the amount by which the arc HN exceeds a quarter of a circle and it is the difference between the arc NM and a quarter of a circle; it follows, for this position, that the arc IHN is greater than a semicircle and the arc IMN is smaller than a semicircle.

If the motion of the Sun on the circle $HIMN$ is uniform, the Sun must traverse the arc IHN in a time longer than the time in which it traverses the arc NMI. But if it traverses the arc IHN, in the circle $ABCD$, it traverses the arc BAC, which is a semicircle, and if it traverses the arc NMI, in the circle $ABCD$, it traverses the arc DCB, which is a semicircle, so the motion of the Sun in the semicircle BAD is slower than its motion in the semicircle DCB; now the latter is the motion that is perceived by the observer.

Then they also determined by analysis the magnitude of the amount by which the arc IHN exceeds the arc NMI, starting from the magnitude of the amount by which the time in which the Sun traverses the arc IHN exceeds the time in which it traverses the arc NMI, because the ratio of the time to the time is equal to the ratio of the distance to the distance if the motion is uniform.

They also determined, from the magnitude of the amount by which the arc IHN exceeds the semicircle, the magnitude of the straight line EG and its ratio to the straight line GH. And in this way they determined by analysis the positions of the orbs of all the wandering stars, as well as the magnitudes of these orbs and the [linear] eccentricity. These are enough as examples of analysis in astronomy.

As for the notions that pertain to music and to the problems that can be devised in that art, they all reduce to numerical problems.

<16> As an example for that, our statement is: the interval of an octave is composed of the interval of a fourth and the interval of a fifth.

Let there be an interval of an octave between the two notes *A* and *B*, let there be an interval of a fourth between the two notes *C* and *D*, let there be an interval of a fifth between the two notes *E* and *F*.

I say that the ratio between the two notes *A* and *B* is composed of the ratio between the two notes *C* and *D* and of the ratio between the two notes *E* and *F*. We suppose that this is so. Let there be an interval of a fourth between the two notes *A* and *H*, the interval of a fifth is then between the two notes *H* and *B*, the interval between the two notes *A* and *B* is composed of the interval between the two notes *A* and *H* and the interval between the two notes *H* and *B*. Since the interval of a fourth is [expressed] in the ratio of [one to] one plus a third, the interval between the two notes *A* and *H* is [expressed] in the ratio of one plus a third; since the interval of a fifth is [expressed] in the ratio of [one to] one plus a half, the interval between the two notes *H* and *B* is [expressed] in the ratio of [one to] one plus a half. It is accordingly necessary that the interval between the notes *A* and *B* is composed of the ratio of [one to] one and a third and <the ratio> of [one to] one plus a half. But the ratio composed of the ratio of [one to] one plus a third and <the ratio> of [one to] one plus a half is the ratio of doubling. It is accordingly necessary that the interval between the two notes *A* and *B* are in the ratio of doubling. But this is so because the interval of an octave is [expressed] in the ratio of doubling.

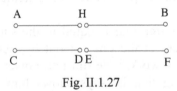

Fig. II.1.27

So the analysis has arrived at the given notion, that is: the interval of an octave is [expressed] in the ratio of doubling, and this is the manner in which analysis is carried out for all problems of the composition [of intervals].

Synthesis of this problem: the interval of an octave is found in the ratio of doubling and the ratio of doubling is composed of the ratio of [one to] one and a third and <the ratio> of [one to] one plus a half. The interval of a fourth is [expressed] in the ratio of [one to] one plus a third. The interval of a fifth is [expressed] in the ratio of [one to] one plus a half and the interval of an octave is composed of the interval of a fourth and the interval of a fifth. This is what we wanted to prove.

We have completed [giving] examples of analysis for all the propositions that serve to characterise the parts of all the mathematical sciences. In all these examples, we have deliberately looked for simplicity, in order to make it easy to understand for someone who is studying the art of analysis.

<SECOND CHAPTER>

It remains for us to set out problems of analysis involving some difficulties, so that analysis becomes a tool to be used by anyone who works through this treatise and a guide for anyone who is trying to acquire the art of analysis; so that this analysis may be directed by the propositions that are used in it and by the complementary results that are added to its objects so that he can exercise the art of analysis: to continue the search for premises that is in fact carried out through the complementary results that are added and through the properties that emerge from them. In these examples we limit ourselves to dealing only with numerical problems and with geometrical problems; they are clearer and all other problems reduce to them.

<17> For one of these problems, our statement is: to find perfect numbers.

A perfect number is one that is equal to the sum of its parts,[49] which measure it. This problem is the one set out by Euclid at the end of the arithmetical books of his work.[50] He did not, however, mention treating it by analysis, and nothing that he says shows how he found the perfect number by analysis. He proposes the number only through synthesis, as for the other problems that he included in his work. We show here how to find a perfect number by analysis, and then proceed to set up the synthesis.

Method for the analysis of this problem: we assume that the perfect number has been found, let it be for example the number AB, and that the parts that measure it are the numbers C, D, E, G, H, I, L, M, N. Let us take AB to be equal to the sum of the numbers C, D, E, G, H, I, L, M, N. We then examine the properties of the numbers that have [aliquot] parts. If we examine the properties of the numbers that have parts, we find, as has been

---

[49] For the Euclidean origin of this term, see footnote 10.
[50] Euclid, *Elements*, IX.35.

shown in Proposition 36 of the ninth book,[51] that if any number of numbers follow one another with the same ratio, and if we subtract from the second and the last a number equal to the first, then the ratio of the remainder of the second to the first is equal to the ratio of the remainder of the last to the sum of all the numbers that precede it.[52] It follows that, given successive proportional numbers in double ratio, if we subtract from the second of them and from the last of them a number equal to the first, then the remainder of the second will be equal to the first and the remainder of the last will be equal to the sum of all the numbers that precede it. But for successive numbers in double ratio, each of them measures the greatest number and each of them is a part of the greatest number. From all this it follows that if the numbers *AB, C, D, E, G, H, I, L, M, N* are in continued double proportion, then each of the <numbers> *C, D, E, G, H, I, L, M, N* is a part of *AB*, and if we subtract from *AB* <a number> equal to *N*, the remainder of *AB* is equal to the sum of the remaining numbers, which are parts of *AB*. But the whole of *AB* is equal to the sum of the parts, so the number *AB* is not in double proportion with all the remaining successive numbers.

Fig. II.1.28

In the same way, one of the properties of successive proportional numbers whose ratio is that of doubling and which begin with one, is that if we subtract one from each of them, <each> remainder will be equal to the sum of the numbers that precede it, because if we also subtract one from the second, which is two, the remainder will be equal to the first, which is one. It follows that for the numbers *C, D, E, G, H, I, L, M, N*, if some of them are successive [numbers] in double ratio beginning with *AB*, and if the last <of those> which is the smallest of them is less by one than twice

---

[51] This is Proposition 35 in Heiberg's edition of the *Elements*.
[52] Proposition 4 of this treatise.

the one that immediately precedes it, then all the numbers that succeed AB are parts of AB, and AB is equal to their sum.

Let the numbers AB, C, D, E, G be in succession in the ratio of doubling, and let the number G be one less than twice the number H; from AB we take away KB equal to G, then AK will be equal to the sum of the numbers C, D, E, G; but G is one less than twice H; so it is equal to the sum of H, I, L, M, N; now KB is equal to G and KB is equal to the sum of H, I, L, M, N, so the whole of AB is equal to the sum of the numbers C, D, E, G, H, I, L, M, N.[53] But the numbers H, I, L, M, N are successive numbers in the ratio of doubling, beginning with one and N is one. But since C is half AB, C measures AB by the units of M, D measures AB by the units of L and E measures AB by the units of I; and the same holds for the remaining numbers. So if the numbers C, D, E, G are the same in number as the numbers H, I, L, M, then each of the numbers that follow AB measures AB by the units of one of the numbers that follow N and each of the numbers that follow N measures AB by the units of one of the numbers that follow AB. Thus, all the numbers are parts of AB and no other number measures AB. If the numbers that follow AB are more numerous than the numbers that follow N, if some of the numbers that follow AB measure AB by the units of the numbers that follow N, and if the remaining numbers that follow AB measure AB by the units of other numbers, then these other numbers are parts of AB. But AB has no parts other than the numbers C, D, E, G, H, I, L, M, N. So the numbers that follow AB are not more numerous than the numbers that follow N. If the numbers that follow AB are less numerous than the numbers that follow N, then some of the numbers that follow N measure AB by the units of the numbers that follow AB, and the remaining numbers that follow N measure AB by the units of other numbers; these other numbers are thus parts of AB. But AB has no other parts than the given numbers. The numbers that follow AB in the ratio of doubling are of the same number as those that follow N. Thus, the numbers C, D, E, G are of the same number as the numbers H, I, L, M.

In the same way, if the number G has one part or several parts, then this part or these parts measure AB, since G measures AB. This part or these parts are one part or several parts of AB, and none of them is one of the numbers C, D, E, G, H, I, L, M, because none of the numbers C, D, E is a part of G, since each of them is greater than G, and none of the numbers H, I, L, M measures G, because if we add [the number] one to G, then the

---

[53] The piece of text between *...* should have appeared several lines earlier, that is before the sentence that begins 'it follows that'. It seems to us that this passage corresponds to a break in the text that we noted in manuscripts B and S.

numbers H, I, L, M measure it and none of the numbers H, I, L, M measures the number one that has been added, because each of them is greater than one, so none of the numbers H, I, L, M measures the number G, so none of them is a part of the number G; so if the number G is one part or several parts other than unity, then this part – or these parts – is a part of AB, and each of them is different from any of the numbers C, D, E, G, H, I, L, M. But AB has no parts other than these numbers and unity, so the number G is a prime number.

The analysis has arrived at there being, between the number AB and the number G, numbers that are all successive and whose ratio is that of doubling; among these, the number G is prime, and the number G is one unit less than double one of the numbers in continued proportion beginning with one and in the ratio of doubling. This concept is possible, that is the existence of a number among the numbers in continued proportion begins with one and in the ratio of doubling, [a number] such that if we subtract one from it we have a prime number.

The synthesis of this problem is carried out as we shall describe: using induction, we consider the numbers that are evenly even numbers, that is those that are in the double ratio starting with one. We subtract unity from each of them; the one that becomes prime we then double as many times as we need to make the number of the numbers in the series that are doubled in this way equal to the number of proportional numbers in the series that precede this number, including the unity that is the first number. The greatest number that the doubling produces is a perfect number.

Example: The numbers A, B, C, D, E, GH are successive <numbers> whose ratio is that of doubling. Of these A is equal to unity, and if we subtract unity from GH, the remainder will be a prime number. We subtract from GH the number unity which is SH, there remains GS a prime number; we double GS as many times as we need to make the number of the doublings equal to the number of the numbers A, B, C, D, E; let the numbers be GS, I, K, L, NO.

I say that the number NO *is a perfect number.*

Proof: We cut off OP equal to GS, so NP is equal to the sum of the numbers L, K, I, GS; but the number PO is equal to the sum of the numbers E, D, C, B, A, so the number NO will be equal to the sum of the numbers A, B, C, D, E, GS, I, K, L. But each of the numbers L, K, I, GS measures NO by the units of one of the numbers E, D, C, B, and each of the numbers B, C, D, E measures NO by the units of one of the numbers GS, I, K, L. All the numbers B, C, D, E, GS, I, K, L are parts of NO; but it has been shown that NO is equal to the sum of these numbers plus A which is equal to unity. It remains for us to prove that no number other than these numbers measures *NO*.

Let the number *M* measure *NO*; I say that *M* is one of the numbers *B*, *C*, *D*, *E*, *GS*, *I*, *K*, *L*. Let the number *M* measure *NO* by the units of the number *Q*. If we then multiply it by *Q*, we have *NO*; but the number *GS* measures *NO* by the units of the number *E*; so if we multiply *GS* by *E*, we have *NO*. So the product of *E* and *GS* is equal to the product of *M* and *Q*. So the ratio of *GS* to *Q* is equal to the ratio of *M* to *E*. Either *GS* measures *Q*, or it does not measure it. If *GS* measures *Q*, then *M* measures *E*; but the numbers *A*, *B*, *C*, *D*, *E* are in continued proportion from one, the one that follows unity is a prime <number>, since it is two; so no number measures the greatest except one of them, as has been proved in Proposition 13 of the ninth book; so the number *M* is one of the numbers *B*, *C*, *D*, *E*. If the number *GS* does not measure *Q*, then they are prime to one another, as has been proved in Proposition 31 of the seventh book; if they are prime to one another, then they are the two smallest numbers in the proportion, as has been proved in Proposition 22 of the seventh book, and if the two numbers *GS* and *Q* are the two smallest numbers in their proportion, then they measure the numbers that follow them in their proportion, as has been proved in Proposition 20 of the seventh book. So if the number *GS* does not measure *Q*, they are the two smallest numbers in the continued proportion and they measure the numbers that follow in their proportion. But the ratio of *GS* to *Q* is equal to the ratio of *M* to *E*. So the number *Q* measures *E*; so the number *Q* is one of the numbers *B*, *C*, *D*. So the number *Q* measures *NO* by the number of units of one of the numbers *GS*, *I*, *K*, *L*. But *Q* measures *NO* by the number of units of *M*, so the number *M* is one of the numbers *GS*, *I*, *K*, *L*.

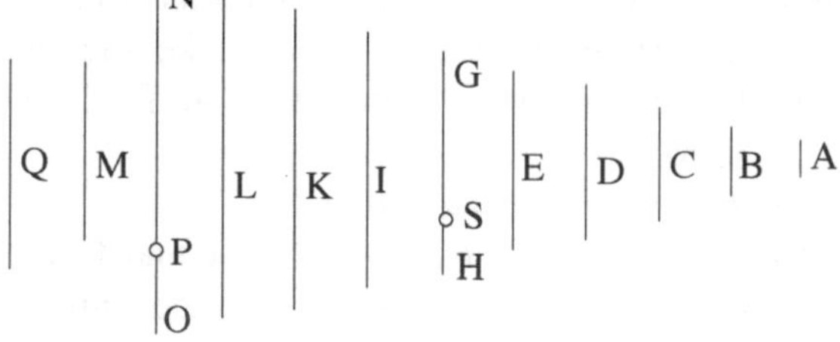

Fig. II.1.29

So every number that measures *NO* is one of the numbers *B*, *C*, *D*, *E*, *GS*, *I*, *K*, *L*. And no part, other than the numbers *B*, *C*, *D*, *E*, *GS*, *I*, *K*, *L* and *A*, which is unity, measures *NO*. But the number *NO* is equal to the sum of these numbers, so the number *NO* is a perfect number. This is what we wanted to prove.

<18> As an example, our statement: to find three numbers such that if to two thirds of the second we add half the first, if to three quarters of the third we add one third of the second and if to one half of the first we add one quarter of the third, these three <sums> are equal.

The analysis of this problem is carried out as we shall describe: let the numbers be *AB*, *CD* and *EG*. Let us cut off a half of *AB*, which is *AH*. We add it to *CD*; let [the result] be *CI*. From *CD* we cut off one third of it, let it be *KD*, and we add it to *EG*; let [the result] be *EL*. We cut off a quarter of *EG*, let [the result] be *MG*, and we add it to *AB*, let [the result] be *AN*. So the three numbers *HN*, *IK* and *LM* are equal. We suppose this is so and we examine what is necessary for the magnitudes of the numbers after the additions [have been made]. Since *HN* is equal to *IK* and *IK* is *IC*, which is half *AB*, plus *CK*, which is two thirds of *CD*, and *HA* is half *AB* and is thus equal to *IC*, there remains *AN* equal to *CK*. But *AN* is a quarter of *EG*, so two thirds of *CD* is equal to a quarter of *EG*; so the whole of *CD* is equal to a quarter of *EG* plus an eighth.

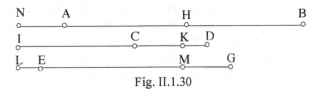

Fig. II.1.30

In the same way, since *HN* is equal to *LM* and *LM* is equal to *LE*, which is a third of *CD*, plus *EM* which is three quarters of *EG*, and *AN* is a quarter of *EG*, we take away from *EM* a quarter of *EG* and from *HN* <we take away> *AN*; so there remains *LE* plus half of *EG* equal to *HA*. But *HA* is half of *AB*, so half of *AB* is a third of *CD* plus half of *EG*; so the whole of *AB* is equal to two thirds of *CD* plus the whole of *EG*. But two thirds of *CD* is a quarter of *EG*, so the whole of *AB* is equal to *EG* plus a quarter of it. So the ratio of *AB* to *EG* is equal to the ratio of five to four; it is thus equal to the ratio of ten to eight. But the ratio of *CD* to *EG* is the ratio of three to eight, and the ratio of *AB* to *CD* is the ratio of ten to three.

So the analysis has resulted in showing that the ratios of the required numbers one to another are known ratios and that it is possible for them to exist.

The synthesis of this problem is carried out as follows: we take a number that has a quarter and an eighth, whatever this number is, let it be the number *AB*; we add a quarter of itself to it, so that it becomes the number *CD* and we also take a quarter of it and an eighth of it, and let <their sum> be *EG*.

I say that the number CD *is the first number we require, that* EG *is the second number and* AB *is the third number.*

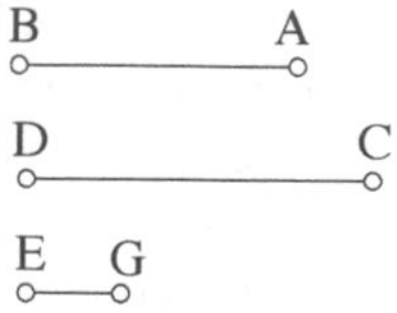

B   A
o————————————o

D   C
o————————————————o

E   G
o————o

Fig. II.1.31

Proof: The number *CD* has ten parts, the number *EG* has three parts and the number *AB* has eight parts. If to the second, which is three parts, we add half of the first, which is five parts, this makes eight parts and there remain five parts of the first. If to the third, which is eight parts, we add a third of the second, which is one single part, the third will be nine parts, and of the second there remain seven parts. If <to what remains of the> first, which is five parts, we add a quarter of the third, which is two parts, the first becomes seven parts, but the third has become seven parts. The numbers become equal after the additions. This is what we wanted to prove.

The synthesis indicates that this problem is indeterminate, since it works for any number that has an eighth. This is what we wanted to prove.

<19> As an example, our statement: to divide two known numbers in three ratios that are equal to known ratios.

Let the two numbers be *AB* and *CD*; the given ratios are the ratio of *E* to *G*, the ratio of *H* to *I* and the ratio of *K* to *L*. Let the greatest ratio be the ratio of *E* to *G* and the smallest ratio be the ratio of *K* to *L*.

The method for the analysis for this problem will consist of supposing that the two numbers have been divided in these ratios; we then examine the properties of these two numbers once they have been divided. Let the two numbers be divided at the points *N*, *M*, *P*, *Q*. Let the ratio of *AM* to *CP* be equal to the ratio of *E* to *G*, let the ratio of *MN* to *PQ* be equal to the ratio of *H* to *I* and let the ratio of *NB* to *QD* be equal to the ratio of *K* to *L*. If we examine the properties of these two numbers after they have been divided, we find that the ratio of *AN* to *CQ* is smaller than the ratio of *E* to *G* and greater than the ratio of *H* to *I*; we find that the ratio of *MB* to *PD* is smaller than the ratio of *H* to *I* and greater than the ratio of *K* to *L*, and we find that the ratio of the sum of *AM* and *MB* to the sum of *CP* and *PD* is smaller than the ratio of *E* to *G* and greater than the ratio of *K* to *L* – this result has in fact been proved in Proposition 6 of this treatise. If the ratio of

AN to *CQ* is greater than the ratio of *H* to *I*, the ratio of *NB* to *QD* is equal to the ratio of *K* to *L* and the ratio of *K* to *L* is smaller than the ratio of *H* to *I*, then the ratio of *AN* to *CQ* is greater than the ratio of *NB* to *QD*. So if the ratio of *AN* to *CQ* is greater than the ratio of *NB* to *QD*, then the ratio of *AB* to *CD* is greater than the ratio of *NB* to *QD*, so it is greater than the ratio of *K* to *L*. But since the ratio of *AN* to *CQ* is smaller than the ratio of *E* to *G* and the ratio of *NB* to *QD* is equal to the ratio of *K* to *L* which is smaller than the ratio of *E* to *G*, each of the ratios of *AN* to *CQ* and of *NB* to *QD* is smaller than the ratio of *E* to *G*, accordingly the ratio of *AB* to *CD* is smaller than the ratio of *E* to *G*. So the ratio of *AB* to *CD* is smaller than the ratio of *E* to *G* and greater than the ratio of *K* to *L*. But the ratio of *H* to *I* is smaller than the ratio of *E* to *G* and is greater than the ratio of *K* to *L*, because the ratio of *E* to *G* is the greatest of the three ratios and the ratio of *K* to *L* is the smallest of the three ratios. If this is so, the ratio of *AB* to *CD* can be equal to the ratio of *H* to *I* or can be higher than it or can be lower than it.

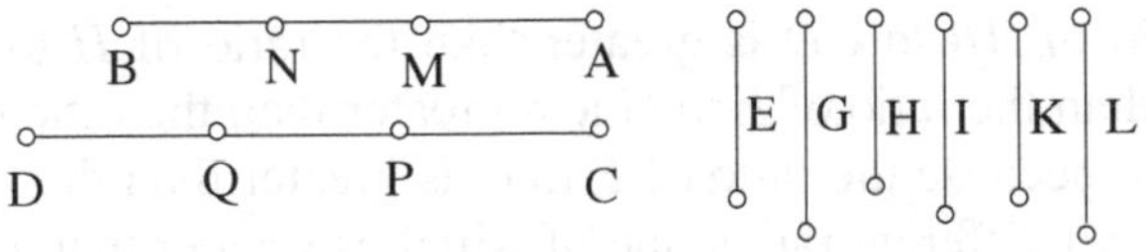

Fig. II.1.32

If the ratio of *AB* to *CD* is equal to the ratio of *H* to *I*, then the ratio of *MN* to *PQ* is equal to the ratio of *AB* to *CD* and is equal to the ratio of the remainder to the remainder, so the ratio of the sum of *AM* and *NB* to the sum of *CP* and *QD* is equal to the ratio of *AB* to *CD* and is equal to the ratio of *H* to *I*.

If the ratio of *AB* to *CD* is greater than the ratio of *H* to *I*, which is the ratio of *MN* to *PQ*, then the ratio of the sum of *AM* and *NB* to the sum of *CP* and *QD* is greater than the ratio of *H* to *I*, while nevertheless being smaller than the ratio of *E* to *G* and greater than the ratio of *K* to *L*.

And if the ratio of *AB* to *CD* is smaller than the ratio of *H* to *I*, then the ratio of the sum of *AM* and *NB* to the sum of *CP* and *QD* is smaller than the ratio of *H* to *I*, while nevertheless being greater than the ratio of *K* to *L*.

So the analysis has established that the ratio of *AB* to *CD* is smaller than the ratio of *E* to *G* and greater than the ratio of *K* to *L*, and either it is equal to the ratio of *H* to *I*, or it is higher than it, or it is lower than it. If the ratio of *AB* to *CD* is equal to the ratio of *H* to *I*, then the ratio of the sum of *AM* and *NB* to the sum of *CP* and *QD* is also equal to the ratio of *H* to *I*. If the ratio of *AB* to *CD* is greater than the ratio of *H* to *I*, then the ratio of the sum of *AM* and *NB* to the sum of *CP* and *QD* is greater than the ratio of *H* to *I*, while nevertheless being smaller than the ratio of *E* to *G*. And if the ratio of *AB* to *CD* is smaller than the ratio of *H* to *I*, the ratio of the sum of *AM* and *NB* to the sum of *CP* and *QD* is smaller than the ratio of *H* to *I*, while nevertheless being greater than the ratio of *K* to *L*.

The synthesis of this problem is carried out as follows: if the ratio of *AB* to *CD* is equal to the ratio of *H* to *I*, from *AB* and *CD* we cut off two numbers such that the ratio of one to the other is equal to the ratio of *H* to *I*; this is possible, albeit indeterminate. Let the two numbers be *AM* and *CP*; it remains that the ratio of *BM* to *PD* is equal to the ratio of *AB* to *CD* which is the ratio of *H* to *I*. But the ratio of *H* to *I* is smaller than the ratio of *E* to *G* and greater than the ratio of *K* to *L*. We divide the two numbers *MB* and *PD* in two ratios equal to the ratios of *E* to *G* and of *K* to *L*, as we have shown in Proposition 6 of this treatise. Let the ratio of *MN* to *PQ* be equal to the ratio of *E* to *G* and the ratio of *NB* to *QD* be equal to the ratio of *K* to *L*, then the two numbers *AB* and *CD* have been divided in the three given ratios.

If the ratio of *AB* to *CD* is greater than the ratio of *H* to *I*, we take a ratio smaller than the ratio of *E* to *G* and greater than the ratio of *AB* to *CD*; this is possible because the ratio of *E* to *G* is greater than the ratio of *AB* to *CD*. And for two different ratios one of which is greater than the other, it is possible to find a third ratio smaller than the greater one and greater than the smaller one; we shall prove that result once this proposition is completed.

Let the ratio of *S* to *O* be smaller than the ratio of *E* to *G* and greater than the ratio of *AB* to *CD*. So the ratio of *S* to *O* is greater than the ratio of *H* to *I* and we have that the ratio of *AB* to *CD* is smaller than the ratio of *S* to *O* and greater than the ratio of *H* to *I*. We divide *AB* and *CD* in two ratios equal to the ratios of *S* to *O* and of *H* to *I*, as has been shown in Proposition 6 of this treatise. Let the ratio of *AN* to *CQ* be equal to the ratio of *S* to *O* and let the ratio of *NB* to *QD* be equal to the ratio of *H* to *I*, then the ratio of *AN* to *CQ* is smaller than the ratio of *E* to *G* and greater than the ratio of *K* to *L*. We divide the two numbers *AN* and *CQ* in two ratios equal to the ratios of *E* to *G* and of *K* to *L*. Let the ratio of *AM* to *CP* be equal to the ratio of *E* to *G* and let the ratio of *MN* to *PQ* be equal to the ratio of *K* to *L*. So the two numbers *AB* and *CD* have been divided in the three given ratios.

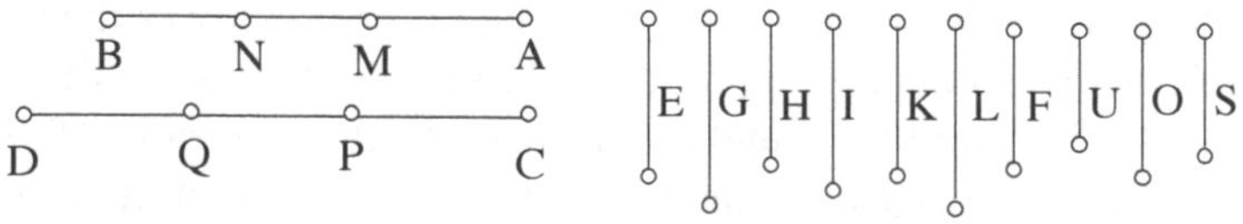

Fig. II.1.33

If the ratio of AB to CD is smaller than the ratio of H to I, we suppose a ratio smaller than the ratio of AB to CD and greater than the ratio of K to L. Let the ratio of U to F be smaller than the ratio of AB to CD and greater than the ratio of K to L. So the ratio of AB to CD is smaller than the ratio of H to I and greater than the ratio of U to F. We divide AB and CD in two ratios equal to the ratios of U to F and of H to I. Let the ratio of AN to CQ be equal to the ratio of U to F and let the ratio of NB to QD be equal to the ratio of H to I; but the ratio of U to F is smaller than the ratio of AB to CD and the ratio of AB to CD is smaller than the ratio of E to G, so the ratio of U to F is smaller than the ratio of E to G. But the ratio of U to F is greater than the ratio of K to L, so the ratio of AN to CQ is smaller than the ratio of E to G and greater than the ratio of K to L. We divide AN and CQ in the two ratios equal to the two ratios of E to G and of K to L. Let the ratio of AM to CP be equal to the ratio of E to G and let the ratio of MN to PQ be equal to the ratio of K to L. So the two numbers AB and CD have been divided in the three given ratios.

So we have proved, from all that we have set out, how to divide two numbers AB and CD in the three given ratios. This is what we wanted to do.

We have proved, further, that this problem is indeterminate, that is to say that it can be solved in several ways: if, indeed, the ratio of AB to CD is equal to the ratio of H to I, then, given any part of the number AB, if we make its ratio to a part of the number CD equal to the ratio of H to I, we have a solution to the problem.

If the ratio of AB to CD is greater or smaller than the ratio of H to I, then we need, in our procedure, to find a ratio smaller than one ratio and greater than <another> ratio; now it is possible to find many ratios greater than a single constant ratio and smaller than a single constant ratio. In this way the problem is again indeterminate. In all cases, it is possible for the two numbers to be divided in the three ratios in many ways.

However, this problem is <subject> to a discussion, because it cannot be solved except when the ratio of the two numbers one to the other is smaller than the greatest ratio and greater than the smallest ratio. If we suppose the ratio of the two numbers is not smaller than the greatest ratio and is not greater than the smallest ratio, what follows is impossible. And the impossibility that follows in this proposition is like the impossibility that followed in Proposition 6 of this treatise. The discussion in this proposition is the same as the discussion in Proposition 6.

As for the way to find a ratio smaller than one ratio and greater than [another] ratio, this is done as we shall describe: let the greater of the two ratios be the ratio of AB to CD and let the smaller be the ratio of EG to HI.

We put the ratio of *EG* to *IL* equal to the ratio of *AB* to *CD*, so the ratio of *EG* to *IL* is greater than the ratio of *EG* to *HI*; so *IL* is smaller than *HI*. We take away from *HL*, an arbitrary [number] *HN*; the ratio of *EG* to *IN* will then be smaller than the ratio of *AB* to *CD* and greater than the ratio of *EG* to *HI*.

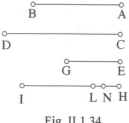

Fig. II.1.34

If the number *HL* is one, we multiply all the numbers by as large a number as we wish until a whole number emerges in place of *HL*. The multiples of the numbers will be in the ratios of the original numbers. Similarly, if one of the numbers given to define the two ratios contains fractions, we multiply all the numbers by the homonymous number of the fraction,[54] then the ratios will be ratios of whole numbers. It is in this way that we can find a ratio smaller than a given ratio and greater than a given ratio. This set of numerical problems is sufficient to provide a basis for practice.

<20> As for geometrical problems, they follow the example in our statement: given a straight line *AB*, on which there are three points that are *A*, *B* and *C*, and a straight line *DG* being known in position and unlimited in length, we wish to draw from the two points *A* and *B* two straight lines that meet one another in a point of *DG* such that if from the point *C* we draw a straight line that passes through this point, it divides the angle formed at this point into two equal parts.

Using the method of analysis we suppose that this has been found and that the straight lines are *AE*, *BE* and *CE*. So the angle *AEC* will be equal to the angle *CEB*. Since the angle *AEB* has been divided into two equal parts by the straight line *EC*, the ratio of *AE* to *EB* is equal to the ratio of *AC* to *CB*.

If *AC* is equal to *CB*, then *CE* is a perpendicular and it has been drawn from the known point *C* to the straight line *AB*. So the straight line *EC* is

---

[54] If one of the given numbers is a sum in which one or more terms are fractions, the homonymous number is the denominator of the fraction we obtain in carrying out the addition.

known in position as has been proved in the twenty-eighth proposition of the *Data*[55] and the straight line *DG* is known in position, by hypothesis. So the two straight lines *DG* and *EC* are known in position and they intersect at the point *E*, so the point *E* is known as has been proved in Proposition 24 of the *Data*.[56]

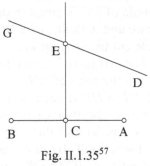

Fig. II.1.35[57]

The synthesis for this case consists of drawing from the point *C* a perpendicular to *AB*, which we extend until it meets the straight line *DG*. From where it meets the straight line *DG* we draw two straight lines to the points *A* and *B* and the problem is completed.

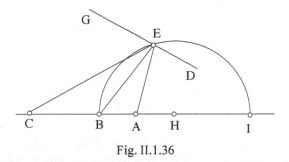

Fig. II.1.36

If the two straight lines *AC* and *CB* are different, let *AC* be the greater, then the ratio of *AE* to *EB* is a known ratio since it is equal to the ratio of *AC* to *CB*, both of which are known, and it is the ratio of the greater to the smaller. So the point *E* lies on the circumference of a circle known in position as has been proved in the first proposition of this treatise. Let the circle be the circle *ECI*; so the circle *ECI* is known in position; but the

---

[55] This is Proposition 29 in Heiberg's edition and in the recension by al-Ṭūsī.

[56] This is Proposition 25 in Heiberg's edition and in the recension by al-Ṭūsī.

[57] In the figures for Problem 20 in the manuscript, we have *DG* ∥ *AB*; but the statement is more general. We have added Figure II.1.35 to fit with the text.

straight line *DG* is known in position, they cut one another at the point *E*, so the point *E* is known as has been proved in Proposition 24 of the *Data*.[58]

The synthesis of this problem is carried out as we shall describe: we put the ratio of *CH* to *HB* equal to the ratio of *AC* to *CB*, thus the ratio of the whole of *AH* to the whole of *HC* is equal to the ratio of *CH* to *HB*. We take *HB*, we take *H* as centre and with distance *HC* we draw a circle, let the circle be *CEI*; let this circle cut the straight line *DG* at the point *E*; we join *AE*, *BE*, *CE* and *HE*; *HE* is equal to *HC*, so the ratio of *AH* to *HE* is equal to the ratio of *AH* to *HC*; but the ratio of *AH* to *HC* is equal to the ratio of *CH* to *HB*, so the ratio of *AH* to *HE* is equal to the ratio of *EH* to *HB*. But the angle *AHE* is common to the two triangles *AHE* and *BHE*, so the triangles *AHE* and *BHE* are similar and the ratio of *AH* to *HE* is equal to the ratio of *AE* to *EB*; but the ratio of *AH* to *HE* is equal to the ratio of *AH* to *HC* and is equal to the ratio of *AC* to *CB*, so the ratio of *AE* to *EB* is equal to the ratio of *AC* to *CB*, so the angle *AEB* has been divided into two equal parts by the straight line *CE*, as has been proved in the sixth book of the work of Euclid. This is what we wanted to do.

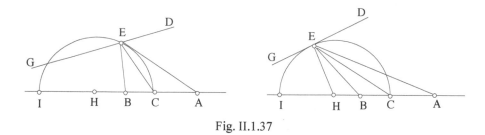

Fig. II.1.37

This problem requires a discussion because the circle *ECI* may not meet the straight line *DG*, and the discussion for this problem is that the straight line *HC* shall not be smaller than the perpendicular drawn from the point *H* to the straight line *DG*, because *HC* is the semidiameter of the circle and the point *H* is the centre of the circle. If the semidiameter of the circle is smaller than the perpendicular drawn from its centre to the straight line *DG*, then the endpoint of the perpendicular will lie outside the circumference of the circle. But the endpoint of this perpendicular is the point of the straight line *DG* that lies closest to the circumference of the circle, so no straight line drawn from the centre of the circle to its circumference reaches the straight line *DG*.

---

[58] This is Proposition 25 in Heiberg's edition and in the recension by al-Ṭūsī.

So if the semidiameter of the circle is smaller than the perpendicular drawn from its centre to the straight line DG, this problem cannot be solved.

If the semidiameter of the circle is equal to the perpendicular drawn from the centre to the straight line DG, then this problem is soluble and there is a single case, as in the example of the first figure. In fact, if the semidiameter of the circle is equal to the perpendicular and if from the centre of the circle, which is the point H, we draw the perpendicular HE, then HE will be the semidiameter of the circle, the straight line DG will be a tangent to the circle at the point E and the circle does not meet the straight line DG in any other point.

If the semidiameter of the circle is greater than the perpendicular, then if from the centre of the circle we draw a perpendicular to the straight line DG that then reaches the circumference of the circle, the straight line DG cuts this straight line which is the semidiameter of the circle and which is perpendicular <to the straight line DG>; so it cuts the circle in two places. If to each of these two points we draw straight lines from the points A, C and B, the angle that is enclosed at this point is divided into two equal parts. The proof relating to each of these two points is the one given earlier.

So if the semidiameter of the circle is greater than the perpendicular drawn from its centre to the straight line DG, then the problem will have two cases involving two different points. If the semidiameter is equal to the perpendicular, then the problem will have a single case. If the semidiameter is smaller than the perpendicular, then the problem has no solution. This is the discussion for this problem.

<21> As an example, our statement: the point A being given, the straight line BC being given in known position and the circle DE being given, we wish to draw from the point A to the circle DE a straight line, inclined at a known angle, so as to reach the straight line BC, in such a way that the ratio of the two straight lines that are generated, the one to the other, is known.

The method of analysis is to suppose that this result has been achieved, that is that we have found the two straight lines AD and DH, that the angle ADH is known and that the ratio of AD to DH in known. We mark the centre of the circle; let it be G. We extend AD and we put DK equal to DH, then the ratio of AD to DK is known; so the point K lies on the circumference of a circle known in position, as has been proved in the second proposition of this treatise. Indeed, we join AG, it will be known in magnitude and in position, because its two endpoints are known, as has been proved in Proposition 25 of the *Data*.[59] We extend *AG* and we put the ratio of *AG* to *GI* equal to the ratio of *AD* to *DK* which is known, so the straight line *GI* will be known in magnitude, as has been proved in Proposition 2 of the *Data*, and it is known in position, so the point *I* is known, as has been proved in Proposition 26 of the *Data*.[60]

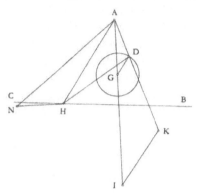

Fig. II.1.38

We join *GD* and *IK*; they will be parallel because the ratio of *AG* to *GI* is equal to the ratio of *AD* to *DK*, so the triangle *AIK* is similar to the triangle *AGD*. So the ratio of *IK* to *GD* is equal to the ratio of *IA* to *AG*; but the ratio of *IA* to *AG* is known, so the ratio of *IK* to *GD* is known. So *IK* is known and the point *I* is known, so the point *K* lies on the circumference of a circle known in magnitude and in position. We join *AH*, the triangle *ADH* is of known shape because the ratio of *AD* to *DH* is known and the angle *ADH* is known, as has been proved in Proposition 39 of the *Data*.[61] The angle *HAK* is known, the ratio of *HA* to *AD* is known and the ratio of *AD* to *DK* is known, so the ratio of *AD* to *AK* is known and the ratio of *HA* to *AK* is known. On the straight line *AI*, at the point *A*, we construct the angle *IAN* equal to the angle *KAH*, which is known, so *AN* is known in position, as has been proved in Proposition 28 of the *Data*.[62] We put the ratio of *NA* to *AI* equal to the ratio of *HA* to *AK*, which is known, so *AN* is of known magnitude, because its ratio to *AI*, which is of known magnitude, is a known ratio, as has been shown in Proposition 2 of the *Data*. We join *NH*. Since the angle *IAN* is equal to the angle *KAH*, the angle *NAH* will be equal to the angle *IAK*; but the ratio of *NA* to *AI* is equal to the ratio of *HA* to *AK*,

---

[59] This is Proposition 26 in Heiberg's edition and in the recension by al-Ṭūsī.

[60] This is Proposition 27 in Heiberg's edition and in the recension by al-Ṭūsī.

[61] This is Proposition 41 in Heiberg's edition and in the recension by al-Ṭūsī.

[62] This is Proposition 29 in Heiberg's edition and in the recension by al-Ṭūsī.

so the ratio of NA to AH is equal to the ratio of IA to AK. So the triangle NAH is similar to the triangle IAK, so their sides are proportional, so the ratio of NH to IK is equal to the ratio of HA to AK; but the ratio of HA to AK is known, so the ratio of NH to IK is known; now IK is of known magnitude, so NH is of known magnitude, as has been proved in Proposition 2 of the *Data*. But the point N is known, because AN is known in magnitude and in position. Accordingly the straight line NH is known in magnitude and the point N is known in position, so the point H lies on the circumference of a circle known in position, as has been proved in Proposition 3 of this treatise.

So the analysis has arrived at a possible result.

It is possible to analyse this problem using another shorter method, which is this: we join the straight line AH, then the triangle AHD will be of known shape, because the ratio of AD to DH is known and the angle ADH is known, so the ratio of AH to AD is known. But the point A is known and the straight line BC is known in position; now from the point A we have drawn a straight line AH that is inclined at a known angle, which is the angle HAD, in such a way that the ratio of AH to AD becomes known, so the point D lies on a straight line known in position, as has been proved in Proposition 3 of this treatise.

The synthesis of this problem starting from the first analysis is carried out in the following way: let A be the given point, BC the given straight line and DE the given circle; the known angle the angle GHI and the known ratio the ratio of K to L. On one of the two sides[63] of the angle we take the point G and we put the ratio of GH to HI equal to the ratio of K to L. We join GI, we extend GH to M and we put HM equal to HI. We mark the centre of the circle, let it be N; we join AN and we extend it; we put the ratio of AN to NP equal to the ratio of GH to HM and we put the ratio of PQ to the semidiameter of the circle equal to the ratio of PA to AN. At the point A, we construct the angle PAJ equal to the angle MGI, we put the ratio of JA to AP equal to the ratio of IG to GM and we put the ratio of JF to PQ equal to the ratio of JA to AP. We take J as centre and with distance JF we draw a circle, let the circle be FO; let this circle cut the straight line BC at the point O; let us join AO.

I say that if we place the straight line AO at an angle equal to the angle GIH, the inclined straight line reaches the circle DE, and that if we join its endpoint and the point A with a straight line, it encloses with it an angle equal to the angle GHI and the ratio of one of the straight lines to the other is equal to the ratio of GH to HI.

---

[63] Lit.: straight lines.

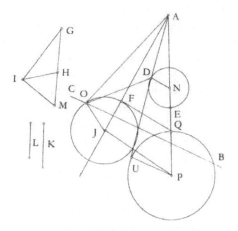

Fig. II.1.39

Proof: We join *JO*, we take *P* as centre and with distance *PQ* we draw a circle, let it be the circle *QU*. We draw from the point *P* a straight line that encloses with the straight line *AP* an angle equal to the angle *AJO*, let the straight line meet the circumference of the circle at the point *U*; we join *AU*. The triangle *APU* is similar to the triangle *JAO*; we draw *ND* parallel to *PU*, so the ratio of *PU* to *ND* is equal to the ratio of *PA* to *AN*. But the ratio of *PA* to *AN* is equal to the ratio of *PU*, which is equal to *PQ*, to the semidiameter of the circle, so the straight line *ND* is the semidiameter of the circle and consequently the point *D* lies on the circumference of the circle.

We join *OD*. Since the triangle *APU* is similar to the triangle *AJO*, the ratio of *JA* to *AP* is equal to the ratio of *OA* to *AU*. But the ratio of *JA* to *AP* is equal to the ratio of *IG* to *GM*, so the ratio of *OA* to *AU* is equal to the ratio of *IG* to *GM*. But the ratio of *UA* to *AD* is equal to the ratio of *MG* to *GH*, because it is equal to the ratio of *PA* to *AN*. So the ratio of *OA* to *AD* is equal to the ratio of *IG* to *GH*. But the angle *OAJ* is equal to the angle *UAP*, so the angle *OAU* is equal to the angle *PAJ*. But the angle *PAJ* is equal to the angle *IGH*, so the angle *OAD* is equal to the angle *IGH* and the ratio of *OA* to *AD* is equal to the ratio of *IG* to *GH*; so the triangle *OAD* is similar to the triangle *IGH*. So the ratio of *AD* to *DO* is equal to the ratio of *GH* to *HI* and the ratio of *GH* to *HI* is equal to the ratio of *K* to *L*, so the ratio of *AD* to *DO* is equal to the ratio of *K* to *L* which is given, and the angle *ADO* is equal to the angle *GHI* which is given. This is what we wanted to do.

This problem requires a discussion and the discussion in this synthesis is like the discussion in the preceding proposition, that is to say that the straight line *JF*, which is the semidiameter of the circle *FO*, should not be smaller than the perpendicular drawn from the known point J to the straight line BC. If it [FO] is equal to the perpendicular, then the problem will have a single case and if it [FO] is greater than the perpendicular, then the problem will have two cases.

The synthesis of this problem starting from the second analysis: let the straight line, the angle and the ratio be given, as they were in the preceding synthesis.

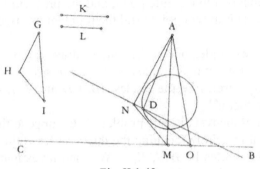

Fig. II.1.40

We put the ratio of GH to HI equal to the ratio of K to L and we join GI. From the point A we draw a perpendicular to the straight line BC, let it be AM. On the straight line AM let us construct an angle MAN equal to the angle IGH. We put the ratio of MA to AN equal to the ratio of IG to GH and from the point N we draw a straight line making a right angle, let it be ND; let us extend it, let it meet the circle at the point D; we join AD and we make the angle DAO equal to the angle NAM. But since the angle DAO is equal to the angle NAM, accordingly the angle NAD is equal to the angle MAO. So the straight line AO meets the straight line BC; let it meet it at the point O. Then the triangle MAO is similar to the triangle NAD. We join OD, then the ratio of MA to AO is equal to the ratio of NA to AD, so the ratio of MA to AN is equal to the ratio of OA to AD. But the ratio of MA to AN is equal to the ratio of IG to GH, so the ratio of OA to AD is equal to the ratio of IG to GH. Now the angle OAD is equal to the angle IGH, so the triangle AOD is similar to the triangle GHI, so the angle ADO is equal to the angle GHI and the ratio of AD to DO is equal to the ratio of GH to HI which is the ratio of K to L. So we have drawn a straight line to the circle, that is the straight line AD, and we have placed it at an angle equal to the angle GHI, which is the angle ADO; the ratio of AD to DO has become equal to the ratio of K to L. This is what we wanted to do.

This synthesis also requires a discussion. The discussion in this synthesis concerns the straight line *ND*. If the straight line *AN* is known in magnitude and in position and a point *N* on it is known and if we have drawn from this point a straight line at a right angle, which is *ND*, then the straight line *ND* is known in position. But the straight line *ND* meets the circle or does not meet it. If the straight line *ND* meets the circle, the problem is soluble, if it does not meet it, then the problem is not soluble. If the straight line *ND* meets the circle, it is a tangent to it or it cuts it. If it is a tangent to it, it meets it in a single point, and the problem has a single case. If it cuts it, it meets it in two points and the problem has two cases.

<22> As an example, our statement: to draw a circle touching three given circles of different magnitudes whose centres do not lie on a straight line. Let the three circles be the circles *AB*, *CD* and *EG*; to draw a circle that is tangent to them.[64]

The method of analysis in this problem is to suppose that this has been achieved and that the circle touching the three circles is the circle *BCE*. Let the centres of the circles be *H*, *I*, *K*, *L*. We then investigate the properties that are necessary in this problem. If the analyst investigates the properties of this problem, it becomes clear to him that any straight line that joins the centres of two of these circles[65] passes through the point of contact, as has been proved in the third book of the work of Euclid.[66] Let us join the centres with the straight lines *HL*, *LK* and *LI*, so they pass through the points *B*, *C* and *E*. We then investigate what is necessary for these straight lines; it emerges that the straight lines *HB*, *EK* and *CI* are each known in magnitude, because these circles are given.[67] But since these circles are of different magnitudes, the differences of these straight lines are known. Let *KE* be the shortest of these straight lines and *CI* the longest. We take away *BM* and *CN*, each equal to *KE*; each of the two straight lines *HM* and *NI* is then known and the straight lines *LK*, *LM* and *LN* are equal. So the points *K*, *M*, *N* lie on the circumference of a circle with centre *L*. We take *L* as centre and with distance *LK* we draw a circle; it passes through the two points *M* and *N*; let the circle be *KMN*, so the point *K* lies on the circumference of

---

[64] Ibn al-Haytham does not specify here that, taken two by two, the given circles lie outside one another, nor that the required circle must touch each of them on the outside. The figures and the argument show that these hypotheses are needed. Moreover, Ibn al-Haytham mentions them in his conclusion.

[65] A straight line joining the required centre to the centre of one of the given circles.

[66] *Elements*, III.12.

[67] They are the radii of the given circles.

the circle *KMN*, and the two points *H* and *I* lie outside it. But since we wish to add something supplementary that will produce properties that did not exist, we join the two straight lines *KH* and *KI*; these two straight lines enclose an angle because, by hypothesis, the three centres are not collinear. But if the two straight lines *HK* and *KI* enclose an angle, then the sum of the two angles *LKH* and *LKI* is less than two right angles, in every case one of these two angles is acute, or both of them are acute.

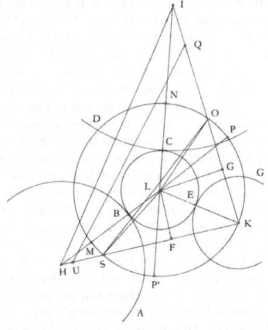

Fig. II.1.41*

First let both of them be acute, then each of the two straight lines *KH* and *KI* cuts the circle *KMN*. Let the straight line *KH* cut the circle at the point *S* and let the straight line *KI* cut the circle at the point *O*. Let us join *SL*, *OL* and *HI* and we draw *HL* [extending it] to meet the circle; let it meet it at the point *P*. So the product of *KH* and *HS* is equal to the product of *PH* and *HM*. So the ratio of *PH* to *HS* is equal to the ratio of *KH* to *HM*. But the ratio of *KH* to *HM* is known, because each of them is known, as has been proved in the first proposition of the *Data*. So the ratio of *PH* to *HS* is known; let it be equal to the ratio of *MH* to *HU*, so the ratio of *MH* to *HU* will be known. Now *MH* is known, so *HU* is known, as has been proved in

---

* The letter ﺟ (transcribed without a diacritical point ﺣ) has been used to designate two different points (G) in the figure.

Proposition 2 of the *Data*. There remains the ratio of *MP*, which is the diameter of the circle, to *US*, [which is] equal to the ratio of *PH* to *HS*, which is known and which is equal to the ratio of *KH* to *HM*; so the ratio of *KH* to *HM* is equal to the ratio of *MH* to *HU*.

In the same way, if we draw the straight line *IL* [extending it] to meet the circle, the product of the whole of this straight line and *IN* is equal to the product of *KI* and *IO*. The ratio of the whole of this straight line to *IO* is equal to the ratio of *IK* to *IN*, which is known. If we put the ratio of *KI* to *IN*, which is known, equal to the ratio of *NI* to *IQ*, then *IQ* will be known and the ratio of the diameter of the circle to *OQ* will be known. But since, by hypothesis, the two points *K* and *H* are known, the straight line *KH* is known in magnitude and in position, as has been proved in Proposition 25 of the *Data*.[68] But since *HU* is of known magnitude and the point *H* on it is known, the point *U* is known, as has been proved in Proposition 26 of the *Data*;[69] so the point *U* is known. In the same way, we prove that the point *Q* is known. We join *UQ*, then *UQ* is known in magnitude and in position and in the triangle *KUQ* each of the sides is known in magnitude and in position; so it will be of known shape, that is to say that its angles are known and the ratios of its sides one to another are known, as has been proved in Proposition 37 of the *Data*.[70] We join *SO*; it will be a chord of the circle *MKN*. Since the angle *SKO* is known, the angle *SLO* is known, because it is double it, the two angles *LSO* and *LOS* are equal and each of them is known. The triangle *LSO* is of known shape, so the ratio of *OS* to *SL* is known and thus the ratio of *OS* to double *SL*, which is the diameter of the circle, is known. The ratio of the straight line *SO* to the diameter of the circle is known and the ratio of each <of the straight lines> *US* and *QO* to the diameter of the circle is known, so the ratio of the straight line *SO* to each of the straight lines *US* and *QO* is a known ratio, as has been proved in Proposition 8 of the *Data*.[71]

So the analysis has led to our drawing in the triangle *UKQ*, which is of known shape, the straight line *SO* in such a way that its ratio to each of the two straight lines *SU* and *OQ* is a known ratio. But the ratio of *US* to *QO* is known, because the ratio of each of them to the diameter of the circle is known, and the ratio of *UK* to *KQ* is known, so the ratio of *UK* to *KQ* either is equal to the ratio of *US* to *QO*, or is not equal to the ratio of *US* to *QO*. If the ratio of *UK* to *KQ* is equal to the ratio of *US* to *QO*, then the

---

[68] This is Proposition 26 in Heiberg's edition and in the recension by al-Ṭūsī.

[69] This is Proposition 27 in Heiberg's edition and in the recension by al-Ṭūsī.

[70] This is Proposition 39 in Heiberg's edition and in the recension by al-Ṭūsī.

[71] This is Proposition 9 in Heiberg's edition.

straight line *SO* will be parallel to the straight line *UQ*, because the ratio of *SU* to *UK* is equal to the ratio of *OQ* to *QK*. If the ratio of *UK* to *KQ* is not equal to the ratio of *US* to *QO*, then the straight line *SO* is not parallel to the straight line *UQ*.

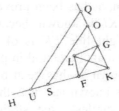

Fig. II.1.41a

If the straight line *SO* is parallel to the straight line *UQ*, the triangle *OSK* is similar to the triangle *QKU*. But the triangle *QKU* is of known shape, as has been proved earlier, so the triangle *OSK* is of known shape. So the ratio of *OS* to *SK* is known. But the ratio of *OS* to *SU* is known, so the ratio of *US* to *SK* is known. But *UK* is of known magnitude, so each of the two straight lines *US* and *SK* is of known magnitude, as has been proved in Proposition 7 of the *Data*.[72] So the straight line *SK* is of known magnitude, and in the same way we prove that the straight line *OK* is of known magnitude; the straight line *OS* will be of known magnitude, because its ratio to *SK* is known. In the triangle *OSK*, each side is of known magnitude and position. But this triangle is inscribed in the circle *MKN*. From the point *L* we draw a perpendicular to the straight line *SK*, let it be *LF*. It divides *SK*, which is of known magnitude, into two equal parts; so the point *F* will be known. From the point *L* we also draw a perpendicular to the straight line *OK*, let it be *LG*; the point *G* is known. We join *FG*; *FG* is known and the triangle *KFG* will be of known shape, because each of its sides is known. The ratio of *GF* to *FK* is known, the angle *KFG* is known and the angle *KFL* is a right angle, so the angle *GFL* is known, because if from a known magnitude we cut off a known magnitude, the remainder will be known, as has been proved in Proposition 4 of the *Data*. In the same way, we prove that the angle *FGL* is known, there remains the angle *FLG* which is known, so the triangle *LFG* is of known shape, as has been

---

[72] *U* and *K* being given, $\frac{US}{SK}$ the known ratio, if $\frac{US}{SK} \neq 1$, defines two points *S* on the straight line *UK*, one on the segment and the other on one of the extensions, and if $\frac{US}{SK} = 1$, a single point *S* that is the midpoint of *UK*. Proposition 7 of the *Data* deals with the point *S* on the segment *UK*, which is the point considered by Ibn al-Haytham.

proved in Proposition 38 of the *Data*.[73] The ratio of *GF* to *FL* is known and the ratio of *FG* to *FK* is known, so the ratio of *KF* to *FL* is known and the angle *KFL* is known, so the triangle *LFK* is of known shape. So the angle *FKL* is known and the straight line *HK* is known in position, so the straight line *KL* is known in position, as has been proved in Proposition 28 of the *Data*.[74] The ratio of *FK* to *KL* is known, because the triangle *FKL* is of known shape. But the straight line *FK* is of known magnitude, so the straight line *KL* is known in magnitude and in position, the point *K* on it is known, so the point *L* is known, as has been proved in Proposition 26 of the *Data*.[75] So the point *L* is known and it is the centre of the circle *BEC* which touches <the given circles>, the straight line *KL* is of known magnitude, the part of it, *KE*, is known because it is the semidiameter of the given circle and the remainder, *EL*, is known and it is the semidiameter of the circle *BEC*. So the circle *BCE* has a semidiameter of known magnitude and its centre is known in position, so the circle *BCE* is known in magnitude and in position. So it can exist, because every magnitude known in magnitude and in position can exist.

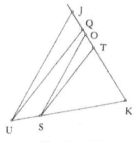

Fig. II.1.41b

If the straight line *SO* is not parallel to the straight line *UQ*, from one of the two points *S* or *O* we draw a straight line parallel to the straight line *UQ*; let it be *ST*. The ratio of *US* to *QT* is known, because it is equal to the ratio of *UK* to *KQ*. But the ratio of *US* to *QO* is known, so the ratio of *OQ* to *QT* is known, as has been proved in Proposition 8 of the *Data*, and the ratio of *QO* to *OT* will be known, as has been proved in Proposition 5 of the *Data*. Now, the ratio of *QO* to *OS* is known, so the ratio of *QO* to each of the two magnitudes *OT* and *OS* is known, so the ratio of *SO* to *OT* is known, as has been proved in Proposition 8 of the *Data*. From the point *U*

---

[73] This is Proposition 40 in Heiberg's edition and in the recension by al-Ṭūsī.

[74] This is Proposition 29 in Heiberg's edition and in the recension by al-Ṭūsī.

[75] This is Proposition 27 in Heiberg's edition and in the recension by al-Ṭūsī.

we draw a straight line parallel to the straight line *SO*, let it be *UJ*. The triangle *UJQ* is thus similar to the triangle *SOT*. So the ratio of *UJ* to *JQ* is equal to the ratio of *SO* to *OT*. But the ratio of *SO* to *OT* is known, so the ratio of *UJ* to *JQ* is known and the angle *UQJ* is known, so the triangle *UJQ* is of known shape, as has been proved in Proposition 41 of the *Data*.[76] So the angle *UJQ* is known and the angle *JUQ* is known; there remains the angle *UJK* which is known.[77] Thus, the angle *JUK* is known, so the triangle *UJK* is of known shape. The ratio of *UK* to *KJ* is known. But the ratio of *UK* to *KJ* is equal to the ratio of *US* to *OJ*, because *UJ* is parallel to *SO*. So the ratio of *US* to *OJ* is known, thus in the triangle *UKJ*, of known shape, we have drawn the straight line *SO* parallel to the straight line *UJ*, in such a way that the ratio of *SO* to each of the two straight lines *SU* and *OJ* is known. The analysis is completed as before, that is to say from the place at which it was assumed that the straight line *SO* was parallel to the straight line *UQ* – which is the base of the triangle of known shape – as far as the place in which it was proved that the circle *BCE* was known in magnitude and in position, which is the point where the analysis is complete.

These two analyses are both based on the fact that the two straight lines *KH* and *KI* cut the circle *KMN*, which happens when each of the two angles *HKL* and *IKL* is smaller than a right angle.

If one of these two angles is not smaller than a right angle, then the other angle is smaller than a right angle. Let the angle *HKL* not be smaller than a right angle, then the angle *LKI* will be smaller than a right angle, so the straight line *IK* will cut the circle *KMN* and the angle *HKL* will be either a right angle or much greater than a right angle.

If the angle *HKL* is a right angle as in the second case of the figure, then the product of *PH* and *HM* is equal to the square of *HK*, so the ratio of *PH* to *HK* is equal to the ratio of *KH* to *HM*. But the ratio of *KH* to *HM* is known, because each of them is known, so the ratio of *PH* to *HK* is known and *HK* is of known magnitude, so the straight line *PH* is of known magnitude. But *HM* is of known magnitude, so the straight line *MP* is of known magnitude and it is the diameter of the circle *KMN*. So the diameter of the circle *KMN* is of known magnitude, so half of it is of known magnitude and the straight line *LK* is thus of known magnitude. But it is

[76] In the manuscripts, the number of the proposition is illegible. Being given the angle *UQJ* and ratio *UJ/JK* does not define a triangle *UJQ* (up to a similarity). We can have 0, 1 or 2 triangles that are solutions to the problem. Proposition 41 of the *Data* treats the case in which an angle and the ratio of the sides enclosing this angle are known.

[77] Depending on the case shown in the figure, the angle *JUK* appears either as the sum or as the difference of known angles.

known in position, because it encloses a right angle with the straight line
HK, which is known in position. So the straight line *KL* is known in
magnitude and in position and the point *K* on it is known, so the point *L* is
known and it is the centre of the circle *BCE*. But the straight line *KL* is of
known magnitude and the straight line *KE* is of known magnitude, so the
straight line *EL* is of known magnitude. Now, we have proved that it is
known in position. So the circle *BCE* that touches <the three circles> is
known in magnitude and in position.

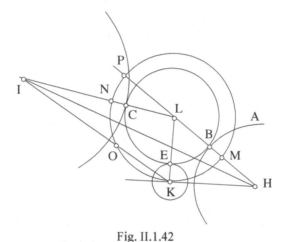

Fig. II.1.42

If the angle *HKL* is greater than a right angle as in the third case of the
figure, then the analysis of this proposition is the same analysis as that we
set out when we supposed that the straight line *SO* was not parallel to the
straight line *UQ*; there is no point of difference between these two analyses.
The analysis of this case of the figure, I mean the third one, has led to the
fact that the circle *BCE* is known in magnitude and in position.[78]

So we prove by this analysis that the required circle that touches the
three given circles is known in magnitude and in position. So it can exist.
To find it, we make use of the lemmas we demonstrated in the analysis and
which have led to the tangent circle being known in magnitude and in
position.

---

[78] Ibn al-Haytham does not mention that in the case where the angle *HKL* = 1 right
angle (Fig. II.1.42), the point *S* coincides with *K* and that in the case where the angle
HKL > 1 right angle (Fig. II.1.43), the point *S* lies on *HK* produced. In the commentary,
we shall see that the argument used to find the straight line *SO* is valid for all cases of
the figure.

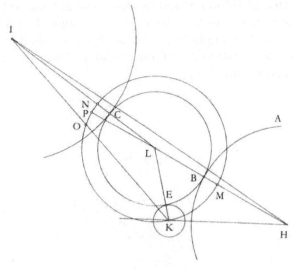

Fig. II.1.43

Among these lemmas arrived at in the analysis, the one that allows us to find the tangent circle is that the triangle *UKQ*, in which the straight line *SO* has been drawn in such a way that the ratio of this straight line to each of the straight lines *SU* and *OQ* is known, is of known shape. It is with the help of this straight line that the problem is completed and it is with the help of these <ratios> that we find the centre of the tangent circle.

The synthesis of this problem is carried out as we shall describe: let the given circles be the circles *AB*, *CD*, *EG* and let the smallest be *EG*; we wish to draw a circle that touches these circles. We mark the centres of these circles; let the points be *H*, *K*, *I*. We join the straight lines *HK*, *KI*, *IH*. Let the straight line *HK* cut the circle *AB* at the point *A* and cut the circle *EG* at the point *E* and let the straight line *KI* cut the circle *CD* at the point *D* and cut the circle *EG* at the point *G*. We take away each of the straight lines *AF* and *DT*, equal to *KE*, and we put the product of *KH* and *HU* equal to the square of *HF*; we put the product of *KI* and *IQ* equal to the square of *IT*. We join *UQ*, then the triangle *UKQ* is of known shape because each of its sides is known in magnitude and in position. We put the arc *EP* equal to the arc *EG* and we join *KP* and *GP*. We put the ratio of the sum of *GK* and *KP* to *PM* equal to the ratio of *KH* to *HF* and we put the ratio of the sum of *PK* and *KG* to *GO* equal to the ratio of *KI* to *IT*. In the triangle *UKQ* we draw a straight line that cuts off from the two straight lines *UK* and *QK* two straight lines such that its ratio to what it cuts off from the straight line *UK* is equal to the ratio of *GP* to *PM* and its ratio to what it cuts off from *QK* is equal to the ratio of *GP* to *GO*; let it be the straight line *SN*. We have shown how to find this straight line using analysis; we shall proceed to the synthesis once the construction of the circle is finished, so that these matters do not become mixed up.

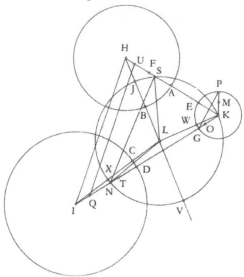

Fig. II.1.44

If in the triangle *UKQ* we draw the straight line *SN* in accordance with the ratio we have mentioned, then the triangle *SKN* will be of known magnitude and each of its sides will be known in magnitude and in position. We draw a circle circumscribed about the triangle *SKN*; let it be the circle *SKN*. The centre of the circle will be known; let it be the point *L*. We join the straight lines *HL*, *KL*, *IL*, *SL* and *NL*. Let the straight line *HL* cut the circle *SKN* at the point *J* and cut the circle *AB* at the point *B*; let the straight line *IL* cut the circle *SKN* at the point *X* and cut the circle *CD* at the point *C* and let the straight line *KL* cut the circle *EG* at the point *W*. The straight lines *LK*, *LJ* and *LX* are equal.

If the angle *HKI* is less than a right angle, then the segment *SKN* is greater than a semicircle, so the straight line *SN* will lie beyond the centre *L*, inside the triangle *UKQ*, as in the first case of the figure; so the angle *SLN* is double the angle *SKN*, so it is equal to the angle *PKG* and thus the triangle *SKN* is similar to the triangle *PKG*. The ratio of the sum of *SL* and *LN* to *SN* is equal to the ratio of the sum of *PK* and *KG* to *GP*. The ratio of *NS* to *SU* is equal to the ratio of *GP* to *PM*, so the ratio of the sum of *SL* and *LN* to *SU* is equal to the ratio of the sum of *GK* and *KP* to *PM* and the ratio of the sum of *PK* and *KG* to *PM* is equal to the ratio of *KH* to *HF*; so the ratio of the sum of *SL* and *LN* to *SU* is equal to the ratio of *KH* to *HF*. We extend *HL* to the point *V*, then *JV* will be the diameter of the circle *SKN*, so it will be equal to the sum of *SL* and *LN*. So the ratio of *JV* to *SU* is equal to the ratio of *KH* to *HF*.

I say first that HJ *is equal to* HF.

Proof: Indeed, it cannot be otherwise.[79] If this were possible, let *HJ* be greater than *HF*. We put the ratio of *HJ* to *HO'* equal to the ratio of *KH* to *HJ*, then *HO'* will be greater than *HU*, because the product of *KH* and *HU* is equal to the square of *HF*; the product of *KH* and *HO'* is equal to the square of *HJ*, but *HJ* is greater than *HF*, so *HO'* is greater than *HU*. But since the product of *VH* and *HJ* is equal to the product of *KH* and *HS*, the ratio of *KH* to *HJ* is equal to the ratio of *VH* to *HS*. But the ratio of *KH* to *HJ* is equal to the ratio of *JH* to *HO'*. So the ratio of *VH* to *HS* is equal to the ratio of *JH* to *HO'*, so the straight line *HO'* is smaller than the straight line *HS*. Now, we have proved that it was greater than the straight line *HU*, so the point *O'* lies between the two points *U* and *S*.

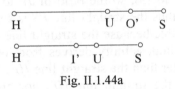

Fig. II.1.44a

In the same way, since the ratio of *VH* to *HS* is equal to the ratio of *JH* to *HO'*, the ratio of *VJ*, the remainder, to *O'S* is equal to the ratio of *VH* to *HS* and is equal to the ratio of *KH* to *HJ*. So the ratio of *JV* to *O'S* is equal to the ratio of *KH* to *HJ*. But the ratio of *KH* to *HJ* is smaller than the ratio of *KH* to *HF*, because *HJ* is greater than *HF*. So the ratio of *JV* to *O'S* is smaller than the ratio of *KH* to *HF*. But the ratio of *KH* to *HF* is equal to the ratio of *JV* to *US*, so the ratio of *JV* to *O'S* is smaller than the ratio of *JV*

[79] To show that $HJ = HF$, we may replace the argument by *reductio ad absurdum* by the following:

    (1) $HF^2 = HK \cdot HU$ by hypothesis,
    (2) $HJ \cdot HV = HS \cdot HK$ (power of *H*),
and we have seen that
    (3) $\dfrac{JV}{SU} = \dfrac{HK}{HF}$.

From (1) et (2), we deduce $HF^2 + HF \cdot JV = HK \cdot HU + HK \cdot SU = HK \cdot HS$; and (2) can be written $HJ^2 + HJ \cdot JV = HK \cdot HS$, hence

$$HF^2 + HF \cdot JV = HJ^2 + HJ \cdot JV \iff (HF - HJ)(HF + HJ + JV) = O,$$

an equation that is satisfied only by $HF = HJ$.

to *US*; so *O'S* is greater than *US*. This is impossible because the point *O'* lies between the points *U* and *S*. Now this impossibility follows from our hypothesis that the straight line *HJ* is greater than the straight line *HF*. So the straight line *HJ* is not greater than the straight line *HF*.

I say that the straight line *HJ* is not smaller than the straight line *HF*. If this were possible, let it be smaller than *HF*. We put the ratio of *HJ* to *HI'* equal to the ratio of *KH* to *HJ*; *HI'* will thus be smaller than *HU* because the product of *KH* and *HU* is equal to the square of *HF* and the product of *KH* and *HI'* is equal to the square of *HJ*. But *HJ* is smaller than *HF*, so *HI'* is smaller than *HU*. But since the product of *VH* and *HJ* is equal to the product of *KH* and *HS*, the ratio of *KH* to *HJ* is equal to the ratio of *VH* to *HS*. But the ratio of *KH* to *HJ* is equal to the ratio of *JH* to *HI'*, so the ratio of *VH* to *HS* is equal to the ratio of *JH* to *HI'* and is equal to the ratio of the remainder, which is *JV*, to the remainder, which is *I'S*. So the ratio of *JV* to *I'S* is equal to the ratio of *KH* to *HJ* and the ratio of *KH* to *HJ* is greater than the ratio of *KH* to *HF*, because *HJ* is smaller than *HF*; so the ratio of *JV* to *I'S* is greater than the ratio of *KH* to *HF*. But the ratio of *KH* to *HF* is equal to the ratio of *JV* to *US*, so the ratio of *JV* to *I'S* is greater than the ratio of *JV* to *US*, and thus the straight line *I'S* is smaller than the straight line *US*. This is impossible because the straight line *HI'* is smaller than the straight line *HU*. This impossibility derives from our hypothesis that the straight line *HJ* is smaller than the straight line *HF*. So the straight line *HJ* is neither smaller than the straight line *HF*, nor greater than it, thus the straight line *HJ* is equal to the straight line *HF*. But *HB* is equal to *HA*, there remain *JB* equal to *FA* and *FA* equal to *KE*, that is *WK*, so the straight line *JB* is equal to the straight line *KW*; now *JL* is equal to *LK*, there remains *BL* equal to *WL*. By an analogous method, we prove that the straight line *IX* is equal to the straight line *IT* and that the straight line *XC* is equal to the straight line *KW*, there remains *CL* equal to *WL*. So the straight lines *LB*, *LW* and *LC* are all three equal. We take *L* as centre and with distance *LB* we draw a circle, let it be the circle *BCW*; this circle touches the three circles because it meets each of these circles in a point of the straight line that joins its centre to the centre of each of these circles. In fact, if from the point *B* we draw a perpendicular to the straight line *HL*, it touches the circle *AB*. It touches the circle *AB* and it touches the circle *BCW*. So the circle *BCW* touches the circle *AB* at the point *B*. In the same way, we prove that it touches the circle *CD* at the point *C* and that it touches the circle *EG* at the point *W*. So the circle *BCW* touches the three circles. This is what we wanted to do.

If the angle *HKI* is a right angle, then the straight line *SN* is a diameter of the circle <*SKN*>, as has been proved in the second case of the figure.

The ratio of *NS* to *SU* is equal to the ratio of *GP*, a diameter of the circle *EG*, to *PM*, and the ratio of *SN* to *NQ* is equal to the ratio of *PG* to *GO*. The remainder of the construction is carried out as before.

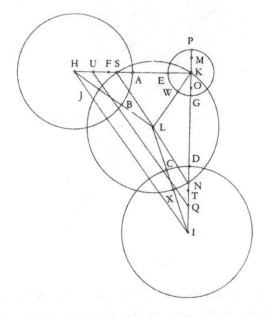

Fig. II.1.45

If the angle *HKI* is greater than a right angle, then the straight line *NS* can lie outside the triangle <*UKQ*>, as in the third case of the figure; it can lie inside the triangle *UKQ* and the centre of the circle will lie outside the triangle *SKN*; the straight line *NS* can itself be the straight line *KQ*, as we shall prove later. The remainder of the proof is carried out as before, that is to say that we shall prove, in both cases of the figure, that the straight line *HJ* is equal to the straight line *HF* and that the straight line *IX* is equal to the straight line *IT*; and this completes the proof.

It remains for us to show how, in the triangle *UKQ* of known shape, to draw a straight line like the straight line *NS*, in such a way that the ratio of *NS* to *SU* is equal to the ratio of *GP* to *PM* and the ratio of *NS* to *NQ* is equal to the ratio of *GP* to *GO*.

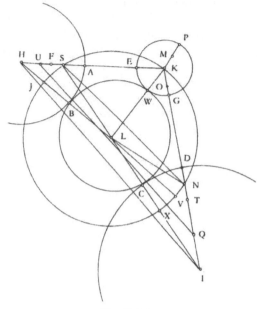

Fig. II.1.46

The analysis for this lemma has been presented in the analysis of the problem; it remains to proceed to the synthesis corresponding to this analysis so that the problem is completed.

We assume the triangle UKQ is given, then we investigate: if the ratio of PM to OG is equal to the ratio of UK to KQ and if the angle QKU is smaller than a right angle, then we put the ratio of PG – which is in the first case of the figure – to GC_a[80] equal to the ratio of QU to UK. We divide the straight line UK at the point S in such a way that the ratio of US to SK is equal to the ratio of PM to GC_a.[81] From the point S we draw the straight line SN parallel to the straight line UQ.

---

[80] We may note that GC_a has been used only as an auxiliary length to define the point S on the segment UK; we could directly consider the ratio $\frac{SU}{SK}$ such that

$$\frac{SU}{SK} = \frac{PM}{PG} \cdot \frac{QU}{UK}.$$

[81] There exist two points S on the straight line UK defined by the ratio $\frac{SU}{SK}$. Ibn al-Haytham considers only the one that lies on the segment UK. But in finding the tangent circle the second point can be used if it lies beyond K on the line UK.

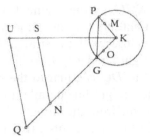

Fig. II.1.47

I say that the ratio of NS *to* SU *is equal to the ratio of* GP *to* PM *and that the ratio of* SN *to* NQ *is equal to the ratio of* PG *to* GO.

Proof: The ratio of *PM* to *GC_a* is a compound of the ratio of *PM* to *PG* and the ratio of *PG* to *GC_a*, but the ratio of *PM* to *GC_a* is equal to the ratio of *US* to *SK*, so the ratio of *US* to *SK* is a compound of the ratio of *PM* to *PG* and the ratio of *PG* to *GC_a*. But the ratio of *PG* to *GC_a* is equal to the ratio of *QU* to *UK*, which is equal to the ratio of *NS* to *SK*, so the ratio of *US* to *SK* is a compound of the ratio of *PM* to *PG* and the ratio of *NS* to *SK*. But the ratio of *US* to *SK* is a compound of the ratio of *US* to *SN* and the ratio of *NS* to *SK*. The ratio compounded of the ratio of *US* to *SN* and the ratio of *SN* to *SK* is equal to the ratio compounded of the ratio of *PM* to *PG* and the ratio of *NS* to *SK*. We eliminate the ratio of *NS* to *SK*, which is common; there remains the ratio of *US* to *SN* equal to the ratio of *PM* to *PG*. So the ratio of *NS* to *SU* is equal to the ratio of *GP* to *PM*. But the ratio of *SU* to *NQ* is equal to the ratio of *UK* to *KQ*. Now the ratio of *PM* to *GO* is equal to the ratio of *UK* to *KQ*, so the ratio of *SU* to *NQ* is equal to the ratio of *PM* to *GO*. By the ratio of equality, the ratio of *SN* to *NQ* is equal to the ratio of *PG* to *GO*. So we have drawn in the triangle *UKQ* the straight line *NS* in such a way that the ratio of *NS* to *SU* is equal to the ratio of *GP* to *PM* and the ratio of *NS* to *NQ* is equal to the ratio of *GP* to *GO*. This is what we wanted to do.

If the ratio of *PM* to *GO* is not equal to the ratio of *UK* to *KQ*, then the ratio of *PM* to *GO* either is greater than the ratio of *UK* to *KQ* or is smaller than it.

If it is smaller than it, then the ratio of *GO* to *PM* is greater than the ratio of *KQ* to *KU*. One of the two ratios of *GO* to *PM* or of *PM* to *GO* is greater than one of the two ratios of *UK* to *KQ* or of *KQ* to *KU*. Let the ratio of *GO* to *PM* be greater than the ratio of *KQ* to *KU*. So we put the ratio of *PM* to *GF* equal to the ratio of *UK* to *KQ*; but the angle *UKQ* is smaller than a right angle, as has been proved in the first case of the figure. On the straight line *PG* we construct a segment of a circle intercepted by an angle equal to the angle *KQU*, let the segment be *PJG*; in it we draw the chord *GJ* equal to the straight line *FO* and we join *FJ*. From the point *U* we draw a straight line that encloses with the straight line *UQ* an angle equal to the angle *GPJ*, let the straight line be *UD*; so it generates the triangle *UQD* such that the ratio of *UD* to *DQ* is equal to the ratio of *PG* to *GJ*. In the triangle *UKD* we draw the straight line parallel to the straight line *UD* such that its ratio to what it cuts off from the straight line *KU* is equal to the ratio of *PG* to *PM* and its ratio to what it cuts off from the straight line *KD* is equal to the ratio of *PG* to *GF*, as we did in the preceding proposition; let the straight line be *SN*.

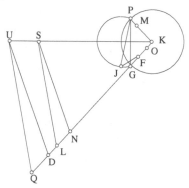

Fig. II.1.48

I say that the ratio of SN *to* NQ *is equal to the ratio of* PG *to* GO.

Proof: From the point *S* we draw the straight line *SL* parallel to the straight line *UQ*, then the triangle *SLN* is similar to the triangle *UDQ*, so the ratio of *SN* to *NL* is equal to the ratio of *UD* to *DQ*. But the ratio of *UD* to *DQ* is equal to the ratio of *PG* to *GJ*, so it is equal to the ratio of *PG* to *FO*, so the ratio of *SN* to *NL* is equal to the ratio of *PG* to *FO*, then the ratio of *LN* to *NS* is equal to the ratio of *FO* to *GP*. But the ratio of *NS* to *SU* is equal to the ratio of *GP* to *PM* and the ratio of *SU* to *LQ* is equal to the ratio of *UK* to *KQ*, which is equal to the ratio of *PM* to *GF*. By the ratio of equality,[82] the ratio of *NL* to *LQ* is equal to the ratio of *OF* to *FG*, so the ratio of *NQ* to *QL* is equal to the ratio of *OG* to *GF* and the ratio of *QN* to *NL* is equal to the ratio of *GO* to *OF*. But the ratio of *LN* to *NS* is equal to the ratio of *OF* to *GP*, so the ratio of *QN* to *NS* is equal to the ratio of *GO* to *GP*. So the ratio of *SN* to *NQ* is equal to the ratio of *PG* to *GO*. But the

---

[82] The product term by term of the three equalities $\frac{LN}{NS} = \frac{OF}{PG}$, $\frac{NS}{SU} = \frac{PG}{PM}$ and $\frac{SU}{LQ} = \frac{OF}{GF}$, and so $\frac{LN}{LQ} = \frac{OF}{GF}$.

ratio of *NS* to *SU* is equal to the ratio of *GP* to *PM*. So in the triangle *UKQ* we have drawn a straight line in accordance with the required ratios.[83] This is what we wanted to do.

This construction is carried out on the basis of <the hypothesis> that the angle *HKI* is smaller than a right angle, as in the first case of the figure.

If the angle *HKI* is a right angle, in the triangle *UKQ* we draw a straight line that cuts off from the straight line *UK* a straight line such that its ratio to the latter is equal to the ratio of *GP*, which is the diameter of the circle *EG*, to *PM*, as in the second case of the figure, and which cuts off from *KQ* a straight line such that its ratio to the latter is equal to the ratio of *PG* to *GO*. We complete the construction as before.

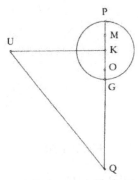

Fig. II.1.49

If the angle *HKI* is greater than a right angle, as in the third case of the figure, then one of the two ratios of *GO* to *PM* or of *PM* to *GO* is smaller than one of the two ratios of *UK* to *KQ* or of *QK* to *KU*. Let the ratio of *GO* to *PM* be smaller than the ratio of *QK* to *KU*. We put the ratio of *GO* to *PF* equal to the ratio of *QK* to *KU*. On the straight line *GP* we construct a segment of a circle intercepted by an angle equal to the angle *QUK*, let the segment of a circle be *PJG*, in which we draw the straight line *PJ* equal to the straight line *MF*, and we join *GJ*. At the point *Q* of the straight line *UQ* we construct an angle equal to the angle *PGJ*; let the angle be *UQC*. This forms a triangle *QKC* and a triangle *QUC*. The triangle *QUC* will be similar to the triangle *PGJ*. So the ratio of *QC* to *CU* will be equal to the ratio of *GP* to *PJ*. In the triangle *QKC* we draw a straight line parallel to the straight line *QC* which cuts off from the straight line *KQ* a straight line such that the ratio of the straight line parallel to it[84] is equal to the ratio of

---

[83] Lit.: the required ratio.
[84] That is, the straight line that has been cut off.

PG to *GO* and which cuts off from the straight line *KU* a straight line such that the ratio of the straight line parallel to it is equal to the ratio of *GP* to *PF*, as we have seen earlier; let the straight line be *NS*.

I say that the ratio of NS *to* SU *is equal to the ratio of* GP *to* PM.

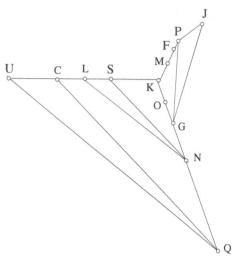

Fig. II.1.50

Proof: We draw *NL* parallel to *QU*; so the triangle *NLS* is similar to the triangle *QUC*. The ratio of *LS* to *SN* is thus equal to the ratio of *UC* to *CQ*. But the ratio of *UC* to *CQ* is equal to the ratio of *JP* to *PG*, so the ratio of *LS* to *SN* is equal to the ratio of *JP* to *PG*, that is of *MF* to *PG*, and the ratio of *LS* to *SN* is equal to the ratio of *MF* to *PG*. But the ratio of *SN* to *NQ* is equal to the ratio of *PG* to *GO* and the ratio of *NQ* to *LU* is equal to the ratio of *QK* to *KU*, which is equal to the ratio of *GO* to *FP*. By the ratio of equality, the ratio of *SL* to *LU* is thus equal to the ratio of *MF* to *FP*, so the ratio of *US* to *SL* is equal to the ratio of *PM* to *MF*. But the ratio of *LS* to *SN* is equal to the ratio of *MF* to *PG*, so the ratio of *US* to *SN* is equal to the ratio of *PM* to *PG*. So the ratio of *NS* to *SU* is equal to the ratio of *GP* to *PM*, and the ratio of *SN* to *NQ* is equal to the ratio of *PG* to *GO*. So we have drawn a straight line *NS* that has the required property. This is what we wanted to do.

But the straight line *QC* can lie outside the triangle *QKU*. It can lie inside the triangle *QKU* and the figure is then as in the case of the figure <that we have given> for the triangle. If *QC* lies outside the triangle, the figure will be as in the third case for the figure.

Now if the straight line QC lies inside the triangle, then the triangle will be similar to the two preceding cases of the figure, because the straight line NS will lie between the point K and the centre L and the circle SKN will cut the straight line UK in a point between the two points U and K. QC can be the straight line QK if the angle PGJ is equal to the angle UQK; it is then that the straight line QK will be divided in a point, such as the point N, so that the ratio of KN to NQ will be equal to the ratio of PG to GO. From the point N we draw the straight line NL parallel to the straight line QU; the ratio of LK to KN is thus equal to the ratio of JP to PG, that is to the ratio of MF to PG. So, by equality, the ratio of LK to NQ is equal to the ratio of MF to GO. But the ratio of NQ to LU is equal to the ratio of QK to KU which is the ratio of GO to FP. So the ratio of KL to LU is equal to the ratio of MF to FP, so the ratio of UK to KL is equal to the ratio of PM to MF. Now the ratio of LK to KN is equal to the ratio of MF to PG, so the ratio of UK to KN is equal to the ratio of MP to PG. So the ratio of NK to KU is equal to the ratio of GP to PM. The straight line NK thus takes the place of the straight line NS and the circle will touch the straight line UK at the point K, as has been proved in the analysis when we divided the investigation[85] of the angle UKQ into three separate cases: an acute <angle>, a right <angle> or an obtuse <angle>.

If the straight line QC lies outside the triangle UKQ and if we take the ratio of US to SK to be compound, which has been explained in detail in the previous proposition, the proof is completed as before.

What we have set out for the triangle UKQ is the entirety of the parts of the investigation of it[86] and the entirety of the cases that can occur for this triangle.

It is in this way that the analysis of this problem and its synthesis are carried out.

But this problem can have several cases. In fact, the circle that touches three circles can touch these three circles with its concave side; it can touch two circles with its concave side and touch only one circle with its convex side; it can touch one circle with its concave side and two circles with its convex side. The analysis and the synthesis then differ. In fact, each of these cases can be analysed in several ways and the arc GJP that we added in the course of the synthesis of the problem, the triangle <SKN> that we determined in this problem and the ratios between these chords that we used do not form part of the body of lemmas that we found by the analysis; however, we added them for the purpose of solving the problem by

---

[85] Lit.: when we have divided the angle.

[86] Lit.: of its parts.

introducing the straight line *NS* in the triangle *UKQ* that resulted from the analysis. We did not carry out an analysis of this result when we arrived at it, because if we had analysed it at that point, the analysis would have been long and difficult, and would have become obscure to most of those who examined it. So in the analysis we stopped at this straight line which we determined only later, by synthesis, trying to make things easier.

All the cases that we have set out derive from <the hypothesis> that the three circles are separate; they can cut one another or touch, one and the same circle can touch them in different positions. We can carry out the analysis of each case in several ways, but our aim is neither to resolve the problem nor to proceed to its solution, our aim is to indicate how the analysis is carried out and to show the method to follow in searching out lemmas that allow us to solve problems. The analysis we have given for this problem and the preceding problems is sufficient for the purpose we set ourselves.

It is here that we end this treatise.

To God the Most High we express our gratitude for the benefits He has bestowed upon us.

II. THE KNOWNS:
A NEW GEOMETRICAL DISCIPLINE

INTRODUCTION

The treatise *The Knowns* is not merely one more text by Ibn al-Haytham. It is a book that its author intended, like his *Analysis and Synthesis*, to be foundational. In such cases, it is not unusual for objectives to multiply and to overlap: one cannot be sure whether what he is doing is a matter of pursuing a line of research already initiated, or of setting a new discipline in place, or of providing new foundations for an established discipline by bringing to completion a contribution that has become classical. All these objectives are interconnected, and while, at first glance, they are different, they in fact prove to be closely linked. It is because of this multiplicity, and not in spite of it, that the book has a place in the history of geometry that is important and unusual in equal measure.

In this treatise, Ibn al-Haytham pursues a line of research initiated a century and a half before, one to which he himself gave powerful impetus and took as far as it will go, that is research concerning motion and trans-formations in geometry: homothety, translation, similarity and even second-degree rational application. Ibn al-Haytham defines the transforma-tions he uses in attacking the different problems that make up the book. In this respect, *The Knowns* belongs to a whole group of writings by Ibn al-Haytham, a group that also includes the texts that are translated, and are the subject of commentaries, in this volume: *The Properties of Circles* and the *Analysis and Synthesis*.

While this research on motion and transformations in geometry does not serve to distinguish *The Knowns* from several other texts, it is a quite different matter when we turn to the second objective of the treatise, which is shared only with the *Analysis and Synthesis*: to invent a new geometrical discipline of 'knowns', whose method is supplied in the book. Two central ideas govern this new discipline: on the one hand, we must no longer think of geometrical objects as being static figures, as they are in Euclid's geo-metry, that is as given once and for all, but rather as figures generated by one or more continuous motions, and thus variable. The problem is then to identify the elements that do not vary in the course of the motion. On the other hand, and this is the second idea, motion must be accepted explicitly not only for use in proofs but also as a legitimate procedure for proof.

This new geometrical discipline imposes new tasks on the geometer. As he starts with figures generated by a motion of some kind he must identify this motion, and in these circumstances must proceed by analysis; and it is analysis that will, further, allow him to search out the elements that are invariant in the course of the motion that generates the figure. But, on the other hand, by starting from definitions of geometrical objects in terms of the motion that generates them – a straight line by rotation about an axis, a circle by rotation of a straight line about a fixed endpoint, and so on – we do not need to bring in additional material to deduce the consequences that follow, notably the properties described in the *Elements*. This approach is obviously a synthetic one. It is in this sense that the *ars analytica* includes the two methods. Surely synthesis is thus also a method of discovery. In its own way, synthesis serves as well as analysis in seeking out the properties that are invariant in the course of the motion that generates a geometrical object, that is an entity perceived by reason. The need for this new discipline becomes clear: it serves to give an account of the geometrical transformations that were then being used more and more; it provides a response to Ibn al-Haytham's new demand for establishing that geometrical objects exist. Through its definitions referring to their generation, this new discipline always provides us with the complete cause of the intellectual entity, and thus of its existence. It is indeed on this account that Ibn al-Haytham makes use of these ideas in, for example, his treatise *The Quadrature of the Circle*.[1] We have already remarked that this geometrical discipline – of *The Knowns* – that Ibn al-Haytham is, as far as I know, the first to have thought of, will re-emerge from the second half of the seventeenth century onwards, under other names and in different places.

The third objective that Ibn al-Haytham sets himself in his treatise *The Knowns* is that of using the new geometrical discipline to provide foundations for Euclid's geometry. It seems that this endeavour forms part of a programme specific to Ibn al-Haytham, one that he has been at pains to put into action in several areas of mathematics, of optics and of astronomy: that of completing what his predecessors have left behind them, either by correcting it, or by providing it with new foundations. There is no lack of examples: the *Conics* of Apollonius, the geometrical constructions of

---

[1] R. Rashed, *Les Mathématiques infinitésimales du IXe au XIe siècle*, vol. II: *Ibn al-Haytham*, London, 1993; English trans. *Ibn al-Haytham and Analytical Mathematics. A History of Arabic Sciences and Mathematics*, vol. 2, Culture and Civilization in the Middle East, London, 2012.

Archimedes,[2] the contributions made by the followers of Archimedes to measuring the paraboloid and the sphere, figures with equal perimeters (isoperimetric), figures with equal areas (isepiphanic) and the solid angle…, the *Optics* of Ptolemy, and so on. This time, he turns his attention to nothing less than Euclid's geometry. To accomplish this final task, Ibn al-Haytham does not work in opposition to Euclid but tries to go further than he did. Thus, the new discipline takes in Euclid's geometry; it justifies it and provides foundations for it, insofar as the new discipline offers means of defining the objects proper to geometry by using the motions that generate them, and at the same time provides geometry with procedures that involve motion and make it possible to prove geometrical theorems. In *The Knowns* Ibn al-Haytham sets out the concepts of this discipline, but it is principally in his *Commentary on the Postulates in the Book by Euclid* and in his *Book for Resolving Doubts in the Book of Euclid*, that he fulfils his purpose of providing foundations for Euclidean geometry. This programme, first proposed by Ibn al-Haytham, and whose significance has not so far been well understood, reappears six centuries later in the writings of Hobbes, but less well understood and in a weaker form.[3]

Seen in this light, the book *The Knowns* belongs to another group of writings by Ibn al-Haytham, which for example includes the two commentaries we have mentioned. So we shall find this treatise central in two respects – for the geometrical work of Ibn al-Haytham and, more generally, for the history of geometry – when we come to study these two commentaries. At the beginning of the second chapter of this volume we have considered the ideas found in this new geometrical discipline. It now remains to examine the geometrical content of *The Knowns*.

---

[2] R. Rashed, *Les Mathématiques infinitésimales du IXe au XIe siècle*, vol. I: *Fondateurs et commentateurs: Banū Mūsā, Thābit ibn Qurra, Ibn Sinān, al-Khāzin, al-Qūhī, Ibn al-Samḥ, Ibn Hūd*, London, 1996; English trans. *Founding Figures and Commentators in Arabic Mathematics*, A History of Arabic Sciences and Mathematics, vol. 1, Culture and Civilization in the Middle East, London, 2012.

[3] On this concept in Hobbes, see *Opera philosophica quae latine scripsit omnia…*, ed. Gulielmi Molesworth; especially his *Elementorum philosophiae sectio prima de corpore*, vol. II, London, 1839, pp. 98–9 and his *Examinatio et emendatio mathematicae hodiernae*, vol. IV, London, 1865, p. 76. See also the commentary by Martial Gueroult, as well as the similarities he establishes between the conception in Hobbes and, following him, that of Spinoza; Martial Gueroult, *Spinoza*, vol. II: *L'âme*, Paris, 1974, pp. 480–7.

MATHEMATICAL COMMENTARY

1. *Properties of position and of form and geometrical transformations*

If we are to believe Ibn al-Haytham, the first part of *The Knowns* included concepts and propositions 'that none of the ancients has set out, and they have not set out anything of this kind'.[4] Exactly what does this novelty that is claimed by an eminent mathematician, who is always rigorous and circumspect, consist of? And in fact, before even setting out on a detailed commentary on this part, we should note that Ibn al-Haytham considers two closely connected areas: sets of points and point by point transformations. In this research, Ibn al-Haytham's principal concern is to identify the non-variable elements of the figure, those of geometrical loci, and those that do vary, those of shape, of position or of magnitude. The majority of the propositions in this part concern properties of position and of shape. Ibn al-Haytham is in fact looking for rectilinear or circular loci that provide answers to problems that associate with each point of a locus already given – a straight line or a circle – a new point, found by a trans-formation of the known locus into the locus we seek. These transformations are displayed explicitly in cases where we have a homothety, a similarity or a translation. In other cases, the transformations, although present, are not identified. In fact, some of them are birational transformations of order 2. Moreover, we need to remember, an essential difference between the two types of transformation we have mentioned above: whereas homotheties, similarities and translations can act on all the points in the plane, the qua-dratic birational transformations employed by Ibn al-Haytham act only from a curve to a curve.

This difference is perhaps the reason why Ibn al-Haytham did not explicitly discuss the second type of transformation, despite its being present in his work.

We may also note that it often happens that Ibn al-Haytham does not enter into any discussion about the existence of solutions and their number. It would be very naïve to suppose that such discussions, which are often easy, were beyond Ibn al-Haytham's capacity; their absence, seen also in other books (for example in the *Completion of the Work on Conics*), is simply an indication of the fact that Ibn al-Haytham did not feel obliged, on every occasion, to enter into a discussion and to carry it through to the end.

Let us take the propositions of this part one by one.

---

[4] See p. 385.

Proposition 1.1. — *Any point* B *situated at a given distance* d *from a fixed point* A *lies on the circle with centre* A *and radius* d.

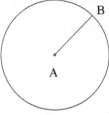

Fig. 2.1.1

Ibn al-Haytham devotes all this part of his work to plane geometry and deliberately concerns himself only with straight lines and circles; he begins by characterising the circle as the locus of a point equidistant from a fixed point.

He stresses the fact that a circle is determinate in position and magnitude when the centre and the radius are known. He distinguishes the invariant elements – a fixed point, a known distance – from the variable elements – positions of the point *B*. Here we see the beginnings of the idea of drawing the figure in a continuous motion; this idea will reappear throughout the text.

The following three propositions are designed to show the characteristics of homothety and similarity.

Proposition 1.2. — *Let there be a given circle* (C, R) *and a given point* B $\in$ (C, R). *The locus of the point* D *such that* $\dfrac{BD}{BC}$ = k, *a given ratio, and that* $C\hat{B}D$ = α, *a given angle, is a circle* (C, r) *concentric [with the given one]. This latter circle is the transform of* (C, R) *in a similarity with centre* C, *ratio* k_1 *and angle* α_1 *which can be deduced from* R, k *and* α.

Ibn al-Haytham shows that the point *D* is the image of the point *B* in this similarity with centre *C*, ratio $k_1 = \dfrac{CD}{CB}$ and angle *BCD* = α_1. The given values k_1, α_1 are known because the triangle *BCD*, in which the angle at *B* and the ratio of the sides enclosing this angle are known, is of known shape.

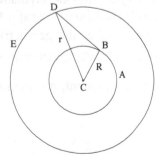

Fig. 2.1.2

Proposition 1.3. — *Let there be a circle* (E, R), *an arbitrary given point* C ≠ E *and an arbitrary point* A ∈ (E, R). *The locus of the point* D ∈ [C, A] *such that* $\dfrac{CA}{AD}$ = k, *a given ratio, is a circle* (G, R₁). *This circle is the transform of the circle* (E, R) *in the homothety* h (C, $\dfrac{k+1}{k}$).

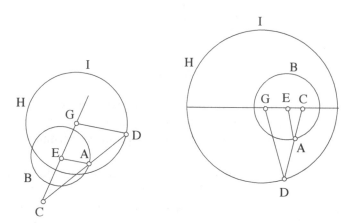

Fig. 2.1.3

Ibn al-Haytham proves the converse and defines the homothety.

Note: Al-Qūhī had studied this problem in his text *Two Geometrical Problems (Mas'alatayn handasiyyatayn)*[5] and it is highly likely that Ibn al-Haytham knew this text by al-Qūhī, as well as many others written by his eminent predecessor. A comparison of the two texts is of interest.

In this third problem of *The Knowns*, Ibn al-Haytham studies the homothetic figure of a circle with given centre *E* and radius *r*. The problem has two parts:

First part: Ibn al-Haytham takes a point *C* that can lie inside or outside the circle and chooses a ratio *k*. A point *A* describes a circle (*E*, r_1). Ibn al-Haytham studies the geometrical locus of the point *D* of the straight line *CA* defined by the given ratio $\dfrac{CA}{AD}$.

In the argument, it is the ratio $k = \dfrac{CD}{CA}$ that is used. Now

---

$$\frac{CD}{CA} = \frac{CA}{AD} + 1,$$

so the ratio $\frac{CD}{CA} = k$ is known.

Ibn al-Haytham draws $DG \parallel EA$ where G lies on CE and deduces

1) $\frac{CG}{CE} = k$, so G is a known point; $CG = k \cdot CE$.

2) $\frac{DG}{EA} = k$, so the length DG is known; $DG = k \cdot r_1$.

So the point D lies on the circle (G, r_2) where $r_2 = k \cdot r_1$.

Second part: Let there be two circles (E, r_1) and (G, r_2) and a point C on the straight line EG such that $\frac{CE}{CG} = \frac{r_1}{r_2}$.

For any straight half line from C that cuts (E, r_1) in A and (G, r_2) in D, we have

$$\frac{DG}{AE} = \frac{r_2}{r_1} = \frac{CE}{CG}.$$

From this we deduce that $DG \parallel AE$; hence

$$\frac{CD}{CA} = \frac{CG}{CE} = \frac{r_2}{r_1}.$$

The ratio $\frac{CD}{CA}$ is the same for any straight half line from C; the same is true for the ratio $\frac{CA}{AD}$.

When the point C lies outside the circle with centre E, the position of the circle with centre G depends on the value of the given ratio. The two circles may cut one another, as in the figure, or touch, or lie entirely outside one another. The argument is the same in all these cases for the figure.

The two parts of Ibn al-Haytham's proposition correspond to the first two propositions of the treatise by al-Qūhī, but in reverse order. However, whereas Ibn al-Haytham presents a single argument in regard to the given point C, whether it lies inside or outside the given circle, in his first two propositions al-Qūhī supposes that C lies inside the given circle; the homothetic circles then lie one inside the other. In a third proposition in his text, he considers the case in which the given point lies outside the given circle. The proof is, however, unchanged. The order of the argument presented by Ibn al-Haytham is more 'natural' and his proof is more concise than those of al-Qūhī. On his proof, Ibn al-Haytham does not make use of the endpoints of the diameters on the straight line *CE*, whereas al-Qūhī considered them in his first proposition in order to introduce the second circle. The reasoning is slightly different. Ibn al-Haytham draws *DG ∥ EA* and from this deduces the equalities between ratios that lead to the conclusion, whereas al-Qūhī starts from an equality between ratios from which he deduces that the two straight lines are parallel and there are other equalities between ratios, and he reaches the conclusion in the same way.

Finally, we may note that in the case where the centre of homothety lies outside the given circle, al-Qūhī proves that the tangent drawn from the centre of homothety to the first circle is also a tangent to the second circle. Ibn al-Haytham, for his part, studies the problem of the common tangents in a general manner in Proposition 2.24 of *The Knowns*. The point of intersection of a common tangent with the straight line through the centres is a centre of homothety.

Proposition 1.4. — *Let there be a given circle* (C, R), *an arbitrary given point* D ≠ C *and a variable point* E *lying on the circle* (C, R). *The locus of a point* G *such that* $\dfrac{DE}{EG} = k$, *a given ratio and that* $D\hat{E}G = \alpha$, *a given angle, is a circle. This last is the transform of the circle* (C, R) *in a similarity with centre* D, *with ratio* k_1 *and angle* β *which can be deduced from the data for the problem.*

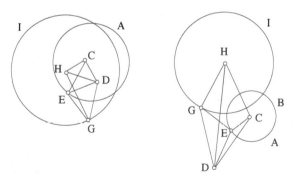

Fig. 2.1.4

Ibn al-Haytham proves the converse proposition and thus defines the similarity.

Proposition 1.5. — *Let there be a given straight line* BC, *a given arbitrary point* A, A $\notin$ BC, *and a variable point* D *on* BC. *The locus of the point* E *such that* $\dfrac{DA}{DE}$ = k, *a given ratio, and* A$\hat{D}$E = α, *a given angle, is a straight line. That straight line is the transform of the straight line* BC *in a similarity with centre* G *such that* AG $\perp$ BC, *with ratio* k_1 *and angle* α. *The ratio* k_1 *is deduced from what is given.*

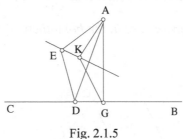

Fig. 2.1.5

Proposition 1.6. — *Given two points* A *and* B *and an angle* α, *the locus of the point* C *such that* A$\hat{C}$B = α *in a half-plane defined by the straight line* AB *is an arc of a circle. This arc is called the subtending arc that contains the angle* α.

Let C be a point that has the properties in the problem and let D be the centre of the circle circumscribed about the triangle ABC, then $A\hat{D}B = 2\alpha$, $D\hat{A}B = D\hat{B}A = \dfrac{\pi}{2} - \alpha$, $DA = \dfrac{AB}{2\sin\alpha}$. The point D and the length DA are determined by what is given. So $C \in \mathscr{C}(D, DA)$.

Fig. 2.1.6a

Notes:

1) The straight line AB cuts the circle into two arcs:

$C \in$ (I) $\Rightarrow$ A$\hat{C}$B = α,

$C \in$ (II) $\Rightarrow$ A$\hat{C}$B = $\pi - \alpha$;

only the first arc is a solution.

2) If a point C has the properties required in the problem, the point symmetrical to it with respect to AB is a solution. The point C lies on the arc AEB or on the arc AE_1B.

3) Conversely, any point C on the arc AEB or on the arc AE_1B satisfies A$\hat{C}$B = α.

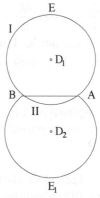

Fig. 2.1.6b

4) This proposition is a preparation for the following proposition, in which we investigate homotheties of a circle in a homothety whose centre is the centre of that circle.

We may note that in this property Ibn al-Haytham makes use of the relation between the angle at the centre and the inscribed angle.

Proposition 1.7. — *Given the subtending arc obtained in Proposition 1.6, the locus of the point* $D \in [AC]$ *such that* $\dfrac{AC}{CD} = k$, *a given ratio, is the arc homothetic to the subtending arc in the homothety* $h\left(A, \dfrac{k+1}{k}\right)$.

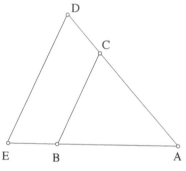

Fig. 2.1.7

Note: The point C lies on the subtending arc of the angle α constructed on AB, and $\dfrac{AC}{CD} = k \Rightarrow \dfrac{AD}{AC} = \dfrac{k+1}{k} = k_1$, so D is the image of C in the homothety $h(A, k_1)$. So D lies on the subtending arc of the angle α constructed on the segment AE such that E is the image of B in the homothety $h(A, k_1)$.

On the extension of AB Ibn al-Haytham constructs the point E such that $\dfrac{AB}{BE} = k$ (so $\dfrac{AE}{AB} = k_1$) and, with the help of homothetic triangles, he proves that the angle $ADE = \alpha$; so he has reduced this to the previous proposition, that is to considering the subtending arc.

Proposition 1.8. — *The locus of a point equidistant from two given points* A *and* B *is the perpendicular bisector of* AB.

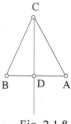

Fig. 2.1.8

Here Ibn al-Haytham is using the third case for the equality of angles in triangles, *Elements* I.8.

Proposition 1.9. — *Let* A *and* B *be two given points and* k $\neq$ 1 *a given ratio. The locus of a point* C *such that* $\dfrac{CA}{CB}$ = k *is a circle the endpoints of whose diameter are the points that divide* AB *in the given ratio* k *and thus form a harmonic range with* A *and* B.

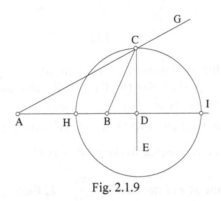

Fig. 2.1.9

Ibn al-Haytham also proves the converse: any point of the circle we have found gives a solution to the problem.

The circular locus that we find is usually called Apollonius' circle.

Proposition 1.10. — *Let* A *and* B *be two given points such that the segment* AB = *l, and let* S *be a given area. The locus of a point* C *such that the area* (ABC) = S *is formed by two straight lines parallel to* AB *whose distance from* AB *is equal to* $\dfrac{2S}{l}$.

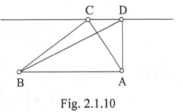

Fig. 2.1.10

Proposition 1.11. — *Let there be two equal circles with centres* E *and* G, *then the circle with centre* G *is the transform of the circle with centre* E *under the translation* EG, *from* E *towards* G, *let it be* T($\overrightarrow{EG}$).

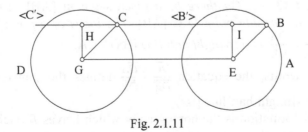

Fig. 2.1.11

Proposition 1.12. — The transform of a circle under a translation is an equal circle.

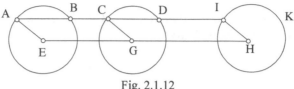

Fig. 2.1.12

Let us return to Ibn al-Haytham's formulation:

Let there be two equal circles (E, R), (G, R) *and an arbitrary straight line parallel to* EG *that cuts the circles in* A *and* C *respectively, such that* AC = EG; *if a point* I *on the extension of* AC *satisfies* $\dfrac{AC}{CI}$ = k, *a given ratio, then* I *lies on a circle equal to the given circles.*

Let H be the point of EG defined by $\dfrac{EG}{GH}$ = k, then H is a known point. We have $AC = EG$ and consequently $CI = GH$; so the quadrilateral ($HICG$) is a parallelogram, hence $HI = GC = R$ and $I \in (H, R)$.

Note: In other words, the hypothesis can be expressed in the form

$$T(\vec{V_2})\ \overrightarrow{CI} = \frac{1}{k}\ \overrightarrow{AC} = \frac{1}{k}\ \overrightarrow{EG} = \vec{V_1}$$

or in the form

$$\overrightarrow{AI} = \left(1+\frac{1}{k}\right)\overrightarrow{AC} = \left(1+\frac{1}{k}\right)\overrightarrow{EG} = \vec{V_2};$$

the point I is derived from C in the translation $T(\vec{V_1})$ or from A in the translation $T(\vec{V_2})$.

Proposition 1.13. — *Let there be a given segment* [AB], *a given point* C ∉ AB *and a variable point* D ∈ [AB]. *The locus of a point* E *on* [CD] *such that* $\dfrac{DC}{DE} = \dfrac{DA}{DB}$ *is a straight half line* [Bx) ∥ CA.

In other words, the equation $\dfrac{\overline{DC}}{\overline{DE}} = \dfrac{\overline{DA}}{\overline{DB}}$ defines the transformation of [BA] into the straight half line [Bx).

This transformation is the homography which leaves B unchanged and transforms A into the point at infinity of CA and the point at infinity of

(*BA*) into the point where (*Bx*) meets the straight line parallel to (*BA*) that passes through *C*.

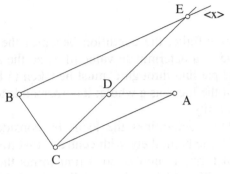

Fig. 2.1.13

Let us use coordinate geometry for this transformation by taking (*BA*) as the axis for the abscissa and (*Bx*) as the axis for the ordinate. The coordinates of the points concerned are: $B(0,0)$; $A(a,0)$; $C(a, c)$; $D(x,0)$; $E(X,Y)$.

The equation of *CD* can be written

$$\frac{X-a}{x-a} = \frac{Y-c}{-c}$$

and the condition

$$\frac{\overline{CD}}{\overline{ED}} = \frac{\overline{AD}}{\overline{DB}}$$

projected onto the axis (*BA*) can be written

$$\frac{x-a}{X-x} = \frac{a-x}{x},$$

that is $X = 0$, which defines the straight line (*Bx*). We then have

$$\frac{Y-c}{c} = \frac{a}{x-a},$$

that is

$$Y = \frac{cx}{x-a},$$

an expression of the homographic transformation. In the case considered by Ibn al-Haytham, since $0 \leq x \leq a$, $y \in [0, \infty[$.

Notes:

1) If x becomes infinite, the condition becomes the identity $-1 = -1$ which does not lead to a determinate value of X; so the straight line $Y = c$ parallel to (BA) and passing through C must be taken to be a singularity as a component part of the locus as a whole; it corresponds to a single point of (BA), the point at infinity.

2) Ibn al-Haytham determines the locus by considering, for a fixed position of D on BA, the homothety with centre D that transforms A into B; this homothety transforms C into E, so it transforms the straight line AC into the straight line BE, which is thus parallel to AC. Since AC is known and B is known, the straight line BE is known. We have a variable homothety, different for each point D.

Let us return to the equation $Y = \dfrac{cx}{x-a}$; if we fix x while leaving c variable, we effectively define a homothety with centre D, and ratio $\dfrac{x}{x-a}$, transforming the straight line CA into BE.

Since Ibn al-Haytham provides no explanations in his short proof, and since the preceding propositions concern homotheties, it is reasonable to suppose that this latter interpretation is the most appropriate one for the text and that Ibn al-Haytham did not think about homographic transformation.

Proposition 1.14. — *Let there be a given segment* [AB], *a given point* C $\notin$ AB *and a variable point* D $\in$ [AB]. *The locus of a point* E $\in$ [C, D) *such that* CD · DE = AD · DB *is an arc of the circle circumscribed about the triangle* ABC.

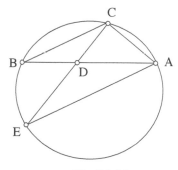

Fig. 2.1.14

Ibn al-Haytham transforms the condition for the problem into a relation of proportions $\dfrac{CD}{DB} = \dfrac{AD}{DE}$, which makes the triangles ADE and CBD similar to one another; thus the angle AEC is equal to the known angle CBD and E lies on the corresponding subtending arc.

We may note that with each point D of AB there is associated a point E of the circle that is the locus; thus we define a correspondence between the straight line AB and this circle. The converse is also true: with each point E of the subtending arc, we can associate the point D in which CE cuts AB; the triangles AED and CBD remain similar because they have two corresponding equal angles, so the equation $CD \cdot DE = AD \cdot DB$ is satisfied.

Let us use coordinate geometry for the relationship between AB and the circle. We take as axes the straight line AB and the perpendicular to AB that passes through C; the coordinates of the points we are concerned with can be written: $A(a, 0)$; $B(b, 0)$; $C(0, c)$; $D(x, 0)$; $E\,(X, Y)$. We have $\dfrac{X}{x} + \dfrac{Y}{c} = 1$ because C, D, E are collinear and the condition for the problem can be written:

$$(x^2 + c^2)\,[(X - x)^2 + Y^2] = (a - x)^2\,(x - b)^2.$$

We have

$$Y = \frac{c}{x}(x - X),$$

so

$$(X - x)^2\,(x^2 + c^2)^2 = x^2\,(a - x)^2\,(x - b)^2,$$

which gives

$$X = x \pm \frac{x(a - x)(x - b)}{x^2 + c^2} = \begin{cases} x\dfrac{(a + b)x + c^2 - ab}{x^2 + c^2} \\[2mm] x\dfrac{2x^2 - (a + b)x + c^2 + ab}{x^2 + c^2}. \end{cases}$$

We next have

$$Y = \mp \frac{c(a - x)(x - b)}{x^2 + c^2}.$$

We note that this transformation is a rational application of degree 2 or 3 depending on which sign we choose. The case considered by Ibn al-Haytham corresponds to the upper sign.

Moreover,

$$x = \frac{cX}{c-Y}, \; X - x = -\frac{XY}{c-Y}$$

and

$$(X - x)^2 + Y^2 = Y^2 \frac{X^2 + (c-y)^2}{(c-Y)^2}, \; x^2 + c^2 = c^2 \frac{X^2 + (c-Y)^2}{(c-Y)^2}.$$

The condition for the problem thus reduces to

$$cY \, [X^2 + (c-Y)^2] = \pm (ac - aY - cX)(cX - bc + bY).$$

In the case considered by Ibn al-Haytham, the first member is negative $(c > 0, \; Y < 0)$ while the product in the second member has the sign of $(a - x)(x - b)$, which is positive; so we must choose the lower sign.

When we make $Y = c$ in the equation, we note that the equation is satisfied identically; so we can take $Y - c$ as a factor. We write the equation as:

$$cY(Y-c)^2 + cX^2Y \quad = [a\,(Y-c) + cX]\,[cX + b\,(Y-c)]$$
$$= (Y-c)\,[(a+b)\,cX + ab\,(Y-c)] + c^2X^2;$$

or

$$(Y-c)\,[cY\,(Y-c) + cX^2 - (a+b)\,cX - ab\,(Y-c)] = 0,$$

$$c\,(Y-c)\,[X^2 + Y^2 - (a+b)X - \frac{ab + c^2}{c}Y + ab] = 0.$$

The first factor $Y - c$ corresponds to the line parallel to AB that passes through C; this straight line is a singularity in the locus, the image of the unique point at infinity of the straight line AB. The second factor gives the equation of the circle circumscribed about the triangle ABC.

In the transformation in which AB becomes this circle, the points A and B are fixed and the point at infinity of the line AB becomes the point $(a + b, c)$ where the line parallel to AB that passes through C meets the circle.

The upper sign would have given a cubic curve that does not correspond to any of the cases considered in this treatise.

If we extend the construction by taking a point D that does not lie on the straight line AB, we find an irrational transformation of the plane into itself.

Propositions 1.15 and 1.16. — *Let there be a given circle* 𝒞 *and two given points* A *and* B *lying outside* 𝒞. *If two straight lines from* A *and* B *cut one another in a point* H *inside* 𝒞 *such that* $\overline{HA} \cdot \overline{HG} = \overline{HB} \cdot \overline{HE}$, *then* CD ∥ AB *and* (A, B, G, E) *are concyclic.*

In fact, in Proposition 1.15, Ibn al-Haytham proves that if A, B, G, E lie on the same circle, then $AB \parallel CD$; in Proposition 1.16, he establishes that if $AB \parallel GE$, then A, B, C, D lie on the same circle.

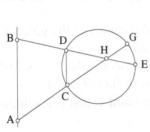

Fig. 2.1.15

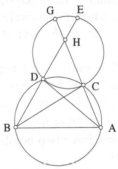

Fig. 2.1.16

Proposition 1.17. — *Let there be two given points* A *and* B *lying outside a given circle; two straight lines from* A *and* B *cut one another in* C *on the circle and cut it again in* D *and* E. *If* $\dfrac{CA}{CD} = \dfrac{CB}{CE}$, *then we either have*

$AD \cdot DC = BE \cdot EC$, *or the ratio* $\dfrac{AD \cdot DC}{BE \cdot EC}$ *is known.*

Let P_A and P_B be the powers of the points A and B with respect to the circle, $P_A > 0, P_B > 0$; we have

$$AC \cdot AD = P_A \text{ and } BC \cdot BE = P_B.$$

Let us write k for the ratio $\dfrac{P_B}{P_A}$; by hypothesis we have

$$\frac{AC}{CB} = \frac{CD}{CE},$$

hence

$$\frac{CD}{CE} = \frac{EB}{k \cdot AD},$$

that is

$$k \cdot AD \cdot CD = CE \cdot EB.$$

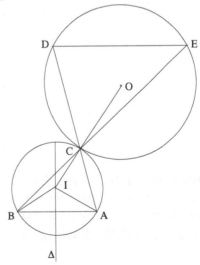

Fig. 2.1.17a

Note: The hypothesis $\dfrac{CA}{CD} = \dfrac{CB}{CE}$ implies $DE \parallel AB$. But if a line parallel to AB cuts the circle in D and E, the point of intersection C of the straight lines AD and BE is not in general a point on the circle. From Proposition 1.16, it can lie inside the circle.

Proposition 1.17 thus appears to be a special case of Proposition 1.16 in which the points D, C, H are coincident.

For a point C to satisfy the conditions in Proposition 1.17, it must be the point of contact of a circle that passes through A and B and is tangent to the given circle. In the text, Ibn al-Haytham supposes we have an external contact.

If I is the centre of the required circle, O the centre of the given circle and R its radius, we have $IO - IA = R$ and $IA = IB$.

If I exists, it is at the intersection of Δ, the perpendicular bisector of AB, and $\mathcal{H}_A$, the branch around A of the hyperbola with foci O and A.

So we can have two points, one point or no point C that is a solution to the problem.

We may also note that in the case where Δ, the perpendicular bisector of AB, passes through the centre of the circle ($P_A = P_B$ and $k = 1$), Δ cuts the circle in C_1 and C_2. The point C_1 is a solution to the problem, whereas the point C_2 is not.

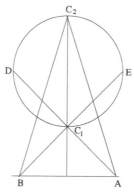

Fig. 2.1.17b

Proposition 1.18. — *Let there be two circles touching internally at* A. *A straight line that passes through an arbitrary point* D *of the smaller circle cuts the greater circle in* B *and* G; *if* D *is a variable point on the smaller circle, then* $\dfrac{DB \cdot DG}{DA^2}$ *is known.*

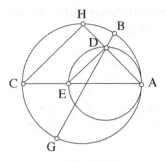

Fig. 2.1.18

The straight line AD cuts the greater circle in H. We have $DB \cdot DG = DA \cdot DH$, hence

$$\frac{DB \cdot DG}{DA^2} = \frac{DA \cdot DH}{DA^2} = \frac{DH}{DA};$$

but

$$\frac{DH}{DA} = \frac{CE}{EA},$$

hence

$$\frac{DB \cdot DG}{DA^2} = \frac{CE}{EA} = \frac{R - r}{r},$$

where R and r are the radii of the greater circle and the smaller circle respectively.

Note: The two circles correspond to one another in the homothety $h\left(A, \frac{R}{r}\right)$, hence

$$\frac{AH}{AD} = \frac{AC}{AE} = \frac{R}{r}$$

and consequently

$$\frac{DH}{AD} = \frac{R - r}{r}.$$

Proposition 1.19. — *Let there be two circles touching internally at* A. *The tangent to the smaller circle at an arbitrary point* D *cuts the greater circle in two points, let* B *be one of them. When* D *varies, the ratio* $\frac{BA}{BD}$ *is constant.*

Let AGC be the common diameter and E the point of intersection of AB and the smaller circle. We have

$$\frac{AE}{EB} = \frac{AG}{GC},$$

hence

$$\frac{AB}{BE} = \frac{AC}{GC} = \frac{R}{R-r} = \frac{AB^2}{AB \cdot BE};$$

but

$$BD^2 = BE \cdot BA,$$

hence

$$\frac{AB^2}{BD^2} = \frac{R}{R-r} \quad \text{and} \quad \frac{BA}{BD} = \sqrt{\frac{R}{R-r}}.$$

If BD cuts the greater circle again in H, we prove in the same way that

$$\frac{HA}{HD} = \sqrt{\frac{R}{R-r}}.$$

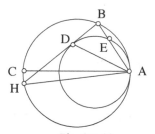

Fig. 2.1.19

So we have

$$\frac{HA}{HD} = \frac{BA}{BD}$$

and consequently

$$\frac{AB + AH}{BD + HD} = \frac{AB + AH}{BH} = \sqrt{\frac{R}{R-r}}.$$

We may note that the two circles correspond to one another in the homothety $h\left(A, \dfrac{R}{r}\right)$.

Note: In Propositions 1.18 and 1.19, we consider a variable point on a given circle and two straight lines that pass through this point, in each case we study a ratio associated with these straight lines and we prove that it is constant and can be expressed in terms of what we are given.

Propositions 1.20 and 1.21 each concern two circles that touch internally. These circles again correspond to one another in a homothety which has as its centre the point of contact and as its ratio the ratio of the radii. These propositions shed light on the reasons for including the preceding propositions. Thus, Proposition 1.20 is a corollary to Proposition 1.19, while Proposition 1.21 is a consequence of Proposition 1.18.

Proposition 1.20. — *Let there be two circles touching internally at the point* A, AGC *their common diameter, the straight line* BDH *touching the smaller circle at* D; *let us join* AD *and let us extend it to* I *on the greater circle. The point* I, *the point of intersection of* AD *and the greater circle, is the mid point of the arc subtended by the tangent at* D.

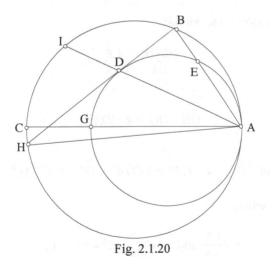

Fig. 2.1.20

From Proposition 1.19, we have

$$\frac{BD}{BA} = \frac{DH}{AH} \text{ or } \frac{DB}{DH} = \frac{AB}{AH},$$

so AD is the bisector of the angle BAH inscribed in the greater circle, so I is the mid point of the arc HB.

Proposition 1.21. — *Let there be two circles touching at* A, AEC *their common diameter; we put* $k = \dfrac{CE}{EA}$ *and* $k_1 = EC \cdot EA$. *If a variable straight line that passes through* E *cuts the smaller circle in* D *and the greater one in* B *and* G, *then*

$$DB \cdot DG + k\, DE^2 = k_1.$$

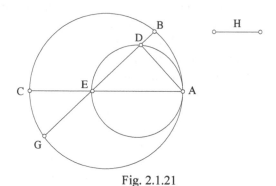

Fig. 2.1.21

From Proposition 1.18, we have

$$\frac{DB \cdot DG}{DA^2} = \frac{CE}{EA} = k,$$

so

$$DB \cdot DG = k \cdot DA^2,$$

hence

$$DB \cdot DG + k \cdot DE^2 = k\,(DA^2 + DE^2) = k \cdot AE^2,$$

hence the result, where

$$k = \frac{CE}{EA} \quad \text{and} \quad k_1 = \frac{CE}{EA} \cdot AE^2 = EC \cdot EA\,;$$

so k and k_1 are known and we can express them in terms of R (the radius of the greater circle) and r (radius of the smaller circle):

$$k = \frac{R-r}{r} \quad \text{and} \quad k_1 = 4r\,(R-r).$$

Propositions 1.22 and 1.23 concern loci of points. In Proposition 1.22, we use a metric property to determine the locus of the point; this proposition serves as a lemma for Proposition 1.23.

Proposition 1.22. — *Let there be a circle with centre* G *and diameter* AC *and two points* E *and* D *on this diameter such that* GE = GD. *For any point* B *lying on the circle, we have*

(*) $BE^2 + BD^2 = AD^2 + DC^2$.

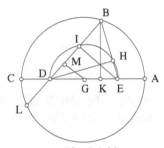

Fig. 2.1.22

In other words, let there be a line segment AC with mid point G, let E and D be two points on this segment such that $GE = GD$; the sum of the squares of the distances from an arbitrary point B on the circle with diameter AC to the points E and D is constant. The converse is true (compare Proposition 1.23); the locus of a point B that satisfies (*) is a circle with centre G and diameter the segment AC.

The proof given by Ibn al-Haytham is valid for any point B on the circle.

Proposition 1.23. — *Let* A *and* B *be two fixed points and* ℓ *a given length. The locus of a point* C *such that* $CA^2 + CB^2 = \ell^2$, *and* $A\hat{C}B$ *is acute, is a circle whose centre is the mid point of* AB *and whose radius is known.*

If the angle C is acute, it is necessary that the given length ℓ satisfies the relation $\ell > AB$ in order for the triangle ABC to exist.

Let us put $\ell^2 - AB^2 = d^2$; let E be a point such that $2EA \cdot EB = d^2$ and let G be such that $AG = EB$. Let us draw the circle with diameter GE and let us prove that it passes through C.

If C does not lie on the circle, the bisector of the angle ACB cuts the circle in H and I, then, from Proposition 1.22, we have

$$HA^2 + HB^2 = AB^2 + 2AE \cdot EB,$$

so

$$HA^2 + HB^2 = CA^2 + CB^2.$$

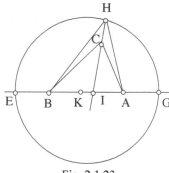

Fig. 2.1.23

But $A\hat{C}I$ being acute implies that $H\hat{C}B$ and $H\hat{C}A$ are obtuse, hence $HB > CB$ and $HA > CA$, hence $HA^2 + HB^2 > CA^2 + CB^2$, which is impossible.

Notes:

1) The argument by *reductio ad absurdum* is valid if the point C is taken to lie inside or outside the circle.

2) From Proposition 1.22, we can calculate the radius of the circle. Let K be the centre, we have seen that

$$CB^2 + CA^2 = GA^2 + GB^2 = 2(GK^2 + KA^2),$$

hence

$$2GK^2 = \ell^2 - 2KA^2 = \ell^2 - \frac{AB^2}{2}.$$

Proposition 1.24. — *Let* AC *be an arbitrary chord in a given circle, if the point* D *of that chord is such that*

(*)   DA · DC = k², *where* k *is given,*

then D *lies on a known circle.*

In other words: the locus of a point D that has a given power k^2 with respect to a given circle (E, R) is part of a concentric circle (E, R'). The radius R' can be found from k and R:

If D lies inside the circle, we have $k^2 = R^2 - R'^2 \Rightarrow R'^2 = R^2 - k^2$.
If D lies outside the circle, we have $k^2 = R'^2 - R^2 \Rightarrow R'^2 = R^2 + k^2$.

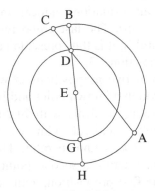

Fig. 2.1.24

Let E be the centre of the circle, ED cuts the circle in B and H, we have

$$DA \cdot DC = DB \cdot DH = EB^2 - ED^2 = k^2,$$

so $ED^2 = R^2 - k^2$. But ED is constant for given k, so D lies on the circle $\left(E, \sqrt{R^2 - k^2} \right)$.

2. Invariant properties of geometrical loci and geometrical transformations

In the second and final part of *The Knowns*, Ibn al-Haytham deals with concepts and propositions, he writes, 'of the same kind as what was set out by Euclid in his book the *Data*, although nothing in this part is to be found in the book the *Data*'.[6] This declaration gives us to understand that in the *Data* Euclid dealt with only a particular class of knowns and that Ibn al-Haytham returns to the study of them to complete Euclid's book with new propositions that Euclid had not thought of. So for Ibn al-Haytham his distant predecessor also considered only a certain class of invariant properties of the figures. What we have is an *a posteriori* interpretation of the presence of this book by Euclid in the 'domain of analysis' as Pappus has set its boundaries in the preamble to the seventh book of his *Mathematical*

---

[6] See p. 410.

Collection. In any case, in this part of his work Ibn al-Haytham investigates certain invariant properties of rectilinear and circular loci.

He employs a multiplicity of methods, some of which are close to those of Euclid. Geometrical transformations appear mainly in the form of homotheties in similar divisions. The last proposition finds the centres of homothety for two given circles; it is thus presented as the procedure that is the converse of that in Proposition 1.3.

He starts with a group of five propositions in which he tries to find a straight line that passes through a known point and has a property *P*. In the first proposition of this group, he deals with finding a straight line that passes through a known point *A* and cuts a known circle between *B* and *C* in such a way that *BA/BC* = *k*, a known ratio. The construction of this straight line reduces to constructing a second point as the intersection of two lines: two circles in the first proposition, a circle and a straight line in the third, two straight lines in the fourth and the fifth. As for the second proposition, it reduces to the first one. We may note that all of these are construction problems, the second one being a *neusis*.

Proposition 2.1. — *From a given point* A *outside a circle, we draw a straight line that cuts the circle in* B *and* C *(where* B *lies between* A *and* C*). If* $\frac{AB}{BC} = k$, *where* k *is a given ratio, then the straight line is known in position.*

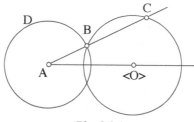

Fig. 2.2.1

So we are concerned with finding a straight line that passes through a known point *A* and cuts a known circle in *B* and *C* such that $\frac{BA}{BC} = k$.

The power of *A* with respect to the circle (*O*, *R*) is known;

$$AB \cdot AC = AO^2 - R^2 = k_1^2, \text{ which is known.}$$

We have

$$\frac{AC}{AB} = \frac{AB + BC}{AB} = 1 + \frac{1}{k} = \frac{AC \cdot AB}{AB^2} = \frac{k_1^2}{AB^2},$$

hence

$$AB^2 = \frac{k_1^2}{1 + \dfrac{1}{k}};$$

the length AB is determined by the data for the problem; let $AB = d$. Consequently $B \in \mathscr{C}(A, d)$, so it lies at the intersection of two circles $\mathscr{C}(O, R)$ and $\mathscr{C}(A, d)$.

Note: Ibn al-Haytham does not prove here that the point B exists. The circles $\mathscr{C}(O, R)$ and $\mathscr{C}(A, d)$ cut one another if and only if

(1) $AO - R < d < AO + R;$

but

$$d^2 = \frac{k}{1 + k} \cdot k_1^2 = \frac{k}{1 + k}\left(AO^2 - R^2\right),$$

hence (1) may be written

$$(AO - R)^2 < \frac{k}{1 + k}\left(AO^2 - R^2\right) < (AO + R)^2,$$

hence

$$(1 + k)(AO - R) < k(AO + R)$$

and

$$k(AO - R) < (1 + k)(AO + R).$$

The second condition is always satisfied, and it remains to satisfy

$$AO < (2k + 1)R.$$

We have

- $AO < (2k + 1)$ R, there exist two straight lines symmetrical with respect to AO that give a solution to the problem.
- $AO = (2k + 1)$ R, there exists a straight line that gives a solution to the problem, that is AO.
- $AO > (2k + 1)$ R, there does not exist any straight line that gives a solution to the problem.

Proposition 2.2. — *From a known point we draw to a circle known in position a straight line that cuts off a known segment from the circle, then the line is known in position.*

This is how Ibn al-Haytham states the second proposition.

We may note that from the definitions in Book III of the *Elements* (Definitions 6, 7, 8 and 11), as well as Proposition 23 of the same book, or the Definitions 7 and 8 and Propositions 88 and 89 of the *Data*, the segment of the circle is known if we know its base and the inscribed angle with its vertex on the arc of the circle that is a boundary of the segment.

In a known circle, a known inscribed angle is associated with a chord of known length. So, to say that a segment of a given circle is known reduces to saying that its base is known. In *Elements* III.34 Euclid gives a construction for this base. Ibn al-Haytham's problem may be rewritten as follows:

From a point A *we draw a straight line that cuts a known circle in* B *and* C; *if the chord* BC *is of known length, then the straight line* BC *is known in position.*

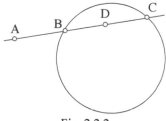

Fig. 2.2.2

The power of the point A with respect to the circle is known, let $AB \cdot AC = k^2$.

Let us put $BC = 2\,BD = 2\,\ell$ (where D is the mid point of BC).

But if A lies outside the circle, we have

$$AB \cdot AC = AD^2 - BD^2 \Rightarrow AD^2 = k^2 + \ell^2.$$

If, on the other hand, A lies inside the circle (a case not considered by Ibn al-Haytham, whose purpose is clearly to get back to the previous proposition), we have $AD^2 = \ell^2 - k^2$.

So the length AD is known. From it we deduce $\dfrac{AD}{DB}$, then $\dfrac{AD-DB}{2DB} = \dfrac{AB}{BC}$ and we have returned to the previous case.

Proposition 2.3. — *Let* A, B, C *be three known points and* D *a point on the segment* BC; *if* $\frac{AD}{DC} = k$, *a given ratio, then the straight line* AD *is known in position.*

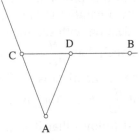

Fig. 2.2.3

By hypothesis we have two known points A and C and $\frac{DA}{DC} = k$, so from Proposition 9, D lies on a circle whose centre is a point of AC. So if D exists, it is at the intersection of this circle and the segment CB. The point D is known, as is the point A; so AD is known.

Proposition 2.4. — *Let there be a known point* A *and two straight half-lines that are parallel and in opposite directions* [CB) *and* [DE). *Let us suppose that a straight line from* A *cuts the straight half-lines in* H *and* G *respectively. If* $\frac{HC}{DG} = k$ *is a known ratio, the straight line* AG *is known.*

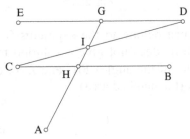

Fig. 2.2.4a

We suppose that A does not lie on the straight line CD. Let I be the point of intersection of AG and DC, we have

$$\frac{IC}{ID} = \frac{HC}{DG} = k \Rightarrow \frac{CD}{ID} = 1 + k,$$

because I lies between C and D.

So the point *I* is known and consequently the straight line *AI* is also known.

Comment: In the statement of this problem, Ibn al-Haytham first makes it clear that *BC* and *DE* are 'two parallel straight lines known in magnitude and in position'. So we are dealing with segments. On the straight lines *CB* and *DE*, the points *H* and *G* such that $\frac{CH}{DG} = k < 0$ correspond to one another in a homothety in which the image of *D* is *C*. So the centre of this homothety is the point *I* such that $\frac{IC}{ID} = k$. So for given *k*, *k* < 0, the point *I* exists and it is unique, $I \in [CD]$. It follows that if $A \notin [CD]$, the straight line *AI* cuts the straight lines *CB* and *DE* respectively in *H* and *G* such that $\frac{CH}{DG} = k$. Note that, in this book, when we write *k* < 0 we always mean that the segments in a ratio have opposite senses.

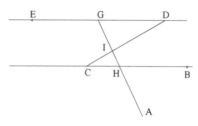

Fig. 2.2.4b

But if *H* and *G* must belong to the segments [*CB*] and [*DE*] respectively, the straight line *AI* does not give a solution to the problem except when *A* is in the smaller of the angles *BIC* and *DIE*, or in the angle that is opposite it at its vertex (the hatched area).

Fig. 2.2.4c

Proposition 2.5. — *Let there be a point* A, *a segment* BC *and an arbitrary point* D *on this segment. If* AD + CD = ℓ, *a known length, then the straight line* AD *is known.*

Let us put $BC = \ell_1$; we have $BD + DC = \ell_1$ and $AD + DC = \ell$.

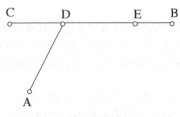

Fig. 2.2.5

If $\ell = \ell_1$, then $AD = BD$, so, from Proposition 1.8, $D \in \Delta$, the perpendicular bisector of AB. The point D must lie at the intersection of the segment BC and Δ.

Note: We may note that if $AB \perp BC$, then $\Delta \parallel BC$, and the point D does not exist. On the other hand, it can happen that the point of intersection of Δ and BC does not lie on the segment BC.

If $\ell_1 > \ell$, then $DB > AD$. Let E be such that $BE = \ell_1 - \ell = BD - AD$; so the point E is known and $ED = DA$. If the point D exists, it thus lies on the perpendicular bisector of EA and on the segment EC; it is then known and the straight line AD is also known.

We may add the same note as before.

If $\ell_1 < \ell$, the argument takes the same form. We then have E lying beyond B.

In Propositions 2.6, 2.7 and 2.8, we go on to the construction of a point by using the intersection of a straight line and a circle. We are in fact dealing with two straight lines that each pass through a known point and have a property P; they are determined by a second point, which is constructed as in the preceding group of propositions by the intersection of two lines – here a straight line and a circle (a subtending arc). We encounter the same procedure later on, in Problems 2.21 and 2.22.

The two Problems 2.6 and 2.7 can be written: *Let there be a straight line Δ and two points* A *and* B, *to find a point* E *on* Δ *such that:*

(6) $A\hat{E}B = \alpha$, *a known angle*

(7) $\dfrac{EA}{EB} = k$, *a known ratio.*

In the two cases, $E \in \Delta$ and $E \in \mathscr{C}$ (a circle). The problem can have 0, 1 or 2 solutions.

In Problem 2.8, we again take a straight line Δ and two fixed points E and G, and we try to find $H \in D$ such that $EH \cdot HG = k$ which is known. The point H lies on a subtending arc and the problem again has 0, 1 or 2 solutions.

In the three Problems 2.6, 2.7 and 2.8, there is a finite number of pairs of straight lines that provide solutions to the problem; whereas in Problem 9, as we shall see, there is an infinite number of such pairs of straight lines.

Proposition 2.6. — *Let there be two fixed points* A *and* B *and a straight line* CD. *Let there be a point* E $\in$ CD *such that* AÊB = α, *a known angle, then the segments* AE *and* BE *are known.*

If the point E exists, it lies at the intersection of the straight line CD and the subtending arc that contains the angle α constructed on AB, from Proposition 1.6; so we can have 0, 1 or 2 solutions.

We are dealing with the intersection of a straight line and two arcs that are symmetrical with respect to AB.

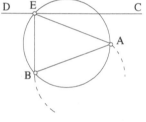

Fig. 2.2.6

Proposition 2.7. — *Let there be two fixed points* A *and* B *and a fixed straight line* CD. *Let there be a point* E $\in$ CD *such that* $\dfrac{AE}{BE}$ = k, *a known ratio. The two straight lines* AE *and* BE *are then known.*

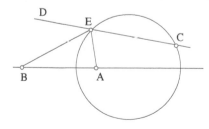

Fig. 2.2.7

If the point E exists, it lies at the intersection of the straight line CD and a circle determined by the data for the problem, from Proposition 1.9; here again we can have 0, 1 or 2 solutions.

Proposition 2.8. — *Let there be* AB *and* CD *two parallel straight lines,* E *and* G *two fixed points on* AB. *Let there be a point* H *on* CD *such that* EH · HG = k, *a known product. The segments* EH *and* GH *are then known in position and in magnitude.*

Let us suppose the point H is known. There exists a point I on AB such that $G\hat{H}I = G\hat{E}H$. The triangles IHG and HEG are similar, hence

$$H\hat{I}G = E\hat{H}G \text{ and } \frac{IH}{HE} = \frac{HG}{GE}.$$

So we have

$$HI \cdot EG = EH \cdot HG = k;$$

since EG is known, we have

$$IH = \frac{k}{EG} = \ell, \text{ a known length.}$$

Let us put in a straight line DB perpendicular to the two known parallel lines; the distance $DB = d$ is known.

If $d = \ell$, then $IH \perp AB$, so $H\hat{I}G$ is a right angle and consequently $E\hat{H}G$ is a right angle.

If $d < \ell$; let $DL = \ell$, the circle (D, ℓ) cuts AB at the point K, $DK = HI = \ell$; then the two straight lines DK and HI are either parallel or antiparallel. In the first case: $H\hat{I}G = D\hat{K}B$, a known angle, and consequently $E\hat{H}G = D\hat{K}B$ a known angle.

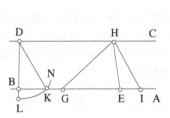

Fig. 2.2.8a

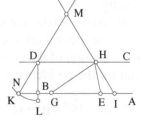

Fig. 2.2.8b

In the second case (when the lines are anti-parallel), the two straight lines DK and HI cut one another in a point M (Figs 2.2.8c and 2.2.8d); we have $M\hat{I}K = I\hat{K}M = D\hat{K}B$, a known angle, so the angles HIG and EHG are again known.

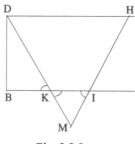

Fig. 2.2.8c   Fig. 2.2.8d

So in all cases, $E\hat{H}G = \alpha$ is known.

If the point H exists, it lies at the intersection of the subtending arc that contains the angle α constructed on EG and the straight line DC; then it is known and the straight lines EH and GH are again known.

Notes:

1) The point H lies on a line parallel to EG, the triangle HEG has a known area S. Now

$$S = \frac{1}{2} EH \cdot HG \sin E\hat{H}G = \frac{1}{2} k \sin E\hat{H}G, \sin E\hat{H}G = \frac{2S}{k},$$

so $E\hat{H}G = \alpha$, an angle determined by the data for the problem.

2) The problem considered is plane because the straight line CD is parallel to EG. If we do not make this hypothesis, we are dealing with a solid problem, as is proved by the following formulation in terms of coordinate geometry:

As coordinate axes we take the straight line EG and the perpendicular bisector of EG. The coordinates of G and E are $(-a, 0)$ and $(a, 0)$ respectively. The equation of the straight line CD is of the form $\alpha x + \beta y = \gamma$ and the condition for the problem can be written

$$GH^2 \cdot HE^2 = ((x + a)^2 + y^2) ((x - a)^2 + y^2) = k^2,$$

or

$$(x^2 - a^2)^2 + 2y^2 (x^2 + a^2) + y^4 = k^2.$$

We eliminate y by using the equation for CD:

$$\beta^4 (x^2 - a^2)^2 + 2\beta^2 (\gamma - \alpha x)^2 (x^2 + a^2) + (\gamma - \alpha x)^4 = \beta^4 k^2,$$

an equation of the fourth degree in x.

In the case considered, where the straight lines are parallel, we have $\alpha = 0$ and the preceding equation reduces to:

$$\beta^4 (x^2 - a^2)^2 + 2\beta^2 \gamma^2 (x^2 + a^2) + \gamma^4 = \beta^4 k^2,$$

or

$$\beta^4 z^2 + 2\beta^2 \gamma^2 z + 4\beta^2 \gamma^2 a^2 + \gamma^4 - \beta^4 k^2 = 0,$$

if we put $z = x^2 - a^2$.

Proposition 2.9. — *Let there be two parallel straight lines* AB *and* CD, *and two known points* E *and* G *on* AB. *From these points are drawn two straight half lines that cut* CD *in* I *and* K *and cut one another in* H *beyond* CD. *If the area of the triangle* HGE *is known, then the segment* KI *is of known length.*

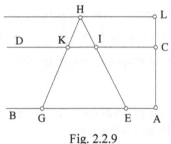

Fig. 2.2.9

Since the triangle HEG has a known area and a fixed base EG, the height from H is of constant length h; so the locus of H is a straight line parallel to AB. Let L be its intersection with the perpendicular AC, the lengths LA and LC are known and we have

$$\frac{AL}{LC} = \frac{EH}{HI} = \frac{EG}{IK} = k,$$

a ratio that does not depend on the position of H.
Then we have

$$IK = \frac{1}{k} EG,$$

where IK is of known length.

Note: Let us put $EG = a$ and let us use d and h respectively to designate the distance between the two parallel lines and the height of the triangle HEG of given area S.

The statement assumes that $h > d$, that is $2S > ad$. We have

$$k = \frac{h}{h-d} = \frac{2S}{2S-ad} \text{ and } IK = \frac{a}{k},$$

hence

$$IK = a\left(1 - \frac{ad}{2S}\right).$$

If $h < d$, $2S < ad$, the locus of the point H lies between the straight lines AB and CD and we have

$$IK = a\left(\frac{ad}{2S} - 1\right).$$

Thus, the locus of the point H is a straight line $\Delta \parallel EG$; the distance between the two straight lines is $h = \dfrac{2S}{EG}$. To each point H of Δ there corresponds a segment IK of constant length on DC.

Proposition 2.10. — In the general case, being given two points A and B and two angles α and β defines a triangle.

Notes:

1) We suppose the angles to be on the same side of AB:
• if their sum is two right angles, the straight lines are parallel;
• if their sum is not two right angles, the straight lines cut one another on one or the other side of AB. The point of intersection C is unique. The data define a unique triangle, because A and B are fixed. The three sides are then known and the ratios of these sides two by two are also known.

2) If a triangle T has known angles, if we take points A and B we can construct a triangle ABC which is similar to it. The ratios of the sides two

by two are then known. The triangle T is determinate, up to similarity. If a side of T is known, T is then determinate, up to isometry.

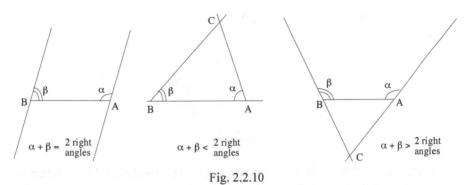

Fig. 2.2.10

In this proposition, Ibn al-Haytham first proves that being given one side of a triangle and two angles adjacent to that side allows us to construct the triangle. He makes two comments on this: one on isometric triangles that he uses in Proposition 2.12, the other on similar triangles that he uses in Proposition 2.11, in which a straight line that passes through a given point and has a property P is characterised by the angle it makes with a given straight line.

We may note that Ibn al-Haytham's investigation in Proposition 2.10 resembles what Euclid did in Propositions 39 and 40 of the *Data*, where we again find the construction of a triangle in which three elements are known; these are three sides in Proposition 39 and three angles in Proposition 40.

Proposition 2.11. — *Let there be a given triangle* ABC *and a given point* D *on the extension of* BC. *If a straight line from* D *cuts* AB *in* E *and* AC *in* G *such that* $\dfrac{GC}{EB} = k$ *is a known ratio, then the straight line* DEG *is known.*

The parallel to *AC* drawn through *B* cuts *DE* in *H*. The points *C, B, D* are known, hence

$$\frac{CD}{DB} = k_1, \text{ a known ratio,}$$

and we have

$$\frac{GC}{BH} = \frac{CD}{DB} = k_1.$$

But $\dfrac{GC}{EB} = k$, by hypothesis, hence

$$\frac{EB}{BH} = \frac{k_1}{k}.$$

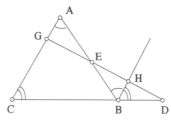

Fig. 2.2.11

Moreover, $E\hat{B}H = B\hat{A}C$, a known angle. The triangle EBH is determinate up to similarity, so the angle BHE is known, and consequently the angle BHD is also known.

But $H\hat{B}D = A\hat{C}B$, a known angle, so the angle BDH is known, so the straight line DHG is known, and the points G and E also.

Note: We are dealing with the construction of a straight line that passes through a given point; the line is defined by an angle.

The problem does not always have a solution. Let us put $BC = a$, $AC = b$, $BA = c$, $BD = d$, $EB = x$, $CG = y$, where $y/x = k$.

In the triangles EBD, CGD and AGE, we have

$$(1) \ \frac{\sin D}{x} = \frac{\sin E}{d}, \qquad (2) \ \frac{\sin D}{y} = \frac{\sin G}{a+d}, \qquad (3) \ \frac{\sin E}{b-y} = \frac{\sin G}{c-x}.$$

From (1) and (2) we obtain

$$\frac{y}{x} = \frac{\sin E}{\sin G} \cdot \frac{a+d}{d},$$

and taking account of (3)

$$\frac{y}{x} = \frac{b-y}{c-x} \cdot \frac{a+d}{d},$$

hence

$$y = kx \iff dk(c-x) = (b-kx)(a+d),$$

hence

$$x = \frac{b(a+d)-kdc}{ak}.$$

We must have $0 < x < c$, which requires $\frac{b}{c} < k < \frac{b}{c} \cdot \frac{a+d}{d}$; these inequalities also imply that $0 < y < b$.

If this double condition is satisfied, the problem has a unique solution.

Proposition 2.12. — *Let there be a given circle, a given straight line* CD *outside this circle and a tangent to the circle at a point* B *which meets* CD *in* E. *If the length* BE *is equal to a given length, then the segment* BE *is known in position.*

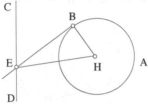

Fig. 2.2.12

Let $\mathscr{C}\,(H,\ BH)$ be the circle, $BE = d$ and $r = BH$ given lengths. The triangle HBE, with a right angle at B, is determinate up to isometry, so $HE = d_1$ is a known length. So the point E lies on $\mathscr{C}\,(H, d_1)$ and it also lies on the given straight line CD. So the angle HEB is known, the straight line EB is thus known.

Notes:

1) There are two tangents that correspond to the point E of the straight line DC.

2) To establish the existence of the point E, we have $HE = d_1 = \sqrt{d^2 + r^2}$. Let h be the distance from H to the given straight line; we have the following cases:

$d_1 < h$, the problem has no solution;

$d_1 = h$, the problem has one solution;

$d_1 > h$, the problem has two solutions.

3) Ibn al-Haytham does not use Pythagoras' theorem in finding HE, but proves that from the data for the problem the triangle HBE, in which we know an angle, the right angle, and the lengths of two sides, is determinate up to isometry. This confirms that this research, just like the *Data* of Euclid, in no way belongs to the domain of algebra.

The construction of point E can be carried out by means of straightedge and compasses.

4) So the problem reduces to constructing E using the intersection of the given straight line and a circle whose centre is known. The radius of the circle is found from what we are given.

The same will be true for a whole group of problems, numbered from 12 to 16.

Proposition 2.13. — *Let there be a given circle* $\mathscr{C}$ (H, HB) *and a given straight line* CD *lying outside the circle. Let us join a point* B *on the circle to a point* E *on* CD *in such a way that* $\hat{BEC} = \alpha$, *a known angle and* BE = d, *a known length; then the straight line* BE *is known in position.*

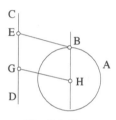

Fig. 2.2.13a

Let H be the centre of the circle, let us put $HK = h$, the distance from H to CD. Let the point G on CD be such that $\hat{HGC} = \alpha$, then the point G is known, as is the length HG,

$$HG = d_1 = \frac{h}{\sin \alpha}.$$

To characterise the point G, here – as in Proposition 2.17 and Proposition 2.23 – Ibn al-Haytham uses Proposition 1.6, that is the subtending arc. It is simpler to note that, with the perpendicular dropped from H on CD, HG makes an angle $\beta = \frac{\pi}{2} - \alpha$ if $\alpha < \frac{\pi}{2}$ or $\beta = \alpha - \frac{\pi}{2}$ if $\alpha > \frac{\pi}{2}$. In both cases, G exists and it is unique.

In regard to the numbers of solutions, there are several cases to consider:

If $d_1 = d$, then $HGEB$ is a parallelogram, because $HG = EB$ and $HG \parallel EB$, so $BH \parallel CD$. The point B lies at the intersection of the circle and the parallel to CD drawn through H. There are two solutions.

If $d_1 > d$, then the straight line HB meets the given straight line; let the point of intersection be C. The point B lies between H and C. We have

$$\frac{HG}{BE} = \frac{HC}{CB} = \frac{d_1}{d} = 1 + \frac{HB}{CB},$$

hence

$$\frac{HB}{CB} = \frac{d_1}{d} - 1.$$

Now $HB = r$, the radius of the circle, so

$$CB = r\frac{d}{d_1 - d} \quad \text{and} \quad HC = \frac{rd_1}{d_1 - d} = R.$$

So the point C lies at the intersection of the given straight line and the circle $\mathscr{C}(H, R)$. The point C exists if and only if

$$R \geq h \Leftrightarrow \frac{rd_1}{d_1 - d} \geq h.$$

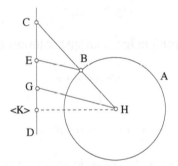

Fig. 2.2.13b

But $d_1 = \dfrac{h}{\sin \alpha}$, so the condition can be written $d \sin \alpha \geq h - r$.

When we find C, we deduce B and $BE \parallel HG$ from it.

If $d_1 < d$, the straight line HB meets the given straight line, but we have that H lies between B and C. In this case we have

$$\frac{d_1}{d} = \frac{CH}{CB} = 1 - \frac{r}{CB},$$

hence

$$CB = \frac{dr}{d - d_1}, \quad HC = \frac{d_1 r}{d - d_1} = R.$$

The condition $R \geq h$ gives

$$d \sin \alpha \leq h + r.$$

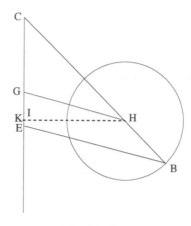

Fig. 2.2.13c

To summarise, this problem has a unique solution if

$$d \sin \alpha = h \pm r;$$

it has two solutions if

$$h - r < d \sin \alpha < h + r;$$

and it has no solution if $d \sin \alpha$ lies outside the interval $[h - r, h + r]$.

This problem also reduces to the construction of a point C using the intersection of the given straight line and a circle with known centre; the radius is deduced from what we were given.

Proposition 2.14. — *Let there be two parallel straight lines* AB *and* CD *and a point* E *lying between them. Let there be a straight line that passes through* E *and cuts* AB *and* CD *in* H *and* G *respectively.*
If EG · EH = k, *then* EG *is known in position.*

The given straight lines correspond to one another in a homothety with centre E. Ibn al-Haytham's argument sets out from a point I on AB, chosen arbitrarily, and uses that property.

If the distances from E to the straight lines AB and AC are called α and β respectively, the ratio of the homothety has the absolute value $\dfrac{\beta}{\alpha}$.

If the straight line EG exists, we have

$$\frac{EG}{EH} = \frac{\beta}{\alpha} = \frac{EG^2}{EG \cdot EH} = \frac{EG^2}{k} \Rightarrow EG^2 = k \cdot \frac{\beta}{\alpha}.$$

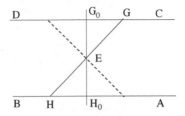

Fig. 2.2.14

The circle with centre E and radius $\sqrt{\dfrac{k\beta}{\alpha}}$ does not cut the straight line CD unless $\dfrac{k\beta}{\alpha} \geq \beta^2$, that is unless $k \geq \alpha\beta$. Thus, if

$k < \alpha\beta$, the problem does not have a solution;

$k = \alpha\beta$, the problem has one solution $G_0 H_0 \perp AB$;

$k > \alpha\beta$, the problem has two solutions symmetrical with respect to $G_0 H_0$.

This problem also reduces to constructing a point G using the intersection of a given straight line and a circle with a known centre; the radius of the circle is deduced from what we are given.

Proposition 2.15. — *Let there be a triangle* ABC, *determinate up to isometry. If a point* D *on the base satisfies*

(1) $$\frac{AD^2}{BD \cdot DC} = k, \text{ a given ratio,}$$

then the straight line AD *is known in position.*

If the straight line AD exists, it cuts the circle $\mathscr{C}$, the circle circumscribing (ABC), again in G; the points G and A lie on opposite sides of BC; hence
$$DB \cdot DC = DA \cdot DG.$$

The condition (1) becomes

$$\frac{DA}{DG} = k.$$

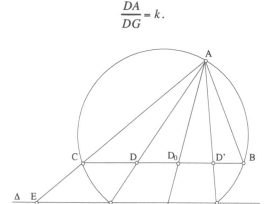

Fig. 2.2.15

Let E be the point of AC beyond C such that $\frac{CA}{CE} = k$, then the required point G lies on Δ, the parallel to BC that passes through E, and it lies on the circle. So G exists if Δ cuts the circle.

Let G_0 be the mid point of the arc CB, the straight line AG_0 is the bisector of the angle A; let D_0 be the point in which it cuts the base and let $\frac{D_0 A}{D_0 G} = k_0$, a known ratio.

If $k > k_0$, the problem does not have a solution.

If $k = k_0$, the problem has one solution, AD_0, the bisector of the angle BAC.

If $k < k_0$, the problem has two solutions. The corresponding points G and G' are such that the arc CG is equal to the arc BG', so the two straight lines AD and AD' are symmetrical with respect to the bisector AD_0.

Note: In the statement of the problem, Ibn al-Haytham writes 'triangle whose sides and angles are known'. The position of the triangle is not given. On the conclusion, he writes 'I say that the straight line AD is known in position'. This refers to the position of AD with respect to the triangle.

This problem is reduced to that of constructing a point G using the intersection of the given circle and a straight line deduced from what we have been given.

Proposition 2.16. — *Let there be two given straight lines* AB *and* AC, *and a point* D *lying inside the salient angle* BAC. *If a straight line that passes through* D *cuts* AB *in* E *and* AC *in* G *such that* $\dfrac{DE}{DG} = k$, *a known ratio, then the segment* EG *is known.*

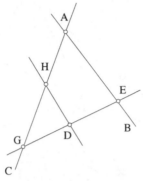

Fig. 2.2.16

The line parallel to *AB* that passes through *D* cuts *AC* in *H* and we have

$$\frac{HA}{HG} = \frac{DE}{DG} = k.$$

So with every value of *k* there is associated a point *G* and consequently a straight line *GD* which cuts *AB* in *E* ($G \neq H$, so *GD* is not parallel to *AB*).

In Propositions 2.13, 2.14, 2.15 and 2.16, Ibn al-Haytham uses parallel straight lines and Thales' theorem; he finds homothetic triangles as well as homothetic straight lines.

Proposition 2.17. — *Let there be two straight lines* AB *and* AC, *and a point* D *in the salient angle* BAC. *If a straight line that passes through* D *cuts* AB *in* E *and* AC *in* G *such that* DE · DG = k², *then the segment* EG *is known.*

Let us suppose that the straight line *GE* exists; let *H* lie on the extension of *AD* and be such that $DA \cdot DH = k^2$; the points *A* and *H* lie on opposite sides of *D*. We have

$$DE \cdot DG = DA \cdot DH \Leftrightarrow \frac{DA}{DE} = \frac{DG}{DH};$$

so the triangles *DAE* and *DGH* are similar and consequently $D\hat{G}H = E\hat{A}D = \alpha$, a known angle. So the point *G* lies on the subtending arc for the angle α

constructed on the segment *DH*. But *G* also lies on the straight line *AC*. So the point *G* exists if the straight line *AC* and the subtending arc cut one another, and we could have 0, 1 or 2 solutions.

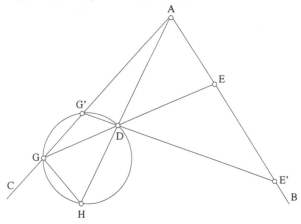

Fig. 2.2.17

With the point *G* that provides a solution to the problem there is associated a segment *GE* known in length and position.

Proposition 2.18. — *Let* A, C *and* D *be three known points on a circle such that* $\overset{\frown}{DC} \neq \overset{\frown}{DA}$. *If a straight line that passes through* D *cuts the arc* AC *which does not contain* D *in a point* B *such that* $\dfrac{BA + BC}{BD} = k$, *a known ratio, then the segment* DB *is known (in position and magnitude).*

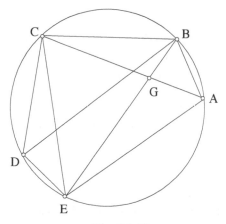

Fig. 2.2.18

Let us suppose the problem has been solved. If E is the mid point of the arc CDA, we have $C\hat{B}E = E\hat{B}A = C\hat{A}E$. Moreover, $B\hat{C}A = B\hat{E}A$, the triangles ABE, GBC and EGA are similar two by two:

$$(ABE) \text{ and } (GBC) \Rightarrow \frac{BE}{EA} = \frac{BC}{CG},$$

$$(ABE) \text{ and } (EGA) \Rightarrow \frac{BE}{EA} = \frac{BA}{AG},$$

hence

$$\frac{BE}{EA} = \frac{BC + BA}{CG + AG} = \frac{BC + BA}{AC},$$

so

$$\frac{AB + BC}{BE} = \frac{AC}{EA} = k',$$

a known ratio (because A, C and E are known points).

So we have

$$\frac{BE}{BD} = \frac{k}{k'},$$

because

$$\frac{BA + BC}{BD} = k.$$

So in the triangle EBD, we know the angle EBD and the ratio $\frac{BE}{BD}$; this triangle is determinate up to similarity and thus its other angles are known.

So the straight line DB makes a known angle with the straight line DE. The point B lies at the intersection of that straight line and the given circle. The points B and D need to lie on opposite sides of AC if B is to provide a solution to the problem. If this is so, the segment BD is determined by its two endpoints.

To summarise, we are concerned with three given points A, C, D that are not collinear and a circumscribed circle (ADC); the problem is to construct a point B on the arc AC such that $\frac{BA + BC}{BD} = k$, a known ratio. Ibn al-Haytham constructs B as the point of intersection of the given circle and a straight line.

Proposition 2.19. — *Let there be a given angle* xAy *and in this a straight half line Az. For any straight line* Δ *that cuts* Ax *in* B, Ay *in* C *and* Az *in* D, *we have*

$$\frac{DC}{DB} = \frac{AC}{k \cdot AB}, \quad where \quad k = \frac{\sin \alpha}{\sin \beta},$$

where α = xÂy *and* β = yÂz, *known angles.*

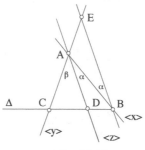

Fig. 2.2.19

First we may note that Ibn al-Haytham does not mention the special case $\alpha = \beta$; in which case *Az* is the bisector and in any triangle *ABC* we then have

$$\frac{DC}{DB} = \frac{AC}{AB} \qquad \text{(in this case } k = 1\text{)}.$$

Let us now look at the general case.

For an arbitrary point *B* on *Ax*, let us draw an arbitrary straight line which cuts *Ay* in *C* and *Az* in *D*, and a straight line parallel to *Az* which cuts the extension of *Ay* in *E*. The triangle *AEB* has two angles known; it is determinate up to similarity. So we have $\frac{EA}{AB} = k$, a ratio which does not depend on the angles that are given. In fact,

$$\frac{EA}{\sin \alpha} = \frac{AB}{\sin \beta}.$$

Moreover,

$$\frac{DC}{DB} = \frac{AC}{AE},$$

hence

$$\frac{DC}{DB} = \frac{AC}{k \cdot AB}.$$

Notes:

1) From this property we immediately deduce: Three concurrent straight lines Ax, Ay, Az determine similar divisions on two parallel lines.

Let there be points D, B, C lying on Δ and D_1, B_1, C_1 on Δ_1. If $\Delta \parallel \Delta_1$, we have

$$\frac{DB}{DC} = \frac{D_1B_1}{D_1C_1}.$$

2) The property established in this proposition is a generalisation of the property of the point of intersection of the internal bisector of an angle of a triangle ABC with the base.

The following proposition is the converse of this.

Proposition 2.20. — *If the angles of a triangle* ABC *are known and if* D *is a point of* [BC] *such that* $\dfrac{DC}{DB} = k$, *a known ratio, then the straight line* AD *makes known angles with* AB *and* AC.

The triangle ABC is determinate up to similarity, so $\dfrac{CA}{CB} = k'$, a known ratio, $k' = \dfrac{\sin \hat{B}}{\sin \hat{C}}$.

Let E be a point on the extension of AC such that $\dfrac{CA}{AE} = k$, a known ratio. The straight line BE is parallel to the required straight line, because $\dfrac{DC}{DB} = \dfrac{CA}{AE} = k$.

But the triangle EAB is determinate up to similarity, because

$$B\hat{A}E = 2 \text{ right angles} - \hat{A}$$

and

$$\frac{AE}{AB} = \frac{AE}{AC} \cdot \frac{AC}{AB} = \frac{k'}{k}.$$

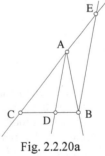

Fig. 2.2.20a

So the angle AEB is known and consequently the angle CAD is also known.

Notes:

1) Starting from Proposition 2.20, we prove that:

if on two parallel straight lines Δ and Δ_1 we have the similar ranges D, B, C and D_1, B_1, C_1 such that $\dfrac{DB}{D_1B_1} = \dfrac{DC}{D_1C_1} \neq 1$, then the three straight lines BB_1, CC_1, DD_1 are concurrent.

This is the converse of Proposition 2.19.

Propositions 2.19 and 2.20 state that if three straight lines are concurrent, they cut off similar ranges on two parallel straight lines, and conversely.[7]

This configuration reminds us of Desargues: in his version the points D and D_1 do not lie on BC and B_1C_1. The straight lines BC and B_1C_1 are parallel and the same holds for BD and B_1D_1; of BB_1, CC_1 and DD_1 are concurrent, DC and D_1C_1 are parallel, and conversely.[8]

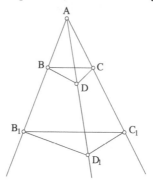

Fig. 2.2.20b

[7] Piero della Francesca (*c.* 1412–1492) gives a version of this theorem in his treatise on perspective, *De prospectiva pingendi*, Book 1, Section 8. See J. V. Field, 'When is a proof not a proof? Some reflections on Piero della Francesca and Guidobaldo del Monte', in R. Sinisgalli (ed.), *La Prospettiva: Fondamenti teorici ed esperienze figurative dall'Antichità al mondo moderno*, Florence, 1998, pp. 120–32, figs pp. 373–5; and J. V. Field, *Piero della Francesca: A Mathematician's Art*, New Haven and London, 2005.

[8] See Girard Desargues, *Exemple de l'une des manieres universelles du S. G. D. L. touchant la pratique de la perspective sans emploier aucun tiers point, de distance ni d'autre nature, qui soit hors du champ de l'ouvrage*, Paris, 1636, pp. 11–12; translation J. V. Field and J. J. Gray, *The Geometrical Work of Girard Desargues*, London and New York, 1987, pp. 158–60 (French text, pp. 200–1).

The case examined by Ibn al-Haytham is a degenerate limit of the configuration in Desargues.

Proposition 2.21. — *Let there be a given circle, a given chord* AB *and a triangle* ABC *inscribed in the circle. If this triangle has a given area, the point* C *is known, then the straight lines* AC *and* BC *are also known.*

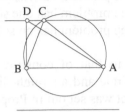

Fig. 2.2.21a

Here Ibn al-Haytham returns to the proof of the second proposition of the first part about triangles with given area and with one given side; he deduces from this that the point *C* lies on a chord parallel to *AB*. As in Proposition 2.10, the point *C* lies on one or other of the two parallel straight lines equidistant from *AB*. Now by hypothesis *C* lies on the circle.

Notes:
The problem can have 0, 1, 2, 3 or 4 solutions.

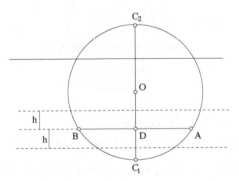

Fig. 2.2.21b

Let *D* be the mid point of *AB* and *O* the centre of the circle. Let us put $AB = 2a$, $OA = R$, $OD = d$; we have $R^2 = a^2 + d^2$. If *S* is the area given for the triangle *ABC*, its height is $h = S/a$.

The perpendicular bisector of *AB* cuts the circle in C_1 and C_2 and we have

$$DC_1 = R - d \text{ and } DC_2 = R + d,$$

hence

$h < R - d$	the problem has four solutions
$h = R - d$	the problem has three solutions
$R - d < h < R + d$	the problem has two solutions
$h = R + d$	the problem has one solution
$h > R + d$	the problem has no solution.

The problem reduces to that of constructing the point C using the intersection of the given circle and a straight line deduced from what was given, using the method that was set out in Proposition 1.10.

Proposition 2.22. — *Let there be a circle and two known points* A *and* B *on this circle. If* C *is a point on the circle such that* CA $\cdot$ CB $= k^2$, *a known ratio, then* C *is known and thus the straight lines* CA *and* CB *are also known.*

This proposition reduces to the preceding one, because

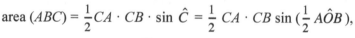

$$\text{area } (ABC) = \frac{1}{2} CA \cdot CB \cdot \sin \hat{C} = \frac{1}{2} CA \cdot CB \sin (\frac{1}{2} A\hat{O}B),$$

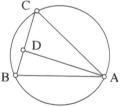

Fig. 2.2.22

where $A\hat{O}B$ is the angle at the centre.

Now $CA \cdot CB = k^2$ is known, and $\sin \frac{1}{2} A\hat{O}B$ does not depend on the position of the point C on the circle, so area $(ABC) = S$ is known. Ibn al-Haytham makes this clear in his proof:

Let us suppose that C is known; let $AD \perp BC$, then the triangle ADC has known angles and

$$\frac{CA}{AD} = k' \left[= \frac{1}{\sin \hat{C}} \right],$$

hence

$$\frac{CA \cdot CB}{AD \cdot CB} = \frac{k^2}{AD \cdot CB} = \frac{CA}{AD} = k',$$

hence

$$AD \cdot CB = \frac{k^2}{k'} \text{ and area } (ABC) = \frac{1}{2}\frac{k^2}{k'} = \left[\frac{1}{2}k^2 \sin \hat{C}\right];$$

so we have returned to the preceding problem and there can be 0, 1, 2, 3 or 4 solutions for the point C.

Note: This problem is like Problem 2.8, with a circle instead of two parallel straight lines. In both cases, the area of the triangle is fixed by the data.

Proposition 2.23. — *We are given a circle and a straight line* CD. *If a straight line cuts the circle in* A *and* B *and the straight line* CD *in* E *in such a way that* $\frac{AB}{BE} = k$, *a known ratio, and* $B\hat{E}C = \alpha$, *a known angle, then the straight line* AB *is known, so the segment* AB *is known.*

Let us suppose that the straight line AB is known. Let H be the centre of the circle and $HI \perp AB$, then I is the mid point of AB and $\frac{IB}{BE} = \frac{k}{2}$.

Let G be a point on CD such that $H\hat{G}C = \alpha$; G exists and is unique, and HG is parallel to the required straight line. Let us put $HG = \ell$.

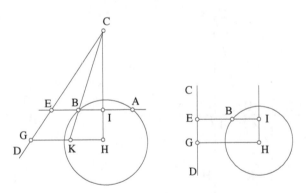

Fig. 2.2.23a   Fig. 2.2.23b

If $HI \parallel DC$, which implies $\alpha = \frac{\pi}{2}$, then $IE = HG = \ell$. But $\frac{IB}{BE} = \frac{k}{2}$,

hence

$$\frac{IB + BE}{BE} = \frac{k+2}{2} = \frac{IE}{BE} \Rightarrow BE = \frac{2l}{k+2} = \ell'.$$

So we return to Proposition 2.13 with $BE = \ell'$ and $B\hat{E}C = \dfrac{\pi}{2}$.

If $\alpha \neq \dfrac{\pi}{2}$, HI cuts the straight line CD at the known point C. The straight line CB cuts HG in K and we have $\dfrac{HK}{KG} = \dfrac{IB}{BE} = \dfrac{k}{2}$; so, from Proposition 2.20, the straight line CK is known in position and the point B lies at the intersection of this straight line and the circle. We then draw the perpendicular from B to CH, which gives us the points E and A. So the points A, B and E are known.

We have just seen that this proposition reduces to Proposition 2.13 or to Proposition 2.20 depending on the given angle, and this reduces to the construction of a point by means of a given circle and a straight line.

Comment: We can restate this proposition as follows: Let there be a circle $\mathscr{C}(H, R)$ and a straight line xy that lies outside it. To find on xy a point E so that a straight half-line Ez cuts the circle in A and B such that $x\hat{E}z = \alpha$ and $\dfrac{BA}{BE} = k$, both known.

The analysis of this problem leads to similar ranges I, B, E and H, K, G if $\alpha \neq \dfrac{\pi}{2}$ and to equal ranges of $\alpha = \dfrac{\pi}{2}$.

The synthesis begins from the data, which allow us to construct K and G.

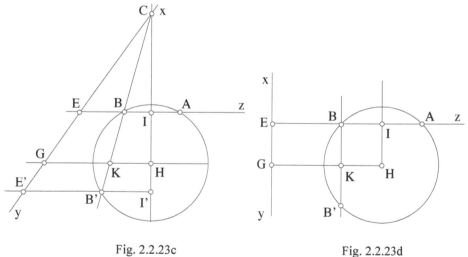

Fig. 2.2.23c   Fig. 2.2.23d

If $\alpha \neq \dfrac{\pi}{2}$, we know C. The required point $B \in \mathscr{C} \cap KC$, from which we deduce Ez, and thus A. The problem can have 0, 1 or 2 solutions.

If $\alpha = \dfrac{\pi}{2}$, the point C does not exist and the required point B is given by the intersection of the circle and the perpendicular to HG at K, hence there are 0, 1 or 2 solutions.

Proposition 2.24. — *Let there be two circles lying outside one another, equal or unequal. If a straight line is a common tangent to the two circles, it is known.*

We may note that the figures provided in the text show circles that lie outside one another; this condition is not necessary in investigating the external common tangent.

1. *External common tangents*

Let there be $\mathscr{C}_1$ (E, EA) and $\mathscr{C}_2$ (G, GD) two circles, and A and D the points of contact. So in this case we have $\overrightarrow{EA} \parallel \overrightarrow{GD}$ and in the same sense.

1.1. *Equal circles*

$AEDG$ is a rectangle that can be constructed immediately, we have $AD = EG$; so AD is known.

What we have is a translation $\overrightarrow{EG}$.

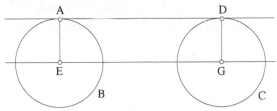

Fig. 2.2.24a

1.2. *Unequal circles*

The straight lines DA and GE cut one another in H which lies beyond E and we have $\dfrac{GH}{HE} = \dfrac{GD}{EA} = \dfrac{R}{r}$. So the point H is known.

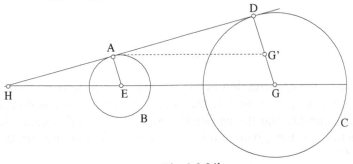

Fig. 2.2.24b

What we have is a homothety $h\left(H, \dfrac{R}{r}\right)$.

2. Internal common tangent
2.1. Equal circles
The reasoning is the same.
This time what we have is a central symmetry $h(H, -1)$.

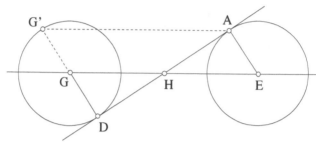

Fig. 2.2.24c

2.2. Unequal circles
$\overrightarrow{EA}$ and $\overrightarrow{GD}$ are parallel and have opposite senses, AD cuts EG in H between E and G and we have $\dfrac{EH}{HG} = \dfrac{EA}{GD} = \dfrac{R}{r}$. So the point H is known.

What we have is a homothety $h\left(H, -\dfrac{R}{r}\right)$.

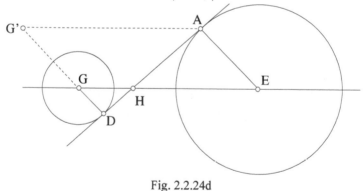

Fig. 2.2.24d

In all cases, the construction of the tangent reduces to the construction of a point, for example D. In 1.1, the point D is at the intersection of the given straight line EG and the perpendicular, and in the other cases, it is at the intersection of the given circle with centre G and the circle with diameter GH.

TRANSLATED TEXT

Al-Ḥasan ibn al-Ḥasan ibn al-Haytham

On the Knowns
Fī al-maʿlūmāt

In the name of God, the Compassionate the Merciful

TREATISE BY AL-ḤASAN IBN AL-ḤASAN IBN AL-HAYTHAM

On the Knowns

Knowledge is an opinion which does not change, and opinion is belief in a certain notion. So knowledge is belief in a certain notion, as it is, and it is, moreover, a belief that does not change, as when we believe that the whole is greater than the part. But belief cannot occur without there being someone who believes, and a notion that is believed. Now belief cannot be unchanging except if the notion that is believed does not change. If this is so, then knowledge is belief in a notion that does not admit of change. A known is the notion in which one believes that does not admit of change, and the person who knows is the one who believes a notion that does not admit of change. On the other hand, believing notions that change cannot be considered to be knowledge, because these latter are not fixed according to one single attribute, as when one believes that Zayd is standing upright: it is possible that he is not standing upright at the moment when we believe [this], and that he is standing upright at other moments. So if we situate it in time, as when we believe that Zayd is standing upright at this time, or that he was standing upright at some particular moment, our belief can be true; if one is certain that it is a true belief, its being called 'knowledge' will come about as a metaphor, since it [the belief] resembles knowledge on account of the correctness of the belief. But knowledge, itself, does not, in truth, admit of change at any moment. If knowledge is a belief, and if belief can occur only for someone who believes, then knowledge can occur only when there is a person who knows.

Further, belief in a notion that does not admit of change may be divided into two kinds[1]: on one hand, when we believe a notion that does not change, while knowing that this notion does not change; on the other hand, when we believe a notion that does not change, without knowing that it does not change. Believing a notion is indeed separate from belief in whether this notion changes or not. Someone who believes a notion which

---

[1] Lit.: parts.

does not change, while knowing that it does not change, has knowledge of that notion. Moreover, he knows that he has knowledge of it, because in believing that notion that does not admit of change, he has knowledge of that notion, and in knowing that it does not admit of change, he knows that he has knowledge of it. But someone who believes a notion that does not change, without knowing that it does not change, has knowledge of that notion, but does not know that he has knowledge of it, because he does not know if this notion admits of change or not. And someone who believes with this kind of belief believes a notion without proof or necessity, but by hearsay and imitation, taking it on trust, or by intuition. It is legitimate to say that he has knowledge of that notion, because he believes a notion that does not admit of change, which is the definition of knowledge.

Knowledge is divided into two parts: knowledge *in actu*, and knowledge *in potentia*. Knowledge is *in actu* when it has become a belief for someone who believes; knowledge is *in potentia* when it can become a belief for someone who believes.

If knowledge is a belief, [and] if belief cannot occur without someone who believes and a notion that is believed, that is to say the known, and if knowledge is divided into two parts, knowledge *in actu* and knowledge *in potentia*, then the known is also divided into two parts: a known *in actu* and a known *in potentia*. A known *in actu* has become known by someone who has knowledge of it; a known *in potentia* is one that can become known by someone who has knowledge of it. But we have shown that a known is a notion that does not admit of change; notions that do not admit of change are thus divided into two parts: some are beliefs for someone who believes; the others can be beliefs for someone who believes. These two parts can be known only if the notions belonging to these two parts do not themselves admit of change. If all of what we have said is true, a known is in truth any notion that does not admit of change, whether or not it is believed by someone who believes.

All known notions are divided up into two kinds: one is concerned with quantity and the other is not concerned with quantity. Our treatise will deal only with known notions that are concerned with quantity.

Quantity is divided up into two parts: one is discrete quantity and the other is continuous quantity. Discrete quantity is divided up into two parts, which are: the letters of words and numbers. Continuous quantity is divided up into five parts, which are: line, surface, solid, weight and time.

The known notions contained in this treatise are those concerning letters of words, those which concern number, those which concern lines, those which concern surfaces, those which concern solids, those which concern weights and those which concern time.

The notions which concern the letters of words are divided into three parts: one is concerned with the essence of the letters; another is concerned with the quantity of the number of the letters – that part reduces to what concerns number; and the third part is that which concerns the arrangements of the letters and the combinations of them among themselves, which are words.

The notions concerning number are divided up into four parts: one concerned with the essence of number; another concerned with the quantity of number; the third concerned with the properties of number, [such] as those that concern numbers that are perfect, abundant or deficient, squares, cubes or numbers like them which are properties of the nature of number; the fourth part is concerned with the association of numbers one with another as in commensurability, ratios, augmentation, diminution and <the relation> of the whole and the part.

Notions concerning lines are divided up into seven parts: one is concerned with the essence of a line; another is concerned with the endpoint of a line which is a point; the third is concerned with the figure of a line; the fourth is concerned with magnitudes of lines; the fifth is concerned with positions of lines, that is to say their situation[2] – that part is divided up into seven parts: one is the position of the line with respect to fixed points, the second is the position of the line with respect to a single fixed point, the third is the position of the line with respect to a moving point or with respect to <several> moving points, the fourth is the position of the line with respect to a fixed line, the fifth is the position of the line with respect to a moving line, the sixth is the position of the line with respect to a fixed surface, and the seventh is the position of the line with respect to a moving surface. The sixth part of the first system of division is concerned with ratios of magnitudes of lines, one to another, and the seventh part is concerned with the composition of a group of these lines when some of them meet with one another.

Notions concerned with surfaces are divided up into parts like those into which the notions concerned with lines are divided up, except the one that is concerned with endpoints, because the boundaries of surfaces are lines.

In the same way, notions concerned with solids are divided up into parts, like those into which notions concerned with surfaces were divided up, except for the last part, which is concerned with composition, because composition of solids arises only from composition of the positions of their

---

[2] We use this term to translate the Arabic term *naṣba* that is reguarly used to translate the Greek θέσις.

surfaces; similarly, positions of solids with respect to anything one can use to situate them are the positions of the surfaces of the solids.

Notions concerned with weights are divided up into three parts: one is concerned with the essence of weight, the next is concerned with magnitudes of weights and the third is concerned with the ratios of weights one to another.

Notions concerned with time are divided up into three parts: one is concerned with the essence of time, the second is concerned with the magnitude of time and the third is concerned with ratios of intervals of time one to another.

The known concerned with the essence of the letters of a word, that is the ordinary letters used in all languages, and whose forms and sound do not change from one language to another. The essence of the letters of a word is, in fact, the separate sounds, used in the words of dialogues and conversations, in all languages; but the words of the languages are different, depending on the difference in character of the people who speak these languages, and all the different words, in all the different languages, there are combined letters. Among the composite letters, some are common to all the languages, the others are peculiar to one language, and not to another. What is common to all the languages has a form and figure that do not change; so this is known because it does not change in all the words there are in all the languages. Among these letters, what is not common [to all languages] has a form that can change in the [different] languages; some of them are to be found in some languages and are not to be found in the other languages, and some are found in one of the languages with a [particular] property and in another language they have another property. Thus, for the letters of a word, the known concerned with the essence of the letters is the letters that are common to all the languages.

As was mentioned earlier, the known concerned with the number of the letters reduces to that concerned with the quantity of the number.

As for the known concerned with the arrangement of letters and the combination of letters one with another, this is the words used in all languages. In fact, words are letters put together, combined with one another, but not every combination of letters is a word used in a language, further, the majority of the combinations of letters do not make a word that is used. The forms and the arrangement in the words that are used do not change, but any word that is used in a language always takes the same shape and does not change in the language in which it is used. As for the known concerned with the combination of letters in a word, this is the words used in all languages.

As for the known concerned with the essence of number, it is merely unity. In fact, the essence of number is unity and what is generated from repeating it. Every number involves nothing more than unity and repetition. In a number the repetition is not [always] one and the same repetition, but a repetition that increases and that decreases, so it is touched by change, whereas unity is not susceptible of change in any way. So the known concerned with the essence of number is only unity.

The known concerned with the quantity of a number is that every number has a finite plurality, which does not change either by increasing or by decreasing. This type of number is subdivided into two parts: one is where the number is limited by necessity and the other is where it is limited by hypothesis. What is limited by necessity is as in the number of the planets, the number of the orbs, the number of the elements, and similar things, that is to say any numbers for which the counted numbers[3] neither increase nor decrease. If the thing that is numbered neither increases nor decreases, then its number does not change either by increasing or by decreasing, but number does not undergo change except by increasing and decreasing. If the numbered does not lend itself either to increasing or to decreasing, its number does not lend itself either to increasing or to decreasing, this part of number is that which applies to a quantity limited by necessity. As for the other part, it is that limited by hypothesis, that is to say that in his imagination, or in a numerical problem that he poses by hypothesis, a human being gives himself a certain number; he supposes that it does not change, or he assumes determinate numbers in the world of sensation and in existence; thus, by way of hypothesis, he will have taken a number that does not change. It is in these two ways that the quantity of a number is known.

The known concerned with the properties of numbers, such as the properties of a square, of a cube, of a surface, of a solid, of a perfect number, of an abundant number, of a deficient number and numbers like them, it is the form of each of these numbers from which its properties have been constituted, as the form of the square from which its properties have been constituted, that is to say as the product of a number with itself. The known notion of a square concerned with the properties of the square is the product of a number with itself. This notion is [to be found] in every square and it is a notion that does not change for any square, despite a change in the sides of the squares and change in the sizes of the squares; every property of every square is, in fact, constituted from the multiplication of a number, which is its side, by itself. Similarly, the form of the cube, from which its properties are constituted, is the multiplication of a number by what is obtained by multiplying it by itself. Similarly, the form of a surface is the

---

[3] See Aristotle, *Physics*, IV, chap. XI, 219b 7–10.

multiplication of a number by a number. Similarly, the form of a solid is the multiplication of a number by what is obtained from the multiplication of a number by a number. The form of a perfect number is that it is equal to the sum of its aliquot parts. The form of an abundant number is that the sum of its aliquot parts exceeds it. The form of a deficient number is that the sum of its aliquot parts is less than it. Similarly, every number like this has a form from which its properties are constituted. The known for any number that has a property or properties is the form it has from which its property or properties are constituted, because this form does not change in any of the numbers of that species, despite a change in its quantity and change in its aliquot parts and its sides.

The known concerned with the combination of numbers with one another is divided up into six parts: one of them – and it is the first – is the equality of every unity, in every number, to every unity, in every number; the second part is that every number is a multiple of every unity that is in it, and a multiple of every unity in every number that is combined with it; the third part is that two arbitrary numbers are commensurable through unity and that unity measures each of them; the fourth part is that any number is [one] of the parts of any number that is combined with it; the fifth part is that two different arbitrary numbers are such that one is greater than the other and that other is smaller than the first. These notions hold for all numbers and do not change for any numbers. As for the sixth part, that is ratios; every numerical ratio is between two numbers; a numerical ratio is the measure of the quantity of the number expressed in terms of the quantity of the number with which its ratio is set up. A known ratio is the ratio of two numbers one to the other, whose quantity is known, as well as the ratio of two numbers that are equimultiples of two numbers whose quantity is known, or of parts, in equal number, of two numbers whose quantity is known, or two homologous parts of two numbers whose quantity is known. Two arbitrary numbers are the two smallest numbers in their ratio or are equimultiples of the two smallest numbers in their ratio; two arbitrary numbers, that are the two smallest numbers in their ratio, in fact measure two numbers in the ratio of equality: the smaller measures the smaller and the greater measures the greater. It is possible that two other numbers that are equimultiples of the two smallest numbers in their ratio measure the two numbers that are measured.[4] If this is the case, then two arbitrary numbers, which are not the two smallest numbers in their ratio, are equimultiples of the two smallest numbers in their ratio, because they are measured by the two smallest numbers in their ratio; it is possible that these two numbers are equimultiples of multiples of the two smallest numbers in their ratio. If

---

[4] That is measured by the smallest.

the two numbers which are the measure are known, then the ratio of one to the other is known and it is the ratio of the two numbers that are measured; the ratio of two numbers that are measured, one to the other, will then be known even if their quantities are not known. But if the quantity of the two numbers that are measured is known, then the ratio of the two numbers that are measures, one to the other, is also known, because the ratio of the parts is equal to the ratio of their equimultiples. So the known ratio is the ratio of two numbers, one to the other, whose quantity is known; it is the ratio of two numbers, which are multiples of two numbers whose quantity is known; it is the ratio of two homologous parts of two numbers of known plurality, and it is the ratio of two numbers which are equal parts of two numbers whose plurality is known. In general, the known numerical ratio is that of two numbers of known quantity, or is equal to the ratio of two numbers of known quantity. So the known in a known numerical ratio is the quantity of each of the two numbers in the ratio one to the other – if each of them is known – or the quantity of two known numbers which are in their ratio.

As for the known concerned with the essence of a line, it is that the line is a length without breadth, because this notion applies to all lines and does not change for any of them. The length of a line and its figure change with the lines, since, for lines, there are straight lines, circular ones, [and] curves with different kinds of curvature. So the known concerned with the essence of a line is that the line is a length without breadth.

The known concerned with an endpoint of a line, which is a point, is made up of two notions: one relates to its essence, that is to say that it is not divisible, and the other to its position, that is to say its distance from another point or points that exist in the imagination, if this distance or these distances do not change. This notion is divided up into three parts: one is that this point, itself of known position, is fixed, and that the point or the points that exist in the imagination are also fixed and that none of them moves with any sort of motion; the second part is that the point that exists in the imagination is fixed, whereas the point of known position is free to move about the fixed point, with a circular motion, and the distance between them [sc. the two points] does not change; and the third part is that the point of known position is at a distance that does not change from a point that exists in the imagination or is at distances that do not change from points that exist in the imagination and are such that the two points, or all the points, are free to move with one equal motion, all together, and that the distances that lie between the point of known position and the points, do not change. These two notions are known and concerned with a point which is the endpoint of a line.

As for the known concerned with the figure of the line, it is the notion which constitutes the essence of the line; for the straight line, it is its two endpoints, plus the fact of being the shortest distance. In fact, a straight line is the distance there is between its two endpoints, on condition that this distance is the shortest distance between its two endpoints; so what constitutes its essence is its two endpoints, because it is its two endpoints which delimit the distance that there is between them. If we make the further demand of the distance that it shall be the shortest, this distance will be a straight line. The known notion concerned with the figure of the straight line which does not change for any of the straight lines, is the two endpoints, plus the fact of being the shortest. As for a circular line, what constitutes its essence is the circular surface of which the line is the boundary; but what constitutes the essence of that circular surface is the centre together with the distance between the centre and the circumference. So what constitutes the essence of the circular line – and is its prime constituent – is its centre and the distance between the line and its centre. The known notion concerned with the figure of the circular line is the centre and the semidiameter. If the magnitude of the semidiameter does not change, the circular line is a whole circle or an arc of a circle, whether the arc or the circumference of the circle is convex or concave. As for curved lines that can have a known figure, these are the ones which are in an arrangement, an order, and a notion which constitutes their essence and which does not change for any line their [various] species. For a curved line, the known notion concerned with its figure is the notion which constitutes its essence. So a line of known figure is a line for which the notion that constitutes its essence is known.

As for the known concerned with magnitudes of lines, it is the value of the length of the line. But we have knowledge of the value of the length of the line from the fact that we have knowledge of the distance there is between its two endpoints, while knowing the figure of the line. Thus, a finite line of known magnitude is one for which the distance between its two endpoints does not change, that is to say does not increase or decrease, and which is such that its figure does not change. In fact, there is an infinity of lines with different figures between two points, each of which [lines] is called a distance, and we cannot imagine any one of them by imagining only its two endpoints, apart from the straight line, because it is the shortest line to join two points. But since the form of linearity is established in the imagination, and the figure of linearity is no different from one straight line to another and it does not change, accordingly a straight line of known magnitude is one for which the shortest distance between its two endpoints does not change.

A circular line is also the distance between its two endpoints if it is an arc; but it is not the shortest distance and, in addition, its magnitude is not limited by its two endpoints because it is possible that, between its two endpoints, there are many circular lines of different magnitudes, none being equal to any other nor having a ratio to it; in consequence, the magnitude of a circular line will not be known unless its semidiameter is known, that is to say unless the magnitude of the latter does not change. But if its semidiameter is known, then its figure will be known because its semidiameter is what constitutes its essence. A finite circular line has a known magnitude only if the distance between its two endpoints is of known magnitude – that is to say the straight line which is its chord – and if, in addition, its figure is known.

Similarly, a curved line is of known magnitude only if its figure is known, but its figure will not be known unless we know the notion which constitutes its essence, because between two points there are many curved lines, none being equal to any other nor having a ratio to it. So, a finite curved line is of known magnitude only if the distance between its two endpoints – that is to say the straight line that is its chord – is of known magnitude and if, in addition, the figure of the curved line is known.

A finite line of known magnitude is one for which the distance between its two endpoints is of known magnitude and one whose figure is also known.

A circular line which is a whole circle and which is of known magnitude is one whose semidiameter is of known magnitude, because, if its semidiameter is of known magnitude, then the magnitude of the circular line does not change and its figure does not change.

A curved line, if it is closed,[5] is of known magnitude only if the distance from each point taken on the line to its centre, or to a fixed point that lies inside it, is of known magnitude, that is to say the straight lines.

As for the known concerned with the position of a line with respect to fixed points, it is the distances from the points, which are on the line, to each of the two [fixed] points, or more than two fixed points. If these distances do not change, and if the line that has this property is a line which does not move with any kind of motion, except for increasing and decreasing,[6] then it does not change its position, but changes only its magnitude. A line which does not move with any kind of motion is known in position with respect to fixed points, because if the distance from each point taken on the line to each of the two [fixed] points, or more than two fixed points, is a distance which does not change, then this line does not move with any

---

[5] Lit.: if it surrounds perfectly.
[6] That is extension and shortening.

kind of motion, whether the line is straight, circular or of an arbitrary figure. If the line moves along a rectilinear path, the distance of each of its points changes with respect to any fixed point, whether the line is straight or not straight. Similarly, if the line moves along a curvilinear path or if it moves along a circular path, then the points that lie on the line can preserve the distance between each of them and a single point if the line is moving round that single point. As for the remaining fixed points, the distances between them and the points that lie on the line change in every case.

A line of known position with respect to fixed points is a line which does not move with any kind of motion, except for increasing and decreasing, and it is one for which the distances from points that lie on it to each of the two [fixed] points, or more than two fixed points, are distances that do not change. The line that has this property is said to be known in position absolutely, without any condition or addition, whether the line is straight or not straight. A straight line known in position absolutely is one that does not move with any kind of motion, except for increasing and decreasing. A circular line known in position absolutely is one whose centre known in position, and whose semidiameter is of known magnitude.[7] As for the known for this line, it is the distances between the points that lie on it and the fixed points, because these distances do not change.

As for the known concerned with the position of a line with respect to a single fixed point, it is the distances between each point taken on the line and the fixed point, if the distances do not change. The line that has this property is said to be known in position with respect to the fixed point, and this line is not known in position absolutely, because the line can preserve the distances that lie between it and the fixed point, even if it is moving; in fact, this line can move round the fixed point in such a way that the distances between the points on it and the fixed point do not change; indeed if we join its two endpoints to the fixed point with two straight lines and if the triangle generated by the line and the two straight lines drawn from its two endpoints to the fixed point is put in motion about the fixed point, then the distances between the points that lie on the line and the fixed point do not change, and the line will, however, be in motion, whether the line is straight or not straight. If the line is a circumference of a circle and is in motion about its centre, then the distances from points that lie on it, to the fixed point, which is its centre, do not change. So a straight line known in position with respect to a single fixed point is the line for which the distances from the points that lie on it to the fixed point are distances which do not change, whether the line is fixed, immobile or mobile, rotating about the fixed point and whether the line is straight or not straight.

---

[7] Compare Euclid, *Data* 6.

As for the known concerned with the position of a line with respect to a moving point or to moving points, it is the distances between each point taken on the line and the moving point or moving points, if the distances between the points are known and if the line moves with a motion equal to the motion of the moving point or moving points, and in the direction in which the point or points move. A line known in position with respect to a moving point or moving points is a line for which distances from its points to the moving point or the moving points are distances which do not change, and which [sc. the line] moves, despite this, with a motion equal to the motion of the moving point or the moving points, and in the direction of their motion, whether the line is straight or not straight.

The known concerned with the position of a line with respect to a fixed line is the angle enclosed by that line and the fixed line, if the two lines intersect, and it is the angle formed if we extend these two lines until they meet one another – if the two lines are ones which can meet one another – and if they do not intersect. A line known in position with respect to a fixed line is – if the two lines are such that they can intersect one another – a line which encloses a known angle with the fixed line, whether the line known in position is also fixed and does not move with any kind of motion, or whether it is moving, but at the same time preserving the form of the angle enclosed by the line known in position itself and the fixed line to which it is compared.

A straight line known in position with respect to a fixed line, if it cuts the fixed line or if it can cut it, is a straight line which encloses a known angle with the fixed line; it can be fixed motionless or it can move as a whole, while preserving the angle, or it can increase or decrease. A straight line that has this property does not, in fact, change in position with respect to the fixed line, because the angle enclosed by the two <lines> does not change, whether the fixed line is straight or not straight. The known for the position of this line is the known angle.

A circular line of known position with respect to a fixed line – if the fixed line cuts it or can cut it if it is extended indefinitely – is the circular line which encloses with the fixed line a known angle; it can be fixed, motionless or in motion about its centre, its centre being fixed, motionless, whether the fixed line is straight or not straight, or in motion along the fixed line, but with the angle enclosed by the two lines not changing. This can take place if the fixed line is straight or circular. The position of the circular line that has this property does not change with respect to the fixed line because the angle between it and the fixed line does not change; the known is the angle.

If the line does not cut the fixed line or cannot cut it, it will then be of known position with respect to the fixed line when, once it has been cut by a straight line which makes a known angle with one of the two lines, this latter encloses a known angle with the other line, whether the line of known position is fixed, motionless or in motion, while preserving the form of the angle formed between it and the line that cuts it, when that is possible for it. What is known for the line that has this property is the two angles formed by the intersection of each of the two lines with the line that cuts them.

A curved line of known position with respect to a fixed line is a line which does not move with any kind of motion, whether the fixed line is straight or not straight, or the curved line that is moving with respect to the fixed line, if the fixed line is straight or circular, and whether the point on the curved line which lies on the straight or circular line does not change, and if the angle enclosed by it and the straight or circular line does not change; this, if the curved line cuts the fixed line. If it does not cut it, it will be of known position if its relation with the straight line which cuts it, as well as the fixed line in accordance with two known angles, is the relation described earlier in regard to the fixed line.

The known concerned with the position of the line with respect to a moving line is the known, in the preceding paragraph, without any difference between them concerning the angles or the division <into cases>. The only difference between this line and the preceding line is that the line to which the position is referred is fixed for the first line, whereas it is moving for the latter one, and that, with respect to that reference line, the line moves with its own motion and in the direction of the motion of the latter line, whether the line of known position is straight or not straight.

The known concerned with the position of the line with respect to a fixed surface is the right angle, if the line is perpendicular to the fixed surface or to the plane tangent to the fixed surface at the endpoint of the perpendicular, if the fixed surface is convex or concave, or the angle enclosed by that line and the perpendicular drawn from a point on the line, perpendicular to the surface or perpendicular to the plane tangent to the fixed surface at the endpoint of the perpendicular, if the angle is known. The line known in position with respect to a fixed surface is the perpendicular erected on the fixed surface or on the plane tangent to the fixed surface at the foot of the perpendicular, or the one that encloses a known angle with the perpendicular, whether the line known in position is fixed, motionless or in motion with respect to the fixed surface, while preserving the right angle or the known [angle]. The known is the angle.

The known concerned with the position of a line with respect to a moving surface is the known in the preceding paragraph, that is to say the angle. The only difference between this line and the preceding line is that the surface to which the position of the line is referred is fixed for the first line, and is in motion for the latter, and that the line, whose position is referred to the surface, moves with a motion equal to its motion and in the direction of its motion, whether the line is straight or nor straight. A line known in position with respect to a moving surface is a line perpendicular to the moving surface or to the plane tangent to the moving surface at the foot of the perpendicular, or a line which encloses with the perpendicular drawn from a point of the line, perpendicular to the moving surface or to the plane tangent to the moving surface at the endpoint of the perpendicular, a known angle, if the line is moving with a motion equal to the motion of the surface and in the direction of this motion.

The known concerned with ratios of magnitudes of lines, the one to the other, includes two notions: one is the figure of the two lines one of which is referred to the other; the second is the quantity of each of the two lines; in fact, of two lines, one cannot have a ratio to the other, and there cannot be a ratio between the two lines, unless they are of the same species and what constitutes their essence is the same notion, as [when there are] two straight lines or two arcs of the same circle or of two equal circles. It is for only these two species of line that there can be ratios between the magnitudes of individuals. As for the species of lines, other than those two, there is no ratio between their magnitudes. So the known ratio between the lines is the one between two straight lines or two circular lines of the same species, and such that the magnitude of each of them is known, or one that is equal to the ratio of two lines of their species such that the magnitude of each of them is known. The known of two straight or circular lines that are of the same species, and whose ratio one to the other is known, is the magnitude of each of the two lines, if each of them is known, or the magnitude of each if the two lines of known magnitude, [lines] for which the ratio of one to the other is equal to the ratio of the two lines for which the ratio of one to the other is known. The two lines for which the ratio of one to the other is known are straight lines or circular <lines>, for which the magnitude of each of them is known, or the magnitude of each of the two lines of known magnitude for which the ratio of one to the other is equal to the ratio of two lines whose ratio one to the other is known. The known ratio between two lines is that between two known lines, because the ratios between the known magnitudes does not change, it being given that, for known magnitudes, their magnitudes do not change, so the magnitude of one does not change when it is measured by the magnitude of the other.

The known concerned with figures made up of lines that meet one another is their form. It is a notion made up of their angles and their magnitudes, measured one by another, <measures> that are the ratios of one to another, if the sides are straight lines or arcs of equal circles. If the angles of a figure are known – that is to say that they do not change – if we know that they do not change and if the ratio of the quantity of each of the sides to each of the sides that remain is a known ratio, then the form of the figure does not change, whether the magnitude of each of the sides is known and does not change or whether the magnitudes of the sides change, while preserving the ratios between them and the angles that they form, whether all the sides are straight or whether all the sides are circular, <belonging to> equal circles, or whether some of them are straight and the others circular, if the ratios of those that are straight to the straight ones do not change and if the ratios of the circular ones to the circular ones do not change. A figure of known form, enclosed by straight lines or arcs of equal circles, is one whose angles are known and one in which the ratios of the sides one to another are known.

Figures of known form, made up of curved lines, are those in which only the angles are known because there cannot be ratios between the magnitudes of the curved lines, except when they are equal, because the parts of a curved line are not measured by a single magnitude, and none of them can be superposed on another and the parts of one of them do not have similar forms, but, on the contrary, two parts of the same curved line always have different forms. A figure enclosed by curved lines or by lines some of which are curved is thus of known form, if only its angles are known.

The known concerned with the essence of a surface is that the surface is only a length and a width, because this notion holds for all surfaces and does not change in any of them. As for the quantity of the length of the surface, of its width and of its shape, it changes from one surface to another, because surfaces have different figures and shapes, in regard to their plane <shape>, their being convex and their being concave. The known concerned with the essence of surfaces is that a surface is only a length and a width.

The known concerned with the figure of a surface, that is to say the shape of the surface, is the notion that constitutes its essence; so for a plane surface the latter is the edges which enclose it, in addition to its smallness because a plane surface is the smallest surface enclosed by its edges. The known notion concerned with the figure of a plane surface that does not change for any plane surface, is the edges of the surface [together] with its smallness.

What constitutes the essence of a spherical surface is the spherical solid; what constitutes the essence of the spherical solid is its centre and its semidiameter, so what constitutes the essence of the spherical surface, which is the primary cause, is its centre and its semidiameter, whether the spherical surface is a complete sphere or a portion of a sphere, convex or concave.

As for non-spherical convex or concave surfaces, which can be of known figure, they are those that have a [spatial] arrangement, ordered and have a notion that constitutes their essence and ones that do not change in any of their species. The known notion for a non-spherical convex or concave surface which is concerned with its figure is the notion that constitutes its essence. So a surface of known shape is one for which the notion that constitutes its figure is known.

The known concerned with the magnitudes of surfaces is the quantity of the area of the surface, if the area of the surface does not change either by increasing or by decreasing. So a surface of known magnitude is the surface such that the quantity of its area does not change. As for what the area of a surface is and how to have knowledge of the area of a surface, we have set it out in our book *On Measurement*,[8] [where] we have explained it in an exhaustive manner, and it is not appropriate to explain, in this book, how to <find> the area.

The known concerned with the position of a surface with respect to fixed points is the distances from each point taken on the surface to two fixed points, or more than two fixed points, if these distances do not change. A surface that has this property is a surface that does not move with any kind of motion, apart from increasing and decreasing; this, in fact, does not change its position, but changes its magnitude because, if the distances from the points that lie on the surface to two [fixed] points, or more than two fixed points, are distances that do not change, then the surface does not move with any kind of motion, whether the surface is plane, convex or concave. In fact, if the surface moves along a rectilinear path or along a curvilinear path, then it is necessary that the distances between its points and the fixed points change; but if it moves along a circular path,[9] then it is possible for it to preserve the distances between its points and only one point among the fixed points, if the surface moves around this single point. A surface of known position with respect to fixed points is thus one which does not move with any kind of motion, except for increasing and decreasing. A surface that has this property is said to be known in position

---

[8] See *Les Mathématiques infinitésimales*, vol. III, chap. IV.

[9] See the commentary in our discussion of the position of a line with respect to fixed points, p. 329.

absolutely, unconditionally, whether the surface is plane, convex or concave.

The known concerned with a position of the surface with respect to a single fixed point is the distances between each point taken on the surface and the fixed point, if these distances do not change. A surface that has this property is said to be of known position with respect to the fixed point; this surface will not be known in position absolutely because the distances between points of this surface and the fixed point can be known without their magnitudes changing, even if the surface moves when its motion is around this fixed point. A surface of known position with respect to a single fixed point is thus a surface for which the distances from its points to the fixed point are distances that do not change, whether the surface is fixed, motionless or in motion along a circular path around the fixed point, [and] whether the surface is plane, convex or concave.

The known concerned with a position of a surface with respect to a moving point is the distances which lie between the points of the surface and the moving point, if the distances are known; the surface moves with a motion equal to the motion of the point, and in the direction of its motion. A surface of known position with respect to a moving point is a surface for which the distances from its points to the moving point are known distances, if, in addition, the surface moves with a motion the same as that of the moving point, and in the direction of its motion, whether the surface is plane, convex or concave. It is the same for a surface of position known with respect to moving points.

The known concerned with the position of a surface with respect to a fixed line is a right angle, if the line is perpendicular to the surface or perpendicular to the plane tangent to the surface at the endpoint of the perpendicular, if the surface is convex or concave, or the angle enclosed by the fixed line and the perpendicular drawn from the point on the fixed line, perpendicular to the surface or to the plane tangent to the surface at the endpoint of the perpendicular. A surface of known position with respect to a fixed line is thus a surface to which the fixed line is perpendicular, or perpendicular to the plane tangent to the surface at the foot of the perpendicular, or the surface such that the fixed line encloses a known angle with the perpendicular, whether the surface is plane, convex or concave, [and] whether the surface is fixed, immobile or in motion along a circular path around the fixed line. So the known is the angle. This case is similar to that of the line [considered] with respect to a fixed surface.

The known concerned with the position of a surface with respect to a moving line is the same known as that in the paragraph before this one, that is to say the angle. The only difference between this surface and the preceding surface is that the reference line for the position of the preceding surface was fixed and immobile, whereas the reference line for the position of this surface is moving, and that, in addition, the surface moves with a motion equal to its motion and in the direction of its motion, whether the surface moves with that motion alone or whether it moves with this motion while [also] moving with a circular motion about the moving line. So a surface of known position with respect to a moving line is the surface to which the moving line is perpendicular, or perpendicular to the plane tangent to the surface at the endpoint of the perpendicular, or the surface such that the perpendicular to the surface or to the plane that is tangent to it encloses a known angle with the moving line and is such that the surface moves with a motion equal to the motion of the moving line and in the direction of its motion, or moves with this motion or also with a circular motion about the moving line; the known is the angle.

The known concerned with the position of a surface with respect to a fixed surface is the angle at which the two surfaces cut one another, if this angle is known, that is to say the angle enclosed by the two lines drawn in the two surfaces from a point on the [line of] intersection, [lines] that cut one another, if these two lines are perpendicular to the [line of] intersection, this when the two surfaces are plane. If the two surfaces are not plane, then the surface will be of known position with respect to the other surface if the surface[10] which cuts them and which is perpendicular to each of them gives rise, at the intersection, to a known angle enclosed by the two intersections generated by the surface perpendicular to the two surfaces, the angle will be at a known point of the intersection, this if the two surfaces cut one another; if the two surfaces do not cut one another and one does not meet the other, then one will be known in position with respect to the other if each of them is known in position with respect to the surface[11] which cuts both of them and which is perpendicular to each of them, the known for each of these surfaces will be the known angle or the two known angles. So a surface known in position with respect to a fixed surface is the surface that encloses a known angle with the fixed surface at the intersection of the two surfaces, or that encloses a known angle at the intersection of the surface of known position and the surface that is perpendicular to it and perpendicular to the fixed surface.

The known concerned with the position of a surface with respect to a moving surface, is like <the known for> a position of the surface we have already mentioned. The only difference between the two is that the refer-

---

[10] Ibn al-Haytham does not indicate explicitly here whether we are concerned with a plane surface.

[11] *Idem.*

ence surface for the position was fixed for the first surface, whereas it is moving for this latter surface, and that the surface of known position moves with a motion equal to its motion and in the direction of its motion. A surface of known position with respect to a moving surface is thus one which encloses a known angle with the moving surface, at the intersection between the two surfaces, or one which encloses a known angle at its intersection with the surface that is perpendicular to it and perpendicular to the moving surface, if the moving surface moves with a motion equal to the motion of the reference surface for its position, and in the direction of its motion.

As for the known concerned with ratios of magnitudes of surfaces the one to the other, there are two notions; one is the figure of the two surfaces, one of which is referred to the other, the second is the quantity of each of the two surfaces, that is to say the area of each of them. In fact, for two arbitrary surfaces, one does not necessarily have a ratio to the other, and there is not necessarily a ratio between the two surfaces, except if they are of the same species and if the notion that constitutes their essence is the same, as for two plane surfaces or two spherical surfaces from the same sphere or from two equal spheres. It is for these two species of surface only that there can be ratios between magnitudes of the individuals belonging to them. The known for two plane surfaces or spherical [ones] of the same species, whose ratio one to another is known, is thus the magnitude of each of the two surfaces – if each of them is known – or the magnitude of each of two surfaces of known magnitude, whose ratio one to another is equal to the ratio of the two surfaces, whose ratio one to another is known. So two surfaces whose ratio one to another is known are two plane or spherical surfaces for which the magnitude of each is known; or the magnitude of each of two surfaces of known magnitude, whose ratio one to another is equal to the ratio of the two known surfaces, one to another. A known ratio between two surfaces is one which is between two surfaces of known magnitude, because the ratios between known magnitudes do not change; it being given that the magnitudes are known, their magnitudes do not change, so the magnitude of one does not change when it is measured by the other.

The known concerned with figures made up of surfaces which meet one another, which are solids, are the shapes of surfaces that meet one another. So if each of the surfaces that enclose the solid is of known shape, then the solid is of known shape. So a solid figure of known shape is one which is enclosed by surfaces of known shape, whether each of these surfaces is of known magnitude or is not of known magnitude, on condition that it preserves its shape.

The known concerned with the essence of a solid is that it has three dimensions, because this notion is to be found in all solids and does not change in any of them. As for the quantity of the length of the solid, of its width and of its depth, it changes depending on the solids. Similarly, the figures of solids change depending on the solids. So the known concerned with the essence of a solid is that it has three dimensions.

The known concerned with the figure of a solid is the notion that constitutes the figure of the solid, that is to say its boundaries which are the surfaces that enclose it. So a solid of known figure is one for which the surface or the surfaces which enclose it are of known figure.

The known concerned with magnitudes of solids is the quantity of the volume of the solid, if the quantity of the volume of the solid does not change, either by increasing or by decreasing. So a solid of known magnitude is one such that the quantity of its volume does not change.

The known concerned with positions of solids with respect to fixed points or to one fixed point, or to a point or points that move, to a fixed straight line or to a moving straight line, to a fixed surface or to a moving surface, is the positions of the surfaces of the solids with respect to these things, so it is the positions of the surfaces which we have already discussed, because if the position of the surface of the solid is known, that is to say that it does not change, then the position of the solid does not change. So a solid of known position is one whose surface or surfaces are of known position, whatever thing it is to which the position has been referred.

The known concerned with ratios of magnitudes of solids one to another, is thus the magnitude of each of the two solids, one of which bears a ratio to the other, if each of them is known, or it is the magnitude of each of two solids of known magnitude, whose ratio one to another is equal to the ratio of the two solids whose ratio one to another is known.

The known concerned with the essence of a weight is the force that moves [it] towards the centre of the Universe, because that notion does not change for any weight. It is this force which is called weight.

The known concerned with magnitudes of weights is the quantity of the weight. But the quantity of the weight is known by its ratio to the size of the measure[12] by which the size of weights is measured is known: the *raṭl*, the *manā*, the *mithqāl* and the weight of *darāhim*[13] and things like them. If the ratio of the quantity of the weight to the size of the weight of the measure is a known ratio, then the magnitude of the weight is known,

---

[12] Measure (*miqyās*): a unit chosen for measuring weight.

[13] *raṭl* = 144 *dirham* = 450 gr.; *manā* = 2 *raṭl* = 2,130 *dirham* = 812,5 gr.; *mithqāl* = 4,464 gr.; *dirham* = 3,125 gr. See Walther Hinz, *Islamische Masse und Gewichte umgerechnet ins metrische System*, Leiden, 1955.

as it does not change because of the fact that the weight of the measure does not change and the known ratio does not change; so this ratio is a numerical ratio. Now, we have proved earlier how a numerical ratio is known. But the ratios of weights can be non-numerical ratios, these are the irrational ratios which are ratios of an actual weight that does not change to an actual weight that does not change, without either of the two weights having a [numerical] ratio to the measure. It remains that what we use for weights is only a numerical ratio. So a weight of known magnitude is one for which the ratio of its quantity to the size of the weight of the measure is a known ratio.

The known concerned with ratios of magnitudes of weights one to another is the quantity of each of the two weights, one of which is compared with the other, if each of them is of known magnitude, or it is the magnitude of each of the two weights of known magnitude, whose ratio to one another is equal to the ratio of the two weights whose ratio one to another is known.

The known concerned with the essence of time is the interval that elapses between two instants, because the essence of the interval does not change for arbitrary times, but it is the magnitudes of the times which differ.

The known concerned with the magnitude of time, that is the quantity of time, and the quantity of time is known in relation to the motion of the celestial sphere, because the revolution[14] of the celestial sphere is the measure[15] by which we measure time. So a time of known magnitude is the time whose ratio to the revolution of the celestial sphere is a known ratio.

The known concerned with ratios of the parts of time one to another is the quantity of each of two times, one of which is compared with the other, if each of them is of known magnitude; or it is the magnitude of each of two times of known magnitude, whose ratio one to another is equal to the ratio of the two times whose ratio one to another is known.

These notions which we have set out are all the knowns which are concerned with quantity in a detailed and precise manner. We are not aware of any predecessor who has given details of them in this way and made them precise in this fashion. These notions are pieces of knowledge each of which stands up in its own right and anyone who seeks to understand the sciences that deal in truths needs to know them. These notions are, in addition, the rules and premises used to solve mathematical problems; the solution of mathematical problems is completed only with their help.

[14] In the plural in the original.
[15] The sidereal day, whose duration is that of a revolution of the celestial sphere, is taken as the unit of time.

For solving these problems, we may need other notions of the same kind as those in *The Knowns* which were not mentioned by Euclid in his book devoted to *Data*, and which were not mentioned by any of [our] predecessors. We set them out in this treatise so that this treatise may gather together all that has not been mentioned by [our] predecessors regarding knowns.

The notions that we set out now are divided into two parts; one of the two parts concerns notions that have not been mentioned by any of [our] predecessors, these [predecessors] have not mentioned anything of the kind; the second part is of the kind of thing that Euclid mentioned in the *Data*, though without anything about them being mentioned in the book of the *Data*.

For this let us introduce premises constructed on what we said earlier in this book of *The Knowns*, so that we could use it later on, following the premises.

We have already reminded ourselves that a known ratio is one which is between two known magnitudes or two magnitudes in the ratio of two known magnitudes. If this is so, then by composition of the separated known ratio, it will be known because the known ratio that has been iso-lated is equal to the ratio of two known magnitudes, one to the other; so by composition of the ratio that has been isolated, it will become equal to the ratio of the sum of two known magnitudes to one of them. But the sum of two known magnitudes is a known magnitude; so by composition of the known ratio that has been isolated, it will become equal to the ratio of two known magnitudes, one to the other; it is a known ratio. Similarly, by separation of a known composed ratio, it will become known, because its separation will be equal to the ratio of two known magnitudes composed, if they are separated. The same holds for a known ratio, if it is inverted.

Similarly, if we have a known ratio between two magnitudes and if one of the two magnitudes is known, it follows that the other is known, because a known ratio is one that exists between two known magnitudes. So the ratio of one known magnitude to the other magnitude is equal to the ratio of two known magnitudes, one to the other. But the ratio that exists between two known magnitudes does not change, thus the ratio of the known magnitude to the other magnitude is a ratio that does not change. So the second magnitude does not change, because if it changed, the ratio of a known magnitude to it would change, because the true nature of a ratio is to measure the quantity of a magnitude against the quantity of a magnitude. So if the quantity of a known magnitude does not change and its ratio to the other magnitude does not change, then the quantity of the other magnitude does not change. Thus, for two magnitudes such that the ratio of one to the other is known, if one is known, then the other is known.

Similarly, if we have two straight lines of known magnitude which enclose a known angle, then the straight line which joins their two end-points is of known magnitude and with each of the two straight lines [it] encloses a known angle. Yet it is of known magnitude, since its two end-points do not change – because they are the endpoints of two straight lines of known magnitude – and since the position of one with respect to the other does not change. But the line encloses a known angle with each of the two straight lines, since the distance to each of the two endpoints from each point of the other straight line does not change, accordingly the position of the straight line which joins the two endpoints does not change with respect to each of the two straight lines, because if the distance from each of the endpoints of the straight line to a single point does not change, then the distance from each point of the straight line to this point does not change. It follows that the position of the straight line that joins the two endpoints with respect to each of the two straight lines does not change. So, if the position of the straight line that joins the two endpoints does not change with respect to each of the two straight lines, then it encloses a known angle with each of the two straight lines. It follows also that the ratio of the sides of the triangle that is formed, one to another, is known, because their magnitudes are known.

Similarly, we have already reminded ourselves that a straight line known in position absolutely is one that does not move with any kind of motion, except increasing and decreasing; and that a circular line known in position absolutely is one whose centre is known and whose semidiameter is known. If this is so, it does not move with any kind of motion. The same holds for any line known in position absolutely which does not move with any kind of motion. It follows that if two lines known in position absolutely cut one another, then the point of intersection is known in position, because it does not undergo displacement, or change, whether the two lines are straight, circular, curved or of two different species.

Similarly, we have already reminded ourselves that a straight line of known magnitude is one whose length does not increase or decrease and its magnitude does not change. It follows that a straight line known in magnitude and in position is one that does not change by any kind of change.

Similarly, we have already reminded ourselves that a straight line of known position with respect to another straight line is one that encloses a known angle with the other straight line.

These notions were proved by Euclid in his book the *Data* using methods different from those we have set out here. We have proved them from knowns that we have given earlier in this book, in order that this book shall have no need of what Euclid set out in his book the *Data*.

Now that we have introduced these premises, let us begin by proving the notions that we have decided to include in this book, and which are needed for solving the problems, which divide up into two parts, as we have shown.

FIRST PART

These are the notions that none of the ancients has set out, and they have not set out anything of this kind.

– **1** – If from a point of known position we draw a straight line of known magnitude, then its endpoint will lie on the circumference of a circle of known position.

Example: Let there be a point *A* of known position; we draw from this point the straight line *AB* which is of known magnitude.

I say that the point B lies on the circumference of a circle known in position.

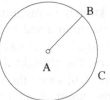

Fig. II.2.1.1

Proof: Let us make point *A* a centre and with the distance *AB* let us draw a circle; let the circle be *BC*. Since the circle *BC* has a centre known in position, the surface of the circle does not move in any way; and since the semidiameter of the circle is of known magnitude, the position of its circumference does not change in any way. So the circumference of the circle *BC* is known in position, so the point *B* lies on the circumference of a circle known in position, which is the circle *BC*. This is what we wanted to prove.

– **2** – If from the centre of a circle known in magnitude and in position we draw, as far as its circumference, a straight line which is then inclined at a known angle, and is such that the ratio of the first straight line to the second is known, then the endpoint of the second straight line lies on the circumference of a circle of known position.

Example: Let there be the circle *AB*, known in magnitude and in position, whose centre is *C*; from the point *C* we draw the straight line *CB* which is turned to lie along the straight line *BD*, so that the angle *CBD* is known and the ratio of *CB* to *BD* is known.[16]

I say that the point D *lies on the circumference of a circle of known position.*

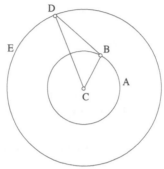

Fig. II.2.1.2

Proof: The circle *AB* is known in magnitude and in position, so the straight line *CB* is of known magnitude and its ratio to *BD* is known, so the straight line *BD* is of known magnitude, as has been proved in the premises. But since angle *DBC* is known, the straight line *BD* is known in position with respect to the straight line *CB*; since the straight line *CB* is of known magnitude, the position of the point *B* with respect to the point *C* is a known position that does not change; the same holds for the position of the point *D* with respect to the point *B*. Now, since the position of the point *D* with respect to the point *B* does not change – that is to say that one does not move away from or towards the other – similarly the position of the point *B* with respect to the point *C* does not change, and the angle *CBD* does not change – that is to say that, with respect to the straight line *BC*, the straight line *BD* does not incline in any direction – and similarly, with respect to the straight line *BD*, the straight line *CB* does not turn in any direction, then the position of the point *D* with respect to the point *C* does not change. We join *CD*; it will be of known magnitude because the position of its two endpoints, one with respect to the other, does not change. But since the point *C* is known in position and the straight line *CD* is of known magnitude, the point *D* will lie on the circumference of a circle

---

[16] In Euclid, *Data* 51, the triangle *CBD* is said to be 'given in species'.

known in position, as has been proved in the preceding proposition. We make *C* a centre and with distance *CD* we draw the circle *DE*, then it will be known in position, and thus the point *D* will lie on the circumference of a circle known in position. This is what we wanted to prove.

– 3 – In the plane of a circle of known magnitude and position, if from a known point, that is not its centre, we draw a straight line to meet the circumference of the circle and if we extend it so that the ratio of the first straight line to the second straight line is known, then the endpoint of the second straight line lies on the circumference of a circle known in position.

Example: Let there be a circle *AB* known in magnitude and in position, and the known point *C* which is in the plane of the circle, but which is not at its centre; from the point *C* we draw the straight line *CA* as far as the circumference of the circle and we extend it to [the point] *D* such that the ratio of *CA* to *AD* is known.

I say that the point D *lies on the circumference of a circle of known position.*

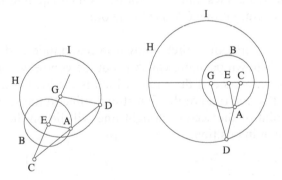

Fig. II.2.1.3

Proof: We mark off the centre of the circle, let it be *E*, and we join *CE*; it will be of known magnitude because its two endpoints are known. We extend it in the direction of *E*, we join *EA* in our imagination, and we imagine *DG* parallel to the straight line *AE*; so the ratio of *GD* to *EA* will be equal to the ratio of *DC* to *CA* and equal to the ratio of *GC* to *CE* and the ratio of *DA* to *AC* will be equal to the ratio of *GE* to *EC*. But the ratio of *DA* to *AC* is known because the ratio of *CA* to *AD* is known, so the ratio of *GE* to *EC* is known; now *EC* is of known magnitude, so *GE* is of known magnitude and *GC* is of known magnitude,[17] as has been proved in the premises, so the ratio of *GC* to *CE* is known as has also been proved in the

---

[17] So the point *G* is known in position.

premises. Now the ratio of *GC* to *CE* is equal to the ratio of *GD* to *EA*, so the ratio of *GD* to *EA* is known and *EA* is of known magnitude, so the straight line *GD* is of known magnitude. We make point *G* a centre and with distance *GD* we draw a circle, let it be the circle *DHI*; the circle *DHI* is then known in magnitude and in position because its centre is of known position and its semidiameter is of known magnitude, so the point *D* lies on the circumference of a circle known in position. This is what we wanted to prove.

Starting out from this proof, we prove that: if an arbitrary straight line from the point *C* cuts two circles *AB* and *HI*, then the ratio of its two parts, one to the other, is equal to the ratio of the two parts of the straight line *CD* one to the other, because for every straight line [drawn] from the point *C* and which cuts the two circles, if from the two centres *E* and *G* we draw two straight lines to the two points of intersection, then the ratio of the two straight lines drawn from the two centres to the two points of intersection, one to the other, is equal to the ratio of *GC* to *CE*; these two straight lines will thus be parallel,[18] and the ratio of the two parts of the straight line which cuts the two circles, one to the other, is then equal to the ratio of the two parts of the straight line *CD* one to the other.

– 4 – In the plane of a circle known in magnitude and in position, if from a point of known position, which is not the centre, we draw a straight line to the circumference of the circle, which is inclined at a known angle, in such a way that the ratio of the first straight line to the second is known, then the endpoint of the second straight line will lie on the circumference of a circle known in position.

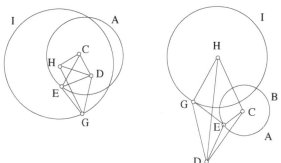

Fig. II.2.1.4

[18] See commentary, pp. 310–12.

Example: Let there be a circle *AB* of known magnitude and position whose centre is *C*, and the point *D* of known position, *DE* comes out from it and is inclined at a known angle which is the angle *DEG*, in such a way that the ratio of *DE* to *EG* is known.

I say that the point G *lies on the circumference of a circle known in position.*

Proof: We join *DC*, it will be known in magnitude and in position because its two endpoints are known in position. We make the angle *DCH* equal to the known angle *DEG*; we make the ratio of *DC* to *CH* equal to the known ratio of *DE* to *EG* and we join the two straight lines *CH* and *DG*, then the two triangles *DCH* and *DEG* are similar, so the angle *CDH* is equal to the angle *EDG*, so the angle *HDG* is equal to the angle *CDE* and the ratio of *CD* to *DH* is equal to the ratio of *ED* to *DG*. But since the straight line *DC* is of known magnitude and position and the angle *DCH* is known, the straight line *DH* will be known in position. Since the ratio of *DC* to *CH* is known and the straight line *DC* is of known magnitude, the straight line *CH* is of known magnitude. Since the two straight lines *DC* and *CH* are known in magnitude and position, and since the angle *DCH* is known, the straight line *DH* will be known in magnitude and in position, because its two endpoints do not change. We join *CE* and *HG*. Since the ratio of *CD* to *DH* is equal to the ratio of *ED* to *DG*, the ratio of *CD* to *DE* will be equal to the ratio of *HD* to *DG*. But the angle *CDE* is equal to the angle *HDG*, so the triangle *CDE* is similar to the triangle *HDG* and the ratio of *DC* to *CE* is equal to the ratio of *DH* to *HG*. But the ratio of *CD* to *DE* is known, because they are both of known magnitude, so the ratio of *DH* to *HG* is known. But *DH* is of known magnitude, so the straight line *HG* is of known magnitude and the point *H* is known, because it is the endpoint of the straight line *CH* of known magnitude and position. With centre *H* and distance *HG* let us draw a circle *GI*, then it will be known in magnitude and in position, because its centre is known in position and its semidiameter is of known magnitude. So the point *G* lies on the circumference of a circle known in position, which is the circle *GI*. By this same proof, we [can] prove this proposition for the case in which the straight line *DC* passes through the point *E*. This is what we wanted to prove.

By an analogous proof, we prove that for every straight line from the point *D* which ends on the circumference of the circle *AB* and which is inclined at an angle equal to the angle *DEG*, in such a way that the ratio of the first straight line to the second straight line is equal to the ratio of *DE* to *EG*, the endpoint of the second straight line will lie on the circumference of the circle *GI*, [at the position] where there is the point *E* with respect to the circle *AB*, because the proof of this results in the straight line which joins the point *H* to the endpoint of the second straight line being equal to the straight line *HG*. It necessarily follows that, for every straight line from the point *D* which ends on the circumference of the circle *AB*, a straight line which is inclined at an angle equal to the angle *DEG* and which ends on the circle *GI*, the ratio of the two straight lines, one to the other, will always be equal to the known ratio of *DE* to *EG*.

‒ **5** ‒ If, from a point of known position, we draw to a straight line known in position a straight line which is inclined at a known angle, in such a way that the ratio of the first straight line to the second is known, then the endpoint of the second straight line will lie on a straight line known in position.

Example: Let there be a point *A* of known position and the straight line *BC* of known position; from the point *A* we draw the straight line *AD* which is inclined at a known angle, say the angle *ADE*, so that the ratio of *AD* to *DE* is a known ratio.

I say that the point E lies on a straight line known in position.

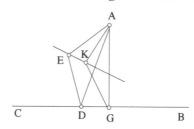

Fig. II.2.1.5

Proof: From the point *A* we draw a perpendicular to the straight line *BC*, let it be *AG*. Since the straight line *BC* is known in position and the point *A* is known in position, the distances between the point *A* and each point of the straight line *BC* do not change and the straight line *AG* is the shortest distance between the point *A* and the straight line *BC*, so the straight line *AG* does not change and the point *G* does not change; so the straight line *AG* is of known magnitude, because it does not change. Now, the point *A* is known in position, the straight line *AG* does not change and the point *G* does not change, so the straight line *AG* is known in magnitude and in position. We make the angle *AGK* equal to the angle *ADE* and we make the ratio of *AG* to *GK* equal to the known ratio of *AD* to *DE*, then the straight line *GK* is of known magnitude because *AG* is of known magnitude. Now, since the angle *AGK* is known, the straight line *GK* is known in position because the straight line *AG* is known in position ‒ in fact if the position of the straight line *GK* changed, the angle *KGA* would change – so the straight line *GK* is known in magnitude and in position. So the point *K* does not change and the point *A* does not change. We join *AK*, it will be known in magnitude and in position and the angle *GAK* will be known, as has been proved in the premises. We join *AE*. Since the ratio of *AG* to *GK* is equal to the ratio of *AD* to *DE* and the angle *AGK* is equal to the angle *ADE*, the triangle *ADE* will be similar to the triangle *AGK*. So their angles are equal, so the angle *DAE* is equal to the angle *GAK* and the ratio of *DA* to *AE* is equal to the ratio of *GA* to *AK*. We join *KE*. Since the angle *DAE* is equal to the angle *GAK*, the angle *GAD* is equal to the angle *KAE*. Since the ratio of *GA* to *AK* is equal to the ratio of *DA* to *AE*, the ratio of *GA* to *AD* is equal to the ratio of *KA* to *AE*. Since the angle *GAD* is equal to the angle *KAE* and the ratio of *GA* to *AD* is equal to the ratio of *KA* to *AE*, the triangle *KAE* is similar to the triangle *GAD*. So the angle *AKE* is equal to the angle *AGD*; but the angle *AGD* is a right angle, so the angle *AKE* is a right angle. But the straight line *AK* is known in magnitude and in position and the angle *AKE* is a right angle, so the straight line *KE* is known in position – in fact, if its position changed, the right angle would change; but since the angle is a right angle, the position of the straight line *KE* does not change. So the straight line *KE* is known in position and the point *E* lies on the straight line *KE*, so the point *E* lies on a straight line known in position, which is the straight line *KE*. This is what we wanted to prove.

– 6 – If, from two points known in position, we draw two straight lines which meet one another in a point and which, at this point, enclose a known angle, then this point lies on the circumference of a circle known in magnitude and in position.

Example: Let there be two points *A* and *B* known in position, from which we draw two straight lines *AC* and *BC*, in such a way that the angle *ACB* is known.

I say that the point C *lies on the circumference of a circle known in magnitude and in position.*

Fig. II.2.1.6

Proof: We join *AB* and we imagine a circle circumscribed about the triangle *ACB*, let the circle be *ACB*; and let its centre be *D*. We join *AD* and *BD*, then the angle *ADB* is known, because it is double the angle *ACB*; there remain the two known angles *DAB* and *DBA* which are equal because the two straight lines *AD* and *BD* are equal, so the angle *BAD* is known and the straight line *AB* is known in magnitude and in position because its two endpoints are known. So the straight line *AD* is known in position because its point *A* is known and the angle *BAD* is known – if its position changed, the angle *BAD* would change. Similarly, we [can] prove that the straight line *BD* is known in position, so each of the straight lines *AD* and *BD* is known; neither of the two straight lines *AD* and *BD* moves with any kind of motion; the point *D*, which is the point of intersection, thus does not change in any way;[19] so the point *D* is known in position and the point *A* is known in position, so the straight line *AD* is known in magnitude, and the straight line *BD* is also. So each of the two straight lines *DA* and *BD* is known in magnitude and in position and the point *D* is the centre of the circle *ACB*. So the circle *ACB* is known in magnitude and in position and the point *C* lies on the circumference of this circle. So the point *C* lies on the circumference of a circle known in magnitude and in position, which is *ACB*. This is what we wanted to prove.

– 7 – If, from two points known in position, we draw two straight lines which meet one another in a point and which enclose a known angle, if we then extend one of the two straight lines in such a way that the ratio of the first straight line to its extended part is a known ratio, then the endpoint of the second straight line lies on the circumference of a circle known in position.

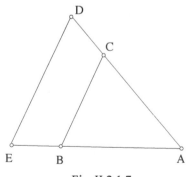

Fig. II.2.1.7

[19] Ibn al-Haytham insists on the fact that the position of *D* is the same irrespective of the point *C* that is proposed as a solution of the problem.

Example: Let there be the two points *A* and *B* known in position, from which we draw two straight lines *AC* and *BC* which meet one another at the point *C*, in such a way that the angle *ACB* is known. We then extend the straight line *AC* to *D* in such a way that the ratio of *AC* to *CD* is a known ratio.

I say that the point D *lies on the circumference of a circle known in position.*

Proof: We join *AB*, thus it will be of known magnitude and position because its endpoints are known; we extend it in the direction of *B* to *E* and we make the ratio of *AB* to *BE* equal to the known ratio of *AC* to *CD*; *BE* will thus be known in magnitude. But it is known in position because it lies on the extension of the straight line *AB*, known in position. So the whole straight line *AE* is known in magnitude and in position; its endpoints, which are *A* and *E*, are known. We join *DE*, thus it is parallel to the straight line *CB* because the ratio of *AB* to *BE* is equal to the ratio of *AC* to *CD*. So the angle *ADE* is equal to the known angle *ACB*, so the angle *ADE* is known. So, from the two points *A* and *E* known in position, we have drawn the two straight lines *AD* and *ED* which enclose a known angle, which is the angle *ADE*. So the point *D* lies on the circumference of a circle known in position, as has been proved in the preceding proposition. This is what we wanted to prove.

– **8** – If, from two points known in position, we draw two straight lines which meet one another in a point, in such a way that they are equal, then the point where they meet lies on a straight line known in position.

Example: Let there be the two points *A* and *B* known in position, for which we draw the two straight lines *AC* and *BC* which meet one another at the point *C* in such a way that they are equal.

I say that the point C *lies on a straight line known in position.*

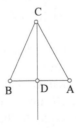

Fig. II.2.1.8

Proof: We join the straight line *AB*, thus it will be known in magnitude and in position because its two endpoints do not change. Let us divide it into two halves at the point *D*; so the point *D* is known because it does not

change. We join *CD*. Since the two straight lines *AD* and *DC* are
<respectively> equal to the two straight lines *BD* and *DC* and the base *AC*
is equal to the base *BC*, accordingly the angle *ADC* is equal to the angle
BDC, so they are right angles. So the straight line *DC* is known in position,
because the two angles which are on either side of this straight line do not
change; and the point *D* does not change, so the point *C* lies on a straight
line known in position, which is the straight line *DC*. This is what we
wanted to prove.

– **9** – If, from two points of known position, we draw two straight lines
which meet one another in a point, in such a way that the ratio of one to the
other is known and there is a ratio of a greater to a smaller, then the point
where they meet lies on the circumference of a circle known in position.

Example: Let there be the two points *A* and *B* known in position, from
which we draw the two straight lines *AC* and *BC* which meet one another at
the point *C*, in such a way that the ratio of *AC* to *CB* is known, and it is a
ratio of a greater to a smaller.

*I say that the point C lies on the circumference of a circle known in
position.*

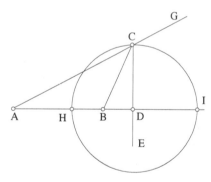

Fig. II.2.1.9

Proof: We join *AB* and we extend it in the direction of *B* to *D*; we
imagine *AC* extended in the direction of *C* to *G* and we imagine the angle
ACE equal to the angle *CBD*. Since *AC* is greater than *CB*, the angle *CBA*
will be greater than the angle *CAB*; now, since the angle *ACE* is equal to
the angle *CBD*, the angle *ECG* is equal to the angle *CBA*, so the angle *ECG*
is greater than the angle *CAB*; now the angle *ACE* is common, so <the
sum> of the two angles *ECG* and *ACE* is greater than <the sum> of the two
angles *CAB* and *ACE*; now <the sum> of the angles *ECG* and *ACE* is equal
to two right angles, so the two angles *CAB* and *ACE* <have a sum> less
than two right angles, so the two straight lines *AB* and *CE* meet; let them

meet at the point D. Thus, the two triangles ACD and CBD are similar because the angle ACD is equal to the angle CBD and the angle CDB is common, and there remains the angle CAD [which is] equal to the angle BCD. So the ratio of AD to DC is equal to the ratio of CD to DB and is equal to the ratio of AC to CB. But the ratio of AC to CB is known, so the ratio of AD to DC is known and the ratio of CD to DB is known. But the ratio of AD to DB is equal to the ratio of the square of AD to the square of DC, and the ratio of the square of AD to the square of DC is known, because the ratio of AD to DC is known, so the ratio of AD to DB is known. We make DH equal to DC, then the ratio of AD to DH is known, the ratio of HD to DB is known, and there remains the ratio of AH to HB [which is] known. But since the ratio of AD to DB is known, the ratio of AB to BD is known; but AB is of known magnitude, so the straight line BD is of known magnitude; now point B on it is known, so the point D is known. Since the ratio of AD to DH is known and AD is of known magnitude, accordingly DH is of known magnitude. But DH is equal to DC. We make D a centre and with distance DH we draw a circle, so it passes through the point C; let the circle be HCI. So the circle HCI is known in magnitude and in position because its centre is known in position, being the point D, and its semidiameter is of known magnitude, being the straight line DH. So the point C lies on the circumference of a circle known in position, which is the circle HCI. This is what we wanted to prove.

Starting out from this proof, we prove that for two straight lines from two points A and B, which meet in a point that lies on the circumference of the circle HCI, the ratio of one to the other is equal to the ratio of AC to CB; and this because if, from the two points A and B, we draw two straight lines which meet one another in [D], an arbitrary point on the circumference of the circle HCI and if, from the point D, we then draw a straight line to this point, two triangles are formed whose [common] vertex is this point, and which are such that the ratio of AD to the straight line extended from the point D to this point is equal to the ratio of this straight line to the straight line DC. Accordingly, the two triangles are similar and the ratio of one of the straight lines to the other is equal to the ratio of AD to DH which is equal to the ratio of AC to CB. So for two straight lines drawn from the two points A and B and which meet one another in a point on the circumference of the circle HCI, the ratio of one to the other is equal to the ratio of AC to CB.

– 10 – If, from two points of known position, we draw two straight lines which meet one another in a point, and if we join the two points with a straight line in such a way that the triangle that is formed is of known magnitude, then the point where they meet lies on a straight line known in position.

Example: Let there be two points *A* and *B*, known in position, from which we draw the two straight lines *AC* and *BC* which meet at the point *C*, in such a way that the triangle *ACB* is of known magnitude.

I say that the point C lies on a straight line known in position.[20]

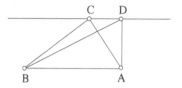

Fig. II.2.1.10

Proof: We join *AB*, it will be of known magnitude, and we draw the straight line *AD* at a right angle [to *AB*]; so the straight line *AD* is known in position because the angle *BAD* does not change and its point *A* does not change. We make the area enclosed by these two straight lines *BA* and *AD* equal to double the triangle *ACB* of known magnitude; this is possible. So *AD* will be of known magnitude, because if its magnitude changed, the area enclosed by the two straight lines *BA* and *AD* would change; but this area does not change, because it is of known magnitude. So the straight line *AD* is known in magnitude and known in position, and its point *A* is known, so the point *D* is known. We join *BD*, so the triangle *BDA* is known in magnitude and it is equal to the triangle *ACB*. We join *DC*, so it will be parallel to the straight line *AB* because the two triangles *ACB* and *ADB* are equal and their two bases are equal,[21] thus the angle *ADC* is a right angle, the straight line *AD* is known in magnitude and in position and its point *D* is known; so the straight line *DC* is known in position and the point *C* lies on a straight line known in position. This is what we wanted to prove.

– **11** – If, between two equal circles, we draw a straight line parallel to the one that joins the centres of these two circles, in such a way that its endpoints are in two similar directions,[22] then this straight line is equal to the straight line between the two centres.

---

[20] Two straight lines parallel to *AB* provide a solution to the problem.

[21] Euclid, *Elements*, I.39.

[22] If a parallel to the line of centres cuts the circles, it cuts each of them in two points. With each of the two points of the first circle we can associate one or other of the two points of the second, and consequently two segments. Ibn al-Haytham makes it precise how to choose which points to associate.

Example: Let there be the two circles *AB* and *CD* whose centres are *E* and *G*; we join *EG* and we draw the straight line *BC* parallel to the straight line *EG*.

I say that the straight line BC *is equal to the straight line* EG.

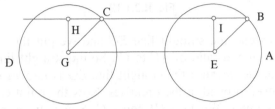

Fig. II.2.1.11

Proof: We join the two straight lines *EB* and *GC*, they will be equal; we erect the two perpendiculars *EI* and *GH*, they will be equal and parallel. But the two straight lines *EB* and *GC* are equal and are in two similar directions with respect to the two perpendiculars *EI* and *GH*; so they are parallel, because the two triangles *BEI* and *CGH* are equal, so the angle *EBI* is equal to the angle *GCH*,[23] and the straight line *BI* will be equal to the straight line *CH*; *IC* is common, so the straight line *BC* is equal to the straight line *IH*; but the straight line *IH* is equal to the straight line *EG*, so the straight line *BC* is equal to the straight line *EG*. This is what we wanted to prove.

– **12** – If, between two equal circles known in magnitude and in position, we draw a straight line parallel to the straight line that joins their centres; if we then extend it in one of the two directions and if we put its ratio to the extended part, a known ratio, then the endpoint of the second straight line lies on the circumference of a circle known in position.

Example: Let there be two equal circles *AB* and *CD*, known in magnitude and in position, and let their centres be *E* and *G*; we join *EG*, we draw the straight line *AC* parallel to the straight line *EG* and we extend it to *I*, in such a way that the ratio of *AC* to *CI* is a known ratio.

I say that the point I *lies on the circumference of a circle known in magnitude and in position.*

---

[23] The following gloss is found in the margin of manuscript [B]: 'Known from the proof of Proposition 7 of Book VI of the *Elements*'.

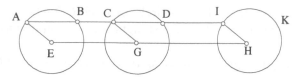

Fig. II.2.1.12

Proof: We extend the straight line *EG* and we put the ratio of *EG* to *GH* equal to the known ratio of *AC* to *CI*. So the straight line *GH* will be known in magnitude because the straight line *EG* is known in magnitude, from what has been proved in the premises. Now the point *G* is known, so the point *H* is known. We join *HI* and *GC*. Since *AC* is parallel to the straight line *EG* and it is equal to it, and since the ratio of *AC* to *CI* is equal to the ratio of *EG* to *GH*, the straight line *CI* is equal to the straight line *GH* and it is parallel to it, so the straight line *HI* is equal to the straight line *GC* and is parallel to it. But the straight line *GC* is known in magnitude, so the straight line *IH* is known in magnitude and the point *H* is known. We take *H* as a centre and with distance *HI* we draw the circle *IK*; it will be known in magnitude and in position, so the point *I* will lie on the circumference of a circle known in magnitude and in position, which is the circle *IK*. This is what we wanted to prove.

– **13** – If, from a known point, we draw to a straight line known in magnitude and in position a straight line which cuts it, [and] if we then extend it in such a way that the ratio of the first straight line to the second straight line is equal to the ratio of the two parts of the straight line known in magnitude and in position, then the endpoint of the second straight line lies on a straight line known in position.

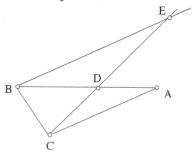

Fig. II.2.1.13

Example: Let there be a straight line *AB* known in magnitude and in position and a known point *C*; from the point *C*, we draw to the straight line *AB* the straight line *CD*, which we extend to *E* in such a way that the ratio of *CD* to *DE* is equal to the ratio of *AD* to *DB*.

I say that the point E *lies on a straight line known in position.*

Proof: We join *AC*, it is known in magnitude and in position. We join *BE*. Since the ratio of *CD* to *DE* is equal to the ratio of *AD* to *DB*,[24] the straight line *BE* is parallel to the straight line *AC*. But the straight line *AC* is known in magnitude and in position and the straight line *AB* is known in position, so the angle *CAB* is known and it is equal to the angle *ABE*, so the angle *ABE* is known. Now, the straight line *AB* is known in position and its point *B* is known, so the straight line *BE* is known in position and the point *E* lies on a straight line known in position, which is the straight line *BE*. This is what we wanted to prove.

– 14 – If, from a known point, we draw to a straight line known in magnitude and in position a straight line which cuts it; if we then extend this straight line, in such a way that the product of the first part and the second is equal to the product of the two parts of the straight line known in magnitude and in position, one with the other, then the endpoint of the second straight line lies on the circumference of a circle known in position.

Example: Let there be the straight line *AB* known in magnitude and in position, and the known point *C*; from the point *C* we draw the straight line *CD* which we extend to *E*, in such a way that the product of *CD* and *DE* is equal to the product of *AD* and *DB*.

I say that the point E *lies on the circumference of a circle known in position.*

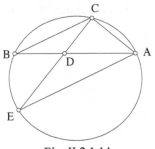

Fig. II.2.1.14

Proof: We join *AC*, *CB* and *AE*; the ratio of *CD* to *DB* is equal to the ratio of *AD* to *DE* and the two angles at the point *D* are equal, so the two

---

[24] The following gloss is found in the margin of manuscript [B]: 'the two angles at *D* are equal, so the two triangles are similar from the Proof 6 of <Book> VI of the *Elements*, so the angle *A* is equal to the angle *B*'.

triangles *AED* and *CBD* are similar and the angle *AEC* is equal to the angle *CBD*. We draw a circle to circumscribe the triangle *ABC*, thus it passes through the point *E*; let the circle be *ACBE*. Since the two points *A* and *C* are known, the straight line *AC* is known in magnitude and in position; and since the points *A*, *B* and *C* are known, the angle *ABC* is known; but since the angle *ABC* is known and the two points *A* and *C* are known, the circle *ACBE* is known in magnitude and in position, as has been proved in Proposition 6 of this treatise. So the point *E* lies on the circumference of a circle known in magnitude and in position. This is what we wanted to prove.

– **15** – If from two known points we draw two straight lines to a circle known in magnitude and in position, if they cut one another in a point inside the circle and if they are extended to end on the circumference of the circle, in such a way that the product of the two parts of one of the two straight lines, one with the other, is equal to the product of the two parts of the other straight line, one with the other; if, with a straight line, we join the first two points in which the two straight lines cut the circle, then this straight line is parallel to the straight line that joins the first two points.

Example: Let there be two points *A* and *B* known in position and the circle *CDEG* known in magnitude and in position; from the two points *A* and *B* we draw two straight lines *ACHG* and *BDHE* which cut one another at the point *H* – the point *H* being inside the circle – in such a way that the product of *AH* and *HG* is equal to the product of *BH* and *HE*. We join *CD* and *AB*.

I say that the straight line CD *is parallel to the straight line* AB.

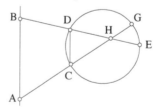

Fig. II.2.1.15

Proof: The product of *AH* and *HG* is equal to the product of *BH* and *HE*, so the ratio of *AH* to *HB* is equal to the ratio of *EH* to *HG*. But the product of *CH* and *HG* is equal to the product of *DH* and *HE*,[25] so the ratio of *EH* to *HG* is equal to the ratio of *CH* to *HD*. In consequence, the ratio of *AH* to *HB* is equal to the ratio of *CH* to *HD*, the triangle *AHB* is thus

---

[25] The power of a point inside a circle (Euclid, *Elements*, III.35).

similar to the triangle *CHD*, so their angles are equal, so the straight line *CD* is parallel to the straight line *AB*. This is what we wanted to prove.

– **16** – If from two known points we draw two straight lines to <the circumference of> a known circle, if they cut one another in a point inside the circle and if, at the point of intersection, they divide one another in the same ratio, then the first two points in which the two straight lines cut the circle lie on the circumference of a circle that passes through the two known points.

Example: Let there be two points *A* and *B* from which we draw, to a circle *CDGE*, the two straight lines *ACHG* and *BDHE*, which cut one another at the point *H*, in such a way that the ratio of *AH* to *HG* is equal to the ratio of *BH* to *HE*.

I say that the two points C *and* D *lie on the circumference of a circle which passes through the two points* A *and* B.

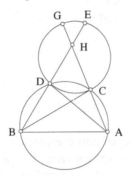

Fig. II.2.1.16

Proof: We join *AD* and *BC*. Since the ratio of *AH* to *HG* is equal to the ratio of *BH* to *HE*, the ratio of *AH* to *HB* is equal to the ratio of *GH* to *HE*. But the ratio of *GH* to *HE* is equal to the ratio of *DH* to *HC* because the product of *CH* and *HG* is equal to the product of *DH* and *HE*, so the ratio of *AH* to *HB* is equal to the ratio of *DH* to *HC*. Now, the angle *AHB* is common to the two triangles *AHD* and *BHC*, so the two triangles *AHD* and *BHC* are similar and the angle *HDA* is equal to the angle *HCB*; so the angle *ADB* is equal to the angle *ACB*. We imagine a circle circumscribed about the triangle *ACB*, it passes through the point *D*, let the circle be *ACDB*. The two points *C* and *D* lie on the circumference of a circle that passes through the two points *A* and *B*. This is what we wanted to prove.

– **17** – If from two points known in position we draw two straight lines to a circle known in magnitude and in position, if they meet one another on the circumference of the circle, if they also reach as far as the circumference of the circle and if they divide one another in the same ratio, then the ratio of the product of one of the two straight lines and the part of it that lies inside the circle to the product of the other straight line and the part of it that lies inside the circle is a known ratio.[26]

Example: Let there be two known points *A* and *B* and the circle *CDE* known in magnitude and in position; from the two points *A* and *B* we draw the two straight lines *ACD* and *BCE* which cut one another at the point *C*, in such a way that the ratio of *AC* to *CD* is equal to the ratio of *BC* to *CE*.

I say that either the product of AD *and* DC *is equal to the product of* BE *and* EC *or its ratio to it is a known ratio.*

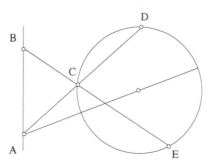

Fig. II.2.1.17

Proof: The point *A* is known and the circle is known in magnitude and in position, so the straight line drawn from the point *A* to the centre of the circle *CDE* and which ends on its circumference is known in magnitude and in position and the part of it that lies outside the circle is of known magnitude because its point of intersection with the circumference of the circle is known; the product of the whole straight line and the part of it that lies inside the circle is thus known, since it is enclosed[27] by two known straight lines. But the product of this straight line and the part of it that lies outside the circle is equal to the product of *DA* and *AC*, so the product of *DA* and *AC* is known. Similarly, we [can] prove that the product of *EB* and *BC* is known. So these two products[28] are known. Either they are equal, or the ratio of one to the other is known. So the ratio of *AC* to *CB* is equal to the ratio of *EB* to *AD* or to a straight line whose ratio to *AD* is known. But the ratio of *AC* to *CB* is equal to the ratio of *DC* to *CE*, so the ratio of *DC*

---

[26] The text seems to assume that the given points lie outside the circle.

[27] The product of the lengths of two segments is the area of the rectangle 'enclosed' by the two segments.

[28] Lit.: surfaces.

to *CE* is equal to the ratio of *EB* to *AD* or to a straight line whose ratio to *AD* is known. If the ratio of *DC* to *CE* is equal to the ratio of *EB* to *AD*, then the product of *AD* and *DC* is equal to the product of *BE* and *EC*; if it is equal to the ratio of *EB* to a straight line whose ratio to *AD* is known, then the ratio of the product of *AD* and *DC* to the product of *BE* and *EC* is known. This is what we wanted to prove.

– **18** – For two known circles that are tangent, one inside the other, if we draw a straight line that cuts the two circles in whatever way, and if we join the point of intersection of the smaller circle to the point of contact with a straight line, then the ratio of the product of the two parts of the straight line that cuts the greater circle, one with the other, to the square of the straight line that joins the point of intersection and the point of contact, is known.

Example: Let there be two circles *ABC* and *ADE* tangent at the point *A*; we draw the straight line *BDG* which cuts the two circles and we join *AD*.

I say that the ratio of the product of BD *and* DG *to the square of* DA *is known.*

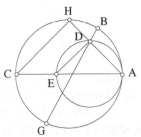

Fig. II.2.1.18

Proof: We draw the common diameter to the two circles, let the diameter be *AEC*; we extend *AD* to *H* and we join *DE* and *HC*. The two angles *ADE* and *AHC* are right angles, so the straight line *DE* is parallel to the straight line *CH*. So the ratio of *HD* to *DA* will be equal to the ratio of *CE* to *EA*. But the ratio of *HD* to *DA* is equal to the ratio of the product of *HD* and *DA* to the square of *DA*, so the ratio of the product of *HD* and *DA* to the square of *DA* is equal to the ratio of *CE* to *EA*. But the product of *HD* and *DA* is equal to the product of *BD* and *DG*, so the ratio of the product of *BD* and *DG* to the square of *DA* is equal to the ratio of *CE* to *EA*. Now the ratio of *CE* to *EA* is known because each of them is known. So the ratio of the product of *BD* and *DG* to the square of *DA* is known. This is what we wanted to prove.

And at this point it becomes clear that if a straight line is drawn from a point of contact and if it cuts the two circles, then it is divided by the smaller circle in a known ratio, let it be the ratio of *AE* to *EC*.

– **19** – For two known circles tangent internally, if we draw a straight line touching the smaller circle, which ends on the greater circle, and if we draw a straight line from the point of contact of the two circles to the end-point of the tangent straight line, then its ratio to the tangent straight line is known.

Example: Let there be two circles *ABC* and *ADG* tangent at the point *A*; we draw the straight line *DB* tangent to the smaller circle and we join *AEB*.
I say that the ratio of AB *to* BD *is known.*

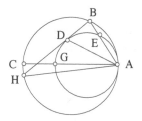

Fig. II.2.1.19

Proof: We draw the common diameter, which is *AGC*, thus the ratio of *AG* to *GC* is known and is equal to the ratio of *AE* to *EB*,[29] so the ratio of *AE* to *EB* is a known ratio. So the ratio of *AB* to *BE* is known and the ratio of the square of *AB* to the product of *AB* and *BE* is known. But the product of *AB* and *BE* is the square of *BD*, so the ratio of the square of *AB* to the square of *BD* is known and is equal to the known ratio of *AC* to *CG*; so the ratio of *AB* to *BD* is known. This is what we wanted to prove.

If we draw *BD* in the other direction to *H* and if we join *AH*, we [can] prove, as has been shown earlier, that the ratio of *AH* to *HD* is known, that the ratio of the square of *AH* to the square of *HD* is equal to the ratio of *AC* to *CG*, and thus that the ratio of *AB* to *BD* is equal to the ratio of *AH* to *HD*. This is what we wanted to prove.

Starting from this, we prove that the ratio of the sum of the two straight lines *BA* and *AH* to the straight line *BH* is known.

– **20** – Let us draw the circle again. We draw the straight line *BDH* tangent to the smaller circle, we join *AD* and we extend it to *I*.
I say that the point I *divides the arc* BIH *into two halves.*

---

[29] From the comment made at the end of Proposition 18.

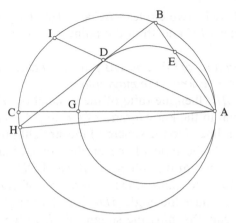

Fig. II.2.1.20

Proof: We join *AB* and *AH*; then the ratio of *AB* to *BD* is equal to the ratio of *AH* to *HD*. By permutation, the ratio of *BA* to *AH* will then be equal to the ratio of *BD* to *DH*. So the straight line *AD* has divided the angle *BAD* into two halves, so the angle *BAD* is equal to the angle *DAH* and the arc *BI* is equal to the arc *IH*. This is what we wanted to prove.

– **21** – Let there be two known circles tangents internally; if from their point of contact we draw the common diameter and if from the endpoint of the diameter on the smaller circle we draw a straight line which cuts the smaller circle, then it is divided into two parts such that the product of one and the other plus a square, whose ratio to the square of the straight line inside the smaller circle is a known ratio, is known in magnitude.

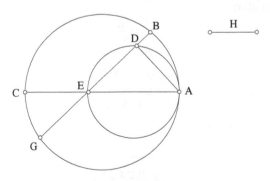

Fig. II.2.1.21

Example: Let there be two circles *ABC* and *ADE* tangent at the point *A*; we draw the diameter *AEC* and from the point *E* we draw the straight line *EDBG*.

I say that the product of GD *and* DB *plus a square, whose ratio to the square of* DE *is known, is known in magnitude.*

Proof: We join *AD*; then the ratio of the product of *GD* and *DB* to the square of *DA* is equal to the known ratio of *CE* to *EA*. Let the ratio of the square of the straight line *H* to the square of the straight line *DE* be equal to the ratio of *CE* to *EA*; the ratio of the product of *GD* and *DB*, plus the square of *H*, to <the sum> of the squares of *AD* and *DE* is thus equal to the ratio of *CE* to *EA*. But <the sum> of the two squares of *AD* and *DE* is equal to the square of *AE* because the angle *ADE* is a right angle. So the ratio of the product of *GD* and *DB*, plus the square of *H*, to the square of *AE* is equal to the ratio of *CE* to *EA* which is equal to the ratio of the product of *CE* and *EA* to the square of *EA*, so the product of *GD* and *DB*, plus the square of *H*, is equal to the product of *CE* and *EA* which is known. But the ratio of the square of *H* to the square of *DE* is known, because it is equal to the ratio of *CE* to *EA*. The product of *GD* and *DB* plus a square, whose ratio to the square of *DE* is known, is thus known in magnitude. This is what we wanted to prove.

– 22 – If in a circle known in magnitude and in position, we draw a diameter known in position on which we take two points on either side of the centre, such that their distances to the centre are two equal distances, and if from these two points we draw two arbitrary straight lines that meet one another in a point on the circumference of the circle, then the sum of their two squares is known and is equal to the sum of the squares of the two parts of the diameter.

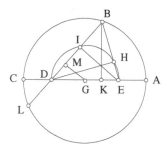

Fig. II.2.1.22

Example: Let there be a circle *ABC* known in magnitude and in position, in which we draw the diameter *AC* known in position; let its centre be

G; on the diameter we take two points E and D, we put EG equal to DG and we draw the two straight lines EB and DB.

I say that the sum of the squares of the two straight lines EB *and* DB *is equal to the sum of the squares of the two known straight lines* AD *and* DC.

Proof: On the straight line ED we draw a semicircle; let the semicircle be HI; let it cut the two straight lines at the points H and I.[30] We join EI, DH and we draw the perpendicular BK. Since the arc $EHID$ is a semicircle, the angle EHD is a right angle and the angle EID is a right angle. But since BK is perpendicular, the angle BKD is a right angle, so the circle circum-scribed about the triangle BKD passes through the point H. So the product of DE and EK is equal to the product of BE and EH. But since each of the angles BKE and BIE is a right angle, the circle circumscribed about the triangle BKE passes through the point I, so the product of BD and DI is equal to the product of ED and DK, so the square of ED is equal to the product of BE and EH, plus the product of BD and DI. We extend BD to L and we draw the perpendicular GM, then M divides DI into two halves and divides BL into two halves. So the straight line BI is equal to the straight line DL and the product of DB and BI is equal to the product of BD and DL. But the product of BD and DL is equal to the product of AD and DC. So the product of DB and BI is equal to the product of AD and DC. But the product of EB and BH is equal to the product of DB and BI. The product of EB and BH, plus the product of DB and BI, is thus equal to twice the product of AD and DC. The sum of the products[31] of BE and EH, of BD and DI, of DB and BI and of EB and BH is equal to the square of ED, plus twice the product of AD and DC. But the sum of these four products[32] is the sum of the squares of EB and of DB. So the sum of the squares of EB and DB is equal to the square of ED, plus twice the product of AD and DC.[33] Twice the product of AD and DC is equal to twice the product of EC and CD. But twice the product of EC and CD, plus the square of ED, is equal to the square of EC, plus the square of CD which is equal to the square of AD, plus the square of DC. So the sum of the squares of EB and BD is equal to the sum of the squares of AD and DC which are known. This is what we wanted to prove.

Whatever the position we have given to the two straight lines EB and DB, the sum of their squares is equal to the sum of the squares of the two

---

[30] The points H and I do not need to lie on the same semicircle of diameter ED. The argument remains valid.

[31] Lit.: surfaces (*suṭūḥ*).

[32] Lit.: surfaces.

[33] The following gloss appears in the margin in manuscript [B]: 'But, by Proof 7 of <Book> II of the *Elements*'.

parts of the diameter, and the proof for all positions is the one that we have set out and differs only in regard to differences in the position of the two points *H* and *I*. It is in fact possible for one of the two straight lines *EB* and *BD* to be a tangent to the smaller circle at the endpoint of its diameter or to cut the other half of the semicircle of the smaller circle. For each of these positions, the sum of the squares of the two straight lines *EB* and *DB* will be equal to the sum of the squares of the two parts of the diameter.

– **23** – If from two points we draw two straight lines which meet in a point and enclose an acute angle in such a way that the sum of their squares is known, then the point where they meet lies on the circumference of a circle known in magnitude and in position.

Example: From two points *A* and *B* we draw two straight lines *AC* and *BC* which meet one another at the point *C* and which enclose an acute angle that is the angle *ACB*, and are such that the sum of their squares is known.

I say that the point C *lies on the circumference of a circle known in magnitude and in position.*

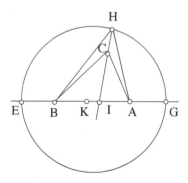

Fig. II.2.1.23

Proof: We join *AB*; it is known and its square will be smaller than the sum of the squares of *AC* and *CB* because the angle *ACB* is acute; the excess <of the sum> of the squares of *AC* and *CB* over the square of *AB* is known. We put twice the product of *AE* and *EB* equal to the excess <of the sum> of the squares of *AC* and *CB* over the square of *AB*. We put *AG* equal to *BE* and on *GE* as diameter we draw a circle; let it be the circle *GHE*.

I say that the circle GHE *passes through the point* C.

If it does not pass through the point *C*, then we divide the angle *ACB* into two halves with the straight line *CI*. We extend *IC* to *H* and we join *AH* and *HB*, then the [sum of the] squares of *AH* and *HB* exceeds the square of *AB* by twice the product of *AE* and *EB*, as has been proved in the previous proposition. But the [sum of the] squares of *AC* and *CB* exceeds the square of *AB* by twice the product of *AE* and *EB*, so <the sum> of the squares of *AH* and *HB* is equal to <the sum of> the squares of *AC* and *CB*. But the angle *ACI* is acute, so the angle *HCA* is obtuse, so the straight line *HA* is greater than the straight line *AC*. Similarly, we [can] prove that the straight line *HB* is greater than the straight line *BC*. So the two straight lines *AH* and *HB* are greater than the two straight lines *AC* and *CB*. <The sum of> the squares of the two straight lines *AH* and *HB* is greater than <the sum of> the squares of *AC* and *CB*; but they are equal, which is impossible. So the point *C* lies on the circumference of the circle *GHE* and the circle *GHE* is known in magnitude and in position, because its diameter, which is *GE*, is known in magnitude and in position. So the point *C* lies on the circumference of a circle known in magnitude and in position. This is what we wanted to prove.

– **24** – If, in a circle known in magnitude and in position, we draw an arbitrary chord which we divide into two parts in such a way that the product of one of the two parts and the other is known, then the point of division lies on the circumference of a circle known in position and in magnitude.

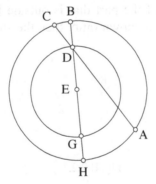

Fig. II.2.1.24

Example: Let there be a circle *ABC* known in magnitude and in position in which we draw the arbitrary chord *AC* which we divide at the point *D*, in such a way that the product of *AD* and *DC* is known.

I say that the point D *lies on the circumference of a circle known in magnitude and in position.*

Proof: Indeed, let us name the centre of the circle, say as the point *E*. We join *ED* and we extend it in both directions to *B* and *H*, then the product of *HD* and *DB* is equal to the product of *AD* and *DC*.[34] But the product of *AD* and *DC* is known, so the product of *HD* and *DB* is known. But the diameter *HB* is known, so half of it, which is *EB*, is known; there remains the square of *ED*, which is known, so the straight line *ED* is known.

We make *E* a centre and with the known distance *ED* we draw a circle, let the circle be *DG*. So the circle *DG* is known in magnitude and in position because its centre is known in position and its semidiameter is of known magnitude. So the point *D* lies on the circumference of a circle known in magnitude and in position. This is what we wanted to prove.

SECOND PART

The work in this part is of the same kind as what was set out by Euclid in his book the *Data*, although nothing in this part is to be found in the book the *Data*.

– **1** – If, from a known point, we draw to a circle known in magnitude and in position, a straight line which cuts the circle; if the point lies outside the circle and if the ratio of the part that is outside [the circle] to the part that lies inside the circle is a known ratio, then the straight line is known in position.

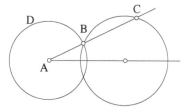

Fig. II.2.2.1

Example: Let there be a known point *A* and a circle *BC* known in magnitude and in position; we draw the straight line *ABC* in such a way that the ratio of *AB* to *BC* is known.

I say that the straight line ABC *is known in position.*

Proof: The point *A* is known and the circle *BC* is known in magnitude and in position, so the straight line drawn from the point *A* to the centre of the circle, and which ends on its circumference, is known in magnitude and in position; the part which lies outside the circle is of known magnitude, so

---

[34] Euclid, *Elements*, III.35.

the product of *CA* and *AB* is of known magnitude and the ratio of *CA* to *AB* is known; but it is equal to the ratio of the product of *CA* and *AB* to the square of *AB*. So the square of *AB* is known and the straight line *AB* is of known magnitude. But the point *A* is known, so the point *B* lies on the circumference of a circle known in position, as has been proved in the first proposition of this book; let the circle be *BD*. The circle *BD* is known in position and the circle *BC* is known in position, so the point *B* is known and the point *A* is known, so the straight line *AB* is known in position, so the straight line *ABC* is known in position. This is what we wanted to prove.

– **2** – If, from a known point, we draw to a circle known in position, a straight line which cuts off from the circle a known segment,[35] then the line is known in position.

Example: Let there be a known point *A* and a circle *BC* known in position; we draw the straight line *ABC* in such a way that the segment *BC* is known.

I say that the straight line ABC *is known in position.*

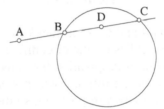

Fig. II.2.2.2

Proof: The point *A* is known, so the product of *CA* and *AB* is known. But since the segment *BC* is known and the circle is known, the straight line *BC* is known.[36] So the straight line *BC* is known and the product of *CA* and *AB* is known. The ratio of twice the product of *CA* and *AB* to the square of *BC* is thus known. We divide *BC* into two halves at the point *D*, then the ratio of the product of *CA* and *AB* to the square of *BD* is known, so the ratio of the square of *AD* to the square of *DB* is known. So the ratio of *AD* to *DB*, that is *DC*, is known. So the ratio of *AC* to *CB* is known, so the ratio of *AB* to *BC* is known. But the point *A* is known and the circle *BC* is known, so the straight line *ABC* is known in position. This is what we wanted to prove.

[35] Euclid, Definitions 7 and 8 of the *Data*.
[36] Euclid, *Data*, 88 and 89.

– **3** – If, from a known point, we draw to a straight line known in magnitude and in position, a straight line such that its ratio to what it cuts off from the <first> straight line is a known ratio, then the straight line that has been drawn is known in position.

Example: Let there be a known point A and a straight line BC known in magnitude and in position; we draw the straight line AD in such a way that the ratio of AD to DC is a known ratio.

I say that the straight line AD *is known in position.*

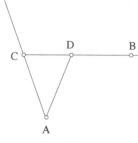

Fig. II.2.2.3

Proof: The two points A and C are in fact known and the ratio of AD to DC is known, so the point D lies on the circumference of a circle known in position, as has been proved in Proposition 9 of this treatise. So the point D lies on the circumference of a circle known in position and it lies on the straight line BC [which is] known in position; so the point D is known and the point A is known, so the straight line AD is known in position. This is what we wanted to prove.

– **4** – If, from a known point, we draw to two parallel straight lines known in magnitude and in position, a straight line which cuts off two alternate straight lines,[37] in such a way that the ratio of one to the other is known, then the straight line that is drawn is known in position.

Example: Let there be a known point A and two straight lines BC and DE known in position[38] and parallel; we draw the straight line AHG in such a way that the ratio of HC to DG is known.

I say that the straight line AG *is known in position.*

---

[37] Here 'alternate' is the translation of *mutabādilayn*, which refers to the two segments DG and CH which lie on either side of HG.

[38] We need to assume that CB and DE are parallel segments and in contrary senses so that CH and DG will be so too. We may note that the points B and E play no part in what follows.

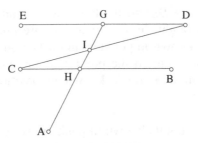

Fig. II.2.2.4

Proof: We join *DC*, it is known in magnitude and in position because its two endpoints are known, and it cuts the straight line *HG*; let it cut it at the point *I*. So the ratio of *CI* to *ID* is known, so the ratio of *CD* to *DI* is known. But *CD* is known, so *DI* is known. But the point *D* is known, so the point *I* is known. But the point *A* is known, so the straight line *AIG* is known in position. This is what we wanted to prove.

– **5** – If, from a known point, we draw to a straight line known in position and in magnitude, a straight line such that, added to what it cuts off from the known straight line, <it gives a sum that is> known, then it is known in position.

Example: Let there be a known point *A* and a straight line *BC* known in magnitude and in position; we draw the straight line *AD* in such a way that *AD* plus *DC* is known.

I say that AD *is known in position.*

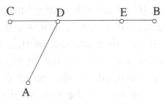

Fig. II.2.2.5

Proof: AD plus *DC* is a known <sum>; *BD* plus *DC* is a known <sum>, so either the two straight lines *AD* and *DB* are equal, or one is greater than the other by a known magnitude.

If they are equal, then from the two known points *A* and *B* we have drawn two equal straight lines *AD* and *DB*, so the point *D* lies on a straight line known in position, as has been proved in Proposition 8 of this treatise.

If one is greater than the other by a known magnitude, let *BE* be that excess; so *BE* is known, so the point *E* is known, and the straight line *AD* is equal to the straight line *DE*; so the two points *A* and *E*, from which we drew the two equal straight lines *AD* and *DE*, are known. So the point *D* lies on a straight line known in position and it lies on the straight line *EC*, which is known in position; so the point *D* is known. But the point *A* is known, so the straight line *AD* is known in position. This is what we wanted to prove.

– 6 – If, from two points known in position, we draw to a straight line known in position two straight lines that enclose a known angle, then they are known in position and in magnitude.

Example: Let there be two known points *A* and *B* and the straight line *CD* known in position; we draw two straight lines *AE* and *BE* which enclose a known angle; let it be the angle *AEB*.

I say that the two straight lines AE *and* BE *are known in magnitude and in position.*

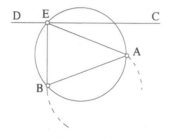

Fig. II.2.2.6

Proof: The two points *A* and *B* are known, from them we have drawn two straight lines *AE* and *BE* which enclose a known angle; then the point *E* lies on the circumference of a circle known in position, as has been proved in Proposition 6 of the first section of this treatise. So the point *E* lies on the circumference of a circle known in position; but it lies on the straight line *CD* known in position, so the point *E* is known. But each of the two points *A* and *B* is known, so the two straight lines *AE* and *BE* are known in magnitude and in position. This is what we wanted to prove.

– 7 – If, from two known points we draw, to a straight line known in position, two straight lines in such a way that the ratio of one to the other is known, then the two straight lines are known in position and in magnitude.

Example: Let there be two known points *A* and *B* and the straight line *CD* known in position; we draw the two straight lines *AE* and *BE* in such a way that the ratio of one to the other is known.

I say that each of the two straight lines AE *and* BE *is known in magnitude and in position.*

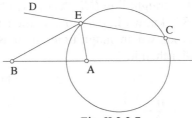

Fig. II.2.2.7

Proof: The point *E* lies on the circumference of a circle known in position, as has been proved in Proposition 9 of the first section of this treatise. But it lies on the straight line *DC*, so the point *E* is known and the two straight lines *AE* and *BE* are known in magnitude and in position. This is what we wanted to prove.

– **8** – If we have two parallel straight lines known in position, if we take two points on one [of them] and if we draw from these two points two straight lines which meet one another in a point on the other parallel straight line in such a way that the product of one of the two straight lines [we have] drawn and the other is known, then the two straight lines are known in magnitude and in position.

Example: Let there be two parallel straight lines *AB* and *CD*, known in position; on the straight line *AB* we take two points *E* and *G* and from these points we draw two straight lines *EH* and *GH* in such a way that the product of *EH* and *HG* is known.

I say that the two straight lines EH *and* GH *are known in magnitude and in position.*

Proof: We imagine an angle *GHI* equal to the angle *HEG*, then the straight line *HI* meets the straight line *GA*, because <the sum> of the two angles *IHG* and *IGH* is smaller than two right angles; let it meet it at the point *I*. The triangle *IHG* is then similar to the triangle *HEG*, so the angle *HIG* is equal to the angle *EHG*, the ratio of *IG* to *GH* is equal to the ratio of *HG* to *GE* and it is equal to the ratio of *IH* to *HE*, so the ratio of *IH* to *HE* is equal to the ratio of *HG* to *GE* and the product of *HI* and *EG* is equal to the product of *EH* and *HG*. But the product of *EH* and *HG* is known, so the product of *IH* and *EG* is known. But *EG* is known, so *IH* is known, because if two straight lines enclose a known area with a right angle and if one of the two straight lines is known, then the other straight line is known, because the magnitude of the area does not change and the angle of the area does not change, so the magnitude of the other straight line does not change, so the straight line *IH* is known. We take an arbitrary point on the straight line *CD*, let it be *D*; from the point *D*, at a right angle, which is the angle *CDB*, we draw the straight line *DB*; thus it will be known in position. But the straight line *AB* is known in position, so the point *B* is known, then the straight line *DB* is known in magnitude, so it is either equal to the straight line *HI* or smaller than it.

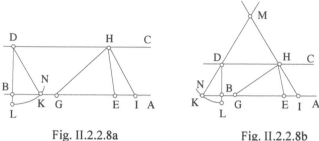

Fig. II.2.2.8a   Fig. II.2.2.8b

If it is equal, then the straight line *HI* is perpendicular and the angle *HIG* is a right angle. But it is equal to the angle *EHG*, so the angle *EHG* is a right angle. If the straight line *DB* is smaller than the straight line *HI*, we make the straight line *DL* equal to the straight line *HI*. We make *D* a centre and with distance *DL* we draw the circle *LKN*. So this circle is known in position. But the straight line *AB* is known in position, so the point *K* is known. We join *DK*, thus it is known in magnitude and in position, because the two points *D* and *K* are known; but the straight line *DK* is equal to the straight line *HI*; either they are parallel or they meet one another. If they are parallel, then the angle *HIG* is equal to the angle *DKB*. But the angle *DKB* is known, because the two straight lines *DK* and *KB* are known in position, so the angle *HIG* is known, so the angle *EHG* is known. And if the two straight lines *IH* and *KD* meet one another, let them meet one another at the point *M*. So the ratio of *IM* to *MK* is equal to the ratio of *IH* to *KD*; but *IH* is equal to *KD*, so *IM* is equal to *MK*, so the angle *MIK* is equal to the angle *MKI*; but the angle *MKI* is known, so the angle *MIK* is known, so the angle *EHG* is known.

In all cases, the angle *EHG* is thus known and the two points *E* and *G* are known, so the point *H* lies on the circumference of a circle known in position, as has been proved in Proposition 6 of the first section of this treatise. But the point *H* lies on the straight line *CD* which is known in position, so the point *H* is known, so each of the straight lines *EH* and *GH* is known in magnitude and in position. This is what we wanted to prove.

– **9** – If we have two parallel straight lines known in position, if we take two points on one of them and if from these two points we draw two straight lines which cut the second straight line, traverse it and meet one another in a point in such a way that the triangle that is formed is of known magnitude, then the straight line traversed by the two straight lines that start on the second parallel straight line is of known magnitude.

Example: Let there be two parallel straight lines *AB* and *CD* known in position; we take two arbitrary points on the straight line *AB*, which are the two points *E* and *G*, and from the two points *E* and *G* we draw two straight lines *EIH* and *GKH* which meet one another at the point *H* in such a way that the triangle *EHG* is known in magnitude.

I say that the straight line IK *is known in magnitude.*

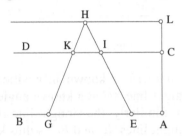

Fig. II.2.2.9

Proof: The two points *E* and *G* are known, from them we have drawn the two straight lines *EH* and *GH*, and thus we have formed the triangle *EHG* which is known in magnitude; so the point *H* lies on a straight line known in position, parallel to the straight line *EG*, as has been proved in Proposition 10 of the first section of this treatise; let this straight line be the straight line *LH*. We draw the perpendicular *ACL*; it will thus be known in magnitude. Since the straight line *AL* is known in position and the straight line *LH* is known in position, accordingly the point *L* is known; but the point *A* is known, so the straight line *AL* is known in position and in magnitude.

Similarly, we [can] prove that the straight line *LC* is known in magnitude and in position; so the ratio of *AL* to *LC* is known and the ratio of *EH* to *HI* is known; the ratio of *EG* to *IK* is thus known and *EG* is known, so *IK* is known. So the straight line *IK* is known in magnitude. This is what we wanted to prove.

– 10 – If, from the endpoints of a straight line known in position, we draw two straight lines [that are] at two known angles and which meet one another in a point, then they are known in magnitude and in position.

Example: Let there be a straight line *AB* known in magnitude and in position; from its endpoints we draw the two straight lines *AC* and *BC* [that are] at two known angles, and which meet one another at the point *C*.

I say that the two straight lines AC *and* BC *are known in magnitude and in position.*

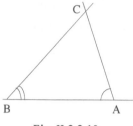

Fig. II.2.2.10

Proof: The straight line *AB* is known in position and its point *A* is known; we draw the straight line *AC* at a known angle, so the straight line *AC* is known in position. Similarly, the straight line *BC* is known in position. Each of the two straight lines *AC* and *BC* is thus known in position, so the point *C* is known. But the two points *A* and *B* are known, so each of the two straight lines *AC* and *BC* is known in magnitude and in position. This is what we wanted to prove.

If each of the straight lines *AB*, *AC* and *BC* is of known magnitude, then the ratio of each of them to one of the two others is known. We shall prove from this proof that for any triangle with known angles, the ratios of its sides, two by two,[39] are known. In fact, if the angles of the triangle are known, if we take a straight line known in magnitude and in position and if from its endpoints we draw two straight lines at angles equal to two of the angles of the triangle whose angles are known, a triangle is formed whose sides are known; the ratios of its sides, two by two,[40] are known, as has been proved in this proposition, and the triangle formed is similar to the triangle whose angles are known; it follows that the ratios of the sides of the triangle, two by two,[41] [in the triangle] whose angles are known, are known.

– **11** – If we extend one of the sides of a triangle whose sides are known in magnitude and in position, if we take a known point on the extension and if from this point we draw a straight line that intersects the triangle and cuts off from its two sides two straight lines on the side

---

[39] Lit.: the ones to the others.
[40] Lit.: the ones to the others.
[41] Lit.: the ones to the others.

towards its base in such a way that the ratio of one to the other is known, then the straight line is known in position.

Example: Let there be a triangle *ABC* whose sides are known in magnitude and in position; we extend one of its sides, which is *BC*; on the extension we take a point *D* and from the point *D* we draw the straight line *DEG* in such a way that the ratio of *GC* to *EB* is known.

I say that the straight line DEG *is known in position.*

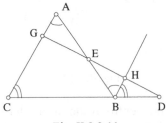

Fig. II.2.2.11

Proof: We draw the straight line *BH* parallel to the straight line *AC*; the ratio of *GC* to *HB* is thus equal to the ratio of *CD* to *DB*. But the ratio of *CD* to *DB* is known, because each of them is of known magnitude; so the ratio of *GC* to *BH* is known. But the ratio of *GC* to *EB* is known, so the ratio of *EB* to *BH* is known, since the ratios of *GC*, *EB* and *HB*, two by two,[42] *are known ratios of three magnitudes, two by two, so the ratio of *EB* to *BH* is the known ratio of two magnitudes one to the other,*[43] so the ratio of *EB* to *BH* is known and the angle *EBH* is known because it is equal to the angle *BAC* which is known, so the angle *BHE* is known as has been proved in the premises. So the angle *BHE* is known and the angle *HBD* is known because it is equal to the angle *ACB*; there then remains the angle *HDB* [which is] known, so the straight line *DG* is known in position and the straight line *AC* is known in position, so the point *G* is known and the point *D* is known, so the straight line *DEG* is known in magnitude and in position. This is what we wanted to prove.

– **12** – If we have a circle known in magnitude and in position and a straight line known in position and if we draw a straight line tangent to the circle, which ends on the straight line known in position, and which is known in magnitude, then it is known in position.

---

[42] Lit.: the ones to the others.

[43] *...* In the text, we may read: 'are the ratios of three known magnitudes, two by two, then the ratio of *EB* to *BH* is the ratio of two known magnitudes one to the other'.

Example: The circle *AB* is known in magnitude and in position and the straight line *DC* is known in position; we draw the straight line *BE* tangent to the circle such that *BE* is known in magnitude.

I say that it is known in position.

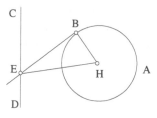

Fig. II.2.2.12

Proof: We mark off the centre of the circle, let it be *H*, and we join *HB* and *HE*. Since the circle is known in magnitude and in position, the straight line *HB* is known in magnitude. Since the straight line *BE* is a tangent, the angle *HBE* is a right angle. Since *BE* is known in magnitude, the ratio of the straight line *HB* to the straight line *BE* is known. Since the angle *HBE* is a right angle, the position of the straight line *BE* with respect to the straight line *BH* is known. Since the straight line *HB* is known in magnitude, and its ratio to the straight line *BE* is known and the angle *HBE* is a right angle, the angle *HEB* is known and the straight line *HE* is known in magnitude, as has been proved in the premises. Since the point *H* is known and the straight line *HE* is known in magnitude, the point *E* lies on the circumference of a circle known in position, as has been proved in the first proposition of this treatise. Since the point *E* lies on the circumference of a circle known in position and it lies on the straight line *CD* known in position, the point *E* is known. But the point *H* is known, so the straight line *EH* is known in position. But the angle *HEB* is known, so the straight line *EB* is known in position. This is what we wanted to prove.

– **13** – If we have a circle known in magnitude and in position and a straight line known in position, if from <the circumference> of the circle we draw a straight line to the straight line known in position which encloses a known angle with the latter and if the straight line we have drawn is known in magnitude, then it is known in position.

Example: Let there be a circle *AB* known in magnitude and in position and a straight line *CD* known in position; we draw the straight line *BE* which encloses with the straight line *CD* a known angle which is the angle *BEC*, in such a way that *BE* is known in magnitude.

I say that it is known in position.

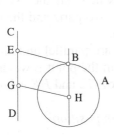
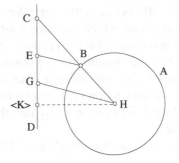

Fig. II.2.2.13a   Fig. II.2.2.13b

Proof: We mark off the centre of the circle; let it be the point *H*. From the point *H* we draw a straight line *HG* which encloses with the straight line *CD* an angle equal to the known angle *BEC*, which is the angle *HGC*. We join *HB*; either it is parallel to *CD* or it meets it. If *HB* is parallel to the straight line *CD*, the area *HBEG* is a parallelogram. But since the point *H* is known and the angle *HGC* is known, the straight line *HG* is known in magnitude and in position, because if we take the point *C* as known, the two points *H* and *C* are known, the point *G* will thus lie on the circumference of a circle known in position. But it lies on the straight line *CD* known in position, so the point *G* is known and the point *H* is known; so the straight line *HG* is known in magnitude and in position. And if *HB* is parallel to *CD*, the angle *GHB* is known because it is equal to the known angle *DGH*; so the straight line *HB* is known in position and the circle *AB* is known in position, so the point *B* is known. But from this latter we have drawn the straight line *BE* at a known angle, which is the angle *HBE*, because it is equal to the angle *HGE*; so the straight line *BE* is known in position.

If the straight line *HB* meets the straight line *CD* – let it meet it at the point *C* – the ratio of *HG* to *BE* is then equal to the ratio of *HC* to *CB*. But the ratio of *HG* to *BE* is a known ratio because each of the two straight lines is known, so the ratio of *HC* to *HB* is known, the ratio of *HB* to *BC* is then known. But *HB* is known in magnitude, so the straight line *BC* is known in magnitude and, in consequence, the straight line *HC* is known in magnitude, so the point *C* lies on the circumference of a circle known in position and it lies on the straight line *CD* known in position; so the point *C* is known; and the point *H* is known, so the straight line *HC* is known in magnitude and in position. But the straight line *HG* is known in position, so the angle *CHG* is known. So the angle *HBE* is known and the straight line *HB* is known in position, so the straight line *BE* is known in position. This is what we wanted to prove.

– **14** – If, between two parallel straight lines known in position, we take a point from which we draw a straight line which cuts the two straight lines in such a way that the product of one of the two parts and the other is known, then the straight line is known in position.

Example: The two straight lines *AB* and *CD* are parallel and known in position; we take a point *E* between them and from the point *E* we draw the straight line *EGH* in such a way that the product of *GE* and *HE* is known in magnitude.

I then say that the straight line GH *is known in position.*

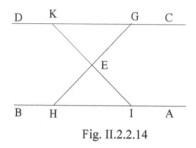

Fig. II.2.2.14

Proof: We take a point *I* on the straight line *AB* and we join *EI*; it will be known in magnitude and in position. We extend *IE* to *K*, so *EK* will be known in position. But the straight line *CD* is known in position, so the point *K* is known. But the point *E* is known, so the straight line *EK* is known in magnitude and in position, the ratio of *IE* to *EK* is then known and is equal to the ratio of *HE* to *EG*. So the ratio of *HE* to *EG* is known; in consequence, the ratio of the product of *HE* and *EG* to the square of *EG* is known; but the product of *HE* and *EG* is known, so the square of *EG* is known. So the straight line *EG* is known in magnitude, the point *G* thus lies on the circumference of a circle known in position; now it lies on the straight line *CD*, which is known in position; so the point *G* is known, so the straight line *GEH* is known in position. This is what we wanted to prove.

– **15** – Let there be a triangle with known sides and angles; if we draw a straight line from its vertex to its base, in such a way that the ratio of the square of the straight line [we have] drawn to the product[44] of the two parts <cut off> on the base is a known ratio, then the straight line [we have] drawn is known in position.

---

[44] Lit.: surface.

Example: The triangle *ABC* has known sides and angles; we draw inside this triangle the straight line *AD* in such a way that the ratio of the square of *AD* to the product of *BD* and *DC* is a known ratio.

I say that the straight line AD *is known in position.*

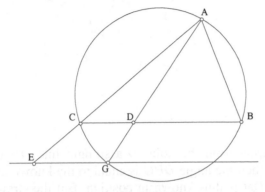

Fig. II.2.2.15

Proof: We circumscribe a circle about the triangle *ABC*; let the circle be *ABC*, and we extend *AD* to *G*; the product of *AD* and *DG* is then equal to the product of *BD* and *DC*. So the ratio of the square of *AD* to the product of *AD* and *DG* is known and is equal to the ratio of *AD* to *DG*. So the ratio of *AD* to *DG* is known; let it be equal to the ratio of *AC* to *CE*; so *CE* is known. We join *EG*, it will be parallel to the straight line *CD* and the angle *AEG* will be equal to the known angle *ACD*, so the angle *AEG* is known and the straight line *CE* is known in magnitude and in position; so the straight line *EG* is known in position; but the circle *ABC* is known in position, so the point *G* is known. But the point *A* is known, so the straight line *AG* is known in position and, in consequence, the straight line *AD* is known in position. This is what we wanted to prove.

– 16 – Let there be two straight lines which cut one another and which are known in position; if we take a point between them and we draw from this point a straight line which cuts the two straight lines known in position in such a way that the ratio of one of the two parts to the other is known, then the straight line is known in magnitude and in position.

Example: Let there be two straight lines *AB* and *AC* known in position and a given point *D*; from the point *D* we draw a straight line *EDG* in such a way that the ratio of *ED* to *DG* is known.

I say that the straight line EG *is known in magnitude and in position.*

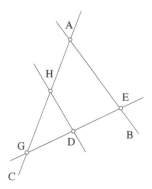

Fig. II.2.2.16

Proof: We draw from the point *D* a straight line *DH* parallel to the straight line *AB*, then the angle *DHG* is equal to the known angle *BAC* and the straight line *DH* is thus known in position. But the straight line *AC* is known in position, so the point *H* is known and the straight line *AH* is known in magnitude. But the ratio of *AH* to *HG* is equal to the ratio of *ED* to *DG* which is known, so the ratio of *AH* to *HG* is known. But *AH* is known in magnitude, so the straight line *HG* is known in magnitude. But the point *H* is known, so the point *G* is known. But the point *D* is known, so the straight line *GD* known in magnitude and in position and its ratio to *DE* is known, so the straight line *DE* is known in magnitude; so the straight line *GDE* is known in magnitude and in position. This is what we wanted to prove.

– **17** – Let there be two straight lines which cut one another and which are known in position; if we take a point between them and we draw from this point a straight line which cuts the two straight lines known in position in such a way that the product of one of the two parts and the other is known, then the straight line is known in magnitude and in position.

Example: The two straight lines *AB* and *AC* are known in position and the point *D* is known; we draw the straight line *DGE* in such a way that the product of *DE* and *DG* is known.

I say that the straight line EG *is known in magnitude and in position.*

Proof: We join *AD*; it will be known in magnitude and in position. We put the product of *AD* and *DH* equal to the product of *ED* and *DG* which is known, then the product of *AD* and *DH* is known. But *AD* is known, so *DH* is known, as has been proved in Proposition 10 of the first section of this treatise. But the point *D* is known, so the point *H* is known. We join *GH*, so we shall have the ratio of *AD* to *DE* equal to the ratio of *GD* to *DH*; so the triangle *DGH* is similar to the triangle *AED* and the angle *EAD* is thus equal to the angle *DGH*; but the angle *EAD* is known, so the angle *DGH* is known. But the straight line *DH* is known in magnitude and in position, so the point *G* lies on the circumference of a circle known in position, as has been proved in Proposition 6 of the first section of this treatise.

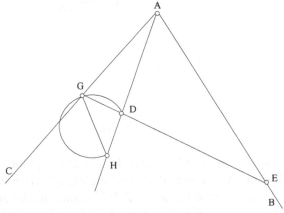

Fig. II.2.2.17

But the point *G* lies on the straight line *AC* which is known in position, so the point *G* is known; but the point *D* is known, so the straight line *DG* is known in magnitude and in position, its product with *DE* is known, so the straight line *DE* is known in magnitude and in position and, in consequence, the straight line *EDG* is known in magnitude and in position. This is what we wanted to prove.

– **18** – Let there be a circle known in magnitude and in position in which we draw a chord which cuts off a known segment from it; if we then take a point on one of the two arcs, not its mid point; if from this point we draw a straight line to the other segment and if we join the two endpoints of the chord to the endpoint of the straight line with two straight lines in such a way that the ratio of the sum of these two straight lines to the first straight line is a known ratio, then the first straight line is known in magnitude and in position.

Example: Let there be the circle *ABCD* known in magnitude and in position; in this circle we draw the chord *AC* which cuts off a known segment from it; we take a point *D* on the arc *ADC* in such a way that the two arcs *AD* and *DC* are different, we draw the straight line *DB* and we join the two straight lines *AB* and *CB* in such a way that the ratio of the sum of the two straight lines *AB* and *CB* to the straight line *DB* is known.

I say that the straight line DB *is known in magnitude and in position.*

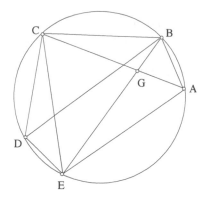

Fig. II.2.2.18

Proof: We cut the arc *ADC* into two halves at the point *E* and we join *AE*, thus it will be known in magnitude and the triangle *ABE* will be similar to the triangle *AEG*, because the angle *EAG* is equal to the angle *ABE*; so the ratio of *BE* to *EA* is equal to the ratio of *BA* to *AG* which is equal to the ratio of the sum of *AB* and *BC* to *AC* and the ratio of *BE* to *EA* is equal to the ratio of the sum of *AB* and *BC* to *AC*; so the ratio of the sum of *AB* and *BC* to *BE* is equal to the known ratio of *CA* to *AE*, so the ratio of the sum of *AB* and *BC* to *BE* is known; but the ratio of the sum of *AB* and *BC* to *BD* is known, so the ratio of *EB* to *BD* is known. But the angle *EBD* is known, because the arc *DE* is known, so the angle *EDB* is known, as has been proved in the premises.[45] But the straight line *ED* is known in magnitude and in position, so the straight line *DB* is known in position; but the circle *ABC* is known in position, so the point *B* is known, so the straight line *DB* is known in magnitude and in position. This is what we wanted to prove.

– **19** – Let there be a known angle of a triangle; from the known angle we draw a straight line which divides the known angle into two known parts, then the ratio of the two parts of the base, one to the other, is equal to the ratio of one of the sides which enclose the known angle to a straight line whose ratio to the remaining side is known.

---

[45] The comment made in the premises (p. 384) applies to a triangle in which we know two sides and the angle between them: so we necessarily know the ratio of these two sides. But here the result: 'so the angle *EDB* is known', derives from the fact that the triangle *BDE* is defined up to similarity (Euclid, *Data*, 51); so its other angles are known (Euclid, *Elements*, IV.6).

Example: Let there be a triangle *ABC* whose angle *BAC* is known; we draw the straight line *AD* in such a way that one of the two angles *BAD* and *DAC* is known.[46]

I say that the ratio of CD *to* DB *is known and that it is equal to the ratio of* CA *to a straight line whose ratio to* AB *is known.*

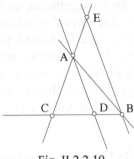

Fig. II.2.2.19

Proof: We make the angle *ABE* equal to the known angle *BAD*, so the straight line *BE* will be parallel to the straight line *AD*. We draw the straight line *CA* to meet it; let it meet it at the point *E*. So the angle *BEA* will be equal to the known angle *DAC*. So each of the angles of the triangle *ABE* will be known, the ratios of its sides, two by two,[47] are thus known, as has been proved in Proposition 10 of the second section of this treatise. So the ratio of *EA* to *AB* is known. But the ratio of *CD* to *DB* is equal to the ratio of *CA* to *AE*, so the ratio of *CD* to *DB* is equal to the ratio of *CA* to a straight line whose ratio to *AB* is known. This is what we wanted to prove.

– **20** – Let there be a triangle with known angles; from one of its angles we draw a straight line which divides its base in a known ratio, then this straight line is known in position.

Example: The triangle *ABC* has known angles and we draw the straight line *AD* in such a way that the ratio of *CD* to *DB* is known.

I say that the straight line AD *is known in position.*

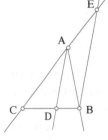

Fig. II.2.2.20

---

[46] If one of these angles is known, the other is also known, because the angle *BAC* is known and *AD* is inside that angle.

[47] Lit.: one to the other.

Proof: We put the ratio of *CA* to *AE* equal to the ratio of *CD* to *DB* which is known. We join *BE*, it will be parallel to the straight line *AD*; so the angle *BEA* will be equal to the angle *DAC*. But since the angles of the triangle *ABC* are known, the ratio of *CA* to *AB* is known. But the ratio of *CA* to *AE* is known, so the ratio of *BA* to *AE* is known, because these two ratios can only be between three magnitudes, so the ratios of each of them to the two others are known. But since the ratio of *BA* to *AE* is known and the angle *BAE* is known, the triangle *BAE* has known angles, as was proved in the premises. So the angle *BEA* is known and it is equal to the angle *DAC*, so the angle *DAC* is known; in consequence, the straight line *AD* is known in position with respect to the straight line *AC* and to the straight line *AB*. So the straight line *AD* is known in position. This is what we wanted to prove.

– **21** – Let there be a circle known in magnitude and in position, we take two points on its circumference; if from these two points we draw two straight lines which meet one another in a point on the circumference of the circle, and if we join the two points with a straight line in such a way that the triangle that is formed is known in magnitude, then each of the two straight lines drawn from the two points is known in magnitude and in position.

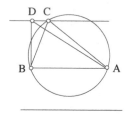

Fig. II.2.2.21

Example: The circle *ABC* is known in magnitude and in position; on its circumference we take the two points *A* and *B* from which we draw the two straight lines *AC* and *BC*, and we join *AB* in such a way that the triangle *ACB* is known in magnitude.

I say that each of the two straight lines AC *and* BC *is known in magnitude and in position.*

Proof: From the point *B* we draw the straight line *BD* at a right angle and we construct the rectangle enclosed by the two straight lines *AB* and *BD* [to be] equal to double the triangle *ACB* which is known in magnitude. The straight line *BD* will thus be known in magnitude, because *AB* is known in magnitude. We join *AD*, so the triangle *ADB* will be equal to the triangle *ACB*. We join *DC*; *DC* will be parallel to the straight line *AB*. So the angle *BDC* is a right angle, so the straight line *DC* is known in position. But the circle *ACB* is known in position; accordingly the point *C* is known. But each of the two points *A* and *B* is known, so each of the straight lines *AC* and *BC* is known in magnitude and in position. This is what we wanted to prove.

– **22** – Let there be a circle known in magnitude and in position; we take two points on its circumference and we draw from these two points two straight lines which meet one another in a point on the circumference of the circle in such a way that their product with one another is known, then each of them is known in magnitude and in position.

Example: The circle *ABC* is known in magnitude and in position; we take the two points *A* and *B* on its circumference, from which we draw the two straight lines *AC* and *BC* in such a way that the product of *AC* and *CB* is known.

I say that each of the straight lines AC *and* BC *is known in magnitude and in position.*

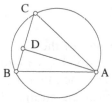

Fig. II.2.2.22

Proof: We draw the perpendicular *AD*. Since the two points *A* and *B* are known, the segment *ACB* is known, so the angle *ACB* is known. But the angle *ADC* is a right angle, so the angles of the triangle *ACD* are known, so the ratio of *CA* to *AD* is known, and the ratio of the product of *AC* and *CB* to the product of *AD* and *CB* is known. But the product of *AC* and *CB* is known, so the product of *AD* and *CB* is known. Now the product of *AD* and *CB* is double the triangle *ACB*; so double the triangle *ACB* is known and, in consequence, the triangle *ACB* is known. But the two points *A* and *B* are known, so the point *C* is known, as has been proved in the preceding proposition; so each of the straight lines *AC* and *BC* is known in magnitude and in position. This is what we wanted to prove.

– **23** – Let there be a circle known in position and a straight line known in position; we draw a straight line which cuts the circle and which ends on the [first] straight line and is such that it is divided by the circumference of the circle in a known ratio and encloses with the [first] straight line a known angle, then this straight line is known in magnitude and in position.

Example: The circle *AB* is known in magnitude and in position and the straight line *CD* is known in position; we draw the straight line *ABE* in such a way that the ratio of *AB* to *BE* is known and the angle *BEC* is known.

I say that the straight line AB *is known in magnitude and in position.*

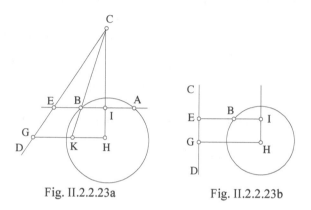

Fig. II.2.2.23a   Fig. II.2.2.23b

Proof: We mark off the centre of the circle, let it be the point *H*, and we draw the perpendicular *HI*; it divides *AB* into two halves, so the ratio of *IB* to *BE* is known. We draw *HG* such that the angle *HGC* is equal to the angle *BEC* which is known, then the straight line *HG* will be known in position, because if we make *C* a known point and we join *HC*, the point *G* lies on the circumference of a circle known in position, so the point *G* is known. But the point *H* is known, so the straight line *HG* is known in magnitude and in position. If the straight line *HI* is parallel to the straight line *DC*,[48] then the straight line *IE* will be equal to the straight line *HG*, so it is known in magnitude. So the straight line *BE* is known in magnitude and the angle *BEC* is known, so the straight line *BE* is known in position, as has been proved in Proposition 13 of the second section of this treatise; in consequence, the straight line *ABE* is known in magnitude and in position.

If the straight line *HI* is not parallel to the straight line *DC*, it meets it; let it meet it at the point *C*. Since the angle *CIE* is a right angle, the angle *CHG* is a right angle and the straight line *HG* is known in magnitude and in position. But the straight line *HI* is known in position and the straight line *CD* is known in position, so the point *C* is known. We join *CB* and we

---

[48] This happens if the known angle *BEC* is a right angle.

extend it to *K*. So the ratio of *HK* to *KG* is equal to the ratio of *IB* to *BE*, which is known; the ratio of *HK* to *KG* is thus known. But the triangle *HCG* has known angles and from its vertex we drew the straight line *CK* which divided the straight line *HG* in a known ratio; so the straight line *CK* is known in position, as has been proved in Proposition 20 of the second section of this treatise. But the point *K* is known because it divides the known straight line *HG* in a known ratio. So from the point *K* we have drawn a straight line *KC* known in position, which cuts the known circle *AB* in a point *B*; so the point *B* is known. So we have drawn the straight line *BE* at a known angle; in consequence, *BE* is known in position since the point *E* lies on the circumference of a circle known in position; the point *E* is thus known. But the point *B* is known, so the straight line *BE* is known in magnitude and in position and its ratio to *BA* is known; so the straight line *ABE* is known in magnitude and in position. This is what we wanted to prove.

– **24** – Let there be two circles known in magnitude and in position; we draw a straight line tangent to the two circles; it is known in magnitude and in position.

Example: The two circles *AB* and *CD* are known in magnitude and in position; we draw a straight line *AD* which touches them.

I say that the straight line AD *is known in magnitude and in position.*

Proof: We mark off the two centres, let them be *E* and *G*, and we join *EG*. The circles *AB* and *CD* are either equal or different.

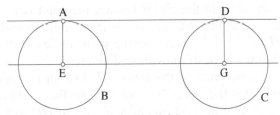

Fig. II.2.2.24a

Let them first be equal, the straight line *AD* touches the two circles either in two similar directions as in the first figure[49] or in two different directions as in the second figure. If the contact is as in the first figure, then we join *EA* and *GD*, so the two angles at the points *A* and *D* are right angles, so the two straight lines *EA* and *GD* are parallel and they are equal.

---

[49] Ibn al-Haytham distinguishes between two cases for equal circles: an exterior common tangent (Fig. II.2.2.24a) and an interior common tangent (Fig. II.2.2.24c).

Then the straight line AD is equal to the straight line EG and is parallel to it, so the angle GEA is a right angle, so the straight line EA is known in position; now, it is known in magnitude, so the point A is known. But the angle EAD is a right angle, so the straight line AD is known in position and it is equal to the straight line EG which is known in magnitude; so the straight line AD is known in magnitude and in position.

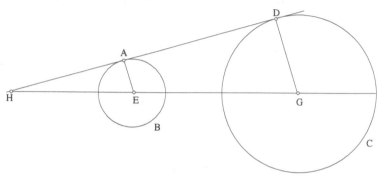

Fig. II.2.2.24b

If the two circles AB and CD are different, then the two straight lines GD and EA are different[50] and they are parallel, so the straight line DA meets the straight line GE in the direction towards the smaller circle – let it be the circle AB; let the straight lines meet at the point H. So the ratio of GH to HE is equal to the ratio of GD to EA. But the ratio of GD to EA is known, because each of these straight lines is known, so the ratio of GH to HE is known, so the straight line EH is known, the point H is consequently known, and the straight line HG is known in magnitude and in position. But the angle HDG is a right angle, so the point D lies on the circumference of a circle known in position, whose diameter is HG, and it [the point D] lies on the circumference of the circle CD known in position. So the point D is known. But the point H is known, so the straight line HD is known in magnitude and in position and the ratio of HD to DA is known, so the straight line AD is known in magnitude and in position. This is what we wanted to prove.

[50] See Fig. II.2.2.24b.

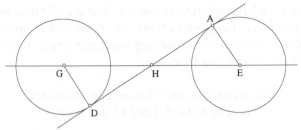

Fig. II.2.2.24c

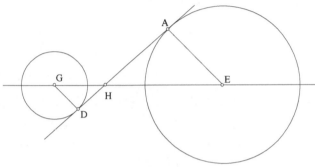

Fig. II.2.2.24d

If the contact is in two different directions, as in this figure,[51] we mark off the two centres; let them be *E* and *G*, and we join *EG*; so it will be known in magnitude and in position. We join *EA* and *GD*, the two angles at the two points *A* and *D* are then right angles and are in two different directions with respect to the straight line *EG*; the two points *A* and *D* will then lie to either side of the straight line *EG*. So the straight line *AD* cuts the straight line *EG*; let it cut it at the point *H*; so the two straight lines *EA* and *GD* will be parallel and the two triangles *EAH* and *HDG* are similar. So the ratio of *EH* to *HG* is equal to the ratio of *EA* to *GD* which is known, because each of the two straight lines *EA* and *GD* is known in magnitude. So the ratio of *EH* to *HG* is known and the straight line *EG* is known in magnitude, each of the two straight lines *EH* and *HG* is known in magnitude and the angle *EAH* is a right angle. So the point *A* lies on the circumference of a circle known in position whose diameter is *EH* and it is the circumference of the circle *AB*, so the point *A* is known. Similarly, we [can] prove that the point *D* is known.

So the straight line *AD* is known in magnitude and in position, whether the two circles are equal or unequal. This is what we wanted to prove.

---

[51] See Figs II.2.2.24c and II.2.2.24d.

The notions we have set out are notions that are of great use in solving geometrical problems. They are notions that none of our predecessors has set out. What we have set out regarding these notions is sufficient for our purpose, and here we end this treatise.

<The treatise on the> *Knowns* is completed.
Praise be to God, Lord of the worlds.

III. ANALYSIS AND SYNTHESIS:
EXAMPLES OF THE GEOMETRY OF TRIANGLES

According to the tenth-century biobibliographer al-Nadīm, Archimedes composed two books devoted exclusively to the geometry of triangles: *The Book on Triangles* and *The Book on the Properties of Right-angled Triangles*.[1] Again according to al-Nadīm, Menelaus too was the author of a *Book on Triangles*, of which 'a small part has been translated into Arabic (*wa-kharaja minhu ilā al-'arabī shay' yasīr*)'.[2] On the basis of this testimony, it seems that ancient mathematicians had singled out triangles by dedicating specialised texts to them, and that at least two of these were translated into Arabic. Given its association with the illustrious name of Archimedes, there was every chance that the subject would be taken up by his successors, in particular by the mathematicians of the tenth century; but the historical research required to establish the details still remains to be done. For the time being, al-Sijzī provides an example that supplies a partial answer to our questions. He too wrote a book called *On Triangles*.[3]

So we would expect that Ibn al-Haytham, who wrote a monograph on the 'properties of circles' and a treatise on the 'properties of conic sections', would also have written a book about triangles. He did not do so expressly, but he did write two short treatises on the geometry of triangles. One has come down to us under the simple title *On a Geometrical Problem* and the other under the more explicit title *On the Heights of Triangles* or *On the Properties of the Triangle in Regard to Height*. In both of these texts, Ibn al-Haytham's work follows on from what was done by his predecessors; in the first treatise, his immediate predecessors – Ibn Sahl and al-Sijzī, and in the second one those he calls the ancients. In the former treatise he proceeds by analysis and synthesis, whereas in the second he gives only the synthesis.

---

[1] Al-Nadīm, *Kitāb al-fihrist*, ed. R. Tajaddud, Teheran, 1971, p. 326.

[2] *Ibid.*, p. 327.

[3] Al-Sijzī refers several times to this book. For instance, in his *Anthology of Problems* (ms. Dublin, Chester Beatty 3652), he refers 'to our book on triangles' (Problem 9, Proposition 20), and similarly in Problem 45, Proposition 72; Problem 50, Proposition 80; Problem 53, Proposition 86. See R. Rashed and P. Crozet, *Al-Sijzī: Œuvres mathématiques*, forthcoming.

1. *On a geometrical problem: Ibn Sahl, al-Sijzī and Ibn al-Haytham*

In his treatise on analysis and synthesis, Ibn al-Haytham makes use of a series of distinctions that he presents as being valid in all the mathematical disciplines of the quadrivium. The principal distinction is that between theoretical analysis and practical analysis (*'amalī*). Theoretical analysis is applicable to propositions and theorems; practical analysis deals with constructions and the determination of an unknown magnitude or number. This distinction, which was already introduced by Thābit ibn Qurra and was taken up by his successors, reappears here. Practical analysis is not in any way to be identified with the 'problematic' analysis of Pappus, and this for two related reasons. On the one hand it in fact also includes geometrical constructions as well as the determination of unknown magnitudes and numbers; on the other hand, it is applicable to all the mathematical disciplines and not only to geometry. This practical analysis can in turn be subdivided into several types: when there is a single solution, when there are several solutions, when there may be an infinite number of solutions; when there is no diorism (that is, no discussion), when there is a diorism, and so on. Now, investigating the conditions for passing from one of these types to another is an interesting logical problem and one that is very fertile in yielding mathematical insights; it in fact requires that we return to the conditions for the problem and those for the construction, in order to change them. This change in the conditions from one type to another in turn constitutes a valuable method of invention; Ibn al-Haytham carries this out, as well as the transformation of a problem of the geometry of triangles that had already been considered, successively, by his direct predecessors Ibn Sahl and al-Sijzī.

One of the problems proposed by Ibn Sahl, and for which he provides the synthesis,[4] is that of the construction of a triangle such that one of its sides is equal to a given straight line $DC = 2c$, and the sum of the other two sides is equal to a given straight line $AB = 2a$. Ibn Sahl imposes a supplementary condition, namely that all the angles of the triangle should be acute.

We know from the outset that it is necessary to have $a > c$. Ibn Sahl accordingly considers the ellipse with major axis $AB = 2a$, centre E, foci C and D, where $EC = ED = c$. Let us put $IK = 2b$ as the minor axis ($EK = b$).

---

[4] See R. Rashed, *Géométrie et dioptrique au Xe siècle. Ibn Sahl, al-Qūhī, Ibn al-Haytham*, Paris, 1993; English version: *Geometry and Dioptrics in Classical Islam*, London, 2005; and 'Ibn Sahl et al-Qūhī: Les projections. Addenda & corrigenda', *Arabic Sciences and Philosophy*, vol. 10.1, 2000, pp. 79–100.

Any point X on this ellipse gives a triangle with base $CD = 2c$, and $XC + XD = 2a$. Thus, the problem has an infinite number of solutions.

The supplementary condition – that the triangle must have only acute angles – leads Ibn Sahl to consider the perpendiculars to the axis AB at C and D and the circle with diameter CD, in order to find the arcs of the ellipse on which to take X so as to ensure the triangle XCD satisfies the three conditions.

By hypothesis $a > c$ and $a > b$; but we can have $b > c$, $b = c$, or $b < c$, which are the three cases Ibn Sahl investigates. The circle C with diameter CD cuts the straight line IK in K'.

• $b > c$, K' lies inside the ellipse.

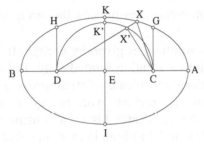

Fig. 2.3.1

For any point X on the arc GKH of the ellipse, X lies outside the circle C, $D\hat{X}'C = 1$ right angle, so $\hat{X}$ is acute. We also have that $\hat{D}$ and $\hat{C}$ are acute and $XC + XD = 2a = AB$, except if X is at G or at H.

• $b = c$ $K = K'$.

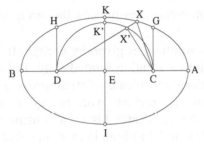

Fig. 2.3.2

Any point X on the arc GKH, except for G, K and H, makes $\hat{X}$ acute, because X lies outside the circle; $\hat{C}$ and $\hat{D}$ are acute.

• $b < c$ K' lies outside the ellipse.

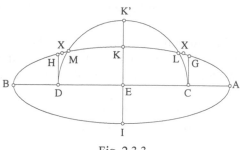

Fig. 2.3.3

The circle cuts the ellipse in L and M. Any point X on the arcs LG and HM, excluding the endpoints, lies outside the circle and makes $\hat{X}$, $\hat{C}$ and $\hat{D}$ acute.

This is Ibn Sahl's solution, as given by al-Sijzī. It is clear that the idea of using an ellipse gives a solution immediately. But one can find a construction for such a triangle by means of straightedge and compasses, as al-Sijzī indeed noted. This is precisely what he says in a letter to Naẓīf ibn Yumn,[5] to whom he communicates his own construction. But before we examine al-Sijzī's solution, let us look in more detail at the one by Ibn Sahl.

If a point X provides a solution to the problem, that is if XCD is the triangle constructed with $XC + XD = 2a$ and $CD = 2c$, which are given lengths, and if we extend CX by a length $XM = XD$, we have $CM = 2a$. The point M lies on the circle with centre C and radius $2a$. Conversely, to any point M of this circle there corresponds a point X, the point of intersection of the straight line CM and the perpendicular bisector of MD,[6] and the triangle CXD that we obtain satisfies the *two* given conditions.

But Ibn Sahl imposes a supplementary condition: the triangle CXD must have all its angles acute. Let us examine its three angles.

1. The angle DCX. It is acute if the straight half line $[CX)$ lies within the angle FCE (E is the point of intersection of the circle $(C, 2a)$ and the straight half line $[CD)$). Consequently, the angle DCX is acute if M is any point on the arc FE, apart from its endpoints (to the endpoint F there corresponds the point X', the point of intersection of CF and the perpendicular bisector of DF).

    [5] See p. 623.
    [6] This is the method used in the point by point construction of the ellipse with foci C and D and major axis $2a$.

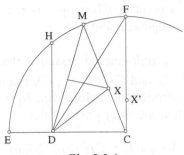

Fig. 2.3.4

2. The angle CDX. The perpendicular to CD at D cuts the circle in H. We shall have $C\hat{D}H = 1$ right angle; if X is at X_1 on the straight line DH, the point M then lies on the arc HE at the point M_1 such that M_1X_1D is an isosceles triangle. In the triangle CM_1D we have $C\hat{D}M_1 = 1$ right angle $+ \hat{M}_1$, so

$$\frac{\sin \hat{M}_1}{2c} = \frac{\sin \left(1 \text{ right angle} + \hat{M}_1\right)}{2a} = \frac{\cos \hat{M}_1}{2a},$$

hence

$$\tan \hat{M}_1 = \frac{c}{a} = \tan D\hat{F}C.$$

So we have $H\hat{D}M_1 = \hat{M}_1 = D\hat{F}C = F\hat{D}H$. The straight line HD bisects the angle FDM_1. When M lies on the arc FM_1, but not at either of its endpoints F and M_1, then both $X\hat{C}D$ and $X\hat{D}C$ are acute.

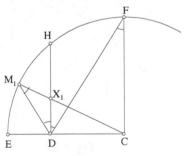

Fig. 2.3.5

3. The angle $C\hat{X}D$. We have $C\hat{X}D = 2\,C\hat{M}D$ (Fig. 2.3.4); so for $C\hat{X}D$ to be acute, we must have $C\hat{M}D < 45°$. Consequently, the point M must lie outside the arc that subtends $45°$ constructed on CD. This arc cuts the

straight line *DH* in a point *K* whose position in regard to the point *H* depends on the given lengths 2*a* and 2*c*.

We have *CDK*, a right-angled triangle that is isosceles (because $D\hat{K}C = 45°$), so *DK* = 2*c* and *CK* = $2c\sqrt{2}$; *CK* is a diameter of the circle, on which we find the subtending arc. Moreover, by hypothesis we have *CH* = 2*a* and *a* > *c*, hence there are three possibilities:

• If $c\sqrt{2} < a$, we have *CK* < *CH* (Fig. 2.3.6). The subtending arc lies completely inside the circle (*C*, 2*a*), so any point on the arc FM_1, apart from the endpoints, gives a solution.

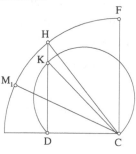

Fig. 2.3.6

• If $c\sqrt{2} = a$, we have *CK* = *CH* (Fig. 2.3.7). The subtending arc and the circle (*C*, 2*a*) touch one another at the point *H* (*K* = *H*), so any point on the arc FM_1, apart from the points *F*, *H* and M_1, gives a solution.

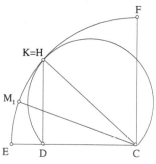

Fig. 2.3.7

• If $c\sqrt{2} > a$, we have *CK* > *CH* (Fig. 2.3.8). The subtending arc cuts the circle (*C*, 2*a*) in two points M_2 and M_3 symmetrical with respect to the straight line *CK*, and it cuts the straight line *CF* in a point *K'* such that *CK'* = *DK* = *DC* = 2*c*. If a point describes the subtending arc from the point *K* to the point K', its distance from the point C decreases from $CK = 2c\sqrt{2}$ to CK' $= 2c$; so at one moment this distance takes the value $2a$ (because $2c < 2a < 2c\sqrt{2}$); so the subtending arc cuts the arc HF of the circle with centre C at the point M_2 and cuts the arc HE of the circle at the point M_3. So we must investigate the positions of the points M_1 and M_3 that both lie on the arc HE. The triangles M_1DC and M_3DC have a common side CD, of length $2c$, and the sides M_1C and M_3C are of equal length $2a$. The angles M_1DC and M_3DC are obtuse; in triangle M_3DC, we have

$$\frac{\sin M_3\hat{D}C}{2a} = \frac{\sin D\hat{M}_3C}{2c} = \frac{\sin\frac{\pi}{4}}{2c},$$

and in triangle M_1DC,

$$\frac{\sin M_1\hat{D}C}{2a} = \frac{\sin D\hat{M}_1C}{2c}.$$

Now, we have seen that $D\hat{M}_1C = D\hat{F}C$ (Fig. 2.3.5) and in the present case $D\hat{F}C < \frac{\pi}{4}$ because F lies outside the subtending arc. So we have $\sin M_1\hat{D}C < \sin M_3\hat{D}C$ and consequently $M_1\hat{D}C > M_3\hat{D}C$ (because the angles are obtuse). We then have that M_3 lies between M_1 and H. So, if $c\sqrt{2} > a$, to any point M on one or other of the arcs FM_2 or M_3M_1 (apart from the endpoints) there corresponds a point X that gives a solution to the problem.

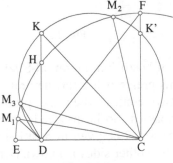

Fig. 2.3.8

Thus, the straightedge and compasses construction is possible and, in all cases for the figure, it leads to an infinity of solutions. The method employed here in the discussion that arises when we have the condition that the triangle should be 'acute-angled' leads us to consider the relative positions of a subtending arc and a circle.

However, Ibn Sahl's method involves the relative positions of a circle and an ellipse, and, in the last case for his figure, the circle cuts the ellipse at the points L and M, for which Ibn Sahl gives no construction. We may also note that the three cases for the figure that are considered here ($a > c\sqrt{2}$, $a = c\sqrt{2}$ and $a < c\sqrt{2}$) correspond to the three cases examined by Ibn Sahl: $b > c$, $b = c$ and $b < c$. In fact, given $a^2 = b^2 + c^2$, if $b = c$, $a^2 = 2c^2$. The point X which lies on the ellipse can be found from the point M which lies on the circle with centre C and radius $2a$. To the points F, M_2, M_3, M_1 that come into play when M lies on the circle there correspond, for X, the points X', X_2, X_3, X_1 that are none other than the points G, L, M, H of the ellipse in the text by Ibn Sahl (Fig. 2.3.3 in this text). In fact, the circle with centre C and radius $2a$ is the 'director circle' of the ellipse; the latter is the locus of the centres of the circles that pass through D and touch the director circle at M.

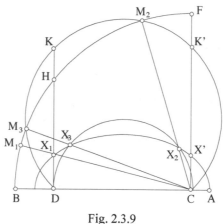

Fig. 2.3.9

So at the request of Naẓīf ibn Yumn, al-Sijzī looks again at Ibn Sahl's problem. He makes his preference clear: he prefers not to have recourse to the intersection of two conics when the problem can be solved with straightedge and compasses. So he proceeds to construct an acute-angled triangle ABG whose base AB is given in position and in magnitude, and in which the sum of the two other sides is known (let $AB = 2c$ and $BG + AG = 2a$). Al-Sijzī's response is synthetic, but we can imagine how his analysis went.

Let ABG be a triangle that provides a solution to the problem; we extend AG by a length $GE = GB$; so we have $AE = 2a$, a known length. Moreover, the triangle GBE is isosceles, hence $G\hat{B}E = B\hat{E}G$. So we have $B\hat{G}A = 2B\hat{E}A$, so angle BGA being acute corresponds to $B\hat{E}A < 45°$. So the point E lies on the circle with centre A and radius $2a$, and outside the arc subtending 45° constructed on AB.

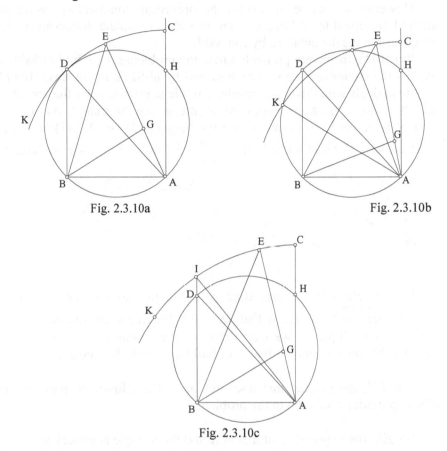

Fig. 2.3.10a   Fig. 2.3.10b

Fig. 2.3.10c

Al-Sijzī does not mention the ellipse with foci A and B and major axis $2a$ on which the point G lies. But it is clear that the whole problem is connected with constructing this ellipse point by point. Using the subtending arc leads to distinguishing the different cases for the figure, but al-Sijzī does not show that these different cases correspond to the equality or inequality of the lengths a and $c\sqrt{2}$.

Al-Sijzī's discussion is incomplete; he does not in fact mention the angles A and B of the triangle ABG (the angles that must be acute) except in the last lines of his letter. He refers to the straight lines from A and B and concludes: 'so let us take the triangle lying between the straight lines AC and BD which are parallel'. The point G obviously indeed lies between these straight lines, but the point E can move beyond the straight line perpendicular to AB at the point B. If we take E between the two straight lines AC and BD, we have a condition that is sufficient but not necessary.

However, as can be seen from the preceding commentary, using the method presented by al-Sijzī, we can provide a complete discussion which produces the results obtained by Ibn Sahl.

Ibn al-Haytham very probably knew the problem set out by Ibn Sahl, as well as the solutions proposed by him and by al-Sijzī. In any case, Ibn al-Haytham deals with this same problem, while substituting for the condition that the triangle has acute angles the condition that the triangle has a given area, S; which reduces to saying that the height XH (Fig. 2.3.11) measured from the base CD is of magnitude $h = \dfrac{2S}{CD} = \dfrac{S}{c}$; so h is thus a known length.

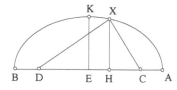 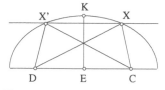

Fig. 2.3.11

Let us return to the preceding solution, that of Ibn Sahl; we have $EK = b = \sqrt{a^2 - c^2}$. We know that $EK = b$ is the maximum ordinate for the points of the ellipse. If we now draw a straight line parallel to AB at a distance h from that straight line, we shall have the following cases:

$h > EK$, the parallel line does not meet the ellipse; no point of the ellipse provides a solution to the problem;

$h = EK$, the required point X is at K and the triangle is isosceles;

$h < EK$, the parallel line cuts the ellipse in two points X and X' which give two triangles that provide a solution to the problem. These triangles are equal.

So the necessary and sufficient condition for X to exist is $h \leq EK$; now

$$\left\lfloor EK^2 = a^2 - c^2 \right\rfloor \Leftrightarrow \left\lfloor h^2 \leq a^2 - c^2 \right\rfloor \Leftrightarrow \left\lfloor 4a^2 \geq 4h^2 + 4c^2 \right\rfloor$$
$$\Leftrightarrow \left\lfloor AB^2 \geq 4h^2 + DC^2 = DC^2 + \frac{16S^2}{DC^2} \right\rfloor.$$

This condition, which is not mentioned by Ibn Sahl or al-Sijzī, is the one that Ibn al-Haytham gives in his diorism.

Ibn al-Haytham seems in fact not to have forgotten al-Sijzī's recommendations. He returns to the problem using the inscribed circle of the triangle or the circumscribed one. So all the constructions he proposes can be carried out with straightedge and compasses.

Ibn al-Haytham gives five analyses of the transformed problem, one after the other. In the first four, he employs the inscribed circle of the triangle and in the fifth the circumscribed circle.

Let us summarise these analyses.

In the first four, let us call the triangle ABC, the centre of the inscribed circle I, let AB, BC, CA be the sides and their points of contact with the circle E, G and D respectively. Let us put $BC = a$ the known side, $AB + AC = l$ a known length, let S be the area of the triangle, a known area.

Analysis 1: The known lengths allow us to calculate $AE = \dfrac{l-a}{2}$ and $IE = \dfrac{2S}{l+a} = r$, the radius of the circle; $\tan I\hat{A}E = \dfrac{IE}{AE}$ is known and the angle IAE is also known; so $C\hat{A}B = 2I\hat{A}E$ is known. We take K on BC such that $A\hat{K}C = B\hat{A}C$; we prove that AK is known and that the triangles ABC and AKC are similar; hence $AK \cdot BC = AB \cdot AC$, so $AB \cdot AC$ is known. So we have that the sum l and the product p of the two sides are known, so the two sides are known.

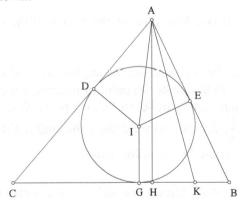

Fig. 2.3.12

Note: Ibn al-Haytham does not carry out the synthesis corresponding to this analysis. We may note that AB and AC are the roots of the equation $x^2 - lx + p = 0$, where $\Delta = l^2 - 4p$. These roots are real if $l^2 \geq 4p$, that is if $l^2 \geq a^2 + 4h^2$, where $h = AH$ the perpendicular to BC at H. In fact, we have

$$S = \frac{ah}{2} \quad \text{and} \quad p = \frac{ah}{\sin B\hat{A}C}.$$

Now

$$B\hat{A}C = 2\,I\hat{A}B \quad \text{and} \quad \tan I\hat{A}B = t = \frac{IE}{AE} = \frac{ah}{l+a} \cdot \frac{2}{l-a},$$

so

$$t = \frac{2ah}{l^2 - a^2}.$$

Moreover,

$$\sin B\hat{A}C = \frac{2t}{1+t^2},$$

hence

$$p = \frac{\left(l^2 - a^2\right)^2 + 4a^2h^2}{4\left(l^2 - a^2\right)};$$

and we have

$$l^2 \geq 4p \Leftrightarrow l^2 \geq l^2 - a^2 + \frac{4a^2h^2}{l^2 - a^2} \Leftrightarrow l^2 \geq a^2 + 4h^2,$$

a condition that Ibn al-Haytham gives in the course of the synthesis for the fifth analysis.

Analysis 2: As in the first analysis the angles *IAC* and *BAC* are known. If we use the letter *K* to designate the point of intersection of the height *AH* and the line drawn through *I* parallel to *BC*, we prove that the angle *IAK* is known (cos $I\hat{A}K = \frac{AK}{AI}$); from this we deduce the angles *KAE* and *ABC*. Ibn al-Haytham draws his conclusion in two ways:

• $AB = \frac{AH}{\cos H\hat{A}B}$, so *AB* is known and $AC = l - AB$ is also known.

• We know that $B\hat{A}C = 2\,I\hat{A}B$ and $A\hat{B}C$; so the triangle *ABC* is of known shape and the ratio $\frac{AB}{AC}$ is known. Thus, the lengths *AB* and *AC* whose sum and ratio are known are themselves known.

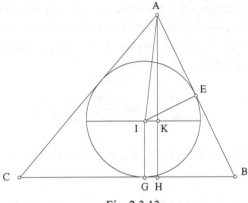

Fig. 2.3.13

Note: As before, $IE = \dfrac{2S}{l+a} = r = \dfrac{ah}{l+a}$ and $AE = \dfrac{l-a}{2}$, hence

$$AI^2 = \frac{4a^2h^2 + \left(l^2 - a^2\right)^2}{4(l+a)^2}.$$

Moreover,

$$AK = AH - IG = h - r = \frac{hl}{l+a},$$

hence

$$\cos I\hat{A}K = \frac{2hl}{\sqrt{4a^2h^2 + \left(l^2 - a^2\right)^2}},$$

hence

$$\cos I\hat{A}K \leq 1 \iff 4a^2h^2 + (l^2 - a^2)^2 \geq 4h^2l^2 \iff l^2 \geq a^2 + 4h^2,$$

a condition that Ibn al-Haytham gives in the course of the fifth analysis.

Analysis 3: If we extend BA by a length $AF = AC$, the length $BF = l$ is known and we have $CF \parallel AI$. We then prove that the triangle BCF is of known shape (that of BAH, in which we know the angles BAH [see Analysis 1] and ABH [see Analysis 2]), so the length CF is known. So the triangle AFC is also of known shape, and the ratio $\dfrac{FC}{CA}$ is known. So we know the length CA and from it we deduce $AB = l - CB$.

Note: Here one recognises a construction similar to that of al-Sijzī. The straight line FC is parallel to AH, the bisector of the angle BAC, a known angle, as we have seen in Analysis 1. If we put $B\hat{A}C = 2\alpha$, we have $A\hat{F}C = F\hat{C}A = C\hat{A}H = \alpha$. Now

$$\frac{BF}{BC} = \frac{l}{a} \quad \text{and} \quad \frac{BF}{BC} = \frac{\sin F\hat{C}B}{\sin A\hat{F}C},$$

so $\sin F\hat{C}B = l\dfrac{\sin\alpha}{a}$ and angle FCB is thus known and we have

$$\sin F\hat{C}B \le 1 \Leftrightarrow \frac{l}{a}\sin\alpha \le 1 \Leftrightarrow \frac{2hl}{\sqrt{4a^2h^2 + \left(l^2 - a^2\right)^2}} \le 1 \Leftrightarrow l^2 \ge a^2 + 4h^2,$$

a condition that Ibn al-Haytham gives in the synthesis of the fifth analysis.

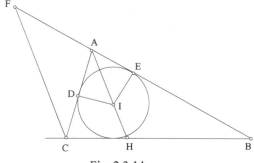

Fig. 2.3.14

Analysis 4: S, the area of the triangle ABC, is given and the angle BAC can be determined as in the first analysis. The product $AB \cdot AC = \dfrac{2S}{\sin B\hat{A}C}$ is known. As, by hypothesis, we also know the sum $AB + AC$, we can find each of the sides, as in the first analysis. In the same way, we can derive the necessary and sufficient condition.

Analysis 5: Ibn al-Haytham considers the circle circumscribed about triangle ABC. The bisector of the angle BAC cuts BC in E and the circle in D. Using the property of E, as the foot of the bisector: $\dfrac{EB}{EC} = \dfrac{AB}{AC}$, and the fact that the triangles ABD and BDE are similar, we prove that the ratio $\dfrac{AE}{ED}$ is known, and from it we deduce that the lengths DG (G is the mid point of

BC; we have $\dfrac{AH}{DG} = \dfrac{AE}{ED}$ and AH is known), DB, DA, AE and ED are known. Now the power of the point E with respect to the circle gives $EA \cdot ED = EB \cdot EC$, so the product $EB \cdot EC$ is known; as we know that $\dfrac{EB}{EC} = \dfrac{AB}{AC}$, we deduce that the product $AB \cdot AC$ is known; but since, by hypothesis, the sum is also known, each of the straight lines AB and AC is thus known.

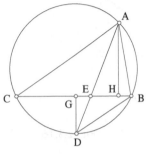

Fig. 2.3.15

Synthesis of 5: The base given for the construction is AB; the required vertex is the point M. The given lengths are GH and CD such that

(1) $GH = MA + MB$
(2) Area $(AMB) = AB \cdot CD$.

So the height from M is $MN = 2\,CD = DE$.

On GH we take a point I such that $\dfrac{AB}{HI} = \dfrac{GH}{AB}$. Let K be the mid point of AB and let L be a point on the perpendicular bisector of AB such that $\dfrac{ED}{KL} = \dfrac{GI}{IH}$. We draw the circumscribed circle of triangle ABL. On the straight half line LA, we take P such that $\dfrac{PA}{AL} = \dfrac{ED}{KL}$. If the line drawn through P parallel to AB cuts the circle at the point M, then the triangle AMB is the triangle we require.

Proof: By hypothesis, we have

$$\frac{GH}{AB} = \frac{AB}{HI} \quad \text{and} \quad \frac{ED}{KL} = \frac{PA}{AL}.$$

We draw MN perpendicular to AB, and ML, which cuts AB in F; we have

$$\frac{MN}{KL} = \frac{MF}{FL} = \frac{PA}{AL} = \frac{ED}{KL},$$

hence $MN = ED$.

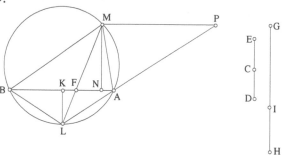

Fig. 2.3.16

So we then have area $(MAB) = \frac{1}{2} MN \cdot AB = CD \cdot AB$, and (2) is satisfied. The point L is the mid point of the arc AB, hence arc BL = arc AL which implies $B\hat{M}L = A\hat{M}L = B\hat{A}L$; so the triangles AML and AFL are similar, and we have

(1) $\qquad \dfrac{ML}{AL} = \dfrac{AL}{LF} = \dfrac{MA}{AF};$

moreover,

(2) $\qquad \dfrac{MA}{AF} = \dfrac{MB}{BF},$

because F is the foot of the bisector of the angle BMA.

From (1) we deduce $LA^2 = ML \cdot LF$, and from (1) and (2) we deduce

$$\left(\frac{MA + MB}{AB}\right)^2 = \frac{ML^2}{LA^2} = \frac{ML}{LF};$$

but

$$\frac{MF}{LF} = \frac{GI}{IH},$$

so

$$\frac{ML}{LF} = \frac{GH}{HI},$$

but

$$GH \cdot HI = AB^2 \text{ and } \frac{GH}{HI} = \frac{AB^2}{HI^2} = \frac{GH^2}{AB^2},$$

hence

$$\frac{MA + MB}{AB} = \frac{GH}{AB}$$

and finally $MA + MB = GH$.

So if the line parallel to AB cuts the circle in M, the triangle MAB provides a solution to the problem.

But, in this line of reasoning, we assume that the line drawn through P parallel to AB cuts the circle in M, which requires a discussion of the existence of this point M, a discussion to which Ibn al-Haytham turns his mind.

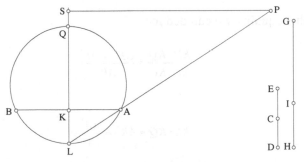

Fig. 2.3.17

Ibn al-Haytham then gives the condition $GH^2 \geq AB^2 + 4ED^2$. He first proves that if $GH^2 < AB^2 + 4ED^2$, the problem is impossible. Next he proves a property which holds for any triangle, a property from which the condition (1) follows. He then returns to the discussion and proves that if $GH^2 = AB^2 + 4ED^2$, we have $ED = MN = KQ$; M is at Q and the required triangle is isosceles.

Finally, he proves that if $GH^2 > AB^2 + 4ED^2$, we have $ED < KQ$.

The straight line drawn through P parallel to AB cuts the straight line LQ between K and Q; so it cuts the circle in a point M of the arc QA and in a point M' of the arc QB; the points M and M' give the triangles MAB and $M'AB$, equal triangles, which are solutions to the problem. Ibn al-Haytham gives only the point M.

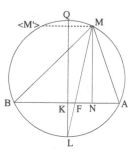

Fig. 2.3.18

Note: In the course of the discussion Ibn al-Haytham notes that any triangle *AMB* can be inscribed in a circle and proves that if we have $GH = MA + MB$ and $ED = MN$ (*MN* being the height of the triangle) and if *LQ* is the diameter perpendicular to the base *AB* at *K*, we have

$$\frac{ED}{KL} = \frac{GH^2 - AB^2}{AB^2}.$$

From this equality we can deduce

$$\frac{ED \cdot KQ}{KL \cdot KQ} = \frac{GH^2 - AB^2}{AB^2}.$$

Now

$$KL \cdot KQ = AK^2 = \frac{AB^2}{4},$$

so we have

$$\frac{4ED \cdot KQ}{AB^2} = \frac{GH^2 - AB^2}{AB^2},$$

hence

$$KQ = \frac{GH^2 - AB^2}{4ED}.$$

Constructing the point *M* is possible only if $MN \le KQ$ (Euclid, *Elements*, III.15). Now

$$MN = ED, \ ED \le KQ \Rightarrow ED \le \frac{GH^2 - AB^2}{4ED} \Rightarrow 4ED^2 \le GH^2 - AB^2,$$

the necessary and sufficient condition.

We may ask ourselves why Ibn al-Haytham did not put the paragraph that begins 'Indeed, in every case a triangle is inscribed in a circle ...' (p. 483) at the beginning of the discussion, and why he has not proved that, from the equality he has established, $\dfrac{ED}{KL} = \dfrac{GH^2 - AB^2}{AB^2}$, we can deduce the condition $GH^2 > AB^2 + 4ED^2$ which he gives at the beginning of the discussion without explaining how it has been obtained. Perhaps it is a simple question of editing. Starting, as we have seen, from a problem proposed by Ibn Sahl, who solved it by using the intersection of conic sections, a problem then picked up by al-Sijzī, who looked for a solution by straightedge and compasses, Ibn al-Haytham transforms the problem and thus moves from one kind of analysis to another: from an indeterminate practical analysis to a determinate practical analysis. The obligation he adopts deliberately and almost systematically that, when it is required, he should give a proof of existence, led him to establish the existence of the point M, and then to prove the necessary and sufficient condition for there to be a solution to the problem. Thus, everything seems to point to the passage from one kind of analysis to another being one of the procedures in mathematical invention. Indeed, this research leads Ibn al-Haytham to discover new properties of the triangle in relation to its inscribed circle and its circumscribed circle.

2. *Distances from a point of a triangle to its sides*

In a second treatise with the title *On the Properties of the Triangle in Regard to Height*, Ibn al-Haytham proposes to study the sum of the distances from a point on one of the sides of the triangle or lying inside it, to the sides of the same triangle. The treatise is purely synthetic. In the preamble Ibn al-Haytham gives a clear explanation of his intention and its development. He starts by referring to the fact that the ancients had considered this problem for the case where the triangle is equilateral, and he gives the two propositions they had established. In regard to other triangles, no results had been found. Ibn al-Haytham accordingly takes up the problem for isosceles triangles, then for arbitrary triangles. He states that he has found, for one and the other variety 'a uniform order' (*niẓām muṭṭarid*, see below, p. 485), that is a formula sufficiently general to define the character of each class. We shall see that this is indeed so. Nonetheless, despite these interesting results, readers familiar with Ibn al-Haytham's writings cannot but be disconcerted by this treatise. Ibn al-Haytham has indeed accustomed us to works that break new ground, always innovative and profound; now this text, although not entirely lacking in interest, nevertheless does not rise to such heights. The fact remains that this relatively modest contribution is shaped by the same principle that governed the composition of other, incomparably more important, works: the purpose is to complete what his predecessors had begun and to exploit all the potentialities to be found in their research. And in fact, this problem of the distance from a point of a triangle to its sides presents itself to Ibn al-Haytham as given prestige by the participation of the 'ancients' – somewhat like the example of the problem of the regular heptagon.[7] But we may ask why Ibn al-Haytham chose the generic term, 'the ancients' (*al-mutaqaddimūn*), whereas he is not usually sparing in using names when he is dealing with authors as highly esteemed as Archimedes. In short, we may wonder which ancients he is referring to.

A manuscript copied at the beginning of the thirteenth century informs us of the existence of a text attributed to Archimedes, with the title *On the Foundations of Geometry* (*Fī al-uṣūl al-handasiyya*), a work translated by Thābit ibn Qurra. The attribution to Archimedes appears in the title, together with the name of the translator, and it is repeated in the colophon.[8] This is a treatise that includes nineteen propositions, the first of which is proved twice. Better still, in the title we find not only the name of the mathematician of Syracuse and that of the prestigious translator, but also the name of the person who commissioned the translation: Abū al-Ḥasan 'Alī ibn Yaḥyā, the friend and protégé of the Caliph al-Mutawakkil and the son of Caliph al-Ma'mūn's astronomer Yaḥyā ibn Abī Manṣūr. So we are given a set of pieces of information that is perfectly coherent and plausible. Now this is the book in which we meet this problem of the distance from a

---

[7] See *Les Mathématiques infinitésimales du IXe au XIe siècle*, vol. III: *Ibn al-Haytham. Théorie des coniques, constructions géométriques et géométrie pratique*, London, 2000, Chap. III; English translation by J. V. Field: *Ibn al-Haytham's Theory of Conics, Geometrical Constructions and Practical Geometry*, A History of Arabic Sciences and Mathematics, vol. 3, Culture and Civilization in the Middle East, London, 2013.

[8] See mss Istanbul, Aya Sofya 4830/5 and Khuda Bakhsh 2519/28 (=2468/28) and below, Appendix, Text 4. This book has been published, but not in a critical edition: *Rasā'il Ibn Qurra*, Osmania Oriental Publications Bureau, Hyderabad, 1947. H. Hermelink has noted that under the title in the manuscript 4830, fols 91^v–92^r (*Kitāb al-mafrūḍāt li-Aqāṭun*) there is a fragment of the text on the foundations of geometry attributed to Archimedes and translated by Thābit ibn Qurra. Further, he discusses Van Schooten's possible knowledge of this text via Golius (H. Hermelink, 'Zur Geschichte des Satzes von der Lotsumme im Dreieck', *Sudhoffs Archiv für Geschichte der Medizin und der Naturwissenschaften*, Band 48, 1964, pp. 240–7).

point of a triangle to its sides in the single case of the equilateral triangle, as it was cited by Ibn al-Haytham as according to the ancients.[9] We are only one easy step away from a claim that it is this treatise that Ibn al-Haytham was using. But on the other hand, no tenth-century source, biobibliographical, historical or mathematical, provides evidence that Archimedes wrote such a book, or that Thābit ibn Qurra translated a work with such a title into Arabic.

But yet another fact contributes to confusing a picture that has so far seemed to be clear. Another manuscript, also dating from the beginning of the thirteenth century, a manuscript of a book called *Kitāb al-mafrūḍāt* (*Book of Hypotheses*), contains all the propositions of the previous one plus twenty-four supplementary propositions; this time the whole thing is attributed to a certain Aqāṭun. The common propositions − nineteen or twenty depending on whether we count the first as one proposition or as two − are the same despite variations in the editing.[10] As for the author, Aqāṭun, not only is he unknown, but there is no proof he ever existed. Furthermore, there is no ancient testimony to confirm that a book with such a title existed or was translated. But such a case is not unique, and other books are translated from Greek without our knowing the translator and without their being cited by early biobibliographers. In this particular case, however, careful examination shows that we in fact have a compilation by a late author, put together from several sources, mainly Greek. This is indeed the conclusion which those who have studied this book seem inclined to draw.[11] So, rather than helping us to know 'the ancients' better, this book tends to complicate the situation.

But we are presented with two other testimonies, which make the situation even more complicated. The first comes from al-Nadīm, the early biobibliographer. He tells us that Thābit ibn Qurra did indeed translate a treatise with exactly this title, a work in three books. This work is not attributed to Archimedes but to Menelaus.[12] Moreover, several other sources inform us of the existence of a work with this title and its translation into Arabic.[13] To add to this precise information we have

[9] See p. 485 and Appendix, Text 4: *Two Propositions of the Ancients*.

[10] See Appendix, Text 4: *Two Propositions of the Ancients*.

[11] Y. Dold-Samplonius, *Book of Assumptions by Aqāṭun*, Thèse de doctorat, Université d'Amsterdam, 1977.

[12] Al-Nadīm, *Kitāb al-fihrist*, ed. R. Tajaddud, Teheran, 1971, p. 327: *Kitāb fī uṣūl al-handasa 'amilahu Ibn Qurra (thalātha maqālāt)*.

[13] Al-Bīrūnī, *Risāla fī istikhrāj al-awtār fī al-dā'ira*, Hyderabad, 1948, p. 49; ed. A.S. Demerdash, Le Caire, 1965, p. 90.

another testimony, from al-Sijzī one of the immediate predecessors of Ibn al-Haytham.

Not only has al-Sijzī had a copy of this book by Menelaus in his hands, but he also informs us about a part of its content that is of interest here: according to al-Sijzī, at the beginning of his book *On the Foundations of Geometry* (*Fī al-uṣūl al-handasiyya*) Menelaus considered the problem of the property of equality working from the perpendiculars drawn in the equilateral triangle to its perimeter. Not satisfied by the proof given by Menelaus, al-Sijzī proposes to work through all possible cases – for the equilateral triangle – when the point lies inside or outside the triangle.[14] In his account, he presents, with his own proof, the propositions that Ibn al-Haytham attributes to the 'ancients'.

This rather complicated situation and the small amount of available evidence merely increases the number of possibilities. For example, we might suppose that the Pseudo-Archimedes, the translation of whose work is attributed to Thābit ibn Qurra, formed part of an authentic Menelaus. we might also suppose that the compilation made by Aqāṭun contained part of the work in three books by Menelaus. Obviously, to become plausible, such a conjecture would require us to start by establishing critical editions of all these texts – something that has not yet been done – and to carry out a rigorous study of the history of the text. For the moment, it is enough to know that Ibn al-Haytham had access to the writings attributed to the ancients – Pseudo-Archimedes or genuine Menelaus – which had raised the question in regard to the equilateral triangle. The text by Menelaus had already been studied by al-Sijzī who had generalised the problem by discussing the case of points lying outside the equilateral triangle. Ibn al-Haytham, who probably knew about this work, wanted to go further still by investigating the case of other triangles – isosceles and scalene – but limiting himself to considering points lying inside the triangles, so as to be able to arrive at what we might call a canonical rule. But why did he not, like al-Sijzī, consider points lying outside the triangle? Even if he did not know al-Sijzī's text, it would have been entirely natural for him to think about such points. Perhaps he wanted to look only at generalising the problem under exactly the conditions found in the ancients, that is only for points lying inside the triangle. We shall see later how easy it is to discuss the case of points that lie outside the triangle.

Thus, having readopted the case investigated by the ancients, Ibn al-Haytham considers the same question for an isosceles triangle, then for a scalene triangle. In this last case, he seems to stop at the investigation of

---

[14] See Appendix, Text 4: *On the Properties of the Perpendiculars*, p. 630.

the distances from an arbitrary point on one side of the triangle to the two other sides. However, Ibn al-Haytham's treatise ends with the case of an arbitrary point that lies inside the scalene triangle. This final proposition contains an error that is, to say the least, surprising: not that Ibn al-Haytham never makes a mistake – that, obviously, happens to him as it does to everyone – but this type of error is one he never commits. The only reasonable suggestion is that a reader took it upon himself to complete Ibn al-Haytham's treatise by adding a proposition, when the eminent mathematician intended to go no further. The single surviving manuscript of this text does not provide a basis for any textual argument that might put such a conjecture to the test. So all we have to go on is our knowledge of Ibn al-Haytham's style and his mathematical works.

Ibn al-Haytham begins by setting out the two propositions established by the ancients with the proofs they provided, which he will use as lemmas:

a) *Let there be an equilateral triangle* ABC, *and an arbitrary point* D *on one of the sides – for example* AB –, *then the sum of the distances* DE *and* DG *to the two sides* BC *and* AC *is constant and equal to the height of the triangle.*

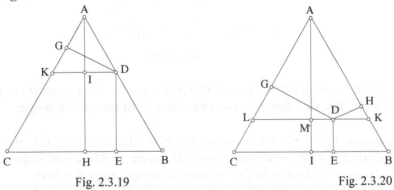

Fig. 2.3.19   Fig. 2.3.20

b) *The sum of the distances from an arbitrary point* D *inside an equilateral triangle* ABC *to the sides* AB, BC, AC *is constant and equal to the height of the triangle.*

According to Ibn al-Haytham, these results were, so far, the only ones known. So the great question is to know how to extend them, with the necessary rigour, to the case of an isosceles triangle, then to an arbitrary one. So it is a matter of finding a similar property, even if it does not involve a constant as in the case of the equilateral triangle; which in fact means finding an expression for the sum of the distances in terms of a parameter. Ibn al-Haytham will first establish that the sum of the distances from a point D on one of the sides of an isosceles triangle, or inside it, to the sides of the triangle is constant for any point lying on a line parallel to the base of the triangle; the sum depends on the distance x from the parallel line to the base of the triangle. Then, as we shall see later, he looks at the scalene triangle.

Ibn al-Haytham begins by establishing two lemmas:

Proposition 1.— *In any triangle, the heights are inversely proportional to the sides on which they stand.*

Proposition 2.— *Let* ABC *be a scalene triangle with a right angle at* A; *we draw the height* AD, *we take a point* E *on* BC *such that* CD = DE *and we draw* AG, *the bisector of the angle* EAB; *then* GD = AD.

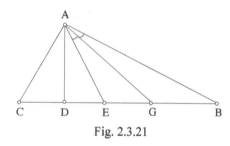

Fig. 2.3.21

After these lemmas, Ibn al-Haytham proves six propositions on the distances. The first four propositions relate to an isosceles triangle.

Proposition 3.— *For any isosceles triangle* ABC *with vertex* A, *the sum of the distances from an arbitrary point* D, *taken on* BC, *to the sides* AB *and* AC, *is equal to the height drawn from the endpoints of the base.*

For this proposition there are three cases for the figure depending on whether angle A is acute, a right angle or obtuse (see figures of the text, p. 489). Here we shall consider only one of the cases for the figure, to make the underlying ideas clear; the reasoning is identical in all the cases.

Let us put $BC = a$, $AC = b$, $AB = c$, $AH = h_A$, $CH = h_C$, $DC = u$, $0 < u < a$. We have

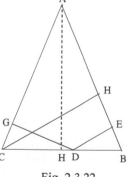

$$DE = (a - u) \sin \hat{B}, \quad DG = u \sin \hat{C} = u \sin \hat{B},$$

Fig. 2.3.22

hence
$$S = DG + DE = a \sin \hat{B} = CH.$$

We can write

(1) $S = h_A \dfrac{a}{b} = h_C = h_B$ (from lemma 1).

In the previous proposition Ibn al-Haytham considered an arbitrary point on the base of the triangle, in the following proposition he considers an arbitrary point on one of the equal sides of the isosceles triangle.

Proposition 4.— *Let* ABC *be an isosceles triangle,* D *an arbitrary point on* AB; *we draw* DG $\perp$ BC *and* DH $\perp$ AC. *Let* AE *be the height from* A, *on which we define the two points* I *and* L *such that*

(1) $\dfrac{AE}{EI} = \dfrac{AB}{BD}$ *and* *(2)* $\dfrac{AI}{IL} = \dfrac{AC}{CB} = \dfrac{AB}{CB}$;

we then have
$$DG + DH = LE.$$

For this proposition there are again three cases for the figure; let us take one of them to make the underlying ideas clear.

The position of *D* on *AB* can be described by $DG = x$, $DG = IE$, so $AI = h_A - x$. Now

$$\dfrac{AI}{IL} = \dfrac{b}{a}$$

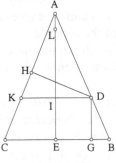

Fig. 2.3.23

implies

$$IL = \dfrac{a}{b}(h_A - x);$$

but

$$EL = EI + IL = x + \frac{a}{b}(h_A - x);$$

now

$$\frac{DK}{BC} = \frac{h_A - x}{h_A},$$

hence

$$DK = \frac{a}{h_A}(h_A - x).$$

But

$$DH = DK \sin \hat{K} = DK \sin \hat{C} = DK \cdot \frac{h_A}{b},$$

so

$$DH = \frac{a}{b}(h_A - x) = IL,$$

and

$$(2) \qquad S = DG + DH = x + \frac{a}{b}(h_A - x) = \frac{a}{b}h_A + x\left(1 - \frac{a}{b}\right) = h_B + x\left(1 - \frac{a}{b}\right).$$

We may note that, if $a = b$, the triangle is equilateral and we then find $DG + DH = h_A$, the result (a) obtained by Ibn al-Haytham's predecessors.

In the following proposition, Ibn al-Haytham considers an arbitrary point inside the isosceles triangle and investigates the sum of its distances from the three sides; he proves that for any point on a line parallel to the base, the sum of the distances can be expressed in terms of the distance of the parallel from the base of the triangle.

For this proposition also there are three cases for the figure (see text, p. 490); let us take one of them to make the underlying ideas clear.

Proposition 5.— *Let there be an isosceles triangle* ABC; *through an arbitrary point* D *inside the isosceles triangle we draw* DE $\perp$ AB, DG $\perp$ AC, DH $\perp$ BC. *Let* AK *be the height of* ABC. *The line through* D *parallel to* BC *cuts* AB *in* M, AK *in* I *and* AC *in* L. *Let* N *be a point on the straight line* AK *such that* $\dfrac{AI}{IN} = \dfrac{AB}{BC} = \dfrac{AM}{ML}$; *we prove that* DE + DG + DH = NK.

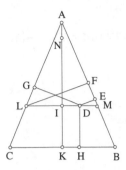

Fig. 2.3.24

We can see at once that the length KN depends on the position of the straight line LM but not on the position of D on this straight line. So let us work out the length of this segment KN.

Let us put $DH = x$ and let us use the same lettering as before. We have $AK = h_A$ and $IK = DH = x$; hence $AI = h_A - x$. But

$$\frac{AI}{IN} = \frac{c}{a} = \frac{b}{a},$$

hence

$$IN = \frac{a}{b}(h_A - x).$$

Now

$$\frac{LM}{BC} = \frac{AI}{AK},$$

hence

$$LM = a\frac{h_A - x}{h_A};$$

moreover,

$$LF = LM \sin \hat{B} = LM \cdot \frac{h_A}{b} = \frac{a}{b}(h_A - x),$$

so

$$LF = IN.$$

Now $DE + DG = LF$, from Proposition 3; so

$$DE + DG = IN$$

and

$$S = DE + DG + DH = IN + IK = KN = \frac{a}{b}(h_A - x) + x = \frac{a}{b}h_A + x\left(1 - \frac{a}{b}\right);$$

and by lemma 1 we have

(3) $S = h_B + x\left(1 - \frac{a}{b}\right).$

We may note that
• if $a = b$ in (3), triangle ABC is equilateral and we obtain the result (*a*) obtained by Ibn al-Haytham's predecessors: $S = h$.
• if $x = 0$, D lies on BC, the base of the isosceles triangle, and we have

$$S = DE + DG = \frac{a}{b}h_A = h_C \quad \text{(the height from } C\text{)}.$$

• if $x \neq 0$, D lies on one of the two sides or on a straight line parallel to the base, and we have the relation (3).

Proposition 6. — This proposition is a corollary to Proposition 5. We consider BE the bisector of the angle B; we have

$$\frac{GA}{GD} = \frac{EA}{EC} = \frac{BA}{BC} = \frac{CA}{BC} = \frac{b}{a};$$

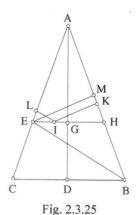

Fig. 2.3.25

If we put $GD = x$, we have $GA = h_A - x = x\,\frac{b}{a}$, hence

$$\frac{a}{b}(h_A - x) = x$$

and from formula (3) we have

(4) $S = 2x.$

In Propositions 3, 4, 5, 6 Ibn al-Haytham derived a formula for finding the sum of the distances from a point on the equal sides of an isosceles triangle or inside it. He has proved that this sum depends on a parameter. It is constant only for the equilateral triangle and in the obvious case where the parameter is zero in the isosceles triangle, when the point lies on its base.

In the following proposition, Ibn al-Haytham deals with another property of the isosceles triangle. Starting from the height CE measured from the side AB, he proves that there are three magnitudes that are in continuous proportion.

Proposition 7. — *Let* ABC *be an acute-angled isosceles triangle with* $AB = AC$; *we draw the heights* AD *and* CE; *then* AB − CE, CE − EB *and* 2EB *are in continuous proportion.*

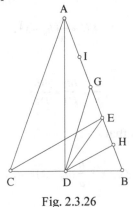

Fig. 2.3.26

Then, next, Ibn al-Haytham deals with an arbitrary triangle. He knows perfectly well that formula (3) does not apply in a more general case. He establishes in what cases this formula still holds, that is when the point is taken on one of the sides.

Proposition 8. — *The sum of the distances from a point on a side of the scalene triangle to the two other sides is given by*

$$S = h_A + x\left(1 - \frac{a}{b}\right),$$

where x = DE, *the distance from* D *to the side* AC.

In the proof, the height used this time is the height from *B*.

Let us put $BH = h_B$; we then have $KB = h_B - x$.

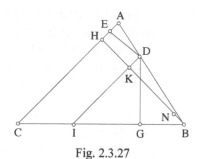

Fig. 2.3.27

Let us put

$$\frac{BK}{KN} = \frac{a}{b},$$

hence

$$KN = \frac{b}{a}(h_B - x).$$

Moreover,

$$\frac{a}{b} = \frac{BI}{ID} = \frac{BK}{DG} \qquad \text{(by Proposition 1)},$$

so $KN = DG$; and it follows that

$$S = DE + DG = x + \frac{b}{a}(h_B - x) = h_A + x\left(1 - \frac{b}{a}\right).$$

If we consider the height from *A* (as in Propositions 4 and 5), we then take *DG* as an unknown line parallel to that height (we interchange the roles of *BC* and *AC* as well as the letters *a* and *b*), and we have

$$S = \frac{a}{b}h_A + x\left(1 - \frac{a}{b}\right), \qquad \text{where } x = DG$$

hence

(5) $$S = h_B + x\left(1 - \frac{a}{b}\right).$$

Following this proposition, there is a ninth proposition in which an effort is made to establish that this property holds for an arbitrary point. Let us first examine the content of this proposition before raising the question of its authenticity.

Let us start by presenting it as it appears in the manuscript.

Proposition 9. — *Through a point* D *inside the scalene triangle* ABC *we draw* DE ⊥ AB, DG ⊥ AC, DH ⊥ BC, LM ∥ BC *which cuts the height* AK *in* I. *We take* N *on* AI *such that* $\dfrac{AI}{IN} = \dfrac{BC}{CA}$; *we prove that*

$$DE + DG + DH = NK.$$

It is true − apart from one correction: $\dfrac{IN}{AI} = \dfrac{CB}{CA}$ instead of $\dfrac{AI}{IN} = \dfrac{CB}{CA}$ −, that

$$LM \parallel BC \Rightarrow \frac{LM}{MA} = \frac{BC}{CA},$$

so

$$\frac{LM}{MA} = \frac{IN}{AI}.$$

Even with this correction, the remainder of the proof is false.

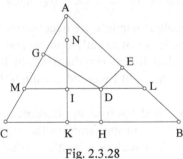

Fig. 2.3.28

In the statement of Proposition 9, the point I is defined by the equality $\dfrac{AI}{IN} = \dfrac{CB}{CA}$; then we have $\dfrac{AI}{IN} = \dfrac{LM}{MA}$ in the triangle ALM with height AI. In Proposition 8 on which Proposition 9 depends in the mind of the editor, the reasoning brings in the scalene triangle BDI with height BK, and the point N on BK defined by $\dfrac{BK}{KN} = \dfrac{BC}{CA}$, and we have $\dfrac{BK}{BN} = \dfrac{BI}{ID}$ (see Fig. 2.3.26). But in Proposition 9 the height AI goes to LM and in Proposition 8 the height

BK goes to *ID*. Further, in 8 the point *D* is a vertex of the triangle *IBD* and in 9 the point *D* is an arbitrary point of *LM*, the base of the triangle *ALM*. One cannot refer back to 8 to prove 9, though this is indeed what is done by the editor of the text.

If we correct the text and put $\dfrac{IN}{AI} = \dfrac{BC}{CA}$, we return to the condition imposed in Proposition 5 in the case of an isosceles triangle *ABC*. Further, the figures for Propositions 5 and 9 are constructed in the same way, with the same lettering, and in 9 the expression 'as it was proved earlier' is undoubtedly a reference to Proposition 5. But the result $DE + DG = IN$ is proved in 5 starting from the conclusion of Proposition 3 in which use is made of the similar triangles *DLG* and *DME* (*DLE* and *DMG* respectively in 9). The fact that these triangles are similar derives from the equality $\hat{B} = \hat{C}$ that implies $\hat{L} = \hat{M}$; now in 9, $\hat{B} \neq \hat{C}$, so in Proposition 9 one cannot have $DE + DG = IN$. Let us find what this sum is.

We have

$$DG \cdot AM = AI \cdot MD \text{ and } AL \cdot DE = AI \cdot DL,$$

hence

$$DG + DE = AI \left(\frac{MD}{AM} + \frac{DL}{AL} \right) \qquad \text{where } AM \neq AL;$$

so this sum depends on the position of *D* on *ML*.

Now if we define *N* by $\dfrac{IN}{AI} = \dfrac{BC}{CA}$, we have $IN = AI \cdot \dfrac{ML}{MA}$. So, in Proposition 9, for a scalene triangle we necessarily have $DE + DG \neq IN$ and $DE + DG + DH \neq KN$. It is not exactly controversial to say that this type of error is not the kind that Ibn al-Haytham might have committed. So it is likely that some editor thought he could complete Ibn al-Haytham's text in this way.

If this is so, we need to find a plausible explanation why Ibn al-Haytham did not present a proposition on the sum of the distances to the sides from a point inside a scalene triangle. Perhaps it is because this time the formula does not involve only a single parameter but two at a time, which makes it considerably less interesting. To illustrate this let us find the sum $DE + DG$ in terms of what is given and the parameters that determine the position of *D*.

As before, let us put $BC = a$, $AB = c$, $AC = b$, $AK = h_A$, $DH = x$, $DL = y$ (see previous figure). We have

$$AI = h_A - x, \frac{AM}{AC} = \frac{AL}{AB} = \frac{LM}{BC} = \frac{h_A - x}{h_A},$$

hence

$$AM = \frac{b(h_A - x)}{h_A}, \quad AL = \frac{c(h_A - x)}{h_A}, \quad LM = \frac{a(h_A - x)}{h_A}.$$

But

$$MD = LM - y = \frac{a(h_A - x) - h_A y}{h_A};$$

we have

$$DG + DE = AI\left(\frac{MD}{AM} + \frac{DL}{AL}\right) = \frac{ac(h_A - x) + h_A y(b - c)}{bc}$$

and

(6) $$DG + DE + DH = \frac{ach_A + c(b - a)x + (b - c)h_A y}{bc}.$$

Thus, the sum depends on the three sides of the triangle, on a height and on two parameters.

Notes:

1. If D lies on AB as in 8, we have $y = 0$ and $DE = 0$, and we return to (5) starting from (6). In fact, (6) can be rewritten

$$\frac{a}{b}h_A + \frac{b - a}{b}x = h_B + x\left(1 - \frac{a}{b}\right),$$

a result that depends on a single parameter.

2. If the triangle is isosceles, we have $b = c$ and we obtain (3) starting from (6), a result that depends on a single parameter.

3. If in (6) we take the parameter x as given, which is the same as taking LM as given, the sum $DG + DE + DH$ depends on the parameter y, that is it depends on the position of D on LM; so it will not be a sum that is constant, and it could not be represented by a segment such as the segment KN that appears in the text.

Perhaps these difficulties encountered in trying to provide a usable generalisation of a formula that describes the sum of the distances, to include the case of a point inside a scalene triangle, prevented Ibn al-Haytham from providing a proposition, which as presented here seems to be the work of an editor who was much less expert than Ibn al-Haytham; unless, perhaps, we should consider the possibility that we are dealing with a rather naïve investigation dating from his youth. Only further manuscript copies, from a manuscript tradition different from that of the copy we have, would allow us to return to this question.

4. There is, however, a linear combination of the three distances DE, DG and DH that remains constant, that is independent of the position of D, an arbitrary point inside the triangle or on one of its sides: this is $c \cdot DE + b \cdot DG + a \cdot DH$ which is always equal to the area of the triangle ABC. If D lies outside the triangle, we need to attach a suitable sign to each of the terms in the preceding sum in order to make it remain equal to the area of ABC.

Finally, let us turn to the question of the distances from exterior points to an equilateral triangle that was examined by al-Sijzī[15] but was not considered by Ibn al-Haytham. The part of the plane *outside* the equilateral triangle ABC can be divided into six parts obtained by extending its sides. We have the three straight lines $XBAX'$, $YCAY'$ and $ZCBZ'$. With the vertex A we associate the region I_A, or $(XBCY)$, lying beyond BC; and the region II_A, enclosed by the angle $X'AY'$. In the same way to the point B there correspond the regions I_B and II_B, and to the point C the regions I_C and II_C.

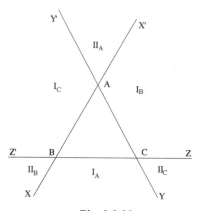

Fig. 2.3.29

[15] *Qawl Aḥmad ibn Muḥammad 'Abd al-Jalīl al-Sijzī fī khawāṣṣ al-a'mida al-wāqi'a min al-nuqṭa al-mu'ṭā ilā al-muthallath al-mutasāwī al-aḍlā' al-mu'ṭā bi-ṭarīq al-taḥdīd*, mss Dublin, Chester Beatty 3652, fols 66r–67r; Istanbul, Reshit 1191, fols 124v–125v. See J. P. Hogendijk, 'Traces of the lost geometrical elements of Menelaus in two texts of al-Sijzī', *Zeitschrift für Geschichte der arabisch-islamischen Wissenschaften*, Band 13, 1999–2000, pp. 129–64, p. 142 *sqq*; and P. Crozet, 'Geometria: La tradizione euclidea rivisitata', in R. Rashed (ed.), *Storia della scienza*, vol. III: *La civiltà islamica*, Rome, 2002, pp. 326–41.

Triangle ABC is equilateral, each of its sides is equal to the length a and the heights are each equal to $h = \dfrac{\sqrt{3}}{2}a$. So it is sufficient to investigate the sum of the distances for any point situated in I_A or II_A.

1. First let M be a point in I_A, that is in the region $(XBCY)$ and let the three distances be ME, MK, MI. Let AH be the height from A $(AH = h)$; the line drawn through M parallel to BC cuts AX in B_1, AY in C_1 and AH in H_1. Let us put $ME = x$, a parameter that determines the position of the straight line B_1C_1; we have $AH_1 = h + x$. Moreover,

$$MK = MC_1 \sin \hat{C}_1 = MC_1 \cdot \frac{\sqrt{3}}{2}$$

and

$$MI = MB_1 \sin \hat{B}_1 = MB_1 \cdot \frac{\sqrt{3}}{2},$$

so

$$MK + MI = B_1C_1 \frac{\sqrt{3}}{2} = AH_1 = h + x.$$

In fact, triangle AB_1C_1 is equilateral with height AH_1 and M is a point on its base; we know from Lemma (b) (p. 457) that $MK + MI = AH_1$.

We have that the sum

(1) $S = ME + MK + MI = h + 2x.$

This sum is the same for any point M of the segment B_1C_1, including the endpoints. If $x = 0$, M is at E on BC and we have $MK + MI = h$, a result proved earlier.

So it follows that for any point M in the region $XBCY$, including the edges, the required sum depends on a parameter that is the distance from M to the straight line BC.

2. Now let there be a point N lying within the angle $X'AY'$, that is in the region II_A and $NE \perp BC$, $NK' \perp AC$, $NI' \perp AB$. The line drawn through N parallel to BC cuts AX' in B', AY' in C' and AH in H'. Let us put $NE = x > h$; we have $AH' = x - h$. Moreover, we have $NK' = NC'\dfrac{\sqrt{3}}{2}$, $NI' = NB'\dfrac{\sqrt{3}}{2}$, so

$$NK' + NI' = B'C' \frac{\sqrt{3}}{2} = AH' = x - h.$$

In fact, triangle $AB'C'$ is equilateral with height AH' and N is a point on its base, so $NK' + NI' = AH' = x - h$, from the lemma mentioned earlier, and

(2) $S = NE + NI' + NK' = 2x - h.$

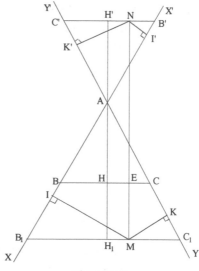

Fig. 2.3.30

This sum is the same for any point N of the segment $B'C'$. If the point N is at A, we have $x = h$, $NK' = 0$, $NI' = 0$ and the sum is equal to $AH = h$.

It is obvious that the same method can be applied for the points in the regions I_B and II_B or I_C and II_C and leads to the same result.

So the sum we are examining can be expressed in terms of h, the height of the equilateral triangle, and a single parameter x that is the distance from the point we are considering to a side of the triangle, that is the distance to BC if M is in one of the regions I_A or II_A, the distance to AC if M is in one of the regions I_B or II_B, and the distance to AB if M is in one of the regions I_C or II_C. So we have two different results for the sum S.

• M in I_A, I_B, I_C, we have $S = h + 2x$
• M in II_A, II_B, II_C, we have $S = 2x - h$.

It is clear that nothing in the above procedure could be unknown to Ibn al-Haytham, nor inaccessible to him. If he did not wish to consider these cases, perhaps it is, as we have said, in order to remain within the framework of the hypotheses of the ancients: considering only interior points. Al-Sijzī, on the other hand, had thought about these exterior points. Ibn al-Haytham's predecessor, al-Sijzī, considered the cases of a point at the vertex of an equilateral triangle, on one of its sides or inside it; in all cases, we have $S = h$, a result established by the ancients and, according to him, by Menelaus. Next he considered the points in the region I_A investigating various positions for the point and he determined S_1, the sum of the distances to the two sides AB and AC of the triangle. In fact, if x is the distance from the point to the base BC, we have in all cases that

$$S_1 = AD + x = h + x.$$

And consequently

$$S = h + 2x.$$

However, al-Sijzī does not examine the case where the point is in the region II and he does not complete the generalisation he began. Perhaps he was interested only in the relation between the three perpendiculars D_1, D_2 and D_3 from the point D. In the case he considers, we have $D_1 + D_2 = D_3 + h$. But he would then have been able to deduce from this that $D_1 + D_2 - D_3$ is constant in the case concerned, which he did not do.

In conclusion, it is clear that the chief concern in these investigations carried out by al-Sijzī, and above all in those by Ibn al-Haytham, is the search for a quantity that remains constant. But the weight of the tradition stemming from the ancients led them to consider only the sum of the distances from a point to the sides of a triangle instead of, in a more algebraic way, thinking about a suitable linear combination of the distances.

3. *History of the texts*

3.1. *On a Geometrical Problem*

Ibn al-Haytham's first treatise, *On a Geometrical Problem* (*Fī mas'ala handasiyya*), appears in two old lists of his writings: we find it mentioned by al-Qifṭī and by Ibn Abī Uṣaybi'a.[16] The treatise itself has come down to us in two manuscripts. It in fact forms part of two particularly important manuscript collections, each of which includes several treatises by Ibn al-

---

[16] See *Les Mathématiques infinitésimales*, vol. II, pp. 526–27.

Haytham. The first is the collection in the Oriental Institute in St Petersburg, whose former number is B 1030 and its present one 89. This collection contains twelve treatises, eleven by Ibn al-Haytham and the twelfth by al-'Alā' ibn Sahl. The treatise that is of interest to us here is the eighth. This collection is one we have come across several times;[17] it was copied about 750/1349 – the date when it was revised by comparing it with its original. It is all in the same hand, in *nasta'līq* script. The treatise on a geometrical problem, which covers folios 102^r–110^v, has neither additions nor glosses. The only addition (fol. 107^v) has been made in the hand of the copyist – in the course of making the copy or during its revision against the original – so as to remedy the omission of a term in a proposition. The figures are drawn by the copyist. The only peculiarity in the copy is the repetition of the text of a page, but not of the figure that accompanies it. The copyist noticed this error, and at the top of the page wrote the word *mukarrara* (repeated). Here this collection is referred to as L.

The second collection is found in the Bodleian Library, Oxford, Seld. A 32. It too contains eight treatises by Ibn al-Haytham, the one that interests us here being the fifth (fols 115^v–120^r). We have already come across this collection.[18] It is written in *naskhī*. The copyist does not give either the place or the date; he has drawn the figures and compared his copy with its original, as is indicated by additions in the margin in the same hand. We may note, however, that there are some glosses in a different hand (fol. 117^r). Here this collection is referred to as O.

These two manuscripts are fully independent. Thus, three words that are missing in L are present in O, while, in relation to L, O is missing six words and a sentence. If such confirmation were needed, other peculiarities, grammatical errors, errors in letters, and so on, further confirm that the two are independent.

As far as we know, this treatise was edited and translated for the first time in our French edition of 2002 and appears here in English for the first time.[19] Nor do we know of any serious and comprehensive study of its content.

---

[17] *Ibid.*, vol. II, pp. 24, 26, 27.

[18] *Les Mathématiques infinitésimales du IXe au XIe siècle*, vol. III: *Ibn al-Haytham. Théorie des coniques, constructions géométriques et géométrie pratique*, London, 2000, p. 535. English translation: *Ibn al-Haytham's Theory of Conics, Geometrical Constructions and Practical Geometry*. A History of Arabic Sciences and Mathematics, vol. 3, Culture and Civilization in the Middle East, London, 2013.

[19] *Les Mathématiques infinitésimales du IXe au XIe siècle*, vol. IV: *Méthodes géométriques, transformations ponctuelles et philosophie des mathématiques*, London, 2002, pp. 619–33.

3.2. *On the Properties of the Triangle*

The second treatise *On the Properties of the Triangle in Regard to Height* (*Fī khawāṣṣ al-muthallath min jihat al-'amūd*), which is also mentioned by al-Qifṭī and by Ibn Abī Uṣaybi'a,[20] has come down to us in a single manuscript that forms part of the collection 2519 of the Library of Khuda Bakhsh (Patna, India). This important collection, which we have already mentioned,[21] consists of 42 treatises on mathematics (Archimedes, al-Qūhī, Ibn 'Irāq, al-Nayrīzī and others); it has 327 folios (32 lines per page, dimensions 24 × 15 cm, and 20 × 12.5 cm for the text) and it was copied in 631–632 of the Hegira, that is in 1234–1235 AD, at Mosul, in *naskhī*. The text by Ibn al-Haytham was transcribed in 1235 and occupies folios 189ʳ–191ʳ; it has neither additions nor marginal annotations. Here it is referred to as H.

An edition of this text, but a not a critical one, was printed in Hyderabad in 1948; we have referred to it as Kh.

This treatise was edited and translated for the first time in our French edition of 2002[22] and appears here in English for the first time. The inadequate Hyderabad edition was translated into English by F. A. Shamsi, with a commentary, under the title *Properties of Triangles in Respect of Perpendiculars*, in Hakim Mohammed Said (ed.), *Ibn al-Haytham, Proceeding of the Celebrations of 1000th Anniversary*, Karachi, n.d., pp. 228–46.

---

[20] *Ibid.*, vol. II, pp. 522–3.

[21] *Les Mathématiques infinitésimales du IXᵉ au XIᵉ siècle*, vol. I: *Fondateurs et commentateurs: Banū Mūsā, Thābit ibn Qurra, Ibn Sinān, al-Khāzin, al-Qūhī, Ibn al-Samḥ, Ibn Hūd*, London, 1996, p. 680; English trans. *Founding Figures and Commentators in Arabic Mathematics*, A History of Arabic Sciences and Mathematics, vol. 1, Culture and Civilization in the Middle East, London, 2012.

[22] *Les Mathématiques infinitésimales*, vol. IV, pp. 635–53.

TRANSLATED TEXTS

Al-Ḥasan ibn al-Ḥasan ibn al-Haytham

In the name of God, the Compassionate the Merciful

TREATISE BY AL-ḤASAN IBN AL-ḤASAN IBN AL-HAYTHAM

On a Geometrical Problem

Let there be a triangle *ABC* of known magnitude whose side *BC* is known and in which the sum of the two sides *BA* and *AC* is known; we wish to know each of the sides *BA* and *AC*.

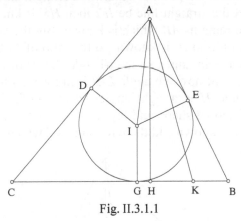

Fig. II.3.1.1

We imagine that in this triangle there is a circle touching its sides; let the circle be *DEG*; let its centre be *I*. We draw from the centre *I* straight lines to the points of contact; let the straight lines be *IE*, *ID*, *IG*. They are perpendicular to the sides of the triangle and they are equal. The product of *IE* and half of the perimeter of the triangle is thus known, because it is equal to the area of the triangle; but the area of the triangle is known and half of the perimeter of the triangle is known, because the perimeter of the triangle is known, so the straight line *IE* is known. But since these perpendiculars make the two straight lines that enclose an angle of the triangle to be equal, the sum of the straight line *BC* and the straight line *AE* is half of the perimeter of the triangle, so the sum of the straight line *BC* and the straight line *AE* is known; now *BC* is known, so the straight line *AE* is known. But *EI* is known, so the ratio of *AE* to *EI* is known. Let us join

AI; the triangle *AEI* is thus of known shape, because the angle *AEI* is a right angle. So the angle *EAI* is known and the angle *DAI* is equal to it, so the angle *BAC* is known. We draw the perpendicular *AH*, it is known, because its product with half of *BC* which is known is the area of the triangle which is known. We draw the straight line *AK* in such a way that the angle *AKC* is equal to the angle *BAC* which is known; the triangle *AKH* is thus of known shape, so the ratio of *HA* to *AK* is known and *AH* is known, so *AK* is known and the product of *AK* and *BC* is known. But the product of *AK* and *BC* is equal to the product of *BA* and *AC*, so the product of *BA* and *AC* is known. But the sum of *BA* and *AC* is known, so each of the straight lines *BA* and *AC* is known. This is what we wanted to prove.

In another way

We return to the triangle and the circle and we draw the perpendicular *AH*; it is known. From the point *I* we draw a straight line parallel to the straight line *BC*; let the straight line be *IK*; then *HK* is known, because *IG* is known, and there remains *AK* which is known. But the straight line *IA* is known, because its ratio to *IE* is known, so the ratio of *IA* to *AK* is known; now the angle *K* is a right angle, so the triangle *AIK* is of known shape, so the angle *IAK* is known; now the angle *IAB* is known, so the angle *HAB* is known. But the angle *H* is a right angle, so the angle *B* is known, the triangle *ABH* is of known shape and the ratio of *BA* to *AH* is known. But *AH* is known, so *BA* is known and there remains *AC* which is known.

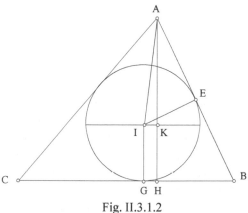

Fig. II.3.1.2

Further, the angle *C* is also known, because each of these angles *B* and *A* is known, so the triangle *ABC* is of known shape and the ratio of *BA* to *AC* is known; but the sum of *BA* and *AC* is known, so each of the straight lines *BA* and *AC* is known. This is what we wanted to prove.

In another way

We return to the triangle and the circle, we extend *BA* and we cut off *AF* equal to *AC*. We join *FC* and we extend *AI* to *H*; then it is parallel to the straight line *FC*, because the angle *BAH* is half the angle *BAC*. But the angle *AFC* is equal to half of the angle *BAC*, so the angle *BFC* is known because the angle *BAH* is known. But the ratio of *BF* to *BC* is known, because each of them is known, so the triangle *BFC* is of known shape and the ratio of *BF* to *FC* is known; but *BF* is known, so *FC* is known. Similarly, the triangle *AFC* is of known shape, so the ratio of *FC* to *CA* is known; but *CF* is known, so *CA* is known and there remains *AB* which is known. This is what we wanted to prove.

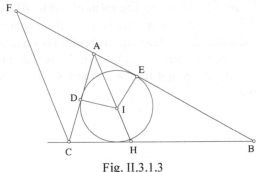

Fig. II.3.1.3

In another way

We return to the triangle and the circle; we prove that the angle *BAC* is known, so the ratio of the product of *BA* and *AC* to <the area of the> triangle is known. But <the area> of the triangle is known, so the product of *BA* and *AC* is known and each of the straight lines *BA* and *AC* is known. This is what we wanted to prove.

In another way

We take the triangle and we draw the circumscribed circle; let it be the circle *ACDB*. We divide the arc *BDC* into two halves at the point *D* and we join *DEA*; it divides the angle *BAC* into two halves. So the ratio of *BA* to *AC* is equal to the ratio of *BE* to *EC*, so the ratio of *AB* to *BE* is equal to the ratio of *AC* to *CE* and it is equal to the ratio of the sum of *BA* and *AC* to the whole of *BC*. But the ratio of the sum of *BA* and *AC* to *BC* is known, because the sum of *BA* and *AC* is known, so the ratio of *AB* to *BE* is known. We join *DB*; the angle *DAC* is equal to the angle *CBD*. So the angle *BAD* is equal to the angle *CBD*, so the triangle *ABD* is similar to the triangle *DBE*, so the ratio of *AD* to *DB* is equal to the ratio of *BD* to *DE* and is equal to the ratio of *AB* to *BE*. But the ratio of *AB* to *BE* is known, so the ratio of

AD to *DB* is known and the ratio of *BD* to *DE* is known, so the ratio of *AD* to *DE* is known and the ratio of *AE* to *ED* is known.[1] We draw the perpendicular *AH*; it is known because its product with *BC* is double <the area> of triangle *ABC* which is known. We draw from the point *D* the perpendicular *DG*; it is parallel to the perpendicular *AH*, so the ratio of *AH* to *DG* is equal to the ratio of *AE* to *ED* which is known and the perpendicular *DG* is known; now *DG* divides the straight line *BC* into two halves, so the straight line *BG* is known and the angle *G* is a right angle; so the straight line *BD* is known. But the ratio of *AD* to *DB* is known, so the straight line *AD* is known; but the ratio of *AE* to *ED* is known, so each of the straight lines *AE* and *ED* is known, the product of *AE* and *ED* is known, the product of *BE* and *EC* is known, the ratio of *AB* to *BE* is known and the ratio of *AC* to *CE* is also known. So the ratio of the product of *BA* and *AC* to the product of *BE* and *EC* is known. Now the product of *BE* and *EC* is known, so the product of *BA* and *AC* is known. Now the sum of *BA* and *AC* is known, so each <of the straight lines> *BA* and *AC* is known. This is what we wanted to prove.

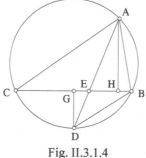

Fig. II.3.1.4

Further, each of the straight lines *BE* and *EC* is known, the ratio of *AB* to *BE* is known, the ratio of *AC* to *CE* is known and each <of the straight lines> *BA* and *AC* is known.

Once this proposition[2] has been proved in several ways, it remains to carry out the synthesis of this problem and to make it [*sc.* the problem] constructible.

We can carry out the synthesis in each of the ways that we have shown; but we shall limit ourselves to carrying out the synthesis in one of the ways so as not to lengthen our account. So we shall carry out the synthesis in the last way. The synthesis in the last way is as we shall describe it.

---

[1] See the justification in the commentary.

[2] The Arabic term is *ma'nā*, which Latin translators render as *intentio*, a word that conveys both the subject matter and the desired result. See also p. 221, n. 1.

We wish to construct on a known straight line a triangle [of area] equal to a given area and such that the sum of its two remaining sides is equal to a known straight line.

Let the known straight line on which we wish to construct the triangle be *AB*, let the known area to which we wish the [area of] the triangle to be equal be the area enclosed by the straight lines *AB* and *CD*, and let the straight line to which the sum of the two remaining sides of the triangle is equal be the straight line *GH*. We put the ratio of *GH* to *AB* equal to the ratio of *AB* to *HI*. We divide the straight line *AB* into two halves at the point *K* and we draw from the point *K* the perpendicular to the straight line *AB*; let it be *KL*. We put *CE* equal to *CD* and we put the ratio of *ED* to *KL* equal to the ratio of *GI* to *IH*. We join *AL* and *BL* and we circumscribe a circle about the triangle *ALB*; let the circle be *ALBM*. We extend the straight line *LA* and we put the ratio of *PA* to *AL* equal to the ratio of *ED* to *KL*. We draw from the point *P* a straight line parallel to the straight line *AB*, let it be *PM*; let it cut the circle at the point *M*. We join *AM* and *BM*.

I say that [the area of] the triangle *AMB* is equal to the area enclosed by the straight lines *AB* and *CD* and that the two straight lines *AM* and *MB* have a sum equal to the straight line *GH*.

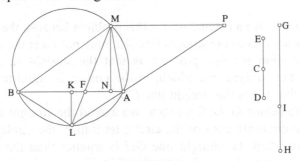

Fig. II.3.1.5

Proof: We join the straight line *LFM* and we draw the perpendicular *MN*; the ratio of *MN* to *KL* is then equal to the ratio of *MF* to *FL*. But the ratio of *MF* to *FL* is equal to the ratio of *PA* to *AL* and the ratio of *PA* to *AL* is equal to the ratio of *ED* to *KL*, so the ratio of *MN* to *KL* is equal to the ratio of *ED* to *KL*, the straight line *MN* is equal to the straight line *ED* and the product of *AB* and half of *MN* is equal to the product of *AB* and *CD*. Now the product of *AB* and half of *MN* is the magnitude of the triangle *AMB*, so the triangle *AMB* is equal to the area enclosed by the straight lines *AB* and *CD*; which is one of the <conditions> we sought.

In the same way, the angle *BML* is equal to the angle *BAL* and the angle *BML* is equal to the angle *AML*, because the arc *AL* is equal to the arc

LB, so the angle *BAL* is equal to the angle *AML*. So the triangle *AML* is similar to the triangle *ALF*, so the ratio of *ML* to *LA* is equal to the ratio of *AL* to *LF* and is equal to the ratio of *MA* to *AF*. But the ratio of *MA* to *AF* is equal to the ratio of *MB* to *BF*, because these two angles which are at the point *M* are equal, so the ratio of the sum of *AM* and *MB* to the straight line *AB* is equal to the ratio of *ML* to *LA*. The ratio of the square of the sum of *AM* and *MB* to the square of *AB* is thus equal to the ratio of the square of *ML* to the square of *LA*, which is equal to the ratio of *ML* to *LF*, which is equal to the ratio of *GH* to *HI*. The ratio of the square of the sum of *AM* and *MB* to the square of *AB* is thus equal to the ratio of *GH* to *HI*. But the ratio of *GH* to *HI* is equal to the ratio of the square of *GH* to the square of *AB*. The ratio of the square of the sum of *AM* and *MB* to the square of *AB* is thus equal to the ratio of the square of *GH* to the square of *AB*, so the ratio of the sum of *AM* and *MB* to *AB* is equal to the ratio of *GH* to *AB*, so the sum of the straight lines *AM* and *MB* is equal to the straight line *GH*.

Now we have proved that <the area> of the triangle *AMB* is equal to the product of *AB* and *CD*. So we have constructed on the straight line *AB* a triangle that satisfies the required condition; let the triangle be *AMB*. This is what we wanted to construct.

It remains to give a discussion for this problem because the straight line *PM*, which is parallel to the straight line *AB*, may not meet the circle.

The discussion for this problem is that the straight line *GH* is not smaller than the straight line which, in the power, is equal to the straight line *AB* plus four times the straight line *CD*.[3]

Proof: We return to the figure and we extend the straight line *LK* until it ends on the circumference of the circle; let it meet the circle at the point *Q*. We suppose that the straight line *GH* is smaller than the straight line which, in the power, is equal to the straight line *AB* plus four times the straight line *CD*. I say that the straight line *PM* parallel to the straight line *AB* does not meet the circle and that the construction of the triangle cannot be completed in accordance with the required condition.

The straight line *PS* meets the straight line *LQ* in all the cases of the figure, if we extend *LQ*; let it meet it at the point *S*. So the ratio of *SK* to *KL* is equal to the ratio of *PA* to *AL* which is equal to the ratio of *ED* to *KL*, so the straight line *SK* is equal to the straight line *ED* and the ratio of *SK* to *KL* is equal to the ratio of *GI* to *IH*. So the ratio of *SL* to *LK* is equal to the ratio of *GH* to *HI*. But the ratio of *GH* to *HI* is equal to the ratio of the square of

---

[3] It is necessary that $GH^2 \geq AB^2 + (4CD)^2$; but $2CD = ED$, so it is necessary that $GH^2 \geq AB^2 + 4ED^2$ (see Note, p. 452).

GH to the square of *AB* which is smaller than the ratio of the square of twice *ED* plus the square of *AB* to the square of *AB*, so the ratio of *SL* to *LK* is smaller than the ratio of the square of twice *ED* plus the square of *AB* to the square of *AB*, so the ratio of *SK* to *KL* is smaller than the ratio of the square of twice *ED* to the square of *AB* which is the ratio of the square of *ED* to the square of *AK* and the ratio of *SK* to *KL* is smaller than the ratio of the square of *ED* to the square of *AK*. But *SK* is equal to *ED*, so the ratio of *SK* to *KL* is smaller than the ratio of the square of *SK* to the square of *AK*. So the ratio of *SK* to *KL* is equal to the ratio of the square of *SK* to a square greater than the square of *AK*. So the product of *SK* and *KL* is greater than the square of *AK* and the product of *QK* and *KL* is equal to the square of *AK*, so the straight line *SK* is greater than the straight line *QK*. So the point *S* is outside the circle and the height *ED* is greater than the straight line *QK* which is the greatest perpendicular that is in the segment *AQB*.[4] So if the square of *GH* is smaller than the square of twice *ED* plus the square of *AB*, then the construction of the triangle cannot be completed in accordance with the given condition.

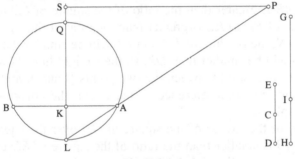

Fig. II.3.1.6

Indeed, in every case a triangle is inscribed in a circle; if a circle is circumscribed about it, if we divide the arc subtended by the straight line *AB* into two halves and if we join its mid point and the vertex of the triangle with a straight line, then that straight line is divided by the straight line *AB* into two parts such that the ratio of one of the parts to the other is equal to the ratio of the perpendicular dropped from the vertex of the triangle to its base – which is the straight line *AB* – to the perpendicular dropped from the mid point of the arc subtended by the straight line *AB* to the straight line *AB* and that the ratio of the two parts of the straight line which joins the mid point of the arc and the vertex of the triangle, the one to the other, is equal to the ratio of the excess of the square of the sum of the two

[4] Euclid, *Elements*, III.15.

adjacent sides at the vertex of the triangle, over the square of *AB*, also to the square of *AB*, because the ratio of the square of the sum of the two sides to the square of *AB* is always equal to the ratio of the straight line which joins the mid point of the arc to the vertex of the triangle to the part of it which is on the side towards the mid point of the arc. So the ratio of *ED*, which is the height of the triangle, to *KL*, which is the other height, is always equal to the ratio of the excess of the square of the sum of the two sides – which in the example is *GH* – over the square of *AB*, to the square of *AB*.

So if the ratio of the excess of the square of *GH* – which is equal to the sum of the two sides – over the square of *AB* to the square of *AB* is equal to the ratio of the square of twice *ED* – which is the height – to the square of *AB*, then the ratio of *ED* to *KL* is equal to the ratio of the square of *ED* to the square of *AK*. So we have the product of *ED* and *KL* equal to the square of *AK*; so the straight line *PS* meets the circle at the point *Q* and the construction is completed.

If the ratio of the excess of the square of *GH* over the square of *AB* to the square of *AB* is greater than the ratio of the square of *ED* to the square of *AK*, the ratio of *ED* to *KL* is greater than the ratio of the square of *ED* to the square of *AK*, the product of *ED* and *KL* will be smaller than the square of *AK* and *ED* will be smaller than *QK*, so the straight line *PS* will meet the straight line *QK* in a point between the two points *Q* and *K* and the straight line *PS* thus cuts the circumference of the circle; the construction of the triangle is completed.

If the ratio of the excess of the square of *GH* over the square of *AB* to the square of *AB* is smaller than the ratio of the square of *ED* to the square of *AK*, then the straight line *PS* meets the straight line *QK* in a point outside the circle, so the straight line *PS* does not meet the circle, as we have established by the proof, and the construction of the triangle cannot be completed.

The discussion of this problem is that the straight line *GH* is not smaller than the straight line which, in the power, is equal to the straight line *AB* plus four times the straight line *CD*. This is what we wanted to prove.

Starting from this discussion we have proved that for two of the sides of a triangle, the square of their sum, if they are made into a straight line, is not smaller than the square of the remaining side plus four times the square of the perpendicular dropped from the angle enclosed by these two sides, onto the side that remains.

The treatise is completed.

In the name of God the Compassionate the Merciful
From Him comes success

TREATISE BY IBN AL-HAYTHAM

On the Properties of the Triangle in Regard to Height

The ancient geometers examined the properties of the equilateral triangle. It thus became clear to them that for any given point on one of the sides of the equilateral triangle, if we draw from that point two perpendiculars to the two remaining sides, then their sum is equal to the height of the triangle. They have written this down and proved it in their books. They examined the heights of other triangles and for them they found neither a complete order, nor an arrangement and thus they did not give any account of them. Since this was how matters stood, necessity encouraged us to investigate the properties of triangles; then we found a uniform order for the heights of the isosceles triangle, and we similarly found an order for the heights of the scalene triangle and a uniform arrangement. When this became clear to us, we composed this treatise.

We first present what the ancients said about the properties of the heights of the equilateral triangle.[1] We then move on with what we ourselves have determined about the properties of the heights of other triangles so that the properties of the heights of all triangles might be gathered together in this treatise.

<I> What the ancients said is as follows. For any equilateral triangle, if on one of its sides we take a point from which we draw the perpendiculars to the other two sides, then their sum is equal to the height of the triangle.

Example: The triangle *ABC* is equilateral; on the side *AB* we take a point *D*, from this we draw the perpendiculars *DE* and *DG* and we draw the height *AH*; then the sum of the two perpendiculars *DE* and *DG* is equal to the height *AH*.

---

[1] See Appendix, Text 4: *Two Propositions of the Ancients.*

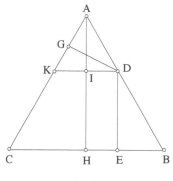

Fig. II.3.2.1

Proof: Let us draw from the point *D* a straight line parallel to the straight line *BC*; let it be *DIK*. The triangle *ADK* is equilateral because it is similar to the triangle *ABC*, so the perpendicular *DG* is equal to the perpendicular *AI* and the perpendicular *DE* is equal to the perpendicular *IH*, so <the sum> of the perpendiculars *DE* and *DG* is equal to the height *AH*. This is what we wanted.

<II> The ancients also said that if we take a point inside an equilateral triangle and if we draw from this point perpendiculars to the sides of the triangle, then the sum of these perpendiculars is equal to the height of the triangle.

Example: The triangle *ABC* is equilateral; inside if we take a point *D* from which we draw the perpendiculars *DE*, *DG* and *DH* and we draw the height *AI*, then the sum of the perpendiculars *DE*, *DG* and *DH* is equal to the height *AI*.

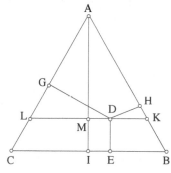

Fig. II.3.2.2

Proof: We draw from the point *D* a straight line parallel to the straight line *BC*; let it be *KML*. Then the triangle *AKL* is equilateral, so the sum of the two perpendiculars *DH*, *DG* is equal to the perpendicular *AM*, as before; but the perpendicular *DE* is equal to *MI*, so the sum of the perpendiculars *DE*, *DG*, *DH* is equal to the height *AI*.

This is what the ancients said about this idea (*ma'nā*).[2] As for what we ourselves have determined, this is what we shall now describe.

<1> For any triangle, if we draw from its angles the perpendiculars to its sides, then the ratio of the perpendicular to the perpendicular is the inverse of the ratio of the side to the side.

Example: In the triangle *ABC* we have drawn the perpendiculars *AD*, *BE* and *CG*.

I say that the ratio of the perpendicular AD *to the perpendicular* BE *is equal to the ratio of* AC *to* CB *and that the ratio of the perpendicular* AD *to the perpendicular* CG *is equal to the ratio of* AB *to* BC.

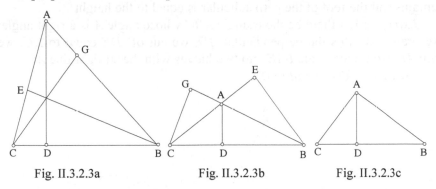

Fig. II.3.2.3a   Fig. II.3.2.3b   Fig. II.3.2.3c

Proof: Each of the angles *D* and *E* is a right angle and the angle *ACD* is common, so the triangle *ACD* is similar to the triangle *BCE*, so the ratio of *AC* to *CB* is equal to the ratio of *AD* to *BE*. In the same way, we prove that the ratio of *AB* to *BC* is equal to the ratio of *AD* to *CG*. If the triangle has acute angles, all three of the feet of the perpendiculars are inside the triangle,[3] as in the first case of the figure.

If the triangle has an obtuse angle, then one of the perpendiculars lies inside the triangle and the other two perpendiculars are outside the triangle, as in the second case of the figure.

If the triangle is right-angled, then the two perpendiculars drawn from the two acute angles are the two sides of the triangle that enclose the right

---

[2] See p. 480, n. 2.

[3] They lie on the sides of the triangle and not on their extensions.

angle, so the feet of the perpendiculars, which are G and E, are at the point A, as in the third case of the figure.

We prove this proposition using a different proof. The product of each of the sides and the perpendicular that falls on it is twice <the area> of the triangle, so the ratio of each of the sides of the triangle to another side is the ratio of the perpendicular that falls on the second side to the perpendicular that falls on the first side. This is what we wanted to prove.

<2> In the same way, if in a scalene right-angled triangle we draw from its right angle the perpendicular to the base, then we cut off from the greater of the two parts of the base a part equal to the smaller one; if we join its endpoint to the right angle with a straight line, then we divide the angle that remains over from the right angle into two halves, then the part cut off from the base between the straight line that divides the angle that remains and the foot of the perpendicular is equal to the height.

Example: Let there be the triangle ABC whose angle A is a right angle; we draw from this the perpendicular AD, we cut off DE equal to DC, we join AE, we divide angle BAE into two halves with the straight line AG.

I say that GD *is equal to* DA.

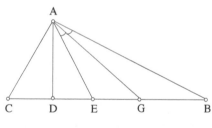

Fig. II.3.2.4

Proof: The angle EAD is equal to the angle DAC, so the angle EAD is half the angle EAC. But the angle EAG is half the angle EAB, so the angle GAD is half the angle BAC. But the angle BAC is a right angle, so the angle GAD is half a right angle; but the angle ADG is a right angle, so the angle AGD is half a right angle, so the straight line GD is equal to the straight line DA. This is what we wanted to prove.

<3> If on the base of an isosceles triangle we take an arbitrary point, from which we draw the perpendiculars to the two [equal] sides of the triangle, then their sum is equal to the perpendicular drawn from one endpoint of the base to the side of the triangle, whether the angle of the triangle enclosed by the two equal sides be acute, obtuse or a right angle.

Example: Let there be the isosceles triangle *ABC* whose sides *AC* and *BA* are equal and whose base is *BC*. On the base *BC* we take a point *D* from which we draw the two perpendiculars *DE* and *DG*.

I say that their sum is equal to the perpendicular CH.

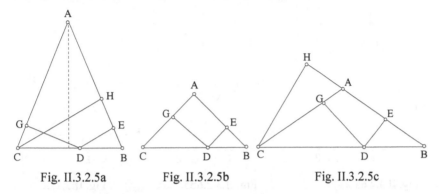

Fig. II.3.2.5a   Fig. II.3.2.5b   Fig. II.3.2.5c

Proof: The two angles *B* and *C* are equal, the two angles *E* and *G* are equal because they are right angles, so the two triangles *BED* and *DGC* are similar and the ratio of *CD* to *DB* is equal to the ratio of *GD* to *DE*. By composition, the ratio of the sum of *GD* and *DE* to *DE* is equal to the ratio of *CB* to *BD*. But the ratio of *CB* to *BD* is equal to the ratio of *CH* to *DE*,[4] so the ratio of the sum of *GD* and *DE* to *DE* is equal to the ratio of *CH* to *DE*, so the sum of the two perpendiculars *GD* and *DE* is equal to the perpendicular *CH*.

This proof is the same whatever the properties of the triangle. This is what we wanted.

<4> In the same way, we return to the figure and the given point on the side *AB*; let it be *D*. We draw the perpendiculars *DG* and *DH*, we draw the perpendicular *AE*, we put the ratio of *AB* to *BD* equal to the ratio of *AE* to *EI* and we put the ratio of *AI* to *IL* equal to the ratio of *AC* to *CB*.

I say that the sum of the perpendiculars DG *and* DH *is equal to the perpendicular* LE.

Proof: We join *DI* and we extend it to *K*, then *DK* is parallel to the straight line *BC*. But since the ratio of *AB* to *BD* is equal to the ratio of *AE* to *EI*, then the ratio of *AC* to *CB* is equal to the ratio of *AK* to *KD*. But the ratio of *AC* to *CB* is equal to the ratio of *AI* to *IL*, so the ratio of *AK* to *KD* is equal to the ratio of *AI* to *IL*. But the ratio of *AK* to *KD* is equal to the

---

[4] Because the triangles *BDE* and *BCH* are similar.

ratio of *AI* to *DH*, as has been proved in Proposition 3 of this treatise. So the perpendicular *DH* is equal to *LI* and *DG* is equal to *EI*, so <the sum> of the perpendiculars *DG* and *DH* is equal to the perpendicular *LE*. This is what we wanted to prove.

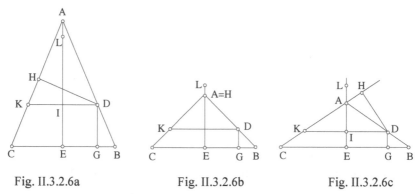

Fig. II.3.2.6a   Fig. II.3.2.6b   Fig. II.3.2.6c

<5> In the same way, we return to the isosceles triangle; let the point lie inside the triangle. Let there be the triangle *ABC* and the point *D* which is inside the triangle; let us draw from this point the perpendiculars *DE*, *DG*, *DH*. We draw from the point *D* a straight line parallel to the straight line *BC*; let it be *MDL*. We draw the perpendicular *AIK* and let us put the ratio of *AI* to *IN* equal to the ratio of *AB* to *BC*, which is equal to the ratio of *AM* to *ML*.

I say that the sum of the perpendiculars DE, DG *and* DH *is equal to the perpendicular* NK.

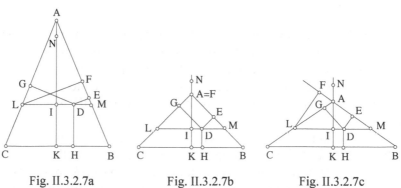

Fig. II.3.2.7a   Fig. II.3.2.7b   Fig. II.3.2.7c

Proof: We draw the perpendicular *LF*. Since the ratio of *AI* to *IN* is equal to the ratio of *AM* to *ML*, we have *IN* equal to *LF*. Now we have proved that the sum of the perpendiculars *DE* and *DG* is equal to the perpendicular *LF*. <The sum> of the two perpendiculars *DE* and *DG* is thus equal to the perpendicular *NI* and the perpendicular *DH* is equal to the perpendicular *IK*, so the sum of the three perpendiculars *DE*, *DG* and *DH* is equal to the perpendicular *NK*. This is what we wanted to prove.

This proof is the same for all isosceles triangles, whether they are acute-angled, or have an obtuse angle or a right angle.

<6> In the same way, we return to the isosceles triangle; let the triangle be *ABC*. We divide the angle *ABC* into two halves with the straight line *BE*, we draw *EH* parallel to the base *BC* and we draw the perpendicular *AGD*.

I say that if from a given point on the straight line EH *we draw two perpendiculars to the straight lines* AE *and* AH, *then their sum is equal to the perpendicular* GD. *On the straight line* EH *we take a point* I *from which we draw the perpendiculars* IK *and* IL. *I say that the sum of* IK *and* IL *is equal to the perpendicular* GD.

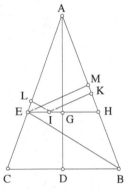

Fig. II.3.2.8

Proof: We draw the perpendicular *EM*. Since *EH* is parallel to the straight line *CB*, the angle *HEB* is equal to the angle *EBC*. But the angle *EBC* is equal to the angle *EBH*, so the angle *HEB* is equal to the angle *EBH*; so the straight line *EH* is equal to the straight line *HB*. So the ratio of *AH* to *HB* is equal to the ratio of *AH* to *HE*; but the ratio of *AH* to *HB* is equal to the ratio of *AG* to *GD* and the ratio of *AH* to *HE* is equal to the ratio of the perpendicular *AG* to the perpendicular *EM*,[5] so the ratio of *AG* to *GD* is equal to the ratio of *AG* to *EM* and the perpendicular *EM* is equal to *GD*; but the perpendicular *EM* is equal to the sum of the perpendiculars

---

[5] Proposition 1.

IK and *IL*, as before.[6] The sum of the two perpendiculars *IK* and *IL* is thus equal to the perpendicular *GD*.

This proof is the same for all isosceles triangles.

<7> For any acute-angled isosceles triangle, the excess of its side – one of the two <equal> sides – over its height – the one that falls on this side – the excess of this height over the segment that it cuts off on the side towards the base and twice this segment are all three in continued proportion.

Let there be a triangle *ABC* whose sides *AB* and *AC* are equal and whose three angles are acute. Let us draw in this triangle the perpendicular *CE*.

I say that the excess of AB *over* CE, *the excess of* CE *over* EB *and twice* EB *are all three in continued proportion.*

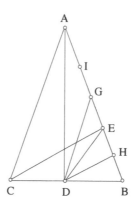

Fig. II.3.2.9

Proof: We draw the perpendicular *AD* and the perpendicular *DH*, which makes[7] *HE* equal to *HB*. We join *ED* and let us divide the angle *ADE* into two halves with the straight line *DG*, then *GH* is equal to *HD*, as we have proved in Proposition 4 of this treatise.[8] Since *CB* is twice *BD* and *EB* is twice *BH*, *CE* is parallel to *DH* and is twice *DH*. Since the product of *AH* and *HB* is equal to the square of *HD*, the product of *AH* and *HE* is equal to the square of *HG*, so the ratio of *AH* to *HG* is equal to the ratio of

[6] Proposition 3.

[7] In the manuscript, we read 'and we make'.

[8] In fact, *ADB* is right-angled and scalene.

HG to *HE* and is equal to the ratio of *AG* to *GE*.[9] But *HG* is greater than *HE*, because *GH* is greater than *HB*. But given that *AD* is greater than *DB*; the angle *BAC* is in fact acute, so the straight line *AG* is greater than the straight line *GE*. We put *GI* equal to *GE*, then the ratio of *AG* to *GI* is equal to the ratio of *GH* to *HE*, so the ratio of *AI* to *IG* is equal to the ratio of *GE* to *EH*, so the product of *AI* and *HE* is equal to the square of *GE*, twice the product of *AI* and *HB* is equal to twice the square of *EG* and twice the product of *AI* and *BE* is equal to the square of *EI*. Since *GE* is equal to *GI* and *EH* is equal to *HB*, thus *IB* is twice *HG*. But *HG* is equal to *HD* and *CE* is twice *HD*, so the straight line *IB* is equal to the perpendicular *CE*, thus *AI* is the excess of *AB* over the perpendicular *CE*. But *IB* is equal to the perpendicular *CE* and *IE* is the excess of *IB* over *EB*. Now *AI*, *IE* and twice *EB*, which is the part cut off by the perpendicular *CE*, are in continued proportion, so the excess of *AB* over the perpendicular *CE*, the excess of the perpendicular *CE* over *EB* – which is the part that it cuts off on the side <*AB*> – and twice *EB* are all three in continued proportion. This is what we wanted to prove.

<**8**> In the same way, let the scalene triangle be *ABC*; on any one of its sides, let us take a point; let it be the point *D*. Let us draw from the point *D* the two perpendiculars *DE* and *DG*. We draw the perpendicular *BH*, we draw *DKI* parallel to the straight line *AC* and we put the ratio of *BK* to *KN* equal to the ratio of *BC* to *CA*.

I say that the sum of the perpendiculars DE *and* DG *is equal to the perpendicular* NH.

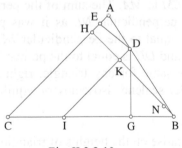

Fig. II.3.2.10

Proof: The ratio of *BI* to *ID* is equal to the ratio of *BC* to *CA*; but the ratio of *BC* to *CA* is equal to the ratio of *BK* to *KN*, so the ratio of *BI* to *ID* is equal to the ratio of *BK* to *KN*. But the ratio of *BI* to *ID* is equal to the

---

[9] See commentary, p. 463.

ratio of *BK* to *DG*, so the perpendicular *DG* is equal to the perpendicular *KN* and the perpendicular *DE* is equal to the perpendicular *KH*. So the sum of the perpendiculars *DG* and *DE* is equal to the perpendicular *NH*. This is what we wanted to prove.

<9> Let us return to the scalene triangle; let it be *ABC*. Let us take an arbitrary point *D* inside it. Let us draw from this point the perpendiculars *DE, DG, DH*. We cause to pass through the point *D* a straight line parallel to the straight line *BC*; let it be *LDM*. We draw the perpendicular *AIK* and we put the ratio of *AI* to *IN* equal to the ratio of *BC* to *CA*.

I say that the sum of the three perpendiculars DE, DG *and* DH *is equal to the perpendicular* NK.

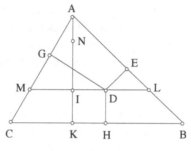

Fig. II.3.2.11

Proof: The ratio of *LM* to *MA* is equal to the ratio of *BC* to *CA*; but the ratio of *BC* to *CA* is equal to the ratio of *AI* to *IN*, so the ratio of *AI* to *IN* is equal to the ratio of *LM* to *MA*. The sum of the perpendiculars *DE* and *DG* is thus equal to the perpendicular *NI*, as it was proved earlier. Now the perpendicular *DH* is equal to the perpendicular *IK*, so the sum of the perpendiculars *DE, DG* and *DH* is equal to the perpendicular *NK*.

This proof is the same for any triangle, right-angled, acute-angled or with an obtuse angle, scalene, isosceles or equilateral. This is what we wanted to prove.

The treatise on the heights of triangles is complete.
Thanks be given to God and blessings upon our Prophet Muḥammad
and those that are His.
I completed transcribing it at Mosul, the Protected city,
in the month Ṣafar 632.

CHAPTER III

IBN AL-HAYTHAM AND THE GEOMETRISATION OF PLACE

As we have seen, the emergence of geometrical transformations – the operations as well as the transformed objects of geometry – spurred Ibn al-Haytham into conceiving a new mathematical discipline: the discipline of the knowns. This discipline was designed to justify the operations, and to provide a basis for the existence of the objects by its introduction of motion. And the same is true of its method: the technique of analysis. In truth Euclid's Common Notions, Postulates and Definitions seem no longer to be enough, being unsuitable for the new representation of the objects of geometrical knowledge that are envisaged. In the *Elements*, this object was simply the *figure*, without any consideration either of its place or, in general, of the space that contained it. From now on, figures are no longer the sole objects of geometry; moreover, a figure can move, undergo translation, dilatation, contraction, inversion and projection. Thus, the figure moves, and motion even comes into the way the object is conceived. For example, it is relevant to the concept of parallelism, and for the process of deriving figures by the transformation of other figures. So it is clear that it was no longer possible to think about the relationships between elements of a single figure, any more than about the relationships between figures and still less about the relationships used in finding these figures, without raising questions about the notion of spatial relationships itself. This is precisely what Ibn al-Haytham is doing in the treatise *On Place*.

But, in the tenth century, reflections on the nature of spatial relationships were common currency among philosophers[1] and philosopher-theologians.[2] It was, clearly, not an idea of space, such as we encounter

---

[1] For example al-Fārābī's text on the void: *Risāla fī al-khalā'*, edited and translated by Necati Lugal and Aydin Sayili in *Türk tarikh yayinlarindan*, XV.1, Ankara, 1951, pp. 21–36. In his lost *Physics*, al-Fārābī must certainly have dealt with this subject. See also the *Physics* of *al-Shifā'* by Avicenna, ed. Ja'far Āl-Yāsīn, Beirut, 1996, Chapters 5 to 9; and *al-Najāt* by Avicenna, ed. M. S. al-Kurdī, Cairo, 1938, pp. 118–24.

[2] Later philosopher-theologians were able to pass on to their own time the theses discussed by their predecessors, notably the discussion in the School of Baṣra from Abū al-Hudhayl al-'Allāf and his nephew al-Naẓẓām, as well as later Abū 'Alī al-Jubbā'ī and his son Abū Hāshim. See for example Ibn Mattawayh, *al-Tadhkira*, ed. Samīr Naṣr Luṭf and Faysal Badīr 'Aūn, Cairo, 1975, especially p. 116. See also Abū Rashīd al-

after Newton, but rather a reflection on place and the void. Indeed, it is in connection with these two concepts that ideas about spatial relationships were formulated. The framework was already to be found in Aristotle's *Physics* and it remained current because of that work's longevity: a theory of place based on the everyday experience that every perceptible body is in some location. In any case, it is in this same intellectual context and in the same language that Ibn al-Haytham returns to the question of space, but he bases himself upon the new concerns generated by a considerable revival in geometry.

The notions of place and of the void had been discussed by Aristotle in several treatises and in some detail in Book IV of the *Physics*.[3] From then on, treatises on physics include a chapter about place and the void in which Aristotle's theory is adopted, improved upon or refuted. Among the ancients, we have Alexander of Aphrodisias, Themistius, Philoponus and Simplicius; in Ibn al-Haytham's time the important names are those of al-Fārābī, Avicenna and the members of the Baghdad School, as well as the philosopher-theologians.[4] The ancients' principal writings on this subject were known in Arabic[5] and were thus, like the works composed by the philosophers of their own time, available to Ibn al-Haytham and his contemporaries. It is no doubt the wide diffusion of these theories, and the interest taken in ideas about place and the void, that allowed Ibn al-Haytham to dispense with setting out their nature in detail. He confines himself with merely referring to them.

Of these numerous theories and their ramifications, Ibn al-Haytham chose only the two principal ones, whose theses he summarises very briefly, without the argumentation that supports them. He begins by returning to

---

Nīsābūrī, *Kitāb al-Tawḥīd*, ed. Muḥammad 'Abd al-Hādī Abū Rīda, Cairo, 1965, pp. 416 *sqq.* See Alnoor Dhanani, *The Physical Theory of Kalām*, Leiden, 1994, pp. 62–89.

[3] It is above all in the first six chapters if Book IV of the *Physics* that Aristotle develops his arguments and his theory of place. See Aristotle, *Physics*, Books III and IV, trans. by E. Hussey, Clarendon Aristotle Series, Oxford, 1983 and trans. P. H. Wicksteed and F. M. Cornford, London and Cambridge (Mass.), 1970, 211a–213a. On the important problem of the place of the Whole, see Marwan Rashed, 'Alexandre et la "magna quaestio"', *Les Études classiques*, 63, 1995, pp. 295–351, especially pp. 303–5. On the problem of place in Aristotle, see the now classic study by V. Goldschmidt, 'La théorie aristotélicienne du lieu', in *Écrits*, Paris, 1984, vol. I, pp. 21–61.

[4] See p. 495, n. 1 and 2.

[5] See the translation of Aristotle's *Physics* with commentaries by Ibn al-Samḥ, by Mattā ibn Yūnus, by Ibn 'Adī and by Abū al-Faraj ibn al-Ṭayyib, edited by 'A. Badawi in *Arisṭūṭālīs, al-Ṭabī'a*, vol. I, Cairo, 1964; vol. II, Cairo, 1965, especially Book IV, vol. I, pp. 271 *sqq.* See also E. Giannakis, 'Yaḥyā ibn 'Adī against John Philoponus on Place and Void', *Zeitschrift für Geschichte der arabisch-islamischen Wissenschaften*, Band 12, 1998, pp. 245–302.

Aristotle's theory, according to which the place of a body is the enclosing surface that envelopes the body.[6] The second thesis considered is that of one of Aristotle's critics, Philoponus, who maintains that the place of a body is the void filled by the body. Having described these theories, Ibn al-Haytham at once goes on to say that the question of place has not yet received the rigorous attention it deserves and to prove, by his critique, that both of these theses seem equally ill adapted to his purpose and his requirements. And indeed, while these theories were an integral part of books and commentaries on physics, Ibn al-Haytham does not intend that his own treatise shall be a work concerned only with physics, but writes above all as a mathematician; and it is a mathematical notion of place that he proposes to develop in his treatise. It is with exactly this purpose that he composes one of the first treatises entirely and exclusively devoted to the notion of place, at the least the first of its kind. Although his use of philosophers' terms – that is of the language of the time – has been a cause of misunderstanding,[7] Ibn al-Haytham nevertheless does not go so far that he disguises the novelty of what he proposes to do, the more so since his text does not contain anything beyond mathematics. Thus, we find no allusion to a substantial discussion of the perception of place that he had already completed, in his famous *Book on Optics*.

In constructing his mathematical theory of place, Ibn al-Haytham begins by criticising the Aristotelian thesis. His intention is, however, less that of pointing out its weaknesses than that of laying the foundations for his own theory. When we come to examine it we see that, even if it sometimes does not genuinely refute Aristotle, this critique allows Ibn al-Haytham, as a mathematician, to free the notion of place from any discussion of material existence, that is to free it of its physical and cosmological connections. In contrast, the refutation of Philoponus' thesis seems to have two purposes. Ibn al-Haytham seems to want to, as it were, put us on our guard against rapidly assimilating his own thesis into that theory. Indeed it is as if he wished to warn us in advance against a mistake that would in

---

[6] See Aristotle, *Physics* IV, 212a; cf. trans. Hussey, p. 28.

[7] In fact, here, as in the *Analysis and Synthesis* as well as in *The Knowns*, that is in the treatises preceded by a theoretical introduction containing a mixture of philosophical and mathematical considerations, Ibn al-Haytham makes use of the language of the philosophy of his time, which has an Aristotelian look. We encounter terms such as 'essence', 'in actu', 'in potentia', 'form', 'place', 'demonstrative syllogism', and so on. Although for a historian versed in Ibn al-Haytham's mathematics, optics or astronomy this terminology does not impede understanding Ibn al-Haytham's actual intentions and ideas, it may happen that it misleads a historian of philosophical theories. Such a reader can indeed see here a trace of Aristotelianism, whereas Ibn al-Haytham thinks very differently. This is exactly the mistake made by the philosopher al-Baghdādī, see p. 500.

reality be made by some commentators, such as 'Abd al-Laṭīf al-Baghdādī[8] and other more recent ones.[9] But thanks to this critique he can also determine the conditions for constructing the geometrical concept of place, and it is at this stage that Ibn al-Haytham's intention becomes clear: we have nothing less than the first known geometrisation of the notion of place. The project is so innovative and so strange in its time that even an Aristotelian philosopher saw it that way, though without being able to understand its exact significance.[10] Let us examine Ibn al-Haytham's procedure.

In Aristotle, place is the place of a body, and its existence is given as something immediately obvious that does not stand in need of proof. To convince oneself of this, it is enough to show what it is not and then go on to investigate what it is, its particular attributes: all of them refer to something that exists. Upwards and downwards are not merely relative to one another, but represent the places towards which certain bodies naturally move. So the true difficulty in the problem of understanding place is connected not with its existence, but with its essence and its definition. Accordingly, we must start by looking for attributes of the essence: all of them arise from a primary relationship 'between what is contained and what contains it, that is between two things united by a relationship of exteriority; so it is this relationship that makes it possible to determine the essence of place'.[11] Thus, Aristotle finds the essence in this primary relationship between the container and that which is contained, what encloses and what is enclosed, and defines place as the first envelope of each body, which does not belong to the body itself but has another body of its own that encloses the former one. Or, as he expresses it: 'the limiting surface of the body continent – the content being a material substance susceptible of movement by transference'.[12] So we are concerned with the inner surface of the containing entity lying next to that which is contained, in which the

---

[8] See Appendix III in *Les Mathématiques infinitésimales*, vol. IV.

[9] Thus, A. Dhanani writes: 'In the end he [Ibn al-Haytham] endorsed this view of space (which derives ultimately from John Philoponus)' (*The Physical Theory of Kalām*, p. 69). This error does not affect the value of A. Dhanani's work, because Ibn al-Haytham is not an example of the theologian philosopher – the true subject of the book.

[10] Al-Baghdādī, like al-Rāzī, has not grasped the essential point of Ibn al-Haytham's theory, namely the one-to-one correspondence between two sets of different distances.

[11] 'entre le contenu et le contenant, c'est-à-dire entre deux choses unies par une relation d'extériorité; c'est donc celle-ci qui permettra de déterminer l'essence du lieu' (V. Goldschmidt, 'La théorie aristotélicienne du lieu', p. 28).

[12] Aristotle, *Physics*, Book IV, 212a, trans. P. H. Wicksteed and F. M. Cornford, p. 313; cf. trans. E. Hussey, p. 28.

body is positioned, in accordance with its nature and in accordance with the order of the cosmos, even if the body can be removed from its place. In short, as Aristotle says, 'Further, place is coincident with the thing, for boundaries are coincident with the bounded'.[13] The image of the inner surface of a vase provides a good illustration of such a representation of place. Thus, place is the whole of the surface adjacent to what encloses the whole of the body whose place it is.

Ibn al-Haytham marshals several arguments against Aristotle's theory – counter-examples, most of them mathematical. We observe that, in all these counter-examples, the only property of the body that the mathematician retains is extension, itself conceived as made up of distances, which is a preparation for a formal idea of place. In short, place becomes ontologically neutral.[14]

Let us begin by examining the least mathematical of all these counter-examples, one that might be found in the writings of commentators on Aristotle and those of his critics. We take a goatskin filled with water; if we press it, the water overflows through the spout and the surface of the goatskin encloses the remainder of the water. If we repeat this several times, the surface of the goatskin will enclose less and less water and thus will be the place of several volumes of water. Thus, we have the same place for differ-

---

[13] Aristotle, *Physics* IV, 212a29–30, ed. and trans. Ross.

[14] We may, moreover, ask ourselves whether one or another of Ibn al-Haytham's mathematical predecessors had already made a move towards stripping the idea of place of ontological elements. In other words, we may wonder whether there was a movement to 'de-ontologise' place, a movement of which Ibn al-Haytham was part. This conjecture stems from a thesis attributed to Thābit ibn Qurra, to be found in a text that is no longer extant.

According to the testimony of the philosopher-theologian Fakhr al-Dīn al-Rāzī, Thābit ibn Qurra, unlike the philosophers, was notable for having a thesis of his own: one that contradicted the Aristotelian theory of natural place. Al-Rāzī wrote as follows: 'the philosophers are in agreement about it (every body has a natural place); nevertheless I have seen in some chapters attributed to Thābit ibn Qurra a surprising theory which he has chosen for himself' (*al-Mabaḥīth al-mashriqiyya*, Tehran, 1966, vol. II, p. 63). Al-Rāzī quotes Thābit ibn Qurra before going on to criticise this theory: 'Thābit ibn Qurra has said: he who believes that the Earth is seeking for the place in which it is to be found, has a false opinion; because there is no need to imagine in any place whatsoever a state that is proper to it making it unlike others. But, on the contrary, if one had imagined all places to be empty and then that the whole Earth arrives at any one of them, it necessarily stops there and does not move away towards another [place], because this one and all the places are equivalent' (*ibid.*, p. 63).

On the theory of natural place and the attraction of the Earth, see Marwan Rashed, 'Kalām e filosofia naturale', in R. Rashed (ed.), *Storia della scienza*, vol. III: *La civiltà islamica*, Rome, Istituto della Enciclopedia Italiana, 2002.

ent volumes. Even though an Aristotelian philosopher can always reply that, in this case, the form of the goatskin changes, the argument is not entirely lacking in force and, at the least, refers us back to a difficulty in the theory of combined matter and form.

All the other counter-examples are geometrical in nature and reduce to the fact that a body can have a change in the surface that bounds it without changing in volume, or can even increase in its outer surface while diminishing in volume.

The first example is that of a parallelepiped that we cut into slices with faces parallel to two of its original faces; we rearrange these slices so that the parallel faces make up the faces of a new parallelepiped. The volume remains unchanged, whereas the area of the outer surface that encloses it, and thus the place, has greatly increased.

Furthermore, if we consider a body with plane faces, which we hollow out so as to give its interior, for example, the form of a concave sphere its volume diminishes whereas its enclosing surface increases. If on the other hand we consider a wax cube that we model into a sphere, its surface area diminishes without its changing its volume, in accordance with the properties of bodies with the same surface area (isepiphanic bodies) established by Ibn al-Haytham in another treatise.[15]

If, again, we model the cube into a regular polyhedron with twelve faces, then this polyhedron has a surface area – and thus a place – greater than that of the initial cube. Ibn al-Haytham had in fact proved that if there are two regular polyhedra with similar faces that have the same total area, then the polyhedron that has the greater number of faces has the greater volume.[16] So if the cube and the polyhedron with similar faces have the same surface area, the volume of the polyhedron would be greater than that of the cube, which is contrary to what was assumed.

An Aristotelian would certainly not find himself at a loss as to how to reply to Ibn al-Haytham's criticisms. He could indeed object that the 'individual' body is no longer the same since in one case it is the form that has been altered and in the other case it is the matter that has been changed. This is indeed the way that the philosopher and physician 'Abd al-Laṭīf al-Baghdādī replies to the mathematician.[17] But this reply would not have affected Ibn al-Haytham's opinion, which was based on other grounds,

---

[15] *Les Mathématiques infinitésimales du IXᵉ au XIᵉ siècle*, vol. II: *Ibn al-Haytham*, London, 1993, Chap. III; English trans. *Ibn al-Haytham and Analytical Mathematics*. A History of Arabic Sciences and Mathematics, vol. 2, Culture and Civilization in the Middle East, London, 2012, pp. 289–95.

[16] *Ibid.*, vol. II, p. 339 and pp. 444–51; English trans. pp. 249 and 336–9.

[17] See Appendix III in *Les Mathématiques infinitésimales*, vol. IV.

lying outside Aristotelianism. We have seen that he gives a different meaning to the word 'body'; he also gives another meaning to the expression 'adjacent surface'. Like the body, this entity has in fact no other quality except extension in three dimensions. The body and the adjacent surface have now been stripped of any physical or cosmological quality. So there is every indication that in his critique of the Aristotelian theory, Ibn al-Haytham is setting out not so much to mount an effective attack but rather to prepare the ground for a deliberately more abstract conception of the idea of place. It is in the course of his critique of the theory of place of the type put forward by Philoponus that Ibn al-Haytham set about constructing his own concept of place.

We may first observe that it is in the light of this theory, but also predominantly against it, that Ibn al-Haytham constructs his concept of place. Nevertheless, Ibn al-Haytham does not approach the theories as a historian and it can happen that in criticising certain notions he introduces into them meanings slightly different from their original ones. All the same, since he does not cite any names or any book titles, we need to be prudent.

In his commentary on Aristotle's *Physics* and especially in his *Corollaries on Place and the Void*,[18] Philoponus develops the theory that place is an extension in three dimensions, empty by definition, and thus distinct from the bodies that may occupy it. He expresses his idea as follows:

> That place is not the limit enclosing a body is adequately clear from what has just been said; that it is a certain three-dimensional interval, distinct from the bodies that are to be found in it (because place and the void are in reality the same in regard to their substance), we might show this by elimination of the other possibilities: indeed if it is not the matter nor the form nor the limit of the enclosing body, there remains only that place is the interval.[19]

As to the meaning of this key concept of extension, Philoponus writes:

> And I certainly do not say that this interval has ever been or could be empty of all body. Absolutely not, but I state that it is something other than the bodies that are to be found in it, that it is empty as regards its own definition, but that it is never separated from a body, rather as we say that matter is something other than forms, but that it can nevertheless never be separated from a form. So in this way we agree that an interval is something other than any body, empty as regards its own definition but there are, continually, new bodies which come to be found in it, itself remaining immobile, as a whole and in regard to its parts, as a whole because the cosmic interval which

---

[18] See *Ioannis Philoponi in Aristotelis Physicorum libros quinque posteriores commentaria*, ed. H. Vitelli, *CAG* XVII, Berlin, 1888.

[19] Philoponus, *In Phys.* 567, 29–568, 1.

admitted the body of the whole Universe can never be moved, in regard to its parts because it is impossible that the interval, incorporeal and empty by its own definition, can move.[20]

For Philoponus, extension exists ('an interval is something other than the bodies that are to be found in it, but it is never without bodies');[21] it is empty by definition. To sum up, what Philoponus means by 'place' is extension in three dimensions, empty but possessing existence, even if one might say that existence is not '*in actu*'.

There remains the question of knowing how, starting from dimensions that are empty and of necessity abstract, we can observe a variety of different bodies. It is this question and the difficulty that it raises which, it seems, persuaded Ibn al-Haytham to move away from Philoponus' theory. This theory is, indeed, incapable of explaining how an extension, defined in this way, is the place even of a body – if not of a family of different bodies – unless we suppose that we are concerned simply with extension conceived in relation to the body. A static theory, if one may call it so, that Ibn al-Haytham makes an effort to turn into a dynamic one, but at the cost of subjecting it to considerable modifications.

From the theory of place put forward by Philoponus, Ibn al-Haytham retains the idea of empty extension and that of the existence of place independently of any body to be found in it. But, as a mathematician, Ibn al-Haytham gives these two ideas a sense different from that given by the philosopher of nature. He begins by assigning empty extension a level of existence, that of mathematical concepts: it is 'imagination' which, as we have already seen, for Ibn al-Haytham is an act of thinking by which, stating from the traces left by objects, we separate out intellectual forms that are unchanging.[22] So we are concerned with an 'imagined void', apprehended by this act that starts from the traces of bodies that move from one position to another. After that we can take this position to be empty, even if it is never empty because it will at once be filled by another body. The act of imagination separates out unchanging intellectual form from this void: the distances between all the imagined points, distances that are themselves imagined because they are not material entities; they are in fact the imagined distances between all the points of the surface of a region of space. This manner of conceiving extension presents two advantages: Ibn al-Haytham does not need to give a purely conventional definition of the void; on the other hand he is in a position to present a mathematical notion of the void without having to believe in the existence of a physical void. So,

---

[20] Philoponus, *In Phys.* 569, 7–17.
[21] Philoponus, *In Phys.* 569, 19–20.
[22] See Introduction, pp. 11–12.

by employing the adjective 'imagined', Ibn al-Haytham ensures that the mathematical notion of place has a level of existence.

But we need to know how this imagined void becomes the place of a body, or of a variety of bodies. In this, Ibn al-Haytham clearly departs from all his predecessors. He does not propose a single set of imagined distances, but two. First, the distances that are 'fixed, intelligible, imagined' (al-thābita al-ma'qūla al-mutakhayyala)[23] in this void-extension, in this region of space. On the other hand, the set of imagined distances between all the points of an arbitrary body. For Ibn al-Haytham these distances, those in both sets, are segments of straight lines. So we shall say that an imagined void is the place of a certain body if and only if the imagined distances from this body 'can be superimposed on and can be identified with' the distances from the imagined void.

These two sets of distances and this 'perfect superposition' are the essential elements of this new conception of place. The end result of the superposition is another set of distances, since we are dealing with segments of straight lines, and thus with lengths without breadth; or, as we shall read later:

> But if on every imagined distance we superimpose an imagined distance, together they will be a single distance, because the imagined distance is simply the straight line which is a length without width. Now if on the straight line which is a length without width we superimpose a straight line which is a length without width, together they become a single straight line, because from their superposition there results no width, nor a length which exceeds the length of one of them. If one of the two imagined straight lines is superimposed upon the other, they become a single straight line which is a length without width. The imagined void filled by the body is thus imagined distances, on which there are superimposed the distances of the body, and which have become the single and same distances.[24]

Ibn al-Haytham's conception is unequivocal, and is sharply different from that of Philoponus. We now know why, from the beginning of the treatise, he has made a point of warning us against hastily assuming the two are the same. Let us assess Ibn al-Haytham's ideas on this matter using words other than his own, to try to make clear both Ibn al-Haytham's intentions and the nature of his contribution.

Ibn al-Haytham simply jettisons the idea found in all his predecessors, that of regarding a body as a whole, and substitutes for it a view of a body as a set of points joined up by segments of straight lines. Thus, of all the qualities a body can have, he retains only its extension, itself understood as

[23] See p. 515.
[24] See p. 513.

a set of line segments. Moreover, the imagined void is also a set of unchanging line segments joining the points of a region of three-dimensional space, independent of any body. Thus, the imagined void, that is place, is conceived from the first as a region of Euclidean space with its intrinsic metric. In other words, let C be the body concerned; with it we associate an abstract construct – place – which is V, the set of the distances (V is the imagined void), with a one-to-one correspondence $C \rightarrow V$. The distances which define V do not depend on the body C that fills them: they are unchanging in magnitude and in position. This place is called the place of the body C if and only if we establish the isometric one-to-one correspondence mentioned above. Place has a reality that is independent of any body: it is the family of imagined distances. This latter is obviously conceived *more geometrico*, within the framework of Euclidean geometry. Finally, the place of a body is defined, as we shall see later, as being the metric of the part of Euclidean space occupied by the body, which is itself conceived in the same way, and the two are connected by an isometric one-to-one correspondence. In such a scheme, it is clear that Euclidean space, a universal void, serves as a substrate for the unchanging distances between all the points, even though that is not stated in so many words. This substrate is indispensable for the coherence of the fixed distances that are considered in one or another region of the space, that is at the point, and thus to the conception of places as regions, or parts, of this space. It was not until Descartes, it seems, that it was to be stated, this time explicitly, that space is logically anterior to points.[25] Although it is succinct, Ibn al-Haytham's treatise geometrises the notion of place and mathematises the notions connected with it. It is, as far as I know, the first treatise that includes an attempt of this kind, and this is the direction that will later be followed by seventeenth-century mathematicians, notably by Descartes and Leibniz.[26]

---

[25] Descartes writes in his *Discours de la méthode*: '[...] the geometers' object, which I conceived as being a continuous body, or a space indefinitely extended in length, width and height or depth, divisible into various parts, which could have various shapes and magnitudes, and could be moved or transposed in any way' (*Œuvres de Descartes*, publiées par C. Adam and P. Tannery, Paris, 1965, vol. VI, p. 36).

[26] One simply cannot avoid noticing that it is this direction that the seventeenth-century mathematicians will take, each in his own way, and with differences that need to be pointed out on each occasion. For example, let us look at what Leibniz wrote in *Geometrical Characteristic*, where he represents place as a fragment of geometrical space. Place is a *situs* in Leibniz's sense, that is a relation between different points of a configuration (of an object) and Leibniz indicates it with a '.'. As for example A.B: 'A.B represents the mutual situation of the points A and B, that is an *extensum* (rectilinear or curvilinear, it does not matter which) that connects them and remains the

Now this conception allows Ibn al-Haytham to do what was forbidden to his predecessors: he can compare different geometrical solids, and also diverse figures, which occupy the same place, as well as the places that they occupy. From now on it is legitimate for him to think about their observable relationships, their positions, their shapes and their magnitudes, as he planned to do in *The Knowns*. It is now possible to make a rigorous comparison between a solid – for example a sphere – or an arbitrary figure, such as a circle, and so on, and its transformed version, as well as to compare their respective places, as it is possible to compare each to the other or to a third entity in a different place. He needed something like this new conception of place in order to investigate geometrical transformations.

So Ibn al-Haytham's treatise has close links with the new discipline of the Knowns. It is a book on geometry, or, we might say, on the philosophy of geometry. It consciously positions itself outside the tradition of investigations of place as they appear in Aristotle's *Physics*, or in the works of his critics and commentators, Greek and Arabic. So we should be risking serious misunderstanding if we were to discuss Ibn al-Haytham's theses without being aware that we need to see him as deliberately setting out to conceive place in terms that are mathematical and abstract. This kind of mistake was, however, made by, for example, 'Abd al-Laṭīf al-Baghdādī.

HISTORY OF THE TEXT

The treatise *On Place* (*Fī al-makān*) by Ibn al-Haytham appears on the lists of his writings drawn up by al-Qifṭī and Ibn Abī Uṣaybi'a.[27] In this same treatise, Ibn al-Haytham refers to his book on isoperimetric figures. Further, in a treatise we have edited in *Les Mathématiques infinitésimales* (vol. IV, Appendix III), al-Baghdādī quotes at length from this text by Ibn al-Haytham. The philosopher-theologian Fakhr al-Dīn al-Rāzī also refers to it more than once. That is to say that we have a superabundance of evidence for the authenticity of the attribution of this text.

The treatise itself has come down to us in five manuscripts.

The first, which we shall call C, belongs to the collection 2823 in Dār al-Kutub in Cairo, fols 1ᵛ–5ᵛ. This same collection includes another treatise by Ibn al-Haytham, on the direction of the Qibla (*Fī samt al-qibla*). This latter is copied in the same hand; we read in the colophon 'copied from a

---

same as long as that situation does not vary' (*La Caractéristique géométrique*, text edited and with introduction and notes by Javier Echeverría; French translation with notes and postface by Marc Parmentier, coll. Mathesis, Paris, 1995, p. 235).

[27] See *Les Mathématiques infinitésimales*, vol. II, p. 524; English trans. p. 408.

copy in the hand of Qāḍī Zādeh' (fol. 5ᵛ), that is the hand of the famous astronomer and mathematician in the employ of Ulugh Beg, during the first half of the fifteenth century. The writing is in *nasta'līq*. We can find four omissions of a word and two omissions of a sentence of more than three words.

The second manuscript, which we shall call T, belongs to the collection 2998, fols 166–74, in the library of Majlis Shūrā in Tehran. This collection also includes several other treatises on optics by Ibn al-Haytham: *On Light, On Shadows, On the Light of the Heavenly Bodies*. This collection is in the same hand, in *nasta'līq* script.

We note five omissions of a word and one omission of a sentence of three words, with a relatively large number of errors. C and T have two omissions of a word in common.

The third manuscript forms part of the collection 2196, fols 19ᵛ–22ʳ, in the Salar Jung Museum at Hyderabad (India), we shall call it H.

The fourth manuscript – called L – belongs to the collection no. 1270, fols 25ᵛ–27ᵛ in the India Office Library in London. We do not know the date of the copy, which could be the tenth century of the Hegira. Examination shows that it has one omission of a word and six errors. Apart from these omissions peculiar to each of the manuscripts H and L, they share three omissions of a word and twenty errors.

The fifth manuscript belongs to the collection Fātiḥ 3439, fols 136ᵛ–138ʳ, in the Süleymaniye library in Istanbul – we have called it F. This collection includes several treatises by Ibn al-Haytham. The manuscript was copied in 806/1403–4. It is difficult to read because the ink has faded and it contains a significant number of omissions.

Comparing these five manuscripts two by two, we can divide them into two groups: H and L on the one hand and C and T on the other, while F, on account of its omissions and its errors, remains separate. The probable stemma is particularly simple merely on account of our drastic lack of information.

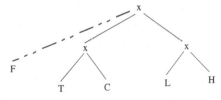

This text by Ibn al-Haytham was published at Hyderabad, by the Oriental Publications Bureau, in a non-critical edition by Osmania based only on manuscript L.

TRANSLATED TEXT

Al-Ḥasan ibn al-Ḥasan ibn al-Haytham

On Place
Fī al-makān

In the name of God, the Compassionate the Merciful

TREATISE BY AL-ḤASAN IBN AL-ḤASAN IBN AL-HAYTHAM

On Place

Men of learning, concerned to attain certainty in the search for truth about things that exist, are in disagreement on the essence of 'place'. Some have said that the place of a body is the surface that encloses the body; others have said that the place of a body is the imagined void filled by the body. However we have not found in any of our predecessors an exhaustive discussion of the essence of place, nor clear evidence to reveal the reality of place. Since this is how things stand, we thought we should search for the essence of place in an exhaustive manner, a search which shows what it is and reveals its reality, and which wipes away the disagreement and abolishes the ambiguity.

We say: place is a common name which is applied to several things each of which is called a place. In fact, the 'place' is the answer one gives to someone who enquires after the place of a body. The answer to someone who enquires after the position of a body can be one among many things. If someone in fact enquires after a man, among others, and says: 'so-and-so, in what place is he?' – and if that man is absent from his country – the answer is that he is in such-and-such a country. That is indeed an indication that a country can be called a place. Similarly, if someone enquires and says: 'so-and-so, in what place does he live?', the answer is: he is in such-and-such a district. That is indeed an indication that the district, which is a part of the city, can be called a 'place'. Similarly, if someone enquires after a man when he is in the house of that man, and says: 'so-and-so, in what place is he?', the answer is: he is in such-and-such a room or in such-and-such a chamber. That is indeed an indication that a room can be called a 'place' and that a chamber can be called a 'place'. And, for each of these locations, no one disputes that one can call them places, whether the object of the enquiry is a man or any other body that is not a man. A single point can remain a matter of disagreement, it is the place of the body whose dimensions do not exceed the dimensions of that body: this is the notion that we must investigate.

We say: for any body, there are two things that can be called 'place'. One is the surface enclosing the body. That is the surface of the air enclosing a body which is in the air, the surface of the water enclosing a body which is in the water and the surface of any body inside which there is a body which is distinct from it. It is for this that one of the two groups which are in disagreement has opted. The other notion is the void imagined as filled by the body. If indeed a body moves from the position it occupies, the enclosing surface by which it was surrounded can be imagined as empty, without a body in it, even if it has been filled with air or water, or with a body different from the one that was in it. By 'position' I mean one of the places mentioned before, each of which is by convention called 'place'.

The imagined void, these are the imagined distances, without matter in them, between the opposite points that belong to the surface enclosing the void. It is for this that the other group has opted. It is not inadmissible to call each of these two notions 'place', but it remains to examine them, along with the properties of each of them, to make it clear whether one deserves this name more than the other, or whether neither deserves it.

The method used for this investigation consists of considering each one individually, and examining whether it necessarily leads to abhorrent ambiguities and irremovable doubts. If one of them is free of ambiguities and doubts, it will be more credible than its companion; but if, for each of them, there arise ambiguities and doubts, the one which carries fewer ambiguities and doubts deserves the name 'place' more than the other one does.

Among the ambiguities that appear in connection with the surface, there is this one: if the shape of the body changes, the shape of the enclosing surface changes.

Among bodies, there is the one which is such that, if its shape changes, the shape of its enclosing surface changes; and in addition, the area of the enclosing surface increases whereas the volume of the body remains as it was, unchanged.

Among those, there is the parallelepiped. If we divide it up into slices with faces parallel and parallel to two of its faces,[1] and we order and then recombine its parts in such a way that each of its parts is placed alongside the other so that the parallel faces become two parallel surfaces; if we join the parts of the body, the one to the other, then the surface enclosing the

---

[1] Lit.: if we divide it up with parallel surfaces.

body will be greater than the first surface which enclosed the body before it was divided up. By its being divided up, numerous surfaces are in fact produced, each of them being equal to each of the two faces which were parallel and parallel to the surfaces generated, and one part of the two surfaces perpendicular to the two parallel faces is removed from the surfaces of the body. So the place of the body will be the surface of the air which encloses the body, superimposed on the surface of the body, which is a multiple of the first surface. Thus, in the second state the place of the body is a multiple of its first place, whereas the body in itself has not been increased at all. This is an unacceptable notion: the place of the body becomes larger whereas the body has not become larger, and has not increased at all.

Among such [bodies] again, there is water: if it is in a goatskin, the inner surface of the goatskin is the place of the water. If we press the goatskin, the water overflows from the spout of the goatskin, and the surface of the goatskin encloses what remains of the water. Then, as we press on the goatskin, the water comes out, and the surface of the goatskin encloses what remains of the water. Thus, the body does not cease becoming smaller, and the place of what remains of it is its original place. It follows that the same place, the inner surface of the goatskin will be the place for bodies of different magnitudes, and considerably different ones: the surface of the goatskin will sometimes enclose the greater, sometimes it will enclose the smaller, and sometimes it will enclose something in between. This is an unacceptable absurdity.

Similarly for any body enclosed by plane surfaces: if in each of its surfaces we dig out a concave hole – be it spherical or cylindrical, or a circular pyramid or a rectilinear pyramid – then each of the concave surfaces that are generated is greater than its plane base which has been removed; thus, what remains of the body, once pieces have been dug out of it, is much smaller than the original solid itself; and the place of this remainder would be greater than the place of the original body. So the body would have become smaller and its place greater. But this is one of the most unacceptable of absurdities.

From all this it follows necessarily that the same body would be able to occupy many places of different magnitudes, while the magnitude of the body does not change. In fact, a body that is such as to suffer action from another body, such as wax, lead, water and any liquid body, can take different shapes, without anything being added to it or removed from it. For example, if wax – or something like it – has the shape of a cube, its enclosing surface will be its place; if we then fashion this same body into a sphere, its place will be the spherical surface which encloses it. But the spherical surface is always smaller than the sum of the faces of the cube, if the body of the sphere is equal to the body of the cube. We have proved this result in our book *On the Sphere which is the Largest of all Solid Figures having Equal Perimeters.*[2] Similarly, we give this body twenty [regular] faces, the sum of its faces will be smaller than the sum of the faces of the cube, because if the one that has twenty faces has the sum of its faces equal to the sum of the faces of the cube, its body will be greater than the body of the cube, since we have also proved that in the book we have just mentioned.

If, similarly, we make it into a body with twelve faces or with eight, or cylindrical, or pyramidal with a circular base or a rectilinear pyramid, then the magnitude of the body will be the same, while the enclosing surfaces will be different. If this is so, the same body, of known magnitude, for which the measure[3] of its magnitude does not change, could be enclosed, at different moments, by surfaces of different magnitudes. So if the place of the body is the surface enclosing this body, then the place of the body will be places of different magnitudes, of infinite number, none of which deserves more than all the others to be a place of this body. And nevertheless there are not several places for a single body.

None of the ambiguities that we have noted can be removed in any way at all. So it is not necessary that the surface enclosing the body shall be the place of the body. And if we call it 'place', it is by way of metaphor and not in an exact manner, as for example when one calls the house, the dwelling, the district, the city, the 'place of a body'.

As for the void imagined as filled by the body, in this case an ambiguity appears when we say that no void exists in the Universe. So if we say that the place of the body is the void, it follows necessarily that the place of the body is a thing which does not exist. And yet the body exists, and any body that exists is in a place. If what is in a place exists, its place exists. It follows necessarily that the void exists; now this is an unacceptable statement in the mouth of someone who claims that the void does not exist. We remove this ambiguity by means of what we are about to describe.

If, to reply to this statement, we say: the void is nothing but distances lacking matter; the imagined void which has been filled by the body is the imagined distances equal to the distances of the body, if we imagine them lacking matter, the imagined void which has been filled by the body is thus imagined distances equal to the distances of the body, on which have been

---

[2] See *Les Mathématiques infinitésimales du IXe au XIe siècle*, vol. II: *Ibn al-Haytham*, London, 1993, p. 384, n. 1; English trans. *Ibn al-Haytham and Analytical Mathematics*. A History of Arabic Sciences and Mathematics, vol. 2, Culture and Civilization in the Middle East, London, 2012, p. 305, n. 1. We might also translate this as 'the greatest of solid isepiphanic figures'.

[3] Lit.: the quantity.

superimposed the distances of the body imagined in the body. But if on every imagined distance we superimpose an imagined distance, together they will be a single distance, because the imagined distance is simply the straight line which is a length without width. Now if on the straight line which is a length without width we superimpose a straight line which is a length without width, together they become a single straight line, because from their superposition there results no width, nor a length which exceeds the length of one of them. If one of the two imagined straight lines is superimposed upon the other, they become a single straight line which is a length without width. The imagined void filled by the body is thus imagined distances, on which there are superimposed the distances of the bodies, and which have become the single and same distances. The imagined void filled by the body will not be anything but the distances of the body, so that if the person who imagines forms in his imagination distances equal to the distances of the body, similar to the figure of the body – the figure that is in the imagination, separate from the body, will not be the place of the body. But the place of the body is no more than the distances on which are superimposed the distances of the body, with which they have been united, and to which the figure that is in the imagination is similar. And if the distances filled by the body do not exist in isolation, when empty of matter, before being filled by the body, it does not follow necessarily that the body does not fill imagined distances, because the distances can be imagined as being in isolation, lacking matter, even if they are never empty of a body which fills them. We shall explain this concept with the help of an example which reveals the form of place.

We say: for any hollow body, such as a goblet, a bowl, a pitcher and things like them, between two opposite points of the surface inside it, which is a concave surface, there is an imagined distance, conceived as unchanging. In the same way, it contains imagined distances, perpendicular to the base of its concavity or oblique. The set of the distances for the inner surface of a goblet, between opposite points of it, are fixed distances, which do not change. So if there is air in the goblet, which fills the inside of the goblet, these distances are the distances of the air that is inside the goblet; if we then fill the goblet with water, the distances between the opposite points of the inner surface of the goblet are the distances of the water which is inside the goblet. If we then pour the water out of the goblet, and we fill the goblet with a drink, the distances between the opposite points of the inner surface of the goblet will be the distances of the drink that is in the goblet. In the same way, for any body with which we fill the goblet, the distances between the opposite points of the inner surface of the goblet will be the distances for it. The distances between the opposite points of the inner surface of the goblet will sometimes be the distances of the air, sometimes the distances of the water, sometimes the distances of the drink, and they will be the distances of all the bodies which fill the goblet, which are bodies of different substances and different qualities. But the distances inside the goblet are intelligible distances, understood, and are fixed in a single state, and do not change; and their magnitude does not increase or decrease. But each of the bodies which fill the goblet has distances that are proper to it, which do not leave it, and whose magnitude does not increase or decrease while the body keeps the form of its substance, even if the figure of the distances changes, and some of them increase while the others decrease. The distances of each of the bodies which fill the goblet are distinct from the distances of the other bodies. If we empty one of the bodies out of the goblet, its distances depart with it, while the distances inside the goblet remain as they were and do not depart with the departing body. If another body comes into the goblet, it comes with distances other than the distances inside the goblet. If it then enters into the goblet, the distances inside the goblet become distances for it. This shows clearly that the distances of any body which fills the goblet are superimposed upon the distances inside the goblet, are identified with them, and become distances of the body which fills the goblet; but the distances inside the goblet are identical with themselves and do not vary.

In the same way, any body such as to suffer action from another, such as air, water, drinks or other bodies such as to suffer action from another, is susceptible of having a different figure and of varying forms – however, the distances are inseparable from the body and their figure and their form do not change except by a decrease of some of their distances and an increase in some others, because their measure, that is the measure[4] of their magnitude, does not vary with the variation in their figure and in their form, as long as their substance preserves its form. If a particular liquid body such as to suffer action from another, such as water and the like, is in vessels of different shapes; if we then pour successively from each of them into a goblet enough to fill the goblet; then the shapes of what comes from them into the goblet, before it comes into the goblet, are different shapes; then, once each of them has come into the goblet, in succession they all take the same shape, whose configuration does not differ in any way. From this we see clearly that there is something which maintains the forms of all these bodies and which moulds them all into the same shape and the same form. And the same form that will have become the form of each of these bodies that come into the goblet is the form of the inside of the goblet; but

---

[4] Lit.: quantity.

the form of the inside of the goblet is the form of the distances of the inside of the goblet; so that is the form of the distances of the inside of the goblet which maintains the forms of all the bodies which fill the goblet in one and the same form. This indeed manifestly shows that inside the goblet there are fixed distances which do not change, and that the distances of the bodies which succeed one another in the goblet, which are bodies that are different in regard to their substance, in regard to their shape and their form before coming into the goblet, the distances of each of them are superimposed onto these fixed distances and take their form; and each of the distances of the bodies becomes one with the distance inside the goblet, [the distance] onto which that distance has been superimposed.

If we say that what maintains the shape of the body and its form is the inner surface of the goblet and not the distances between the opposite points of the surface, then the answer is that the body which comes into the goblet comes in between the opposite points of the inner surface of the goblet; so its distances are superimposed upon the distances between the opposite points of the inner surface of the goblet or[5] upon the set of them. And for any body which comes into the inside of the goblet, in all cases its distances are superimposed upon the distances of the inside of the goblet, distances which are fixed and unchanging.

But the fixed distances, which are inside the goblet, are the imagined voids filled by each of the bodies which fill the goblet, even if these distances do not lack a body which fills them; but they are in the imagination lacking matter, and in sensible existence they are associated with one [specific type of] matter, and the [types of] matter succeed one another.

But every body is enclosed by a body, so the surface of the body which encloses the body which is inside it encloses the imagined distances, known and fixed, which do not change, onto which the distances of the enclosed body are superimposed, and with which they become one. So if we take the enclosed body out of this position, and if another body comes into its place, the distances of the second body are superimposed upon the distances fixed, intelligible, imagined, onto which the distances of the first body had been superimposed.

It is clear, from all that we have shown, that these imagined distances between the opposite points of the surface enclosing the body, which are the imagined void filled by the body, are more deserving of being [considered as] the place of the body than is the surface which encloses the body, given that <the hypothesis of> the surface necessarily implies unacceptable

[5] 'or' in the sense of 'that is to say'.

ambiguities and abominable horrors;[6] whereas the imagined distances between the opposite points of the surface enclosing the body – which are the imagined void filled by the body – do not imply anything abhorrent and do not suffer from any ambiguity. The imagined distances between opposite points of the surface enclosing the goblet are the place in which the body has its position, and which in magnitude does not exceed that of the body. But since these distances – once the body is in position in them and once the distances of the body have been superimposed upon them – become one with the distances of the body and become distances for the body, the imagined void equal to the body which has been filled by the body is the distances of the body themselves. If this is so, then the place of the body is the distances of the body.

If we say that the void is a body, that the body in position in the place is a body, and that it is not permitted for a body to interpenetrate another body and that both become a single body, then the answer is that a body does not interpenetrate a body, if each of them is material – now in matter there is a resistance and an impediment; each of the two impedes the other from coming into its place, while it is fixed in its place. But a void has no matter, and has no resistance. [As for] the void, it is only distances, assembled to receive matter. A physical body is matter that the imagined distances are assembled to receive with the distances. Now all the distances are assembled to receive any matter and any distance; so they have nothing with which to prevent the distances from being superimposed upon the void. So nothing prevents the distances of the physical body that the void is configured to receive being superimposed upon the distances of the void which are lengths without width, and which have nothing that resists. Given that this is so, then the assertion that a physical body does not penetrate into the void, because these are two bodies, becomes null.

Since all that we have shown has been explained, accordingly the place of a body, that is the distances of the body which, abstract in the imagination, are a void without matter, equal to the body, with a shape similar to that of the body, which is what we wanted to prove in this treatise.

The treatise by al-Ḥasan ibn al-Ḥasan ibn al-Haytham on place
is finished.
Thanks be given to God, Lord of the worlds, and blessing upon
Muḥammad his prophet and all his family.

---

[6] The colourful phrase of the original has been translated literally. The horrors in question are logical, that is propositions that are logically untenable.

THE ARS INVENIENDI: THĀBIT IBN QURRA AND AL-SIJZĪ

It is because of Ibn Sinān's research on analysis and synthesis in geometry, but also in opposition to it, that Ibn al-Haytham conceived of the art of analysis. It is also, as we have indicated, following on from the *ars inveniendi* proposed by al-Sijzī, and in opposition to it, that Ibn al-Haytham developed this art. We have seen that Ibn al-Haytham's novel conception can be understood in the light of new demands in geometry.

But Ibn Sinān, and his successor al-Sijzī, both started out from a short essay by Thābit ibn Qurra. Thus, a historical picture emerges, or at least the parts of it that have survived the vicissitudes of time; it suggests that if we wish to get a clearer idea of where Ibn al-Haytham's contribution fits in, we need to make a careful examination of the writings of Thābit ibn Qurra, Ibn Sinān and al-Sijzī. Studies of Ibn Sinān's book and of his other works have already been carried out.[1] So there remain the essay by Thābit ibn Qurra and the treatise by al-Sijzī. It is to these last two works that the following pages will be dedicated, with the purpose of understanding how a new area of study emerged and how it developed before Ibn al-Haytham, so that we can take the measure of the step forward that he made.

I. THĀBIT IBN QURRA: AXIOMATIC METHOD AND INVENTION

We might say that in the beginning there was the translation of Euclid's *Elements*. In the eyes of the mathematicians of the ninth century, as in those of their successors, this book was a model of organisation, but it rather rapidly became the source of a multitude of topics for reflection. There is a whole book to be written on this double part the *Elements* played in Arabic mathematics. Let us remind ourselves that hardly had the *Elements* been translated before they became the object of numerous commentaries that tried to describe the author's intentions, to discuss the organisation of the work, to correct some of the propositions and to rewrite

---

[1] R. Rashed and H. Bellosta, *Ibrāhīm ibn Sinān. Logique et géométrie au Xe siècle*, Leiden, 2000.

some of the proofs. This is how it happens that in the mid ninth century the famous philosopher al-Kindī wrote two books whose titles are particularly eloquent: *On the Rectification of the Book by Euclid* and *On the Intentions of the Book by Euclid*. Others, such as al-Jawharī, wrote commentaries on Euclid's book, and were interested in some of the difficulties it presented, notably the question of the fifth postulate. Others again, such as al-Māhānī, wanted to replace proofs by *reductio ad absurdum* with direct proofs. We could indeed cite many more authors and book titles that bear witness to the central position occupied by the *Elements*, not only at the heart of mathematical activity, but also, more generally, in the intellectual life of the time. Mathematicians who were geometers, but also some who were algebraists, philosophers and intellectuals, quickly began to take an interest in the treatise, in its organisation, its order and its style. Among the intellectuals, we find a state official, a member of a dynasty of great administrators of the Empire: Ibn Wahb.[2] He was, no doubt, an alert reader of the *Elements*, and as such he raises the major question regarding the axiomatic method and discovery. In Ibn Wahb's terms, the question takes the form: in his order of exposition of the propositions, Euclid has regard only to the

[2] Thābit ibn Qurra addresses this letter to Ibn Wahb; but which one? The family of the Banū Wahb, a family of ministers, secretaries of state and men of letters, had moved in the circles of power in Baghdad for at least a century. Except for the founder, Wahb himself, secretary of the famous Barmecide and a minister of Hārūn al-Rashīd, Ja'far (d. January 803), all the others, his sons, grandsons and great grandsons, could have been the Ibn Wahb to whom Ibn Qurra refers. Ibn Qurra does nothing to help us: he does not give any date for his letter, nor the first name or title of his correspondent. A first possible candidate is Sulaymān ibn Wahb (d. 885). Thābit was already in Baghdad and, with his protectors and teachers, the Banū Mūsā, moved in the circles of power. Two other possible candidates are the two sons of Sulaymān: Aḥmad, secretary of state for taxations, celebrated literary expert and poet, the subject of a brief biography by Yāqūt in his *Mu'jam al-Udabā'* (ed. Būlāq, Cairo, n.d., vol. III, pp. 54–63); or 'Ubayd Allāh, a minister of the Caliph al-Mu'taḍid for about a decade, who died in 900, and was thus also a contemporary of Thābit ibn Qurra. It is also possible that Thābit is addressing this letter to al-Qāsim, the son of 'Ubayd Allāh, with whom he shared ministerial responsibilities before himself becoming a minister on his father's death. It was exactly for this same al-Qāsim that Ibn Qurra wrote his *Talkhīṣ* (summary) of Aristotle's *Metaphysics*. This latter essay has the title *Treatise by Thābit ibn Qurra on the Summary of what Aristotle Presents in his Book on* Metaphysics ... *written for the Minister Abū al-Ḥusayn al-Qāsim ibn 'Ubayd Allāh*. It is possible that this same al-Qāsim, who was interested in metaphysics, might be the person who was interested in Euclid's *Elements*, and above all in the method of discovery. In any case, in the present state of knowledge, this last suggestion seems the best one (see for example D. Sourdel, *Le Vizirat abbaside*, Institut Français de Damas, Damascus, 1959–60, vol. I, pp. 300–1, pp. 329–57; vol. II, p. 745).

requirements of proof, so he brings forward the position of certain propositions while he puts back others, irrespective of their significance. So, in the
Elements, Euclid opted simply for syntactic order, ignoring semantics. This
order, Ibn Wahb admits, is suitable for training an apprentice in geometry;
but when it is a matter of carrying it through into research, what one has
learned shows itself to be inadequate: we need another order, that of discovery. This question formulated by Ibn Wahb in the ninth century is
brought up again centuries later by Petrus Ramus,[3] Antoine Arnauld and
Pierre Nicole,[4] among many others.

But the fact that this topic is a perennial one is an indication of the
intimate links it has with the actual style of the *Elements*, that is the axiomatic method (in Euclid's sense, of course) that governs the work. The
Elements, a model of organisation for mathematicians, not only in the ninth
century but for more than two millennia, is also a model in the sense of
being something to be imitated as well as representing an ideal. This double validity as establishing a norm derives from the application of the axiomatic method. But, in the context of Euclid's work, this application is
itself possible only insofar as the geometrical object – the figure – is the
object of knowledge that is subordinate to hypotheses and to construction
processes that take place in the imagination. The question of discovery
might thus, in such a context, seem to be a meta-geometrical one. Discoveries would be at most a matter of circumstance, for the most part resulting
from applying the axiomatic method to verifying a proof or to assessing the
limits of its validity.

In any case, Ibn Wahb is writing to Ibn Qurra to ask him to devise a
method, different from the axiomatic method, that is capable of meeting
the requirements for discovery. So the intention is clear: to provide the
reader, presumed to be familiar with the axiomatic method, with a second
method that will allow him to discover new propositions and to carry out
new constructions. It seems that Ibn Wahb's decision was not actuated
solely by Thābit ibn Qurra's eminence as a geometer; Thābit also had first
hand knowledge of the *Elements* because he had made a revised version of

---

[3] Ramus (Pierre de la Ramée) had questioned the order propositions in Euclid's
Elements: 'Ordo Euclidis displicuit Petro Ramo, quemadmodum ex iis intelligitur, quae
in Scholis Mathematicis lib. 6 et sqq., contra Euclidem passim disputat', quoted in
Antoine Arnauld and Pierre Nicole, *La Logique ou l'art de penser, contenant, outre les
règles communes, plusieurs observations nouvelles, propres à former le jugement*, a
critical edition with an Introduction by Pierre Clair and François Girbal, coll. Le
mouvement des idées au XVIIᵉ siècle, Paris, 1965.

[4] *La Logique ou l'art de penser*, reference in the previous note.

the third Arabic translation of the work, the translation by Isḥāq ibn Ḥunayn.

By way of replying to Ibn Wahb, Thābit ibn Qurra composed a short text – which is translated here.[5] We may note, briefly, that this short work has a simple structure. In the first section, which is introductory, the author takes up the question of the axiomatic style of exposition employed in the *Elements* and that of the order which should be followed for discovery, and presents a classification of geometrical concepts. The second part, consisting of examples that provide illustrations for the first part, might thus be said to supply 'exercises in invention'.

The first part of the essay opens with two interesting observations. Ibn Qurra's immediate purpose is to think out the rules for a method that will lead to the discovery of new propositions and constructions, this for a mathematician who has a grasp of the axiomatic method and an adequate knowledge of mathematics. But Ibn Qurra does not stop there: the method must be applicable in 'any apodeictic science' (*fī kull 'ilm burhānī*). So it is obvious that we are looking at a pragmatic approach. Moreover, it requires us to proceed by classifying concepts so as to be able to distinguish between different species, in order to group them together later into species and thus bear them all in mind when required. Now, to identify the different species, Ibn Qurra begins by distinguishing three kinds of geometrical investigation: geometrical constructions making use of instruments – for example using ruler and compasses for constructing an equilateral triangle; propositions referring to an unknown magnitude or state – for example to determine the area of a triangle whose sides are known, or to find a perfect number; and finally general statements concerning the nature of the object – or concerning a specific property of the object – for example, when the object is a triangle: the sum of its angles is equal to two right angles. Ibn Qurra points out that the first type requires a knowledge of the two others, but the converse is not true.

The first rule of the method is more or less self-evident: it consists of beginning by finding out which type or which grouping the concept we are looking for belongs to. But each of these three groupings includes principles and results established by means of these principles, together with some supplementary ones. By principle (*aṣl*), Ibn Qurra means something in the tradition of the *Posterior Analytics* (I, 10). It is, as he says, a matter of 'common notions', of postulates and definitions. For the latter, these are only the definitions that refer to the essence of the concept we seek. Once

---

[5] For the *editio princeps* of the Arabic text together with a French translation, see *Les Mathématiques infinitésimales du IXe au XIe siècle*, vol. IV: *Méthodes géométriques, transformations ponctuelles et philosophie des mathématiques*, London, 2002.

the mathematician has distinguished, for each of the preceding types, between axioms, postulates and definitions on the one hand and propositions on the other, he is in a position to 'bring to mind' all the ideas that are necessary for forming a conception of the required object: this is the second rule for the method.

The third rule, which Ibn Qurra does not identify by name, is analysis: to start from the necessary conditions for the required object, then work from the necessary conditions for these conditions, and so on. To illustrate this analysis, he examines three constructions in which, in each example, we see how to carry out the analysis. This choice of examples is influenced by a somewhat didactic purpose, as well as a particular interest in the form of analysis that is called 'problematic' in geometry. We may, however, note that by singling out the particular class of propositions that deal with determining a magnitude or a number, Ibn Qurra distances himself from the classic opposition between 'theoretical analysis' and 'problematic analysis'.

As the first text on the method of discovery, the context of Ibn Qurra's short essay sheds light on the emergence of the topic of discovery in mathematics which, in the work of his successors, will move away from its origins and will acquire a quite different scope. But the importance of this essay does not relate only to the fact that it contains the first discussion of this topic; it was also responsible for encouraging its readers to carry out further research in the light of the new mathematical insights it provided.

II. AL-SIJZĪ: THE IDEA OF AN *ARS INVENIENDI*

1. *Introduction*

To establish 'a method intended for students, which includes all that is necessary for the solution of problems in geometry'.[6] This is the expression used by Ibrāhīm ibn Sinān (296/909–335/946) when committing himself to fulfilling his grandfather's promise, Thābit ibn Qurra. But the mathematical context has been in continual change since his time, following the path already mapped out by Thābit's teachers, the Banū Mūsā. The effect of new research in the geometry of measurement and in the geometry of position and of shape, the emergence of a new *mathesis* from the impact of algebra, all these had, as Ibn Sinān himself tells us, spurred mathematicians

---

[6] R. Rashed and H. Bellosta, *Ibrāhīm ibn Sinān. Logique et géométrie au Xe siècle*, p. 96.

into returning to the traditional question of analysis and synthesis, and, more generally, that of the philosophy of mathematics. As we have seen, Ibn Sinān's intervention in this matter proved to be crucial: in the first known substantial treatment of analysis and synthesis, he works out a philosophical logic that allows him to combine an *ars inveniendi* with an *ars demonstrandi*.[7] His contribution was soon extended to provide a proper theory of proof, in which questions of logic take centre stage: the reversibility of implications, auxiliary constructions, classification of propositions according to the number of variables and according to the number of conditions.

Ibn Sinān's successors repeatedly took up the question of analysis and synthesis, either in the course of their technical mathematical works, or, following Ibn Sinān's example, though from a different point of view, dedicating complete treatises to it. Such is the case for Ibn Sahl, al-Qūhī and Ibn al-Haytham, among many others. Nor was this line of research neglected by philosophers who took an interest in mathematics: al-Fārābī discusses it and Muḥammad ibn al-Haytham also pays attention to it. Which is to say that, at least from the mid tenth century onwards, we see the development of a new field of research in the philosophical logic of mathematics, or more generally in the philosophy of mathematics, a meeting place for professional mathematicians and philosophers of mathematics – a field of which 'analysis and synthesis', in their different forms, makes up the core. Now it is in exactly this domain that we come across al-Sijzī, and this is above all the context in which we must situate the book by him that concerns us here: *To Smooth the Paths for Determining Geometrical Propositions*.

Al-Sijzī lived almost a generation after Ibn Sinān; he knew the latter's writings well, in particular the treatise on analysis and synthesis, which he had himself transcribed.[8] He was also familiar with the writings of Thābit ibn Qurra,[9] notably with Thābit's short work on methods for determining geometrical problems, which he had addressed to Ibn Wahb: one of the manuscripts of this short work was indeed written out by al-Sijzī.[10] So

---

[7] *Ibid.*, pp. 21–56.

[8] Ibn Sinān, *Maqāla fī ṭarīq al-taḥlīl wa-al-tarkīb fī al-masā'il al-handasiyya*, ms. Paris, Bibliothèque nationale, no. 2457, fols A^v–18^v.

[9] See R. Rashed and H. Bellosta, *Ibrāhīm ibn Sinān. Logique et géométrie au Xe siècle*, p. 89.

[10] *Kitāb Thābit ibn Qurra ilā Ibn Wahb fī al-ta'attī li-istikhrāj 'amal al-masā'il al-handasiyya*, ms. Paris, Bibliothèque nationale, no. 2457, fols 188^v–191^r; see *Les Mathématiques infinitésimales*, vol. IV, pp. 742–65, and below, pp. 581–9.

Thābit ibn Qurra and Ibrāhīm ibn Sinān provide us with the two clearest points of reference for assessing the place of al-Sijzī's contribution.

Al-Sijzī's text is much more fully developed than that of Ibn Qurra and its purpose is different. It remains true, of course, that they have in common elements of their vocabulary, their intention and their organisation, which makes it reasonable to suggest these books were the first source of al-Sijzī's inspiration. His own book is also made up of two parts: a first part that is introductory, followed by a second that is devoted to examples. In addition to this similarity in form, we have another similarity: like Ibn Qurra, al-Sijzī deals only with geometry, to the exclusion of all other branches of mathematics – and in fact both of them set up geometry as the model for any apodeictic science. Finally, the intention that animated both of them was at once didactic and logical. This intention is, of course, not absent in Ibn Sinān, but in his work the didactic element took second place to the task of developing a theory of proof. In al-Sijzī's case, and in some sense thanks to his having read Ibn Sinān, this project is transformed into a true *ars inveniendi*, something that did not occur in the work of Ibn Qurra. Here we need to look at the direction al-Sijzī is taking and the novelty of what he proposes to do. To make this clear, we must comment on his treatise in detail.

2. *A propaedeutic to the* ars inveniendi

The first part of al-Sijzī's treatise opens with a propaedeutic to research into the methods that will form the main body of the *ars inveniendi*. This propaedeutic is itself made up of two short parts, in accordance with the author's purposes for the work as a whole, which are both didactic and logical. Al-Sijzī starts with a theory of mathematical discovery, a sketchy one, certainly, but nevertheless one that contains the genesis of the psychology of mind (to use a later term) that is connected with the *ars inveniendi*. According to this theory, invention in geometry is the offspring of a 'natural power', of an innate gift and arduous training in principles as well as in methods and theorems. This latter is more important than having a gift, in the sense that, when the natural power is not at its height, it may happen that training will serve to remedy this relative weakness. However, the reverse is not true, since a natural power without training leads nowhere. Without training there is no invention. Under such conditions, there is room for a discipline to guide the geometer from training to discovery: this discipline is the *ars inveniendi*. That is to say that the need for this art is built into the didactic side of the discipline. So we are seeing a first manifestation of its being necessary.

The second part of the propaedeutic is concerned with the preliminary procedures for any method, the operations one must carry out before choosing one method or another. In taking this approach, al-Sijzī in fact starts out from the conception of training in geometry. As he conceives it, the part training plays in invention makes it necessary for the beginner in geometry to start by learning the theorems (al-qawānīn) proved in the Elements. This very natural demand nevertheless raises some questions, which al-Sijzī considers. It is, moreover, in connection with these that he touches on certain logico-philosophical problems some of which he will return to in a later text.[11] The first relates to logical ordering of proofs, the order adopted by Euclid in the exposition in the Elements. If we allow this same order for simple training, which al-Sijzī does, following the example of Thābit ibn Qurra, can we accept it as an ordering in training for research, that is training that leads to discovery? In either case, if we adopt a deductive system like that of Euclid, we should begin with the axioms (the common notions) rather than with theorems. Surely it is indeed more natural and more consistent to start with what is most basic. The more so because the theorems, as such, form part of what we are trying to establish. So the risk is that, if we start with the theorems, we may take the end for the means. To avoid this, it might be preferable to employ methods that begin from the axioms alone so as to determine the objects of research. In this treatise, once the question of the ordering appropriate for training in preparation for discovery has been raised, and once the didactic problem has been reduced to the logico-philosophical problem of the relations between axioms and theorems, al-Sijzī dismisses the possibility of following the order of proof, and recommends starting with the theorems. There are three reasons for this, with distinct origins. First of all, starting out with the axioms alone would lengthen the journey towards discovery to an unreasonable degree. In the second place, if we limit ourselves to only the axioms, it will be difficult, without theorems, to proceed to discovery. Finally, in his system, Euclid combined axioms and theorems in a balanced manner, which allows us to start out from the theorems he established. These arguments, which are summarily listed by al-Sijzī, essentially belong to pro-

---

[11] R. Rashed, 'Al-Sijzī et Maïmonide: Commentaire mathématique et philosophique de la proposition II–14 des Coniques d'Apollonius', Archives internationales d'histoire des sciences, vol. 37, no. 119, 1987, pp. 263–96. See also P. Crozet, 'Al-Sijzī et les Éléments d'Euclide: Commentaires et autres démonstrations des propositions', in A. Hasnawi, A. Elamrani-Jamal and M. Aouad (eds), Perspectives arabes et médiévales sur la tradition scientifique et philosophique grecque, Paris, 1997, pp. 61–77.

grammatic and pragmatic logic. The problem itself is, as we have seen, much more significant, since it concerns the status of the theorem in a deductive system, when we are considering training designed to lead on to research and invention. As regards al-Sijzī's preoccupations, this problem seems to lie in the background, sufficiently present for him to come back to it, but seen in a different light.

Al-Sijzī notes that, in a deductive system, a theorem is at once a lemma and a consequence. He describes this state as *mushtabah*, that is as ambiguous. There is also an additional difficulty: the chain of implications can be unlimited. So the question is to find how, under these conditions, there can be training in regard to theorems; it might perhaps be better to confine ourselves to axioms. Now it is at this very moment that al-Sijzī refers 'to the equilibrium' in Euclid's exposition.

In this treatise, al-Sijzī gives only very brief answers to the questions that he himself raises. He refers to a fear of excessive length, to the difficulty and the equilibrium of Euclid's exposition, all of which is no doubt significant, nevertheless leaves the discussion open. Further, he permits himself to relaunch the debate, since he returns to the question of axioms and theorems, more fully and in greater depth, in a later treatise, where he mentions the one we are concerned with here. This later treatise is his work on the asymptote,[12] where with the help of the pair 'conceive – prove' he works out a classification for mathematical propositions thus distinguishing and illuminating the connections between axioms and theorems. He distinguishes five classes, in order: (1) Propositions that can be conceived directly from axioms; (2) Propositions that can be conceived before we proceed to prove them, that is ones that are close to the axioms; (3) Propositions that can be conceived when we form the idea of how to prove them; (4) Propositions that can be conceived only once they have been proved; (5) Propositions that are difficult to conceive even after they have been proved.[13] Thus, in returning to his own work, al-Sijzī shows us what he was interested in when he was writing his first treatise.

Thanks to this training, from the outset the geometer already has a store of theorems and lemmas, knows how to proceed, and is ready to embark upon research. The important question is to know how to put that readiness into action in such a way that it leads to discoveries. But, before deciding on a specific method, it is necessary to master a mass of faculties and knowledge that underlies all the methods. This underlying mass also

[12] *Ibid.*

[13] He considers the example of Proposition II.4 of the *Conics*, which concerns the asymptotes of a rectangular (equilateral) hyperbola.

includes, without distinction, many elements that are psychological and logico-philosophical.

Once he has recognised what kind of object he requires, as well its specific properties, the geometer must first of all bring to mind the lemmas and theorems that apply to that kind of object, or to a kind of object connected with it. This effort of imagining the lemmas and theorems is necessary for both the classes of object that together constitute geometry: constructions and propositions. Al-Sijzī accordingly proposes some rules to guide us in this search for lemmas and theorems. To establish these rules, he begins by distinguishing two classes of propositions. The first class consists of the propositions that are in themselves possible but which, because of a lack of preliminary results (lemmas), it is impossible for us to prove. This is the class to which the squaring of the circle belongs. In other words, as al-Sijzī will say later, for us these propositions are conceivable but not demonstrable. In the other class, there are propositions that can be proved, and in this case we can proceed according to the following rules:

1. For any proposition we think we can establish from a particular lemma, we can try to establish it by lemmas of the same class – that is ones that, at least, apply to the same objects – or to some of them.

2. Any proposition we think we can establish from a lemma, or from lemmas, we can establish from lemmas of that lemma, or of these lemmas.

3. Any proposition that we cannot establish from a series of successive lemmas could be established from many compound lemmas. Al-Sijzī later examines an example of this.

These rules are applicable to all the methods that the geometer should adopt, and around which al-Sijzī's treatise is organised. However, the application of one or other of them is by no means automatic; it in fact demands preliminary work which involves the faculties of the intelligence. This is, indeed, the way that these faculties are introduced into the *ars inveniendi*. Now there are two characteristics that distinguish these faculties in al-Sijzī's work: they belong to the intellect, and intelligence itself is not natural intelligence, but a power exercised in the art of geometry. All its faculties are, moreover, as it were formed by this art. Al-Sijzī lists them, and presents them at the head of the list of preliminary procedures. They are skill, intelligence informed by practice, and the faculty for calling to mind, instantly and simultaneously, the necessary conditions for the proposition we wish to establish. These three faculties are concomitants in al-Sijzī's actual presentation.

Next in this preparation for applying the rules and methods comes learning all the theorems and lemmas necessary for the object we require, and to learn them exhaustively. The third demand requires that these truths already established become things counted among the intellectual faculties. At this stage, alongside the skill that comes directly from educated intelligence, al-Sijzī mentions, both intuition and cunning. By 'intuition', al-Sijzī here once again means intuition informed by the art. No doubt what we have is an act of thought which immediately grasps the object of knowledge, an act that does not require the intermediate deductions. Cunning, or the faculty for finding ingenious procedures, is itself also obviously the fruit of informed intelligence, of long practice and intense application.

These first three preparatory rules call upon all these considerations regarding intellectual faculties, thus introducing these faculties into the *ars inveniendi*. Exactly like his successors many years later, al-Sijzī obviously could not avoid these elements of a psychology of reason, the logic being what it was. The fourth and final rule is logical in nature: we are concerned with classifying theorems and lemmas according to what they have in common, their differences and their specific properties, that is the properties of the objects they apply to.

Once this propaedeutic work has been done, the geometer, knowing the rules that apply to all methods, is now in a position to make use of one of the following three methods: (1) the method of transformation; (2) 'analysis and synthesis'; (3) ingenious procedures.

These methods are neither of the same nature, nor do they have the same power, but they can be combined. We may note that in this part, as in the remainder of the treatise, al-Sijzī comes up against a difficulty part of which we have already pointed out, and one which he could not surmount. He in fact formulates the logical problems in a mixed language, that of traditional modal logic and that of the theory of proportions; thus he speaks of necessary and impossible propositions. On the other hand, as we shall see later, when he speaks of the logical implications among mathematical propositions, he compares them to ratios in the theory of proportions. This is not a matter of criticising al-Sijzī for not knowing the work of George Boole, or of regretting that he did not think of using the language of algebra, but merely a matter of noting that his writing has an irremovable ambiguity, whose cause is to be found in the mixed form of expression. It seems that a psychology of reason being involved to form the basis of the *ars inveniendi* results in the absence of the logical language appropriate to discussion of the art of analysis. Later, Ibn al-Haytham would invent a new geometrical discipline, with the intention, among other things, of meeting this difficulty: *the knowns*.

3. *The methods of the* ars inveniendi *and their applications*

As we have seen, al-Sijzī proposes three methods. The four methods
that he presents before them are recommended as propaedeutic to these
final three: transformation, analysis and synthesis and ingenious procedures. We observe, however, that he presents all seven as on the same
level, as if all of them carried the same weight. Further, his account of
them is very brief, not to say incomplete. Transformation is merely named,
'analysis and synthesis' are cited in ordinary, that is standard, terms; as for
the ingenious procedures, they appear in connection with a rather vague
reference to Hero of Alexandria. All the same, al-Sijzī gives a clear
explanation of his reason for this choice of exposition, distinguishing
between two possible styles for presenting this material, styles which for
him are complementary: one is general, indicative and not demonstrative;
the second consists of examining examples and giving detailed proofs. It is
in the course of carrying out this second process that al-Sijzī presents these
methods, as it were in action. All the same, this does not completely
explain why his earlier account was short and somewhat allusive.

If we examine al-Sijzī's text more closely, we see that there is in fact
only one method that is truly worthy of its title: 'analysis and synthesis'. In
this respect, al-Sijzī belongs to the tradition established by his predecessors
and his contemporaries; but the task he has deliberately set himself is that
of elaborating on this principal method by adding a group of special procedures, mathematical procedures, theoretical and practical. These special
methods are intended to increase the power of the principal method in the
making of discoveries and thus to make its application easier. Thus, transformation is a mathematical procedure that is theoretical in nature, ingenious procedures are of a technical and practical nature, and all of them are
means of bringing analysis to its conclusion and making it easier to carry
out. It is the combining special methods with the principal method, for the
purpose we have just described, that marks a break with tradition and is
due specifically to al-Sijzī. Let us explain.

As we have said, al-Sijzī's purpose is to extend the power of analysis
and synthesis by introducing theoretical and technical procedures. This is
the approach he chose for instituting the *ars inveniendi*. It is in this context
that he fully grasped the importance of point-to-point transformations in
geometry, which had been used since al-Ḥasan ibn Mūsā,[14] by giving them

---

[14] *Les Mathématiques infinitésimales du IXe au XIe siècle*, vol. I: *Fondateurs et
commentateurs: Banū Mūsā, Thābit ibn Qurra, Ibn Sinān, al-Khāzin, al-Qūhī, Ibn al-
Samḥ, Ibn Hūd*, London, 1996, pp. 885 ff.; English trans. *Founding Figures and*

a name: *al-naql.*[15] It is with this same aim that he develops various procedures that spring from the idea of one element varying while all the others remain fixed. Al-Sijzī in fact notices that there are two ways to investigate the properties of geometrical objects. The first consists of looking for what remains fixed when all the other properties vary – an investigation carried out by imagination based on the senses. In following the second approach, we take the required property as given and look for the lemmas that are necessary for it. The first approach, that of variation, is one that al-Sijzī not only finds interesting, but also one he goes on to employ in its various forms. The most obvious form is that where the *varia* include an element of a figure, in which all the other elements remain constant. It is, so to speak, the paradigmatic form of the procedure. There is also the variation of constructions using a fixed figure; then there is the variation of methods for establishing a fixed property, and, finally, the variation of lemmas for a fixed proposition. As for the second approach, it is none other than that of analysis, as a theoretical procedure. In fact, for al-Sijzī, analysis in turn can be looked at in two, inseparable, ways, but not always in an explicit way: a method of discovery on a par with other special methods, and in this sense it is a mathematical procedure; that is, it is *the* method of discovery, supported by all other procedures, such as transformation, ingenious procedures, variation, and so on. The boundaries that separate the two aspects of analysis are obviously not set solidly in place once and for all, but vary according to the complexity of what we seek to discover: for instance the number of lemmas, or the number of auxiliary constructions. Moreover, in the matter of appreciating the degree of this complexity, al-Sijzī calls upon not only the geometer's skill and intelligence, but also his intuition. From this point onwards, intuition continues to play a central role, either alone,

---

Commentators in Arabic Mathematics, A History of Arabic Sciences and Mathematics, vol. 1, Culture and Civilization in the Middle East, London, 2012.

[15] The term *naql*, from the verb *naqala* (to displace), which was already present in the lexicon in the ninth century. Thābit ibn Qurra uses it, first, to mean a displacement to superimpose two figures, and goes on to modify its meaning so that it signifies a displacement of a magnitude in a continuous motion, that is a transformation. See the text by Thābit ibn Qurra with title *If we extend two straight lines at two angles less than two right angles, they meet one another* (see R. Rashed and C. Houzel, 'Thābit ibn Qurra et la théorie des parallèles', in R. Rashed [ed.], *Thābit ibn Qurra. Science and Philosophy in Ninth-Century Baghdad*, Scientia Graeco-Arabica, vol. 4, Berlin, 2009, pp. 27–73, especially pp. 42–7). Al-Sijzī, who knew this text (since he made a copy of it), in turn modifies the meaning of the term, as in this treatise it is used for a similarity, thus confirming the tendency, already found in Thābit ibn Qurra, who uses the term *naql* for a geometrical transformation: translation or a similarity and so on.

or in combination with reasoning, to assess the degree of difficulty and judge what will be the shortest way of finding the necessary lemmas.

Like his predecessors from Pappus and Proclus onwards, al-Sijzī distinguishes between two applications of analysis, depending on whether we have geometrical constructions or propositions that concern geometrical properties. Behind his brief explanations, we may recognise echoes of Ibn Sinān. Al-Sijzī does in fact point out, in passing, some problems raised and discussed in full by Ibn Sinān: the number of conditions or of lemmas; the number of solutions. But he passes over these logical problems, and does not discuss them. This hint confirms his intentions and also reflects the deliberate choice of a style of exposition that privileges explanation through, and with the help of, studies of examples of the methods, of combinations and applications of them. All we now need to do is to follow al-Sijzī in his choice.

3.1. *Analysis and point-to-point transformation*

The first example al-Sijzī discusses concerns a geometrical construction, in which he proceeds by analysis. He shows how employing point-to-point transformations makes the method simpler and discovery easier. So with the method of analysis he combines the geometrical procedure of transformation, obviously with the aim of increasing the power of the analysis.

It is, al-Sijzī writes, a matter of constructing a figure:

> How to find two straight lines proportional to two given straight lines one of which is a tangent to a given circle and the other meets the circle and is such that, if it is drawn inside the circle, it passes through its centre?[16]

Let us proceed by analysis and let us suppose that the figure has been constructed. So we must find the necessary lemmas.

The problem is: given a circle CED with centre H and diameter CD, and a ratio $\dfrac{A}{B}$, to construct EG, a tangent to the circle, in such a way that the ratio $\dfrac{GE}{GC}$ is equal to $\dfrac{A}{B}$.

---

[16] See p. 596.

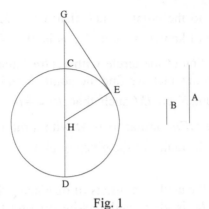

Fig. 1

The difficulty in this construction, pointed out by al-Sijzī himself, lies in the fact that the angle *EGH* of the triangle is unknown. To determine this angle, we construct an auxiliary figure *IKMN* similar to the figure *GEHC*. In other words, we try to find a triangle *IKM* with a right angle at *K* and a point *N* on the hypotenuse such that *MN* = *MK* and satisfying

$$\frac{IN}{IK} = \frac{GC}{GE} = \frac{B}{A} \qquad \text{(a given ratio)}.$$

Since the straight line *IK* is known in position and magnitude, and *B* and *A* are two known magnitudes, the straight line *IN* is of known magnitude. So the point *N* lies on a circle with centre *I* and radius *IN*. The first figure in the manuscript illustrates this situation.

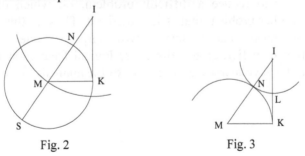

Fig. 2   Fig. 3

To construct the point *N*, we choose a point *L* on the circle and we rotate the straight line *IL* about *I* until the distance from *N*, the new position of *L*, to *M*, the point of intersection of the extended line *IN* with *KM*, the perpendicular to *IK* at *K*, is equal to *KM*. The point *L* is then at the point *N* and we have *NM* = *KM*.

If we extend IN to the point S such that $MS = MN = MK$, we have $IS \cdot IN = IK^2$, so IS is of known magnitude, as is $IM = \dfrac{IS + IN}{2}$. So the point M lies at the intersection of the circle with centre I and radius IM and the perpendicular KM; from which we find the points N and S.

We take the point N on IM such that $IN = \dfrac{B}{A} IK$; it only remains to construct the triangle GEH similar to IKM with a ratio of similarity equal to $\dfrac{HE}{MK}$ (where HE is the radius of the given circle).

Thus, in essence the method consists of displacing the problem by supposing the segment IK is given and seeking to find the circle NKS that touches IK at K. We then return to the problem of the similarity. This is the sense of the term 'transformation' in the context of analysis.

Many mathematicians of the time had made use of a point-to-point transformation, when working by analysis and synthesis, among them some scholars well known to al-Sijzī, such as al-Qūhī. We may cite the example of the construction of the regular heptagon: we construct a triangle of type $(1, 2, 4)$ or $(1, 1, 5)$, or other types; then in the given circle we construct a triangle homothetic with one of them.[17] This technique was to be employed several times by al-Sijzī's successor, Ibn al-Haytham, and we meet it again later on, for example when Fermat finds a tangent to Descartes' *folium* at 45° to its axis.

The essential point here is that of employing a point-to-point transformation in order to reduce a difficult problem, on which one is using analysis, to another problem that is less difficult. Thus, in this context, the point-to-point transformation serves two purposes: one mathematical (transformation of a figure) and the other logical (reduction to an easier problem). So we have increased the power of the analysis in two ways.

[17] R. Rashed, *Les Mathématiques infinitésimales du IXe au XIe siècle*, vol. III: *Ibn al-Haytham. Théorie des coniques, constructions géométriques et géométrie pratique*, London, 2000; English translation by J. V. Field: *Ibn al-Haytham's Theory of Conics, Geometrical Constructions and Practical Geometry*. A History of Arabic Sciences and Mathematics, vol. 3, Culture and Civilization in the Middle East, London/New York, 2013, Chap. III.

3.2. *Analysis and variation of one element of the figure*

Being always in search of techniques to make analysis more powerful, al-Sijzī proposes continuous variation of one element of the figure while all the other elements remain fixed. He illustrates this search for new techniques with a simple example: to establish a specific property of triangles, that of the sum of their angles being the same; to prove in a second stage that this sum is equal to two right angles.

Al-Sijzī takes an arbitrary triangle ABC; the side AB of the angle BAC is fixed, while the vertex C can vary continuously on the straight line AC.

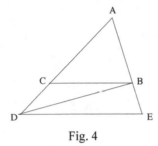

Fig. 4

If C comes to D where $AD > AC$, we have $A\hat{D}B < A\hat{C}B$ and $A\hat{B}D > A\hat{B}C$. We wish to establish that $A\hat{D}B + A\hat{B}D = A\hat{C}B + A\hat{B}C$, or that $C\hat{D}B + C\hat{B}D = A\hat{C}B$, which comes to the same thing, since $A\hat{B}D = A\hat{B}C + C\hat{B}D$.

We draw DE parallel to CB, and by *Elements* I.29 (a necessary premise at this point), $A\hat{D}E = A\hat{C}B$, $A\hat{E}D = A\hat{B}C$ and $B\hat{D}E = D\hat{B}C$. Thus,

$$A\hat{C}B = C\hat{D}B + B\hat{D}E = C\hat{D}B + D\hat{B}C$$

and consequently

$$A\hat{D}B + A\hat{B}D = A\hat{C}B + A\hat{B}C.$$

So the three angles of the two triangles ABC and ABD have the same sum. So al-Sijzī has established that if two triangles have a common angle, then the sum of their three angles is the same. From that we can establish that two arbitrary triangles have the same sum for their three angles; al-Sijzī does not even mention this.

Legendre gives an analogous argument, but without using I.29, to establish that if the sum of the angles of a single triangle is equal to two right angles (respectively, the sum is more than/less than two right angles), the same is true for any triangle.[18]

Al-Sijzī uses this same technique of the variation of one element of the figure to show that this sum is equal to two right angles. This time it is sufficient to consider a special triangle, here the isosceles right-angled triangle *ABC*. We draw *BD* parallel to *AC*; by Proposition I.29 of the *Elements*, we have

$$C\hat{B}D = A\hat{C}B \text{ and } A\hat{B}D \text{ is a right angle.}$$

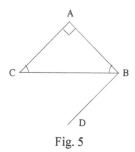

Fig. 5

We see that

$$A\hat{B}C + A\hat{C}B = A\hat{B}D = 90°.$$

It is clear that we can get to this result in the initial figure if we make the point *D* move away to infinity on *AC*, so that the side *BD* becomes parallel to *AC*. In the limit $A\hat{D}B$ collapses and $A\hat{B}D$ becomes the supplement of $D\hat{A}B$ (as before by *Elements* I.29). This is very probably the approach that al-Sijzī took.

3.3. *Analysis and variation of two methods of solution of a single problem*

This time al-Sijzī gives an example in which the problem is kept fixed and the methods of solution are allowed to vary. Thus, there are many roads to discovery, and they are not equivalent: not only are some easier than others, but there are also some that are more elegant. To illustrate this procedure, al-Sijzī considers the division of a triangle into three parts in given ratios.

---

[18] A. M. Legendre, 'Réflexions sur les différentes manières de démontrer la théorie des parallèles ou le théorème sur la somme des trois angles du triangle', *Mémoires de l'Académie des sciences*, 12, 1833, pp. 367–410.

The problem here is to divide a given triangle ABC into three triangles ABH, ACH and BCH whose areas are in given ratios:

$$\frac{ABH}{ACH} = \frac{D}{E}, \qquad \frac{ACH}{BCH} = \frac{E}{G}.$$

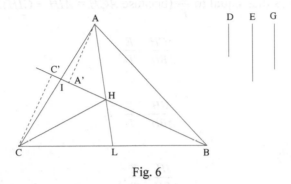

Fig. 6

We first note that the ratio of the areas of ABH and BCH is equal to the ratio $\frac{AI}{CI}$ [19] where I is the point of intersection of AC and the extension of BH. So we construct the point I on AC such that $\frac{AI}{CI} = \frac{D}{G}$. We know that H must lie on BI. It remains for us to choose H on BI in such a way that the ratio of the areas of ACH and ABH is equal to $\frac{E}{D}$. Now this ratio is equal to the ratio $\frac{CL}{BL}$ where L is the point of intersection of BC and the extension of AH. So we construct L on BC such that $\frac{BL}{CL} = \frac{D}{E}$, where H is the point of intersection of BI and AL. We are carrying out an analysis of the problem; the synthesis is self evident and the text passes over it in silence.

Al-Sijzī proposes another method: we construct the ratio $\frac{IH}{BH}$ in which the required point, H, divides BI; this ratio is equal to the ratio of the areas of triangles AIH and AHB. Now

$$\frac{AIH}{CIH} = \frac{AI}{CI} = \frac{D}{G}.$$

[19] The triangles ABH and CBH have a common base BH, so their areas are in the ratio of the heights AA' and CC' from A and C. Now $\frac{AA'}{CC'} = \frac{AI}{IC}$ in the homothety with centre I. So $\frac{ABH}{CBH} = \frac{AI}{IC} = \frac{D}{G}$.

Let us divide the magnitude E into two parts X and Y such that the ratio

$$\frac{X}{Y} = \frac{D}{G};$$

the ratio $\dfrac{AIH}{ACH}$ is thus equal to $\dfrac{X}{E}$ (because $ACH = AIH + CIH$). Since

$$\frac{ACH}{ABH} = \frac{E}{D},$$

we see that

$$\frac{AIH}{ABH} = \frac{X}{D}.$$

Accordingly

$$\frac{IH}{BH} = \frac{X}{D}.$$

Which is what it was required to prove. Here again, the text presents only the analysis; the same holds for the third method, which follows.

The third method is more elegant because it uses the theory of proportions only for dividing one of the sides of the triangle, let it be AB, in the ratio of the three segments D, E, G:

$$\frac{AI}{IH} = \frac{D}{E} \text{ and } \frac{IH}{HB} = \frac{E}{G}.$$

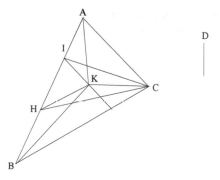

Fig. 7

The three triangles *ACI*, *ICH* and *HBC* have areas in the required ratio.

Through *I* we draw *IK* parallel to *AC* and through *H* we draw *HK* parallel to *BC*; then the area of the triangle *AKC* is equal to the area of the triangle *AIC*, and in the same way the area of the triangle *BKC* is equal to the area of the triangle *BHC*. The areas of the triangles *AKC*, *AKB* and *BKC* are now in the ratios of the segments *D*, *E*, *G*.

Al-Sijzī thus provides three methods – and he is at pains to point out that they are three among many – for constructing an object that has a given property.

3.4. *Analysis and variation of lemmas*

In the first part of the treatise, al-Sijzī recommended, as a useful rule in analysis and synthesis, that one should look back to the lemmas of the lemma that allowed the proposition to be proved. This rule is in turn based on the idea that the lemmas can be varied, at the least by working back up the chain of lemmas that are necessary for establishing the proposition. Here we are obviously concerned with an *indirect* approach, that is intended to increase the number of routes for making discoveries. Here al-Sijzī gives an example that illustrates this rule, Proposition III.20 of the *Elements*:

> In a circle the angle at the centre is double of the angle at the circumference, when the angles have the same circumference as base.

Euclid had proved this proposition from a lemma that required two other preliminary lemmas, that is I.32 (the exterior angle of a triangle), itself based on I.29 and I.31. Al-Sijzī proves the proposition directly from these last lemmas (see text).

3.5. *Analysis and variation of constructions carried out using the same figure*

In the first part of the treatise, al-Sijzī recommends that, as a preliminary approach, one should try to recognise the element which is common to the propositions that will be used in carrying out the analysis and synthesis. He also advises us to grasp what it is that distinguishes these propositions one from another. An investigation that is the more necessary because these 'contribute to one another' (p. 607). Here al-Sijzī illustrates this phenomenon by means of an example, from which he develops a supplementary procedure to make the analysis more powerful: We use the same figure to arrive at different constructions. That is, we fix the figure and we vary the constructions carried out with it. Let us return to al-Sijzī's approach, whose text has been heavily corrupted by the copyist of the manuscript.

Al-Sijzī starts with several propositions which he does not prove on division in extreme and mean ratio, which 'contribute to one another'. Al-Sijzī notes that they have in common the number five, it being given that 'the construction of the regular pentagon in fact involves the division of a straight line in mean and extreme ratio'.[20] Let us return to al-Sijzī's exposition.

Proposition 1: We know from Euclid's *Elements* that the construction of the regular pentagon starts with making a division in extreme and mean ratio. Thus, Book XIII of the *Elements* opens with some propositions on this division, on which al-Sijzī comments[21] in Euclidean style.

Using trigonometry, we find the side of the regular decagon inscribed in a circle of radius r:

$$c = 2r \, \sin\frac{\pi}{10} = r\frac{\sqrt{5}-1}{2}.$$

For the side of the regular pentagon we have

$$C = 2r \, \sin\frac{\pi}{5} = c\sqrt{4 - \frac{c^2}{r^2}} = r\frac{\sqrt{5}-1}{2}\sqrt{\frac{5+\sqrt{5}}{2}}.$$

Proposition 2: The sum $r+c = r\dfrac{\sqrt{5}+1}{2}$ is divided in extreme and mean ratio by r and c: $\dfrac{r+c}{r} = \dfrac{r}{c} = \dfrac{\sqrt{5}+1}{2}$.

Proposition 3: The ratio of D, the diagonal of the regular pentagon, to its side C is again equal to $\dfrac{\sqrt{5}+1}{2}$. In other words, the diagonal PN is divided in extreme and mean ratio at the point Q such that $PQ = PM = C$.

---

[20] See p. 607.

[21] See apparatus criticus in *Les Mathématiques infinitésimales du IXe au XIe siècle*, vol. IV, on p. 799, 16.

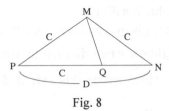

Fig. 8

The isosceles triangles *PMN* and *MQN* are similar, so

$$\frac{MN}{PN} = \frac{QN}{PM},$$

or

$$\frac{C}{D} = \frac{D-C}{C}.$$

Proposition 4: If a straight line of length $2a$ is divided in extreme and mean ratio, the greater part is $a(\sqrt{5}-1)$. So if we add to it a, half the whole straight line, the sum is $a\sqrt{5}$ and its square is five times the square of half the whole straight line.

Proposition 5: If we divide a straight line into two parts in this ratio, and if to the greater part we add twice the smaller part, then the square of the whole straight line is five times the square of the first part.[22]
Let a be the greater part; the smaller part accordingly has the value $a\dfrac{\sqrt{5}-1}{2}$. Twice this added to a gives $a\sqrt{5}$, whose square is five times the square of a.

Proposition 6: Let us continue to consider a straight line divided in extreme and mean ratio, whose greater part is $2a$; and the smaller part is then $a(\sqrt{5}-1)$. If to this part we add a, half the greater part, we obtain $a\sqrt{5}$, whose square is five times that of half the greater part.

Proposition 7: Propositions 4 to 6 show how, starting from a division in extreme and mean ratio, we can construct two segments such that the square of one is five times the square of the other. Now, we carry out the

[22] See p. 607.

inverse procedure, starting from a square divided into five equal squares and proceeding so as to obtain a division in extreme and mean ratio.

For example, if we are given a and $a\sqrt{5}$, the sides of two squares, their sum $a\left(1+\sqrt{5}\right)$ ('composition'), when divided by two, is in the ratio $\frac{1+\sqrt{5}}{2}$ with the first side a; in other words this side a divides half the sum in extreme and mean ratio.

In the same way, the difference $a\left(\sqrt{5}-1\right)$ ('separation'), when divided by two, is in the ratio $\frac{\sqrt{5}-1}{2}$ with the first side a; in other words, half the difference divides this side in extreme and mean ratio.

Here we propose a construction of the division in extreme and mean ratio that has several stages ('propositions that participate in one another'). In the first stage, we construct two segments such that the square of one is equal to three times the square of the other; in the second stage, we use this construction to obtain the division in extreme and mean ratio. The two constructions are done with the same figure.

Example: In this example, we consider a triangle AEB, right-angled at E, and we carry over the smallest side EB to be EG on the larger side EA.

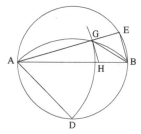

Fig. 9: $2AD^2 = AB^2$; $3AE^2 = (AG + AE)^2$

We have
$$AB^2 = AE^2 + EB^2 \text{ (Pythagoras' theorem)}$$
$$= AG^2 + 2AG \cdot GE + EG^2 + EB^2 = AG^2 + 2EG \cdot AE.$$

As the first stage, we take AG such that $2AG^2 = AB^2$; so we have

$$2\,AG^2 = AB^2 = AG^2 + 2EG \cdot AE,$$

hence

$$AG^2 = 2EG \cdot AE$$

and

$$(AG + AE)^2 = AG^2 + 2AG \cdot AE + AE^2 = 3AE^2,$$

as we wished for the first part.

The construction of G is carried out stating from AB as follows: we find AD such that $2AD^2 = AB^2$ by dividing the semicircle ADB of diameter AB into two equal arcs. We must have $AG = AD$ and angle $AGB = \dfrac{\pi}{2} + \dfrac{\pi}{4}$ (exterior angle of the right-angled isosceles triangle BEG); accordingly G lies at the intersection of the circle with centre A and radius AD and the arc subtending the angle $AGB = \dfrac{3\pi}{4}$.

We then extend AG to E so as to make angle AEB a right angle; so this point is the intersection of the extension of AG and the circle of diameter AB.

In the second stage, we use the same construction, but this time we suppose that $3AG^2 = AB^2 = AG^2 + 2EG \cdot AE$. We then have $AG^2 = EG \cdot EA$ and the straight line AE is divided in extreme and mean ratio at the point G.

So the method consists of constructing AD such that $3AD^2 = AB^2$ by using the first part, then constructing G as the point of intersection of the circle with centre A and radius AD and the arc subtending the angle $AGB = \dfrac{3\pi}{4}$. We then complete the procedure as in the first stage: we extend AG to E, its point of intersection with the circle of diameter AB. We then have

$$5AE^2 = (AE + 2AG)^2.$$

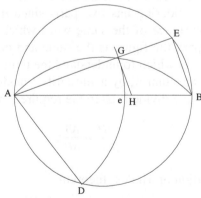

Fig. 10: $3AD^2 = AB^2$; $5AE^2 = (2AG + AE)^2$
($Ae = AE$, $eB = AG$ in Fig. 9)

Note: This construction provides a general method of constructing two segments such that the ratio of their squares is of the form $2^n + 1$. For $n = 2$, this ratio is 5 and the construction is equivalent to division in extreme and mean ratio.

Let us suppose that $(2^{n-1} + 1)AD^2 = AB^2$. If $AG = AD$, as in the figure, we have:

$$(2^{n-1} + 1)\, AG^2 = AB^2 = AG^2 + 2EG \cdot AE,$$

so

$$2^{n-1}\, AG^2 = 2EG \cdot AE.$$

Thus,

$$\begin{aligned}
(2^{n-1}\, AG + AE)^2 &= 2^{2n-2}AG^2 + 2^n AG \cdot AE + AE^2 \\
&= 2^n EG \cdot AE + 2^n AG \cdot AE + AE^2 \\
&= (2^n + 1)\, AE^2.
\end{aligned}$$

So, to obtain a ratio of squares equal to $2^n + 1$, we repeat the construction n times. Each figure is obtained from the previous one:

$$AB_n = 2^{n-1}AG_{n-1} + AE_{n-1} \text{ and } AD_n = AE_{n-1}$$

give

$$AB_n^2 = \left(2^n + 1\right)AE_n^2.$$

We are always dealing with the same figure, with a different choice for the point D.

That is exactly the idea to which this text is trying to draw attention. The case we are considering has two parts since $n = 2$; that explains the jump to a later occurrence of the same word which has caused lacunae in the text. For each part, the figure is the same and the argument is identical and this is the sense in which to understand the term 'participation'.

Al-Sijzī then states that if, by a similarity-transformation, we draw GH parallel to EB, we have divided AB in the required manner, that is in such a way that

$$\frac{AH}{HB} = \frac{AB}{AH};$$

which is an application of Thales' theorem.

3.6. *Variations on a problem from Ptolemy*

Al-Sijzī returns to a problem in the *Almagest*. His investigation is both careful and detailed; he emphasises the difficulties he encountered in carrying out the analysis, providing four proofs, and thus showing that many different methods can be used to solve the problem. In the course of his discussion al-Sijzī demonstrates the part played by auxiliary constructions as well as showing, in each case, the purpose served by their use; but he does not mention, or even hint at, the questions of logic that are raised by there being multiple proofs for the same problem. His principal concern is always the same: to discover the best method of obtaining the required property. This is indeed his motive for providing multiple solutions.

Ptolemy's proposition[23] shows that if in a given circle we take two unequal arcs AC and AB such that arc AC > arc AB, then

$$\frac{\overset{\frown}{AC}}{\overset{\frown}{AB}} > \frac{AC}{AB}.$$

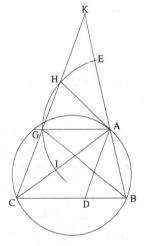

Fig. 11

We take K on BA produced such that $AK = AC$ and take G as the point of intersection of KC and a line through A parallel to BC.

Since the arc AC is cut off by the angle $A\hat{B}C = K\hat{A}G$ and the arc AB is cut off by the angle $A\hat{C}B = G\hat{A}C$, we have

$$\frac{\overset{\frown}{AC}}{\overset{\frown}{AB}} = \frac{K\hat{A}G}{G\hat{A}C}.$$

So we want to show that

$$\frac{G\hat{A}C}{K\hat{A}G} < \frac{AB}{AC} = \frac{AB}{AK} = \frac{GC}{GK} = \frac{\text{tr.}(ABG)}{\text{tr.}(AGK)}.$$

[23] Ibn al-Haytham takes up this proposition and varies the conditions; see R. Rashed, *Les Mathématiques infinitésimales du IXe au XIe siècle*, vol. V: *Ibn al-Haytham: Astronomie, géométrie sphérique et trigonométrie*, London, 2006; English translation by J. V. Field: *Ibn al-Haytham. New Spherical Geometry and Astronomy*, A History of Arabic Sciences and Mathematics, vol. 4, Culture and Civilization in the Middle East, London/New York, 2014. See also *Géométrie et dioptrique*, pp. 248 *sqq.*

Now the first ratio is equal to the ratio of the sectors *GAI* and *GAE*, and we have

$$\text{sect.}(GAI) = \text{sect.}(HAE)^{24} \text{ and sect.}(GAE) = \text{sect.}(AEH) + \text{sect.}(HGA).$$

Moreover,

$$\text{tr.}(AGC) = \text{tr.}(KAH) \text{ and tr.}(KAG) = \text{tr.}(KAH) + \text{tr.}(HAG);$$

since

$$\text{tr.}(HAG) < \text{sect.}(HAG) \text{ and tr.}(KAH) > \text{sect.}(HAE),$$

we thus have

$$\frac{\text{sect.}(GAI)}{\text{sect.}(GAE)} < \frac{\text{tr.}(AGC)}{\text{tr.}(AGK)},$$

hence

$$\frac{G\hat{A}C}{K\hat{A}G} < \frac{\text{tr.}(ABG)}{\text{tr.}(AGK)},$$

which is what it was required to prove.

Note 1: We have presented the argument using Fig. 11, where the angles satisfy $\frac{\pi}{2} > \hat{B} > \hat{C}$.

Let us suppose $A\hat{B}C$ is obtuse, with $\pi > \hat{B} > \frac{\pi}{2} > \hat{C}$ (Fig. 12). The symmetry we pointed out in the first case still exists and the reasoning is identical, because all the equations are satisfied.

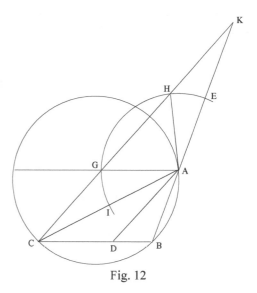

Fig. 12

---

[24] Triangle *KAC* is isosceles, so the perpendicular bisector of *KC* passes through *A*; it is the axis of symmetry of the triangle *KAC* and of the circle (*A*, *AG*); so *KH* = *GC* and arc *HE* = arc *GI* and parts *KHE* and *CGI* (curvilinear triangles) are equal.

Note 2: To express it differently, let us make $\beta = A\hat{B}C$, $\gamma = A\hat{C}B$ and r the radius of the circle *ABC*. The arc *AC* is equal to 2 $r\beta$ and the arc *AB* is equal to 2 $r\gamma$, as for the chords, we have: $AC = 2r \sin \beta$, $AB = 2r \sin \gamma$. Ptolemy's inequality is written as

$$\frac{\beta}{\gamma} > \frac{\sin \beta}{\sin \gamma},$$

that is

$$\frac{\sin \beta}{\beta} < \frac{\sin \gamma}{\gamma} \qquad \text{if } \beta > \gamma.$$

Thus, this inequality indicates that the function $x \to \dfrac{\sin x}{x}$ decreases over the interval $0 < x < \pi$.

Note 3: In another text by al-Sijzī (see below) we find a solution of this problem using a slightly different auxiliary construction, but employing the same reasoning.

Al-Sijzī continues his discussion of this same problem from Ptolemy by proposing a second method of proving the same inequality. For this, he introduces the line *CD*, the bisector of the angle *ACB*,[25] which gives him $\dfrac{BC}{AC} = \dfrac{BD}{DA}$. If the extension of *CD* meets the circle circumscribed about the triangle *ABC* in *H*, we consider the circle with centre *H* that passes through *D*; it meets *HA* in *E* and *HB* in *I*. Since $HA > HD$, the point *E* lies between *H* and *A*. If *I* lay outside *HB*, the proof would be immediate because

$$\frac{BD}{DA} = \frac{\text{tr.}(BDH)}{\text{tr.}(DAH)}, \text{tr.}(BDH) < \text{sect.}(IHD) \text{ and tr.}(DAH) > \text{sect.}(DEH);$$

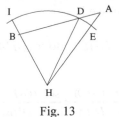

Fig. 13

---

[25] Note that the names of the points *A* and *C* are reversed in this new proof.

So we have

$$\frac{\text{tr.}(BDH)}{\text{tr.}(DAH)} < \frac{\text{sect.}(IHD)}{\text{sect.}(DHE)} = \frac{B\hat{A}C}{A\hat{B}C} = \frac{\overset{\frown}{BC}}{\overset{\frown}{AC}}$$

thus,

$$\frac{BC}{AC} = \frac{\overset{\frown}{BC}}{\overset{\frown}{AC}}.$$

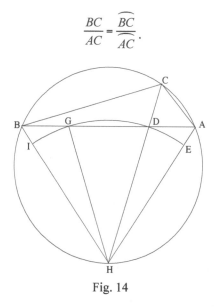

Fig. 14

But Figure 13 is impossible because $HB = HA$ and the point I thus lies between H and B, as the point E lies between H and A. Al-Sijzī then notes that the preceding reasoning is false. But understanding the nature of the error allows us to move forward along the road of discovery and to correct the reasoning. That is probably the point that al-Sijzī wanted to illustrate by this example. This is how he proceeds:

Let G be the point where the circle with centre H and radius HD cuts the side AB. The triangles HDA and HGB are equal and the sectors DHE and GHI are also equal. Now we have

$$\text{tr.}(GHD) < \text{sect.}(GHD) \text{ and tr.}(DHA) > \text{sect.}(DHE),$$

so

$$\frac{\text{tr.}(GHD)}{\text{tr.}(DHA)} < \frac{\text{sect.}(GHD)}{\text{sect.}(DHE)};$$

if we add unity to the two sides, that is by composition of the ratios, we obtain

$$\frac{\text{tr}.(BHD)}{\text{tr}.(DHA)} < \frac{\text{sect}.(IHD)}{\text{sect}.(DHE)},$$

and the proof is completed as above.

We may note that, in another text, al-Sijzī gives the same solution, that is the same apart from some small variants, in the course of the proof. He in fact starts from the same figure with the same lettering and the same assumptions (see below).

Al-Sijzī continues to give more variations on Ptolemy's problem. This time he wants to take account of a special condition that is sufficient as Ptolemy proposes it, namely that $\overparen{AB} < \pi$.

We introduce a point D on the circular arc ACB such that $\overparen{BD} = \overparen{CA}$, and we call the point of intersection of AD with BC the point E. The circle with centre A and radius AE meets the extension of AC in G and the side AB in H.

We have

sect.$(AGE) >$ tr.(ACE) and sect.$(AEH) <$ tr.(AEB),

so

$$\frac{\text{sect}.(AGE)}{\text{sect}.(AEH)} > \frac{\text{tr}.(ACE)}{\text{tr}.(AEB)}.$$

By composition of the ratios, we have

$$\frac{\text{sect}.(AGH)}{\text{sect}.(AEH)} > \frac{\text{tr}.(ACB)}{\text{tr}.(AEB)};$$

the ratio on the left is equal to

$$\frac{\overparen{BC}}{\overparen{BD}} = \frac{\overparen{BC}}{\overparen{AC}}.$$

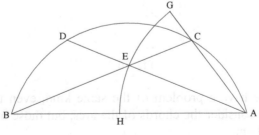

Fig. 15

Further

$$\frac{\text{tr.}(ACB)}{\text{tr.}(AEB)} = \frac{BC}{BE},$$

so

$$\frac{\overparen{BC}}{\overparen{AC}} > \frac{BC}{BE}.$$

Al-Sijzī concludes by saying that $BE = AE > AC$; unfortunately that gives

$$\frac{BC}{BE} < \frac{BC}{AC},$$

and one cannot draw any conclusion as he has inadvertently done.

He also gives a fourth solution to this problem (see below).

Al-Sijzī continues his variations on Ptolemy's problem. He still starts with three points A, B, C on a given circle, with arc $AC >$ arc CB; he draws $CD \perp AB$ and wishes to show that

$$\frac{AD}{DB} > \frac{\overparen{AC}}{\overparen{CB}}.$$

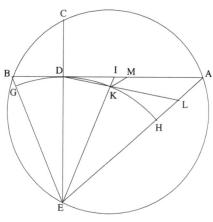

Fig. 16

This time we have a problem of the same kind, even though al-Sijzī does not directly consider the chords of the arcs, but those of the segments associated with them.

We extend CD to the point E on the circle; we take I on BA such that $EI = EB$. The circle (E, ED) cuts the straight lines EB, EI and EA in G, K and H respectively.

Al-Sijzī then says that

$$(1) \qquad \frac{\text{tr.}(ADE)}{\text{tr.}(DBE)} > \frac{\overarc{HKD}}{\overarc{DG}}.$$

He does not prove this inequality, but deduces from it that

$$\frac{AD}{DB} > \frac{\overarc{HKD}}{\overarc{DG}} = \frac{A\hat{E}C}{C\hat{E}B} = \frac{\overarc{AC}}{\overarc{BC}},$$

hence

$$\frac{AD}{DB} > \frac{\overarc{AC}}{\overarc{BC}}.$$

This proof is correct, although it is simply indicative, since al-Sijzī has not proved (1). We can reconstruct the missing proof as follows:

Let us draw DK; it cuts EA in L; through K let us draw a line parallel to EA, which cuts DA in M between I and A. We have

$$\frac{KL}{DK} = \frac{MA}{DM} < \frac{IA}{DI},$$

so

$$\frac{\text{tr.}(KEL)}{\text{tr.}(DEK)} < \frac{\text{tr.}(IEA)}{\text{tr.}(DEI)}.$$

Now

$$\text{tr.}(KEL) > \text{sect.}(KEH) \quad \text{and} \quad \text{tr.}(DEK) < \text{sect.}(DEK);$$

so

$$\frac{\text{sect.}(KEH)}{\text{sect.}(DEK)} < \frac{\text{tr.}(IEA)}{\text{tr.}(DEI)}.$$

By composition of the ratios, we find

$$\frac{\text{sect.}(DEH)}{\text{sect.}(DEK)} < \frac{\text{tr.}(DEA)}{\text{tr.}(DEI)} = \frac{\text{tr.}(ADE)}{\text{tr.}(DBE)};$$

this is what we wanted to prove.

This proof is not only in al-Sijzī's style but is faithful to the way he proceeds. To employ another style, one he did not know, we may let $B\hat{E}D = \alpha$, $D\hat{E}A = \beta$ and $r = ED$. The segments BD and DA are measured by $r \tan \alpha$ and $r \tan \beta$ respectively; so their ratio is $\dfrac{\tan \beta}{\tan \alpha}$. The arcs AC and CB, intercepted by the inscribed angles CEA and CEB, are in the ratio $\dfrac{\beta}{\alpha}$. The inequality that has been proved means that $\dfrac{\tan \beta}{\tan \alpha} > \dfrac{\beta}{\alpha}$ if $\beta > \alpha$. In other words, the function $x \to \dfrac{\tan x}{x}$ increases, for $0 \leq x < \dfrac{\pi}{2}$.

Making this translation might in fact shed light, at least indirectly, on the variations al-Sijzī imposes on Ptolemy's problem.

Al-Sijzī continues his variations on this problem. This time, instead of considering the ratio of the chords of the given arcs, he looks at the ratio of twice the chords of the given arcs. The discussion is, however, not complete. So let us return to the problem.

Let us put $\overset{\frown}{AB} = 2\beta$ and $\overset{\frown}{AC} = 2\gamma$ with $\gamma > \beta$. If AE, the diameter through A, cuts CB in D, we have

$$A\hat{C}B = \beta \text{ and } C\hat{A}D = \frac{\pi}{2} - \gamma,$$

hence

$$A\hat{D}C = \frac{\pi}{2} + \gamma - \beta,$$

an obtuse angle; the same is true for the angle EDB and in consequence we have $EB > ED$.

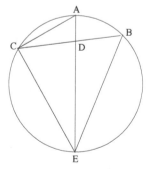

Fig. 17

We draw the circle (E, ED); it cuts EB in H and EC in G.[26] The position of G in relation to the straight line CD depends on the lengths of EC and ED. We have the following cases:

1. G lies beyond C, so it is above CD if $ED > EC$ (Fig. 18).
2. G is at C if $ED = EC$ (Fig. 19).
3. G is below C, so it is on EC, if $ED < EC$ (Fig. 20).

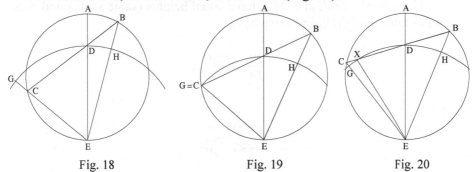

Fig. 18   Fig. 19   Fig. 20

In the triangle EDC, the greatest angle is opposite the greatest side:

$$E\hat{C}D = \frac{\pi}{2} - \beta \text{ and } E\hat{D}C = \frac{\pi}{2} - \gamma + \beta.$$

So we have the following conditions:

1. $ED > EC \Leftrightarrow E\hat{C}D > E\hat{D}C \Leftrightarrow \frac{\pi}{2} - \beta > \frac{\pi}{2} - \gamma + \beta \Leftrightarrow 2\beta < \gamma,$

which is possible because $\gamma > \beta$.

2. $ED - EC \Leftrightarrow \gamma - 2\beta$, which is also possible.

3. $ED < EC \Leftrightarrow \gamma < 2\beta$, which would impose the condition $\beta < \gamma < 2\beta$; a condition that is necessary for G to be below C, and a condition which is not stated by al-Sijzī. His argument applies in cases 1 and 2. Thus, we have

---

[26] If the circle (E, ED) touches the straight line BC, we have $CD \perp AE$, so $A\hat{D}C = 1$ right angle and $\overset{\frown}{AC} = \overset{\frown}{AB}$; which is impossible because $\overset{\frown}{AC} > \overset{\frown}{AB}$.

The point G can lie below the straight line CD without the angle ADC being acute (see case 3); but that requires

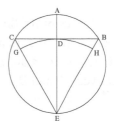

$$\overset{\frown}{AB} < \overset{\frown}{AC} < 2\overset{\frown}{AB}.$$

The condition $\overset{\frown}{AC} < 2\overset{\frown}{AB}$ was not considered by al-Sijzī, and the proof that is given does not apply in this case.

$$\text{sect.}(HDE) < \text{tr.}(DBE) \text{ and sect.}(DGE) > \text{tr.}(DCE),$$

so

$$\frac{\text{tr.}(DBE)}{\text{tr.}(DCE)} > \frac{\text{sect.}(HDE)}{\text{sect.}(DGE)}.$$

The triangles DBE and DCE have equal heights (same vertex and bases on the same straight line), so we have

$$\frac{BD}{DC} > \frac{H\hat{E}D}{D\hat{E}G} \; ;$$

but

$$\frac{H\hat{E}D}{D\hat{E}G} = \frac{\overset{\frown}{AB}}{\overset{\frown}{AC}},$$

hence

$$\frac{BD}{DC} > \frac{\overset{\frown}{AB}}{\overset{\frown}{AC}};$$

which is what al-Sijzī wished to prove.

However, the argument does not apply in case 3, as we can easily check. Let us proceed in the same style as al-Sijzī does:

The triangle EDB is always greater than the sector EDH, but, in this case of the figure (Fig. 20), we do not know whether the triangle ECD is smaller or greater than the sector EGD. We know only that the triangle EXD is smaller than the sector EXD. But the triangle EXC is greater than the sector EXG; which does not allow us to draw any conclusions. We can, however, establish that the ratio $\dfrac{\text{tr.}(EDB)}{\text{sect.}(EDH)}$ is greater than the ratio $\dfrac{\text{tr.}(EXC)}{\text{sect.}(EXG)}$.

Let us prove this inequality: let us construct a point C_1 on DB such that $DC_1 = XC$; the triangle EXC is equal to the triangle EDC_1. As $EC < EB$, since arc $EC <$ arc EB, we have $EC_1 = EC < EB$, so $DC_1 < DB$ and C_1 lies between D and B.

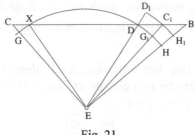

Fig. 21

We draw the circle with centre E and radius $EC_1 = EC$; it cuts EB in H_1 and ED in D_1. We have

$$\text{tr.}(EDC_1) < \text{sect.}(ED_1C_1) \text{ and tr. } (EC_1B) > \text{sect.}(EC_1H_1),$$

so

$$\frac{\text{tr.}(EC_1B)}{\text{tr.}(EDC_1)} > \frac{\text{scct.}(EC_1H_1)}{\text{sect.}(ED_1C_1)};$$

by composition of the ratios, we have

$$\frac{\text{tr.}(EDB)}{\text{tr.}(EDC_1)} > \frac{\text{sect.}(ED_1H_1)}{\text{sect.}(ED_1C_1)} = \frac{\text{sect.}(EDH)}{\text{sect.}(EDG_1)},$$

where G_1 is the point of intersection of EC_1 with the circle $GXDH$.
So we have

$$\frac{\text{tr.}(EDB)}{\text{tr.}(EXC)} > \frac{\text{sect.}(EDH)}{\text{sect.}(EXG)};$$

which is what we wished to establish.
Thus,

$$\frac{\text{tr.}(ECD)}{\text{tr.}(EBD)} = \frac{\text{tr.}(ECX)}{\text{tr.}(EBD)} + \frac{\text{tr.}(EXD)}{\text{tr.}(EBD)} < \frac{\text{sect.}(EXG)}{\text{sect.}(EDH)} + \frac{\text{sect.}(EXD)}{\text{sect.}(EDH)} = \frac{\text{sect.}(EGD)}{\text{sect.}(EDH)}.$$

So finally we have

$$\frac{\text{tr.}(EBD)}{\text{tr.}(ECD)} > \frac{\text{sect.}(EDH)}{\text{sect.}(EGD)},$$

the inequality we wished to prove.

This inequality is, moreover, equivalent to a decrease in the function $x \to \dfrac{\sin x}{x}$, as has been explained in the note on page 545.

Al-Sijzī presents one last variation on Ptolemy's problem, which is stated like this: Let there be two chords AC and BD in a given circle $ADBC$ cutting one another in E; we have

$$\frac{DE}{EB} < \frac{\overset{\frown}{AD}}{\overset{\frown}{CB}}.$$

The ideas in the proof are essentially the same as before. We begin with an auxiliary construction, of a circle (B, BA) that cuts the extension of the chord BD in H; the line through B parallel to AC cuts this circle in I and the extension of DA in G. We have

$$\text{tr.}(ADB) < \text{sect.}(ABH) \text{ and tr.}(AGB) > \text{sect.}(ABI);$$

hence

$$\frac{\text{tr.}(ADB)}{\text{tr.}(AGB)} < \frac{\text{sect.}(ABH)}{\text{sect.}(ABI)},$$

hence

$$\frac{AD}{AG} < \frac{D\hat{B}A}{A\hat{B}G}.$$

But $A\hat{B}G = C\hat{A}B$; $D\hat{B}A$ intercepts the arc AD and $C\hat{A}B$ intercepts the arc BC, so

$$\frac{AD}{AG} < \frac{\overset{\frown}{AD}}{\overset{\frown}{BC}};$$

but

$$\frac{AD}{AG} = \frac{ED}{EB}$$

because $EA \parallel BG$; hence the conclusion.[27]

---

[27] We may note that the figure drawn in the manuscript, which is not very clear, seems to show a semicircle ACB. The argument is the same for a circle. We may have $ACB \le \pi$ or $ACB > \pi$.

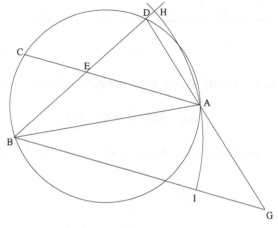

Fig. 22

The preceding argument assumes that $BA > BD$; this is always true if the arc $ADCB \leq \pi$, the points A, D, C, B being in that order. We then have $AB > BD$ and $AB < BG$; we need to have H lying beyond D and I lying between B and G and the argument then applies (see Figs 22.1 and 22.2).

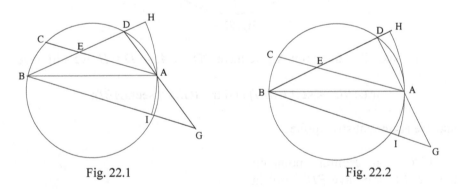

Fig. 22.1   Fig. 22.2

But for the third case, the arc $ADCB > \pi$, we can have: $AB > BD$, $AB = BD$ or $AB < BD$.

In fact, let BB_1 be the diameter $(2r)$ and A_1 such that the arc AB_1 is equal to the arc B_1A_1. If a point D describes the arc AB_1 from A towards B_1, the length BD increases from BA to $BB_1 = 2r$. If D describes the arc B_1A_1, the length BD decreases from $2r$ to $BA_1 = BA$. If D describes the arc A_1B, the length BD decreases from $BA_1 = BA$ to 0; hence

D	A	B_1	A_1	B
BD	AB $\nearrow$	2r $\searrow$	AB $\searrow$	0

So if we take D beyond A_1, we shall have, as in the first two cases,

$$BA > BD \Rightarrow BH > BD,$$

where H lies beyond D and I lies between G and B; and the argument again applies.

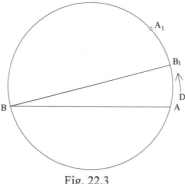

Fig. 22.3

If we take D at the point A_1, we have $BD = BA$, so $D = H$; we still have

$$\text{tr.}(ABD) < \text{sect.}(ABH) \text{ and tr.}(AGB) > \text{sect.}(ABI);$$

and the argument still applies.

If D is an arbitrary point on the arc AA_1, we have $BD > BA$; in this case, H lies between B and D. But the position of the point I depends on the positions of D and C. Thus, if D lies between B_1 and A_1, the angle DAB is acute and the circle (B, BA) cuts AD between A and D (Fig. 22.4) and cuts BG in I between B and G.

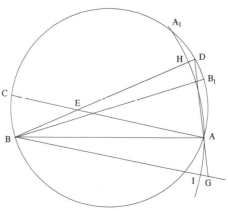

Fig. 22.4

The segment AD cuts the arc AH and we cannot make a comparison between the area of the triangle ABD and that of the sector ABH. The proposed method can no longer be applied.

If D is at the point B_1, the angle DAB is a right angle, the circle (B, BA) touches AD, the point G lies on AD and the point I again lies between B and G; the argument does not apply either, we have

$$\text{tr.}(ABD) > \text{sect.}(ABH) \text{ and tr.}(ABG) > \text{sect.}(ABI),$$

and we thus cannot draw any conclusions (Fig. 22.5).

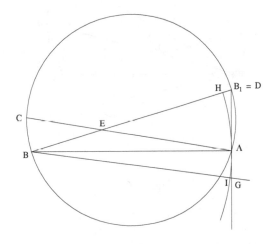

Fig. 22.5

But if D lies between B_1 and A, the angle DAB is obtuse and the straight line AD cuts the circle (B, BA) again beyond A; in this case, depending on the position of D and that of the point C on the arc BC_0 (arc $BC_0 = $ arc AD and arc $BC < $ arc AD), we can have the point G between B and I, $G = I$, or I between B and G (Figs 22.6, 22.7 and 22.8 respectively).

In cases 22.6 and 22.7, the point G lies between B and I or $G = I$; we have

$$\text{tr.}(ADB) > \text{sect.}(ABH) \text{ and tr.}(AGB) < \text{sect.}(ABI),$$

so

$$\frac{\mathrm{tr.}(ADB)}{\mathrm{tr.}(AGB)} > \frac{\mathrm{sect.}(ABH)}{\mathrm{sect.}(ABI)}$$

and in consequence

$$\frac{AD}{AG} > \frac{D\hat{B}A}{A\hat{B}G}.$$

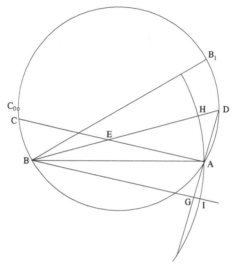

Fig. 22.6

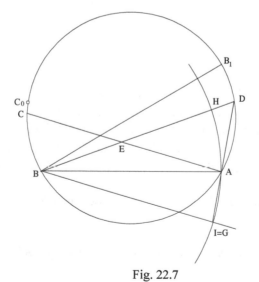

Fig. 22.7

We have $BG \parallel AE$, hence

$$\frac{AD}{AG} = \frac{ED}{EB},$$

and we thus have

$$\frac{ED}{EB} > \frac{\overset{\frown}{AD}}{\overset{\frown}{CB}},$$

which is contrary to the conclusion stated earlier.

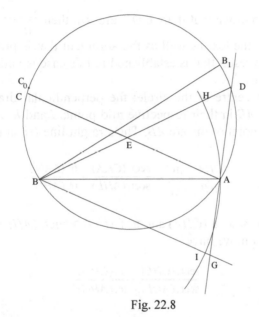

Fig. 22.8

In case 22.8, we cannot compare the area of the triangle ABG with that of the sector ABI.

All this discussion shows that, without stating this in so many words in his text, al-Sijzī seems to assume that he has simply accepted the hypothesis that arc $ADCB \leq \pi$, which indeed corresponds with the figure in the text.

As readers will no doubt have noticed, al-Sijzī's variations on Ptolemy's problem do not relate only to the proofs, but also to deriving different statements and discovering other properties, such as that of the tangent. Moreover, al-Sijzī makes a point, at least at the beginning of the discussion, of raising all the difficulties that the geometer encounters in the course of examining the problem.

3.7. Variations on the same problem from Ptolemy in other writings by al-Sijzī

In the manuscripts of al-Sijzī's writings that have come down to us, we find three additional solutions of this same problem. Two of them differ only very slightly from the two solutions we have given here. The third is simpler, but the argument is based on the same idea. So let us start by looking at this last solution.

1. We wish to prove that if arc CD > arc AB, then $\dfrac{\overparen{CD}}{\overparen{AB}} > \dfrac{CD}{AB}$.

The figure in the text as well as the argument that is presented assume $AB \parallel CD$. But the result that is established in this case is valid whatever the position of the arcs.[28]

Let H be the centre of the circle; the perpendicular drawn from H to CD cuts CD and AB in their respective mid points I and K and cuts the arc AB in L, the mid point of the arc AB. The straight line HL cuts AC in G. We have

$$\frac{C\hat{H}A}{A\hat{H}L} = \frac{\overparen{AC}}{\overparen{AL}} = \frac{\text{sect.}(CHA)}{\text{sect.}(AHL)} > \frac{\text{tr.}(CAH)}{\text{tr.}(AHG)},$$

because tr. (CAH) < sect. (CHA) and tr. (AHG) > sect. (AHL).

By composition, we have

$$\frac{\text{sect.}(CHL)}{\text{sect.}(AHL)} > \frac{\text{tr.}(CHG)}{\text{tr.}(AHG)};$$

now

$$\frac{\text{sect.}(CHL)}{\text{sect.}(AHL)} = \frac{\overparen{CL}}{\overparen{AL}} \quad \text{and} \quad \frac{\text{tr.}(CHG)}{\text{tr.}(AHG)} = \frac{CG}{AG} = \frac{CI}{AK};$$

so we have

---

[28] In fact, let $A'B'$ be an arc of the same circle such that the arc $A'B'$ is equal to the arc AB, $A'B'$ is not parallel to CD; then we have $A'B' = AB$. So

$$\frac{\overparen{CD}}{\overparen{AB}} = \frac{\overparen{CD}}{\overparen{A'B'}} \quad \text{and} \quad \frac{CD}{AB} = \frac{CD}{A'B'},$$

and consequently

$$\frac{\overparen{CD}}{\overparen{A'B'}} > \frac{CD}{A'B'}.$$

$$\frac{\widehat{CL}}{\widehat{AL}} > \frac{CI}{AK};$$

hence

$$\frac{\widehat{CD}}{\widehat{AB}} > \frac{CD}{AB}.$$

This solution is simpler than those we examined before. The diameter *EL* is an axis of symmetry for the given arcs and for their chords and the argument is based on the relations

$$\widehat{LA} = \frac{1}{2}\widehat{AB}, \ \widehat{LC} = \frac{1}{2}\widehat{CD}, KA = \frac{1}{2} AB \text{ and } IC = \frac{1}{2} CD.$$

But, as in all the solutions al-Sijzī proposes, the argument rests on the inequality between the ratio of two sectors and the ratio of the two triangles associated with them and having their bases on the same straight line. In every case, it is from this inequality that the conclusion is deduced. It is only the auxiliary constructions that are different.

The passage that supplies this solution forms part of the text *Reply of Aḥmad ibn Muḥammad ibn 'Abd al-Jalīl al-Sijzī to Geometrical Problems Raised by the People of Khurāsān* (*Jawāb Aḥmad ibn Muḥammad ibn 'Abd al-Jalīl al-Sijzī 'an masā'il handasiyya sa'ala 'anhu Ahl Khurāsān*). See the manuscripts Dublin, Chester Beatty Library 3652, fol. 57, and Istanbul, Süleymaniye Library, Reshit 1191, fol. 118. As we have shown,[29] this last manuscript is a copy of the one in Dublin and from only that one manuscript. For this text, there is an edition comparable to that of the treatise *To Smooth the Paths for Determining Geometrical Propositions*. It is this edition that was translated into English in 1996.[30]

> We wish to show that the ratio of the large arc to the small arc of a circle is greater than the ratio of the chord of the large arc to the chord of the small arc, using a method different from that of Ptolemy in his book, the *Almagest*.

> I have solved this problem by a different method and [using] proofs that can be found in the examples that I have given in the book: *To Smooth the Paths*

---

[29] See pp. 151–2.

[30] *Al-Sijzī's Treatise on Geometrical Problem Solving (Kitāb fī Tashīl al-Subul li-Istikhrāj al-Ashkāl al-Handasiyya)*, translated and annotated by Jan P. Hogendijk, with the Arabic text and a Persian translation by Mohammad Bagheri, Tehran, 1996, Arabic p. 18; English trans. p. 31. Cf. our edition in *Les Mathématiques infinitésimales*, vol. IV, pp. 724–5.

for Determining Geometrical Propositions. Nevertheless, a proof by a method different from those that we have followed in this book is possible; it is as follows.

Let there be two unequal arcs *AB* and *CD*. I say that the ratio of the arc *CD*, the greater [arc], to the arc *AB*, the smaller [arc], is greater than the ratio of the chord *CD* to the chord *AB*.

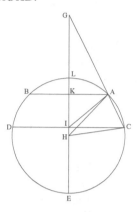

Proof. Let the two chords be parallel. We draw from the centre the line *HIK* perpendicular to the two chords. We extend it on both sides to *E* and *L*. We join *HC* and *HA* and we extend *CA* and *EL* to meet one another in *G*. The ratio of the angle *CHA* to the angle *AHL* is equal to the ratio of the arc *CA* to the arc *AL* and is equal to the ratio of the sector *CAH* to the sector *AHL*. But the ratio of the sector *CAH* to the sector *AHL* is greater than the ratio of the triangle *CAH* to the triangle *AHG*. Thus, by composition, the ratio of the sector *CHL* to the sector *AHL* is greater than the ratio of the triangle *CHG* to the triangle *AHG*. But the ratio of the triangle *CHG* to the triangle *AHG* is equal to the ratio of *CG* to *AG* and to the ratio of *CI* to *AK*. The ratio of the arc *CL* – respectively[31] *CLD* – to the arc *AL* – respectively *ALB* – is thus greater than the ratio of the straight line *CI* – respectively the chord *CD* – to the straight line *AK* – respectively the chord *AB*. Which is what it was required to prove.

2. The two other solutions attributed to al-Sijzī are to be found in a text called *Geometrical Glosses from the Book by Aḥmad ibn Muḥammad ibn ʿAbd al-Jalīl al-Sijzī* (*Taʿlīqāt handasiyya min Kitāb Aḥmad ibn Muḥammad ʿAbd al-Jalīl al-Sijzī*), which has come down to us in two manuscripts, one Dublin, Chester Beatty, no. 3045/14, fols 74ʳ–89ᵛ; and the other Cairo, Dār al-Kutub, no. 699 riyāḍa, 35 pages. The latter is a copy of the former.

---

[31] Lit.: I mean.

The manner in which this book was composed raises an important problem: are we really dealing with a book by al-Sijzī or with a pot pourri made up of parts of his writings, notably from his *Anthology of Problems*? We discuss this question elsewhere,[32] though we regard the matter as settled; here, we give these solutions, which in fact differ from those already given by al-Sijzī in his *To Smooth the Paths* ... only in a slight modification of the auxiliary construction.

Let us start by returning to the first of these two solutions.

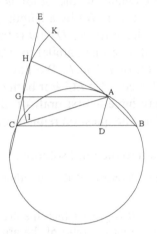

Proof of a proposition from the first book of the *Almagest*, [a proof] that we have established.

The arc *AC* is greater than the arc *AB*; I say that the ratio of the chord of the arc *AC* to the chord of the arc *AB* is smaller than the ratio of the arc *AC* to the arc *AB*.

Proof. We draw *AC*, *AB*, *BC* and we divide the angle *CAB* into two halves with the straight line *AD*.

We draw *CE* parallel to *AD* and we extend *BA* to *E*; we draw *AG* parallel to *BC*, with centre *A* and distance *AG* we draw the arc *IGHK* and we join *AH*. Given that the angle *B* is greater than the angle *ACB*, *AC* will be greater than *AB*. But the triangle *ACE* is isosceles and *AC* is equal to *AE*, so *AE* is greater than *AB*. But the ratio of *CG* to *GE* is equal to the ratio of *BA* to *AE*, so *CG* is smaller than *GE*, and the angle *AGE* is acute.

But the straight line *AH* is equal to the straight line *AG*, so the triangle *AEH* is equal to the triangle *AGC* and the ratio of the triangle *AGC* to the triangle *AGE* is greater than the ratio of the sector *AGI* to the sector *AGK*, because the sector *AGH* is greater than the triangle *AGH* of the segment *HG*. But the ratio of the triangle *AGC* to the triangle *AGE* is equal to the ratio of *GC* to *GE*, and the ratio of the sector *AGI* to the sector *AGK* is equal to the ratio of the angle *CAG* to the angle *GAE*; so the ratio of *CG* to *GE* is greater than the ratio of the angle *CAG* to the angle *GAE*. But the ratio of *GC* to *GE* is equal to the ratio of *BA* to *AC*, so the ratio of *BA* to *AC* is greater than the ratio of the angle *CAG* to the angle *GAE*; but the angle *CAG* is equal to the angle *ACD* and the angle *GAE* is equal to the angle *ABD*, so the ratio of *BA* to *AC* is greater than the ratio of the angle *BCA* to the angle *CBA*. Now the two

[32] R. Rashed and P. Crozet, *Al-Sijzī, Œuvres mathématiques*, forthcoming.

angles *BCA* and *CBA* intercept the arcs *BA* and *AC*, so the ratio of *BA* to *AC* is greater than the ratio of the arc *BA* to the arc *AC*. If we permute, then the ratio of *AC* to *AB* is smaller than the ratio of the arc *AC* to the arc *AB*. This is what we wanted to prove (ms. Dublin, Chester Beatty, no. 3045, fol. 81$^{\text{r-v}}$).

It is clear that this solution is the same as the one given before, the difference being merely that the hypothesis we make here corresponds to the conclusion in the solution given in the treatise; except that the two letters *K* and *E* are exchanged, the proof is the same in both cases. Thus, in the *Treatise*, we extend *BA* by a length *AK* = *AC*; the triangle *KAC* is isosceles and we then have *KC* ∥ *AD* (*AD* is the bisector of the angle *BAC*). Here, we draw through *C* a line parallel to *AD* the bisector of the angle *BAC*; it cuts the extension of *BA* in *E*. We have *CE* ∥ *AD* and we deduce that *EAC* is isosceles; hence *EA* = *EC*. For both constructions, we draw *AG* ∥ *BC*.

We may have a first draft of al-Sijzī's solution, of perhaps a revised version of it. The question remains open.

3. Return to the third solution.

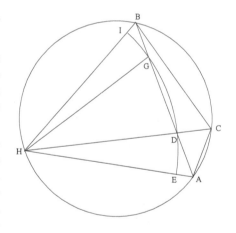

We have determined it in another manner.

The arc *CB* is greater than the arc *CA*. I say that the ratio of the arc *CB* to the arc *CA* is greater than the ratio of the chord *BC* to the chord *CA*.

Proof. We divide the arc *AB* into two halves at the <point> *H*, we draw *AB*, *BH*, *AH* and *CH*, and with centre *H* and with distance *HD* we draw the circle *EDGI* and we draw *HG*. *HG* is thus equal to *DH* (ms. Dublin, Chester Beatty, no. 3045, fol. 82^{r}).

The ratio of the arc *GD* to the arc *DE* is greater than the ratio of the straight line *DG* to the straight line *DA*.[33] By composition, the ratio of the arc *ID* to

---

[33] We have

$$\frac{\overset{\frown}{GD}}{\overset{\frown}{DE}} = \frac{\text{sect.}\,(GHD)}{\text{sect.}\,(DHA)} > \frac{\text{tr.}\,(GHD)}{\text{tr.}\,(DHA)},$$

because

$$\text{sect.}\,(GHD) > \text{tr.}\,(GHD) \text{ and } \text{sect.}\,(DHE) < \text{tr.}\,(DHA).$$

the arc DE is greater than the ratio of the straight line DB to the straight line DA. But the ratio of DB to DA is equal to the ratio of CB to CA, and the ratio of the arc ID to the arc DE is equal to the ratio of the arc CB to the arc CA. This is what we wanted to prove.

In this solution, as in the one in the *Treatise*, we make use of CH, the bisector of the angle ACB defined by the point H the mid point of the arc BA; the straight line CH cuts the chord AB in D. In both solutions – the one in the *Treatise* and the one given here – we make use of the property of the point D, the foot of the bisector of the angle ACB; we have

$$\frac{CA}{CB} = \frac{DA}{DB}.$$

The figure drawn is the same in both cases, with identical lettering. The single slight difference to be found between the two versions is that, in the solution reproduced here, we give the inequality

(1)
$$\frac{\overset{\frown}{GD}}{\overset{\frown}{DE}} > \frac{GD}{DA}$$

without proof. Now it follows from the inequality

(2)
$$\frac{\mathrm{tr}.(GHD)}{\mathrm{tr}.(DHA)} < \frac{\mathrm{sect}.(GHD)}{\mathrm{sect}.(DHE)}.$$

But here we are arguing from (1) by composition, whereas in the *Treatise* we proceeded from (2) by composition:

(2)
$$\Rightarrow \frac{\mathrm{tr}.(BHD)}{\mathrm{tr}.(DHA)} < \frac{\mathrm{sect}.(IHD)}{\mathrm{sect}.(DHE)} \Rightarrow \frac{BD}{DA} < \frac{\overset{\frown}{BC}}{\overset{\frown}{AC}} \Rightarrow \frac{CB}{CA} < \frac{\overset{\frown}{BC}}{\overset{\frown}{AC}}.$$

But
$$\frac{\mathrm{tr}.(GHD)}{\mathrm{tr}.(DHA)} = \frac{GD}{DA},$$

hence
$$\frac{\overset{\frown}{GD}}{\overset{\frown}{DE}} > \frac{GD}{DA}.$$

4. *Analysis and synthesis: variation of the auxiliary constructions*

Without the slightest acknowledgement of making a transition, al-Sijzī then move on to 'analysis and synthesis' of two similar problems of the division of a straight line by a point that has a certain geometrical property. Why do we have these problems? And why here and now? In regard to these reasons and the order of his exposition, al-Sijzī himself, at least in the existing manuscript, says not a word. We may note, however, that, for these examples, he takes the trouble to give a formal exposition, that is he carries out an analysis and then a synthesis of the problems. Finally, the problems themselves are of the same kind as several others that his prede-cessor Ibrāhīm ibn Sinān had considered in order to illustrate different spe-cies of analysis and of synthesis. There is thus every indication that here al-Sijzī is modelling his work on the study by Ibn Sinān. Better still, as is shown by the examples that are investigated, this second look is directed at one of the questions that most interested Ibn Sinān in his researches on analysis and synthesis: auxiliary constructions. Let us now turn to the example studied by al-Sijzī.

Problem 1: Al-Sijzī begins with a straight line *AB*; he wishes to divide *AB* in *D* such that

$$(1) \qquad AB \cdot BD + AD^2 + C = AB^2,$$

where *C* is a given arbitrary square.

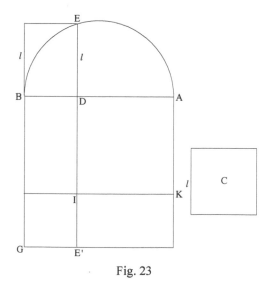

Fig. 23

Let us suppose D is known and satisfies (1). Let us draw $BG \perp AB$, where $BG = AB$ and $DI \parallel BG$. We have area $(DG) = AB \cdot DB$. We construct on AD the square $AKID$. For D to be a solution to the problem, we require area $(KE') = KI \cdot IE' = AD \cdot DB = C$. So it is necessary that l, the side of C, shall be less than or equal to $\dfrac{AB}{2}$.

Problem 2: This time al-Sijzī wishes to divide AB in D such that

(2) $AD \cdot BD + AD^2 + C = AB^2$.

Let us suppose that D is known and satisfies (2). We take $BK = AD$ and we construct the rectangle DK; we have area $(DK) = AD \cdot DB$. We draw $KHI \parallel AD$, so we have area $(AH) = AD^2$; there remains area $(IG) = C$; so we require that $IK \cdot KG = C$, that is that $AB \cdot BD = C$.

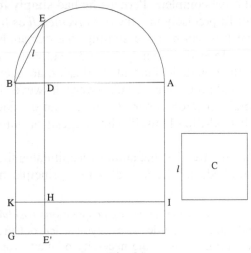

Fig. 24

We draw the semicircle with diameter AB and from the point B we draw the chord $BE = l$; now we have $EB^2 = l^2 = AB \cdot BD$, a construction that is always possible because $l < AB$.

If we put $AB = a$ and $AD = x$, the first problem can be written $ax = x^2 + c$ and the second one $ax + c = a^2$. We may note that al-Sijzī has avoided making this algebraic translation.

The synthesis of these analyses begins with the construction of E on the circle.

5. *Two principal methods of the* ars inveniendi

We recollect that at the beginning of his treatise al-Sijzī gave a list of methods for facilitating invention in geometry; at least seven of them, according to the author. We showed that there is in fact one principal method, 'analysis and synthesis', and several special methods, which will provide the first method with actual means of making discoveries. These special methods all have in common the idea of transformation and of making variations in not only the figures but also the propositions and the procedures used for solution of problems. The set as a whole, including both the principal method and the special ones, suits al-Sijzī's intellectual style, and also the list he proposes, but with the exception of one method that he mentions: that of ingenious procedures, procedures like those of Hero of Alexandria. Having mentioned this method at the beginning of his treatise, al-Sijzī in fact never refers to it again. Perhaps he had introduced it merely to make the list complete. Perhaps he had simply forgotten about it because it does not fit precisely into the *ars inveniendi* as he conceives it. It does not seem that this can be so, assuming that our analysis of al-Sijzī's treatise is correct. Indeed if 'analysis and synthesis' is the principal method, and if all the others, its faithful auxiliaries, are there to serve it, the role of the mechanical procedures for discovery, however important it may be, is of a completely different order: it is that of an outside collaborator of a practical kind. It is al-Sijzī himself who suggests an interpretation along these lines.

Near the end of the treatise, the author recapitulates the list of methods that he has just applied, and he indicates two principal methods. This is what he writes:

> Since an investigation of the nature of propositions (*askhāl*, singular *shakl*) and of their individual properties is always carried out in one of these two ways: either we bring to mind the necessity of their properties by considering a variety of species, [an act of] imagination that draws on sensation or on what is perceived by the senses; or we assume these properties and also the lemmas they make necessary, in succession, by geometrical necessity [...].[34]

Al-Sijzī follows this conclusion with several examples.

So for al-Sijzī the *ars inveniendi* essentially involves only two methods. All the special methods are grouped around the first method, and the second one is simply 'analysis and synthesis'. Now it is exactly this distinction on the one hand, and the nature of the first method on the other

---

[34] See p. 619.

hand and, finally, this intimate relationship between the two of them, which characterise al-Sijzī's conception and shows up the novelty of his contribution.

Moreover, we must note that the first of the two methods splits into two, in accordance with the two meanings of the term *shakl*. This word, which the translators of Greek mathematical writings[35] chose to render as διάγραμμα, designates, as does the Greek, indiscriminately both the figure and the proposition. This double meaning is not too troublesome in causing ambiguity when the figure provides a translation of the proposition into graphic terms, in a static way (so to speak); in other words so long as

---

[35] *Shakl* was in fact the term the Arabic translators used to render διάγραμμα, when they met it, or, more frequently still, καταγραφή and θεώρημα. Thus, when Apollonius writes in *Conics*, I.53, ἐν τῷ μθ´ θεωρήματι, the Arabic translator writes *fī shakl 49*. Examples of this translation are too numerous to merit further discussion here.

It remains that, in contexts like this, it has been possible for the same term θεώρημα to be rendered by the Arabic *ṣūra* (form). For example when Apollonius writes τοῦτο γὰρ δέδεικται ἐν τῷ ια´ θεωρήματι (*Conics*, I.52), the Arabic translator writes *wa-qad tabayyana dhālika fī al-ṣūra* 11 (this has been shown in the *ṣūra* 11); or again when he writes ταῦτα γὰρ ἐν τῷ ιβ´ θεωρήματι δέδεικται (*Conics*, I.54), what we read in the Arabic translation is *kamā tabayyana fī al-ṣūra* 12 (as was shown in the *ṣūra* 12). The term 'theorem' has thus been rendered indifferently as *shakl* and as *ṣūra*. So in regard to terminology the situation is more complex than may at first appear.

We may start by noting that there is some stability in Arabic geometrical vocabulary from about the mid ninth century onwards. This of course does not prevent a certain amount of innovation and a certain amount of adjustment. The terms *shakl* and *ṣūra* provide examples of adjustment. The first term, *shakl*, is stable, and retains its double sense in the geometrical lexicon. As for the term *ṣūra*, it keeps some links with its primary usage, whereas others are modified to apply to a geometrical drawing. Thus, in mathematical writings, *ṣūra* carries multiple meanings:

1. The 'form' of an object, that is its essence; thus we speak of the form of a ratio, or of a number, and so on.

2. The sense of a state of the proposition itself, for example universal or special.

3. The sense of a case of the figure for a proposition which has several.

4. The sense of a possible form of a geometrical object, for example the form of a triangle: isosceles, right-angled and so on.

All these senses apply to propositions or geometrical objects, without any special reference to a figurative representation, that is to a drawing.

5. Finally, there is the sense of *ṣūra*: form, as an image, and expressing the sense of a graphic representation. Thus we speak of *ṣūrat shakl*, the form of a figure, that is the drawing of a figure, or the figure as a representation. It seems, but this is conjectural, that after a time writers gave up using the term *ṣūra* to designate a theorem, but did so while retaining all its other senses. The result is that to the double meaning of *shakl* there is added the multiplicity of meanings of *ṣūra*, without any possibility of setting up an opposition between the two terms. The glossaries of the four previous volumes provide a sufficient number of occurrences of these meanings.

geometry remains essentially a study of figures. But everything becomes more complicated when we begin to transform figures to give variants of the figure, as is already happening in some branches of geometry in the time of al-Sijzī. The double significance then requires an explanation.[36] Let us start with the first sense, that of 'figure'.

In this treatise, al-Sijzī recommends three times that we proceed by varying the figure: when we carry out a point-by-point transformation; when we vary one element of the figure, all the others remain fixed; finally, in the choice of an auxiliary construction. Now several elements are common to these different procedures. The purpose first of all: we are always looking to find, by means of transformation and variation, the invariant properties in the figure associated with the proposition, the properties that are its particular characteristics. Now these are, precisely, the invariant properties that are set out in the figure as a proposition. The second element also relates to purpose: variation and transformation are means to discovery insofar as they lead to these properties that do not change. It is here that the imagination intervenes, as a faculty of the soul capable of pulling out from among the multiplicity offered by the senses, among the variable properties of figures, those properties that do not change, the essences of things. The third common element is concerned with a particular role of the figure, this time as a representation, referred to a number of times by al-Sijzī, that of engaging the imagination, of helping it carry out its task when it is assessing the evidence of the senses. The fourth, no less important, is connected with the figure-proposition duality: there is no bi-univocal relationship. To one and the same proposition there may correspond a variety of figures; similarly to one single figure there may correspond a whole family of propositions. Al-Sijzī chose, moreover, to treat this last case at length. These new relationships between figure and proposition that al-Sijzī, as far as I know, was the first to point out, require that we think about a new area in the *ars inveniendi*: analysis of figures and of their relations to propositions. Now this is precisely what al-Sijzī seems to have initiated.

As an example of the first method for the *ars inveniendi*, al-Sijzī does no more than refer back to one he has already considered: the equality, for any triangle, of the sum of its angles, a property grasped by the imagination on the basis of what is common to the senses. He also mentions another similar example.

---

[36] In this connection, see P. Crozet, 'À propos des figures dans les manuscrits arabes de géométrie: l'exemple de Siǧzī', in Y. Ibish (ed.), *Editing Islamic Manuscripts on Science*, Proceedings of the Fourth Conference of al-Furqān Islamic Heritage Foundation, 29th–30th November 1997, London, 1999, pp. 131–63, esp. pp. 140–3.

For the second method, that of analysis and synthesis in geometry, al-Sijzī adds nothing of substance, but he emphasises the role of apprenticeship and of practice in strengthening the faculty for conceiving properties. He gives only several particularly simple examples to illustrate the procedure.

1. Let there be a circle in which two chords AB and CD cut one another at the point H. We wish to know the origin of the necessity of the equality $AH \cdot HB = CH \cdot HD$. Al-Sijzī begins by drawing CA and DB, and proves that the triangles HCA and HDB are similar; so we have $\dfrac{CH}{AH} = \dfrac{BH}{HD}$, hence the result. But to show that the property is an invariant one, he begins again and draws another circle, EG, passing through H; as before we have that triangles CEH and DHG are similar, hence $\dfrac{HE}{HD} = \dfrac{HC}{HG}$; hence the result. So the property does not depend on the choice of the chord, but on the similarity of the triangles that are formed, and thus on the fact that the angles inscribed in the circle intercept the same arc.

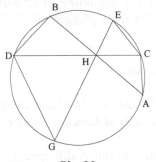

Fig. 25

2. An arc of a circle is intercepted by an inscribed angle equal to the angle formed by the chord of that angle and the tangent to the circle at one end of that chord. Here again al-Sijzī takes his reader by the hand to show him how to find that the property does not vary.

He draws the circle ABC with diameter AB, and BD the tangent at the point B. It is clear that $A\hat{B}D = A\hat{C}B$ because $A\hat{C}B$ intercepts a semicircle; so $A\hat{C}B = 1$ right angle $= A\hat{B}D$. Al-Sijzī then writes:

> we have to look for the variety of the species of that figure, and the necessity of their properties, by an investigation of their nature (p. 620).

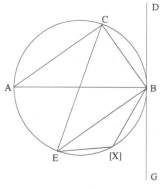

Fig. 26

He in fact proceeds by varying the angle B formed by the chord and the tangent, and proves that it is equal to any inscribed angle such as $E\hat{X}B$ that also intercepts the arc $EACB$, 'visually and geometrically'.

Al-Sijzī concludes his treatise by returning to the didactic intentions he had stated so forcefully at the beginning of it, with an exercise in the way to carry out analysis, written for beginners.

III. HISTORY OF THE TEXTS

3.1. *Book by Thābit ibn Qurra for Ibn Wahb on the Means of Arriving at Determining the Construction of Geometrical Problems*

Thābit ibn Qurra's letter to Ibn Wahb appears in the list of his writings drawn up by Abū 'Alī al-Muḥsin ibn Ibrāhīm al-Ṣābi', and reproduced by al-Qifṭī, with the title *Fī istikhrāj al-masā'il al-handasiyya* (*On the Determination of Geometrical Problems*),[37] it has come down to us under three different titles, the first of which is the closest to that given by al-Ṣābi', and in five manuscripts.[38] Thus, we have:

1. *Fī al-ta'attī li-istikhrāj 'amal al-masā'il al-handasiyya* (*On the Means of Arriving at Determining the Construction of Geometrical Prob-*

---

[37] See *Les mathématiques infinitésimales*, vol. I, p. 145; al-Qifṭī, pp. 116–17.

[38] The multiple titles each of which, as we shall see, expresses one aspect of the letter, have been a cause of confusion for bibliographers. Some have believed we had three different texts by Thābit ibn Qurra and have catalogued them as such. Thus, F. Sezgin in his *Geschichte des arabischen Schrifttums* includes this same letter as numbers 4, 7 and 22, in the belief that he was dealing with three different treatises, pp. 268–70.

lems). This is the title of the letter in the copy in the hand of the tenth-century mathematician al-Sijzī. This copy is part of the famous collection no. 2457 in the Bibliothèque nationale in Paris, fols 188ᵛ–191ʳ, here called B. We have described this collection already.[39] The point to remember here is merely that al-Sijzī checked his copy against his model, as he tells us himself in the colophon.

2. *Fī kayfa yanbaghī an yuslaka ilā nayl al-maṭlūb min al-ma'ānī al-handasiyya* (*How to Conduct Oneself so as to Obtain what one Requires among Geometrical Propositions*). The same letter by Ibn Qurra has come down to us under this title in two manuscripts. The first belongs to the collection Aya Sofya 4832, fols 1ᵛ–4ʳ, in the Süleymaniye Library in Istanbul, here called A; the second is part of the collection 40, fols 155ᵛ–159ᵛ, here called C, of Dār al-Kutub in Cairo. These two manuscripts have also already been described.[40]

3. *Fī al-'illa allatī lahā rattaba Uqlīdis ashkāl kitābihi dhalika al-tartīb* (*On the Cause of Euclid's Ordering of the Propositions in his Book according to this Order*). The letter has come down to us, under this title, in two other manuscripts. The first belongs to the collection Aḥmadiyya 16167, fols 86ᵛ–90ᵛ, in the Library of Tunis, here it is called T; the second is part of the famous collection Leiden, Or. 14, fols 380–388, here called L. We have described this collection and even discovered the models for twelve of the twenty-three treatises it comprises,[41] that is the collection in Columbia University Library, Smith, Or. 45. So the only newcomer is the manuscript collection in Tunis.

This collection of 90 folios – size 13 × 21.5 cm – each with 23 lines of about 13 words, in *nasta'līq* – was copied before 971/1563, the date when it was bought by one of its owners. The copyist himself has not indicated either the date or the place of his copy. The collection comprises the following treatises:

1. *Commentary on Euclid's Postulates by Ibn al-Haytham* (*Sharḥ muṣādarāt Uqlīdis*), fols 1ᵛ–59ᵛ; folio 60ʳ is blank.

2. *Additions by al-'Abbās ibn Sa'īd to Book Five of Euclid* (*Ziyādāt al-'Abbās ibn Sa'īd fī al-maqāla al-khāmisa min Uqlīdis*), fols 60ᵛ–61ʳ.

3. *Remarks from the Commentary on Book Ten of the Work of Euclid, Composed by al-Ahwāzī* (*Kalimāt min sharḥ al-maqāla al-'āshira min kitāb Uqlīdis*), fols 61ᵛ–65ʳ.

---

[39] See for instance *Les Mathématiques infinitésimales*, vol. I, pp. 147–8, 680; English trans. pp. 125, 465.

[40] *Ibid.*, vol. I, pp. 147–9, 679; English trans. pp. 124–9, 464–5.

[41] *Ibid.*, vol. III, pp. 532–4; English trans. pp. 506–8.

4. *Commentary on the Introduction to Book Ten of Euclid by Abū Ja'far Muḥammad ibn al-Ḥasan al-Khāzin* (*Tafsīr ṣadr al-maqāla al-'āshira min Uqlīdis li-Abī Ja'far Muḥammad ibn al-Ḥasan al-Khāzin*), fols 65ᵛ–71ʳ.

5. Anonymous letter on the Fifth Postulate of Euclid, fols 71ᵛ–73ʳ.

6. *Treatise by al-Fārisī who Makes Additions to al-Abharī's Edition on the Famous Problem of the Book of Euclid* (*Maqāla li-al-Fārisī yuḍīfu 'alā taḥrīr al-Abharī fī al-masā'il al-mashhūra min kitāb Uqlīdis*), fols 73ʳ–75ʳ.

7. *Book by Abū Dāwud Sulaymān ibn 'Iṣma on Binomials and Apotomes in Book Ten of the Work by Euclid* (*Kitāb Abī Dāwud Sulaymān ibn 'Iṣma fī zawāt al-ismayn wa al-munfaṣilāt allatī* [ms. *alladhī*] *fī al-maqāla* [ms. *al-maqālāt*] *al-'āshira min kitāb Uqlīdis*), fols 76ᵛ–85ᵛ.

8. A fragment, complementing the preceding book, fols 85ᵛ–86ᵛ.

Now a detailed comparison of T with L allows us to show that treatises 3, 4, 6, as well as the letter by Thābit, are respectively the sole models for treatises 19, 18, 20 and 21 of the Leiden collection, Or. 14. This important finding concerning the history of the text permits us to draw conclusions regarding 17 of the 26 treatises that make up L. We have already identified the single model transcribed by the copyist of L for twelve treatises.[42] Adding in these four extra treatises, we now know the models that were copied for 16 treatises. Further, the model for al-Ṭūsī's edition of the last three books of Apollonius' *Conics*, which is part of this collection, is also known. So we could have set aside the variants of L in establishing the text of Ibn Qurra's letter; if we have noted them, it is because they provide evidence for what we have just said, namely that T is the sole model for L for this treatise, as for the three others mentioned earlier. T and L have in common three omissions of a sentence and of twenty-three words, while L has eleven omissions of a word that are specific to it.

If we now examine the relationships between the other manuscripts, we may note, among other results:

• The copyist of A had two copies at his disposition. Thus, he writes in the colophon, fol. 4ʳ:

قابلت هذه المقالة بالنسخة التي كتبتها منها وبنسخة أخرى غيرها وصححتها بحسب ما كان فيهما.

So manuscript A is copied from two manuscripts A₁ and A₂. However, A is missing a sentence (*kānat al-zāwiyatān al-bāqiyatān mutasāwatayn*),

---

[42] *Ibid.*, vol. III, p. 534.

fol. 2ᵛ. The copyist has indicated with a cross the place where this is missing, but has forgotten to transcribe the missing sentence.

• The copyist of C – the famous Muṣṭafā Ṣidqī – writes in the colophon that he had transcribed the text from a copy in the hand of Avicenna:

وقد استنسخ من نسخة كانت بخط الشيخ الرئيس حجة الحق أبي علي الحسين بن عبد الله بن سينا.

We have discussed this relation of a somewhat legendary tale and shown that C and A have a common ancestor. In any case, for Ibn Qurra's letter alone, they present twelve common omissions of a word, while C has two omissions of a word that are specific to it. As for the sentence omitted in A, Muṣṭafā Ṣidqī could easily have completed it, given his mathematical education.

• Manuscript B in the hand of al-Sijzī contains five omissions of a word that are specific to it.

Examining the variants – additions, errors, etc. – leads us to propose the following stemma:

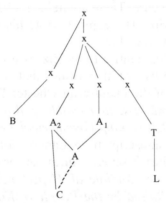

Ibn Qurra's text has been edited from manuscript B alone by A.S. Saidan, in an edition we shall call S in the apparatus criticus; it has never been the subject of a translation or commentary.

3.2. *To Smooth the Paths for Determining Geometrical Propositions*, by al-Sijzī

Al-Sijzī's treatise has come down to us in a single manuscript which belongs to the collection of Nabī Khān and Obaidur-Rahman Khan of

Lahore. This collection includes, among many other treatises by different mathematicians, six writings by al-Sijzī. All the treatises in this collection were copied in the Niẓāmiyya school in Mosul and in the school of Baghdad, between 554 and 557 of the Hegira (1159–62). As for the six texts by al-Sijzī, they were transcribed in Baghdad from the year 556 and in the course of the year 557. The book whose text is translated here appears before two short works by al-Sijzī: a letter to Naẓīf ibn Yumn, copied at the Niẓāmiyya school in Baghdad at the end of the month of Rabīʿ al-ākhir in the year 557 of the Hegira (*bi-tārikh salkh shahr Rabīʿ al-ākhir sanat sabʿ wa-khamsīn wa-khamsimiʾat hijriya*), that is in mid April 1162; and the other one is on the two means and the trisection of an angle, transcribed in the same city and in the same school, at the beginning of Jumādā al-ūlā in the year five hundred and fifty-seven (*bi-tārikh ghurrat Jumādā al-ūlā li-sanat sabʿ wa-khamsīn wa-khamsimiʾat*), that is at the end of April 1162. So it is very probable that the treatise whose text is translated here was transcribed at about that date, that is the end of 556-beginning of 557, in the Niẓāmiyya school of Baghdad. This text covers folios 2–27 in *nastaʿlīq* script; the figures are drawn carefully and the whole thing is done with care. There are neither additions nor glosses in the margins of the manuscript. The copyist wrote his name at the end of the other texts, but it remains illegible.

The attribution of this treatise to al-Sijzī is not in doubt. The work in fact appears on both the lists of his writings that we possess: the first transcribed by the copyist of the manuscript Chester Beatty 3652, fol. 2^r, no. 34, under the title *Fī tashīl al-subul li-istikhrāj al-ashkāl al-handasiyya* (*To Smooth the Paths for Determining Geometrical Propositions*); the second by the copyist of the manuscript from Lahore, fol. 371^v, under the same title. Al-Sijzī himself cites it several times, as for example in his treatise *On the Asymptotes to an Equilateral Hyperbola*,[43] or in his *Reply to Geometrical Problems Raised by the People of Khurāsān* (*Jawāb al-Sijzī ʿan masāʾil handasiyya saʾala ʿanhu Ahl Khurāsān*), Chester Beatty 3652, fol. 57^v.

This text was published for the first time by A. S. Saidan in *The Works of Ibrāhīm ibn Sinān*, Kuwait, 1983, pp. 339–72. Our late friend, no doubt aware of the importance of this work by al-Sijzī, as of that of Thābit ibn Qurra, but in all probability pressed for time, may have wished to draw historians' attention to these texts by getting them published, albeit in a form that is at best provisional (called S). It is this publication of al-Sijzī's

---

[43] R. Rashed, 'Al-Sijzī et Maïmonide: Commentaire mathématique et philosophique de la proposition II–14 des *Coniques* d'Apollonius', p. 288.

text that J. P. Hogendijk reproduced,[44] while, however, adding some corrections, the majority of which are due to comparing A. S. Saidan's text
with the only extant manuscript. This comparison, which was certainly
necessary, would have been welcome had it not left unchanged not only
most of the mistakes, but also the mutilations that the manuscript had
undergone, and if the checking had not itself afforded an opportunity for
new errors. The errors unfortunately remain excessively numerous and
prevent a true understanding of al-Sijzī's text. It is this publication (called
H) which was translated into English, often in a free manner.[45]

3.3. *Letter of al-Sijzī to Ibn Yumn on the Construction of an Acute-angled Triangle*

This text was edited from the autograph version by al-Sijzī, composed
in the month of Ābān 339 of the era Yazdegerd,[46] Bibliothèque nationale,
Paris, no. 2457, fols 136ᵛ–137ʳ, called B. We have also used another copy –
Lahore, fols 28–30, called L – of this same text, although we knew this was
not in any way necessary.
    J. P. Hogendijk has published an edition of this text that is comparable
with the one he produced of the preceding treatise; in our apparatus criticus
it will be called H.

3.4. *Two Propositions from the Ancients on the Property of Heights of an Equilateral Triangle: Ps-Archimedes, Aqāṭun, Menelaus*

The propositions of the ancients taken up by Ibn al-Haytham have
come down to us in two versions, one attributed to Archimedes, translated
by Thābit ibn Qurra (ms. Patna, Khuda Bakhsh 2519, fols 142ᵛ–143ʳ);[47] the
other to a certain Aqāṭun (ms. Istanbul, Süleymaniye, Aya Sofya 4830, fols

---

[44] *Al-Sijzī's Treatise on Geometrical Problem Solving (Kitāb fī Tashīl al-Subul li-Istikhrāj al-Ashkāl al-Handasiyya)*, translated and annotated by Jan P. Hogendijk.

[45] See the apparatus criticus in *Les Mathématiques infinitésimales*, vol. IV, and the review by P. Crozet in *Isis*, 90.1, 1999, pp. 110–11.

[46] The month persan of Ābān 339 of the era of Yazdegerd is between 20 October and 18 November 970 in our era. There are five Thursdays in that interval: the 20 and 27 October and the 3, 10 and 17 November. Since '*day*' can only refer to the 8th, 15th or 23rd day of this month, the date that is indicated can correspond to two days: 27 October or 3 November 970.

[47] See the description of this manuscript p. 473.

91^v–92^r),[48] as well as in a text by al-Sijzī with the title *On the Properties of Perpendiculars Dropped from the Given Point to the Given Equilateral Triangle by the Method of Discussion* (mss Dublin, Chester Beatty 3652, fol. 66^v – called B; Istanbul, Reshit 1191, fols 124^v–125^r – called R).[49] We have discussed[50] the probable relations between these three texts, of which we give translations here.

[48] See p. 454, n. 8.
[49] See section 3.3.
[50] See pp. 453–4.

TRANSLATED TEXTS

1. *Book of Abū al-Ḥasan Thābit ibn Qurra to Ibn Wahb on the Means of Arriving at Determining the Construction of Geometrical Problems*
Kitāb Abī al-Ḥasan Thābit ibn Qurra ilā Ibn Wahb fī al-taʼattī li-istikhrāj ʻamal al-masāʼil al-handasiyya

2. *Book of Aḥmad ibn Muḥammad ibn ʻAbd al-Jalīl al-Sijzī to Smooth the Paths for Determining Geometrical Propositions*
Kitāb Aḥmad ibn Muḥammad ibn ʻAbd al-Jalīl al-Sijzī fī tashīl al-subul li-istikhrāj al-ashkāl al-handasiyya

3. *Letter of Aḥmad ibn Muḥammad ibn ʻAbd al-Jalīl al-Sijzī to the Physician Abū ʻAlī Naẓīf ibn Yumn on the Construction of the Acute-angled Triangle from Two Unequal Straight Lines*
Risālat Aḥmad ibn Muḥammad ibn ʻAbd al-Jalīl al-Sijzī ilā Abī ʻAlī Naẓīf ibn Yumn al-mutaṭabbib fī ʻamal muthallath ḥādd al-zawāyā min khaṭṭayn mustaqimayn mukhtalifatayn

4. *Two Propositions of the Ancients on the Property of the Heights of an Equilateral Triangle: Pseudo-Archimedes, Aqāṭun, Menelaus*

In the name of God, the Compassionate the Merciful

BOOK BY ABŪ AL-ḤASAN THĀBIT IBN QURRA TO IBN WAHB

On the Means of Arriving at Determining the Construction of Geometrical Problems

You have understood everything – Sire, may God give you long life and everlasting renown – when you have grasped what Euclid has truly done in composing the propositions of his work *The Elements* and of his statements, and in the ordering of them in many matters not classified according to their types[1] and without each being connected with one that resembles it; and that his motive in doing this is that he needs to establish the proof of each of the statements and of each of the propositions, and the impossibility of establishing a proof, for many of them, without each being preceded by another which is not of its order and is out of position. To do this he had to bring forward things that should have been postponed and postpone things that should have been brought forward. You have then seen that this approach is necessary for anyone who wants to understand the contents of his book, when he studies it for the first time, that is to say in the state [of mind] he is in up to the point where he has understood, when he has become convinced by what this man says and sets out, and has acquired confidence in him through knowledge of the reliability of his proofs.

If he gets to that point and learns this; if he then attains a second state, more expert than the previous one, and he needs to make use of what he has learned from him and employ it to find what he is trying to find in the different areas of this science and its problems, then he requires a different approach: as he wishes to find one of the propositions or another of the notions mentioned by the geometer,[2] among those he wishes to determine and whose existence and construction he wishes to shed light on, he finds notions like those that are required by what he is looking for, ones he can grasp, gathered together in his soul, present in his understanding, at that

---

[1] Here we are translating the word *jins* as 'type'.
[2] Lit.: the man of this art.

moment. Now this is what will happen if, in his thoughts and reflections, he investigates the notions required in this [particular] type among the types of propositions, or others, or those that necessarily follow from what is specific to this type or what includes it; thus he distinguishes them from the others and accordingly comes to understand them; then he examines them and turns his mind to them, and he takes from them what he needs for the notion he seeks.

Since in this second state that you [Ibn Wahb] have mentioned we need to have recourse to this second approach, that you have described, in regard to the arrangement of notions and the manner in which they are established in the soul according to what is required by each of the types of things we seek, whereas[3] in the first state what was needed was the opposite of this, you ordered me – may God give you glory – to bring this notion to mind and to draw attention to it in a text dedicated to it, to describe it; to attract the attention of anyone who is trying to determine something in areas of this science, still more, in any demonstrative science, on the way to arrive at it and on what he must establish in his soul and call to his mind, from among the principles and notions that are embedded in this science and thanks to which one can prepare the ground for discovery: perhaps all of them, perhaps those of them that are possible, as extensively as he can, in the knowledge that as he reaches further into the notions that are instruments for reaching conclusions in what one seeks to find and to propose to attain what is sought, he will be better able to attain it; and you ordered me to describe by means of examples what method allows one to determine certain notions in geometry and to know them, [a method] to be a guide one imitates and a model one copies in other cases as providing a method of solution, it being given that there is no method that would deal with all of them one by one. I bowed to your command, may God support you.

If we are concerned with one of the notions investigated in geometry or a problem we wish to solve, we need, first of all, to know that everything considered by geometers, what concerns them among the notions in regard to each type of proposition, and others they speak of, consists of three things: one is to describe one of the constructions that employ instruments, by means of which we learn how to shape one of the notions or to find it; the second is to construct a magnitude or a status for one of the notions, itself, with an unknown magnitude or status; and the third is peculiar to their nature or to the properties which define them and are their necessary concomitants or follow them or distinguish them, as for the theorems and the rules that are necessary for them. A description of one of the constructions by which we know how to shape one of the notions or to find it, is for

---

[3] Lit.: in the same way as.

example the construction of an equilateral triangle or a square on a known straight line. Finding a magnitude or a status for one of the notions, itself, whose magnitude or status is unknown, is for example knowing the area of a triangle whose sides or heights are known, or determining a perfect number. To know what is peculiar to their nature or the properties which define them and are their necessary concomitants, or follow or distinguish them, as for theorems and the rules that are necessary for them, as for example knowing that, among rectilinear figures, the triangle is the only <figure> that can have [only] acute angles; that the angles of any triangle, if we take their sum, are equal to two right angles, and that circles that touch one another cannot have the same centre, any more than those that cut one another.

So if one knows what we have set out on the classification of what is of concern to the geometer, one examines the thing one is investigating, whether it is a problem or one of the notions that is investigated, one asks oneself what kind of thing it is, with it one then addresses oneself to the kind of things it belongs with and one looks at the principles and lemmas that one expects in that kind of thing. In addition, one knows that the first of the three kinds we have set out necessarily requires the other two kinds because construction by the art [of geometry] must necessarily be preceded by knowledge of the natures of these things that are constructed. As for the two other kinds, they can almost, in themselves, do without the first kind. We know, also, that each of these three kinds which have been described includes things like its primary fundamentals and the principles by which it is known, as well as things determined from these first principles; and there are often, in addition, [other] principles that are taken into account.

As for the first principles, they are taken as agreed without proof, and among them the definitions that indicate the essence of each of the figures, as well as what will be mentioned, such as the definition of the circle which indicates its essence, the definition of the triangle, and similar statements. Among others, the common notions, which can be called primary knowledge, such as: things equal to the same thing are equal. Among others, the postulates as well as the constructions that are postulated and whose use we agree to, and other things, such as: it is possible to join any point to any point with a straight line, or with any centre and with any distance to construct a circle.

If we do this, and if, for each thing we seek to determine, we direct our attention, as we have said, towards the one of the kinds that we have classified to which it belongs, and if the most significant element in what of this kind we are looking for, in this way we make the lemmas for the thing, then we must next do what I have mentioned: set out the lemmas and the principles which are appropriate to the specific thing, [that is] in seeking it. We then seek to determine the thing starting from a problem or from a geometrical notion, and to distinguish these principles by separating them from the others. The way to achieve this is to assess the thing proposed for investigation: to what type it belongs among figures, and other things, what this type requires in general by way of rules and by way of theorems that are necessary for it, as well as for others, or which are appropriate to it, but not to the others, and which distinguish it; we call them to mind and we present them to our intellect. Next, we examine what is required by each of the conditions in the problem we seek to solve, [conditions] connected with this type [of problem], and each of the specific properties which we add to it; for any problem there is in fact one assumed thing which is sought, and certain conditions by which the [set of conditions in the] discussion is satisfied. Thus, if one of them cannot be employed, the problem will not be solved. So we have to use all the conditions for the problem, and what is made necessary by each condition, so as to specify it completely. If we obtain what we are looking for, so be it; if not, we take things the problem has caused us to arrive at as if they were the result we sought; we put them in the place of the first thing we sought, then in investigating them we follow the same route that we mentioned [before]. We repeat this procedure, time after time, until we get to know the thing we wanted to know, if God wills it.

For what I have written, I propose two or three examples, by which I prove what I have said, and to begin with I make the thing to be found an easy thing, so that our account shall not be a long one.

First, we shall show how to construct a triangle in which one of the angles is double each of the two remaining angles.

For what we are trying to do in this, we must direct our attention to the first of the three types we have described, which is one of constructions. But, as you must begin by knowing the status and nature of the thing constructed, as we have said, we must prepare our minds and organise within them the theorems and the rules that are demanded by the nature of the thing we seek and the type to which it belongs. The type of the thing proposed for investigation is that of triangles. We first call to mind what this triangle makes absolutely necessary in regard to its sides, its angles and other things, such as: in any triangle, the sum of any two of its sides is greater than the third side; an angle exterior to it is greater than each of the interior <angles> opposite to it – but it is equal to the two if they are added together; <the sum> of any two of its angles is smaller than two right angles – but the sum of its three angles is equal to two right angles; any straight line that divides one of its angles and ends on the straight line that subtends it divides it [the triangle] into two triangles whose bases lie in a straight line, and their analogues. But since in this proposition it is the angles that interest us, we must concentrate on them, as well as on what has been mentioned in connection with them.

We next say that we require for the triangle that we seek, and have called to mind, all that is needed for everything of its type. Now this is not enough, [and] does not fulfill all the conditions of the problem, conditions without which the problem cannot be either discussed or made precise. So it remains for us to use the conditions it involves, namely that one of the angles of the triangle we seek is double each of the remaining angles. So we examine what that condition makes necessary. Thus, it makes necessary many things relating to comparing the angles each with the others and with their sum; among other things, that the sum of the three angles of the triangle is double the large angle which we wish to be double the associated one; that half the large angle we have mentioned is equal to each of the two remaining angles; that the sum of its three angles is four times each of the remaining angles; that the two remaining angles are equal, it being given that each of them is half the large angle; that two sides of the triangle must be equal, and other similar things. Next we add and put together the things made necessary by this condition with the things made necessary by the type as a whole, that is to say the type of a triangle. We investigate which of these things is such that if we add it to the former, it will be of use to us in what we are aiming to do. We then find that more than one of them is such that if we add them each one to the others they give and produce what we wanted, or bring us nearer to discovering it. Among other things: when to the first statements about the triangle we add our statement that the sum of its angles is equal to two right angles to one of the statements made necessary by the condition, namely that the sum of the angles of the triangle is equal to double its large angle, then we stop at these two statements and we know that its large angle is a right angle. So we know that in the triangle we need to construct a right angle; but since we want it to be equal to double each of the remaining angles, each of them is thus equal to half a right angle. Accordingly we know that if we can construct a right-angled triangle, in which each of the two remaining angles is half a right angle, then we have learned what we wanted to know.

Now this is something that it is possible for us [to do], it being given that it has been shown in Euclid's *Elements* how to construct a right angle, and that if from the two sides that enclose the right angle we cut off two equal straight lines, the two remaining angles will be equal and each of them will be half a right angle. We shall then have constructed the triangle we sought.

On the other hand if to the first statement that we mentioned – namely that the sum of the angles of a triangle is equal to two right angles – we add another of the statements required by the condition, namely that the sum of the angles of the triangle is four times each of the remaining angles,[4] we stop and we conclude from these two statements that each of the two remaining angles is half a right angle. In consequence we must construct a triangle that includes two angles such that each is half a right angle. Now this is something that is possible for us [to do] starting from constructions set out by Euclid. It in fact remains for us to construct[5] a right angle and divide it into two halves.

So if we draw a straight line and erect two straight lines perpendicular to it at its two endpoints, if we divide each of the two angles that are formed into two halves with two straight lines which we extend to meet one another, we then obtain from them the triangle that we seek, by a construction different from the first one.

If, alternatively, we take a third statement from among those required by the condition, namely that half of the large angle that we have mentioned is equal to each of the remaining angles, it is as if we were saying that the large angle is the angle ABC; if we divide it into two halves with the straight line BD, each of the two halves ABD and DBC is equal to each of the angles BAC and BCA. If we wish to add to the latter the statement: if we divide the angle of the triangle with a straight line and from that we form two triangles into which the triangle is divided, then their bases will lie in a straight line; from these two statements a consequence then follows. But if we examine what makes necessary this thing which has been required in the triangle, we find that it requires that the angle ADB, an exterior angle of the triangle BDC, is equal to <the sum of> the two interior angles DBC and DCB. In the same way, the angle CDB is equal to the sum of the two angles DAB and ABD. This thing would thus have caused us to arrive at a thing such that, if we add it to the preceding one, it gives rise to a consequence. In fact, because of this and the preceding, it must be that the two angles ADB and CDB are equal. So on this account there will be two right angles. So if we draw AD in an arbitrary position and if we extend it to C and put CD equal to AD; if at the point D we erect a perpendicular DB and if we cut off from it a part equal to each of AD and CD, then we have what we wanted. But we have all these constructions at our disposal, and we can accordingly construct what we wanted with a third construction. We follow the same path for everything we wish to find.

---

[4] We are indeed dealing with angles for which each is half the third one.

[5] Lit.: find (najid).

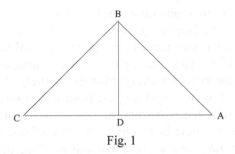

Fig. 1

In the same way, we present another example of something we wanted to find, namely: to show how to construct a triangle such that one of its angles is half of one of the two remaining angles and a third of the other angle.

The method for finding this is similar to the preceding one. In fact, here what the triangle requires, absolutely, is the same thing that was required in what preceded this. As for the two conditions that we imposed here, we have required something different from what preceded this. The conditions have in fact required that the sum of the three angles is six times the first angle we mentioned, three times the second and double the third one. If we add each of these statements to the statement required by the type of every triangle, namely that the sum of its angles is equal to two right angles, it necessarily results and follows from these statements that the first angle is one third of a right angle, the second two thirds of a right angle and the third one is a right angle. So if we construct a triangle in which one of the angles is a right angle and the next is two thirds of a right angle or a third of a right angle, then we have what we wanted. The third, remaining, angle is in accordance with what we demanded, it being given that the three angles [taken together] are equal to two right angles. But the construction of a right angle is possible for us from what is described in the book by Euclid for drawing a perpendicular; and the construction of two thirds of a right angle is possible for us wherever we want, because it is equal to an angle of an equilateral triangle. So it remains for us to construct, at the two endpoints of a straight line, two angles in accordance with what we have mentioned; we extend their sides until they meet one another; we obtain the triangle that we wanted.

In the same way, we propose a third example of what we wanted to find: namely to show how to construct a triangle such that one of its angles is three times each of the remaining angles. We follow a similar method; the lemmas and theorems which are required by the chosen type, which is the triangle, are those we mentioned before. In this problem, the condition itself requires something else, namely that the sum of the three angles is five times each of the two remaining angles, and that it is also equal to one and two fifths times the large angle; and that each of the thirds of the large angle, if that is divided into three equal parts, is equal to each of the two remaining angles. If we add each of the first two statements to the statement required by the type for every triangle, namely that the sum of its three angles is equal to two right angles, from this it results and is necessarily true that each of the two small angles is two fifths of a right angle and that the remaining angle is a right angle plus a fifth. So we have: if on an arbitrary straight line we construct two angles of the magnitude we have mentioned, and if we extend their sides until they meet one another, the third angle remains in accordance with what we demanded; and we have constructed the triangle that we wanted.

But that is possible for us if we can divide a right angle into five equal parts. We shall then have reduced the problem to another problem, we return to investigating it, as if it were our objective.

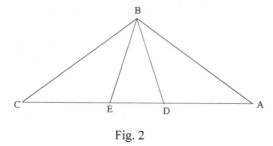

Fig. 2

In the same way, if we add the third statement required by the condition, namely that each of the thirds of the large angle, if this latter is divided into three equal parts, is equal to each of the two remaining angles of the triangle, it is as if we said that the large angle is the angle *ABC*, that the thirds of it are *ABD*, *DBE* and *EBC* and that each of these thirds is equal to each of the angles *BAC* and *ACB*. And we add to this what is required by the shape[6] of the triangle as a whole, namely that it has been divided into three triangles whose bases lie in a straight line, which is *AC*. So it necessarily results that the bases of these triangles have been extended, and that thus each of the two angles *BDE* and *BED* is double each of the two angles *ABD* and *CBE*, which are equal to the angle *DBE*. And it is because of this that each of the two angles *BDE* and *BED* of the triangle *DEB* is double the angle *CBE*. So we have reduced the problem to another problem, we must return to investigating it, and it is to construct a triangle in which one of the angles is equal to double each of the two remaining

---

[6] The original Arabic word is *khilqa* which translators used for the Greek μορφή.

angles – were it not that Euclid has saved us the trouble and has shown the construction in Book IV of the work *The Elements*; and similarly what we have reduced this problem to by the preceding method; and it is easily determined from what is [said] in that place.

What we have arrived at, thanks to these examples chosen by way of illustration, is sufficient for what we had in view, except that we have elected to add a notion to which we draw attention: it must not be allowed to escape notice that, for certain conditions that occur in problems, they can appear to be a single condition, whereas what they result in takes the place of two conditions; and similarly perhaps we think that the construction we have carried out includes a single condition for us, whereas two conditions are required in it and are involved in it. Example: what we have said in the first problem, that one of the angles of the triangle is double each of the two remaining angles; the result of this is two conditions. Similarly, in the construction of the second problem: if we construct a right angle and two thirds of a right angle on a straight line, and if we draw the sides of the two angles so that they meet one another, then we shall have constructed an angle of the triangle equal to a third of another angle of the triangle. That is one of the two conditions of the problem, but we did not carry out a construction [using] the other condition, namely that it should be equal to half of the other angle. But that condition is included in what we constructed, it being given that it necessarily results from it. So it is necessary to examine this and its homologues.

The book by Thābit ibn Qurra on the means of arriving at determining geometrical problems is completed.

In the name of God, the Compassionate the Merciful
May God give us His help

THE BOOK BY AḤMAD IBN MUḤAMMAD IBN 'ABD AL-JALĪL AL-SIJZĪ

To Smooth the Paths for Determining Geometrical Propositions

In this book of ours, we wish to list the theorems the knowledge and possession of which make it easier for the scholar to determine those geometrical constructions he wishes to determine, and to set out the methods and approaches which allow the scholar who employs them to improve his grasp of the ways of determining propositions.[1] Some think there is not any means of knowing theorems in the course of determining them by repeated deductions, by practicing it, by learning it and by studying the fundamentals of geometry, if one does not have an innate natural power that allows one to grasp the deduction of propositions, because learning and practice do not provide what is needed. Now this is not so. Among mankind there are, indeed, some who are naturally gifted and have a good capacity for determining propositions, without much learning or having applied themselves to acquiring knowledge of these things; others are those who have applied themselves and acquire knowledge of the fundamentals and the methods, without being endowed with a good natural capacity. But when someone has an innate natural capacity, and applies himself to learning and one practices, then he is successful and distinguished. When, on the contrary, someone has not got that perfect power, but applies himself and learns, he can do exceptionally well thanks to his learning. As for someone who has the [natural] power and does not learn the fundamentals or practice doing geometrical constructions, he will not acquire any kind of profit without learning. If this is so, if anyone believes that geometrical deduction cannot be carried out except through an innate power, without learning, his belief is mistaken.

---

[1] The word is *al-ashkāl* (sing. *al-shakl*). Depending on the context, we translate it either as 'figure', or as 'proposition'. See p. 569, n. 35.

What is required, first of all, by the beginner in this art, is to know the theorems which are set out order following the common notions, even if that is part of what he is aiming for, that is to say the propositions he seeks to deduce; our objective in this is indeed [finding] methods such that the path leading to them starts out from the theorems and not only from the common notions that precede the theorems. One may indeed speak at great length about these common notions, and Euclid has spared us this in his book the *Elements*, thanks to the theorems that he introduced, [and] that we have mentioned.

As for the theorems preliminary to the objectives,[2] it is difficult to separate them out, because they form part of those we describe as lemmas and necessary consequences, because of the fact that in geometry parts are interconnected one with another: the first of them are lemmas for the later ones, one after the other, as if they were linked to those that follow them up to a certain limit. Now here we are dealing with something equivocal; however we summarise what there is to say in an adequate way, following what Euclid set out in the *Elements*. If someone says: if this is so, how is it possible to obtain the theorems, while the question of deducing propositions has no end to it? Or: why does one not limit oneself to the common notions? I tell him that Euclid applied himself to obtaining these propositions, in a balanced way, because if he had limited himself to the common notions, it would have been difficult for the scholar to proceed to deduction starting from the common notions, without lemmas among the geometrical theorems such as those Euclid has put in order after the common notions. Nor did he seek to make them especially numerous. Anyone whose objective is [to acquire] this art must obtain every one of the theorems Euclid produces in his book *The Elements*, because there is a considerable distinction between the thing and acquiring the thing; and he must get clear in his mind the types of the theorems and their properties[3] in a way that is certain, so that, if he needs to seek for their properties, he will be equipped to find them, and if there is anything he needs to deduce, he will need to investigate and to use his imagination to see the lemmas and the theorems which belong to the same type or to a type that is common to them all.

Example: If we want to determine a figure of the type of the triangle, we shall have to think of all the properties of triangles, the theorems mentioned by Euclid and what necessarily follows for the properties of triangles, in regard to angles, arcs, sides, parallel straight lines, so as to

---

[2] That is, the propositions we seek.

[3] Throughout, we are concerned with specific properties.

make that easy for us[4] and for us to be ready for determining them. Indeed, among figures, some have a property or properties in common, some have none in common, for some this community [of properties] is closer and for others it is more distant, depending on their degree of similarity, on their proportionality and on their homogeneity.

If we are seeking to determine one of the propositions by means of a lemma – by 'lemma' we mean the proposition which precedes it and which is the starting point for determining it – and if it is difficult for us to determine it by means of this lemma, then we must try to do it with the help of lemmas associated with this lemma, assuming what we are looking for by means of this lemma is true. From this assertion it necessarily follows that every proposition determined by starting from one of the lemmas can be determined, in the way we have mentioned, starting from the lemmas associated with it or from some of them, according to the degree of correspondence [with the lemma or lemmas]. Among the properties of the propositions, there are ones it would be easy to determine by means of many lemmas, different ones, and by proceeding in many ways; some one can determine by means of a single lemma; ones that have no lemma, even if the truth of the proposition is imagined and defined as in accordance with nature. This is a necessary result of the closeness of the correspondence between the properties of the lemmas – or of their distance from them – in regard to the properties of the propositions.

It is possible, in the same way, that the propositions have lemmas, and that their lemmas themselves also have lemmas and that one can determine these propositions starting from the lemmas of the lemmas. This property too results from what the propositions have in common, which we have mentioned. It is possible, again, that it is difficult to deduce the propositions – given that they require one to deduce successive lemmas – starting from a theorem or theorems, for which we shall later give some examples, God willing. Perhaps they need numerous theorems and numerous lemmas which are not successive, but composite, as we shall also mention, God willing. Perhaps there will occur to the scholar a method thanks to which it will be easy for him to determine many difficult propositions, [a method] which might be the transformation that we shall explain and of which we shall give examples, God willing.

Another method will be easy for the scholar if he follows it: he assumes the objective he has in view as if it had been constructed, if what is sought is a construction; or established, if he is seeking a property. He then analyses it in successive lemmas or composite lemmas, until he arrives at

---

[4] Lit.: him.

lemmas [already] established as true or false. If he arrives at lemmas that are true, it necessarily follows that he will find what he seeks; if he arrives at lemmas that are false, it necessarily follows that he will not find what he seeks. We call this method 'analysis by inversion' (al-taḥlīl bi-al-'aks).

This method is one of those most commonly used, compared with the other methods. We shall give an example of this later, God willing.

Synthesis (tarkīb) is the inverse of analysis; synthesis in fact consists of employing the method leading to the result through the lemmas, whereas analysis consists of employing a method that will lead to the lemmas which produce what one seeks.

It is a fact of geometry that, through it, the unknown becomes something either constructed or becomes known. From then on one has either constructions or properties. So the scholar must begin by thinking deeply about the question and the things that are sought. In fact, either the question is possible in itself and in accordance with nature but is not [possible] for us; or it is impossible for us to seek to decide it, from the lack of lemmas, as for the quadrature of the circle; or the responses to it are indeterminate and the examples for it innumerable – indeterminate things, that is to say ones lacking complete determinacy which would separate them out from other things; or something one can deduce, but nevertheless only through many lemmas, such as the final propositions of the book of the Conics, which are not easy without the lemmas that Apollonius introduced, and like the final propositions of the book On Circles;[5] or something for which one needs intellectual skill, because one has to imagine, at the same instant, many figures as constructed, in addition to theorems and lemmas; in general, that takes place when we are seeking properties. The man who carries out research in this way is called Archimedes in the language of the Greeks, we are referring to the Geometer. If his purpose is to deduce one of the propositions, the scholar must make the beginning of thought is the end of action and conversely, and inversely,[6] as we have mentioned earlier, and

---

[5] This clearly refers to the Tangent Circles of Archimedes cited by al-Nadīm as among the writings of Archimedes (Kitāb al-Fihrist, p. 326); there is a non-critical edition in Rasā'il Ibn Qurra, Osmania Oriental Publications Bureau, Hyderabad, 1947.

[6] This expression (bi-al-'aks), obviously an idiomatic one, seems to draw its original inspiration from Aristotle. In fact, in the production of a phenomena Aristotle distinguishes between νόησις (fikr), which concerns thinking about necessary conditions, and ποίησις ('amal), which refers to carrying out the construction (Metaphysics, Z, 7, 1032 b 15–30; The Motion of Animals, 7, 701 to 31). Al-Sijzī uses this expression to describe analysis and synthesis; which puts a considerable distance between him and the Aristotelian inspiration. Later, al-Samaw'al returns to this same expression in the sense in which it is used by al-Sijzī (see Al-Bāhir en algèbre d'As-Samaw'al, ed.

that by giving oneself the thing that is sought at the start of the procedure and by drawing out a necessary result from it, starting from the lemmas through which it has been analysed.

Among the ancient geometers, some had recourse to subtle ingenious procedures, when it was difficult for them to deduce what they sought, such as the one whose objects of research were related to proportion, and in investigating them he made use of numbers and multiplication; or the one whose object of research was the area of a figure or equality, and for it he used a drawing on silk or paper, or weighing; he uses other ingenious procedures which are similar to these. Such are the methods of discovery in this art. We shall describe them one by one, so that the scholar may grasp them in his mind and gain possession of them by the will of God the All-High and the goodness shown in His help.

The first one is skill, thought, and the act of calling to mind the conditions that have to be set up.

The second is to acquire the theorems and the lemmas in a comprehensive way.

The third is to persist in using their[7] methods in a comprehensive way, correctly, so as not to rely only on the theorems, the lemmas, their constructions and their ordering, which we have mentioned, but to combine with them skill, intuition and ingenious procedures.[8] The nub of this art is in fact to do with the nature of the ingenious procedures and not only with the mind, but also the natural disposition of those who work, become practiced, become ingenious.

The fourth is to know what they[9] have in common, their differences and their properties. In fact, the properties, their resemblance and their opposition, are, according to this theory, different from listing the theorems and the lemmas.

The fifth is the use of transformation (*naql*).

The sixth is the use of analysis.

The seventh is the use of ingenious procedures (*ḥiyal*), as they were used by Hero.

S. Ahmad and R. Rashed, Damascus, 1972, p. 73 of the Arabic text). It is not as yet known whether this same expression had been used by any of the commentators on Aristotle.

[7] That is of the theorems and the lemmas.

[8] This term is used to describe the use of mechanical devices. See al-Fārābī, *Iḥsā' al-'ulūm*, ed. 'Uthmān Amīn, Cairo, 1968.

[9] See n. 7.

Given that we have completed our account of these things, and we have set them out freely, let us now come to some examples for each of them, so that the scholar may grasp their substance. In fact, in this art we speak in one of these two ways: one is to use words freely, in the way of fantasy and imagination (*takhayyul*); the second is to set things out in full, in the way of exhibiting and presenting examples, so that one experiences them and grasps them perfectly.

But since in this art one speaks only in these two ways, and we have finished [what we wanted to do] in one of them, and that in a general and sketchy way, we must now move on to the second style, the one that consists of exhibiting things, of transmitting items of knowledge, of presenting examples in a comprehensive manner. It is God the All-High who aids us in the search for the truth and guides us on the path of rectitude.

Examples

Question regarding the construction of a figure: How to find two straight lines proportional to two given straight lines one of which is a tangent to a given circle and the other meets the circle and is such that, if it is drawn inside the circle, it passes through its centre?

We suppose, by analysis, that the figure has been constructed so that we are looking for its lemmas. Example: we suppose the ratio is the ratio of *A* to *B*, and the circle is the circle *CD*, and the two straight lines *GE* and *GC* are in the ratio of *A* to *B* – these are those that we seek; as if the figure were constructed and found and as if we had its construction from what we had set out, namely that if we draw *GC* in the circle to *D*, *CD* will be a diameter of this latter. We next seek to know from what construction and from what lemma we have found the construction for it.

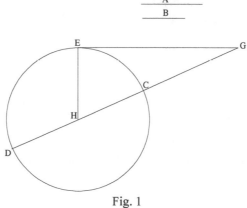

Fig. 1

Since the point G and the straight lines GC, GE and the position of the contact with the circle DC at the point E, are all unknown to us, and since the magnitude[10] of the angle at G is equally unknown, thus there will be a difficulty when it comes to determining the figure. This is the intuition (hads) that I mentioned earlier in regard to knowing the degree of difficulty of the problem: in fact if the figure includes many unknowns, it will be difficult to find it starting out from things that are known, particularly if it appears in a form in which no relationship can be set up between its figures, as we have mentioned. In this figure, there is no accessible relation between the straight lines GC and GE and between the circumference of the circle, nor [is there one] between the angle G and the arc CE. Next we use intuition and also thought, then we proceed towards constructing it by means of transformation (naql), from what we have mentioned: it is easier to determine difficult figures by means of the latter.

We say: how to set out two straight lines GC and GE in a form such that if we draw a circle, it will have GE as a tangent and will meet GC? That will be possible only if we make the angle G and if we know it.[11] So it is necessary for us to seek to know the angle G. Now knowledge of it will not be accessible to us except by seeking another thing knowledge of which is of the same type: we are dealing with angles. How should we try to find that, starting from putting together the straight lines GC and GE, or GE and GH, or GE and GD? Because that is not possible for us in this figure starting from combining other straight lines.[12] It is here that intuition and thought come in. If we join E to C, then perhaps it will be difficult for us to find that, and perhaps we cannot get it by employing this method, because the angles formed there, in that figure, are also unknown from these lemmas. We join E to H; it is here that among the three angles we find the known angle E. Next we must look for the shape of the triangle GEH by combining straight lines and angles; we proceed to a second investigation, once this latter [i.e. angle E] is found. If we find what we have just been seeking, then what we were seeking will be established for us, that is to say that the shape of the triangle GEH is identified as that of a right-angled triangle, such that the ratio of one of its sides to the hypotenuse minus the remaining side is equal to a given ratio. Our original question is thus reduced to this question.

The method we have followed now leads to what the question requires. As it has become our custom to do, we suppose that the triangle is con-

---

[10] Lit.: convexity, curve (ḥāl inḥidāb).
[11] See commentary, pp. 530–1.
[12] Lit.: of another straight line.

structed; it is the right-angled triangle *IKM*, whose right angle is the angle *K*. But *NM* is equal to *KM*, so the ratio of *IN* to *IK* is equal to the ratio of *B* to *A*. It is then here that one employs skill and thought; because the more primitive what we are looking for, [the more] we must employ thought and intuition and not learning. We need to find out how to take *IK* to be such that the ratio of *IK* to *IN* is equal to the ratio of *A* to *B*. Let us draw *KM* of indefinite length, then, in our imagination, let us draw *IL*; if you extend it to [meet] the straight line *KM*, the difference between the movable straight line *IL* and what reaches as far as the straight line *KM* is equal to the straight line which lies between the point *K* and the point of intersection with the straight line *KM*;[13] in consequence, we have here a search for two unknowns.

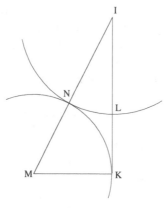

Fig. 2

We draw a circle with centre *I* and distance *IL*, using the fact that we have imagined the straight line *IL* as movable – [rotating] about the point *I* – so as to establish that, in the course of the imagined motion, *L*, the end-point of the straight line *IL*, does not fail to fall on the circumference of the circle. But the basic shape of the triangle is placed before us so that we perceive the figure correctly, by seeing it, at the moment of construction. We then look for the centre of a circle, <a centre> that lies on both the straight lines *IM* and *KM*. It is here that in consequence one uses intuition and thought in a correct way, but that will not be possible either, unless by means of a supplementary construction. We imagine that construction:

How to draw *IN* to *S* in a shape in which the straight line *KM* divides it into two halves and with the condition that the whole of *NS* is double *KM*. The problem reduces to another figure which is the following one.

---

[13] See commentary, pp. 531–2.

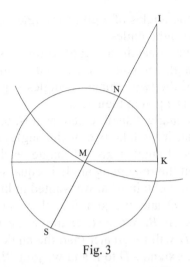

Fig. 3

Next, here we employ reflection; as usual we think of the objective as achieved, and that by supposing *INS* satisfies the condition that *NS* is double *KM* and *NM* is equal to *KM*. With centre *M* and distance *MK* we draw the circle *KS* – it is clear that *IK* is a tangent to the circle – so as to employ intuition and thought. Once this is done, we must look for the property of this figure starting from the fact that we have a tangency, [a subject] first treated by Euclid in the *Elements*. The most immediate property of this figure is that the power of *IK* is equal to *IS* by *IN*. In consequence, we have found this construction provides help in starting out from this property, and that in making *NS* a straight line on the extension of *IN* such that the power of *IK* is equal to *IS* by *IN*. If we do that, then our construction will be even easier; in fact, we have found the straight lines *IN*, *IK* and *IS*. Consequently it remains for us to find the shape of *IS* according to how *KM* divides *NS* into two halves. We first divide *NS* into two halves in *M* and, in imagination, we rotate *IS* about the point *I*, then *KM* cuts *NS* into two halves; which is easy by means of construction, from the fact that with centre *I* and distance *IM* we draw a circle that *KM* cuts at the point *M*. We draw *IMS* and we cause *SKN* to rotate; then we have constructed this figure as we wanted it. We next transform it into the given circle using similarity and the ratio, and we have proved it. This is what we wanted to prove.

Since the exercise of intelligence in deducing properties is more satisfying than constructions, accordingly we give an example of looking for properties of figures. We suppose [we have] a triangle *ABC* and we look for the property of its angles, namely: the sum of the three angles is equal to the sum[14] of the angles of a given triangle, before it is known that the sum is equal to two right angles.

The method for our research in regard to this first objective is to suppose that one of the angles remains as it is and to make the sides round it vary so that the sum of the two remaining angles is greater or smaller than that of the first two or is equal to their sum.

We choose that the angle A shall remain as it is, unlike the other angles. But, it being given that, if we take two of the angles of a triangle given to be equal to two angles of another given triangle, each to its homologue, it necessarily follows that the remaining angle is equal to the other remaining angle, and thus we do not obtain what we wanted to know.

We extend AC to D and we join BD, then the angle ADB will be smaller than the angle ACB. We next examine the two angles ABC and ABD, so the angle ABD will be greater than the angle ABC. We repeat this process once more, we extend AD to E and we join BE. The angle E will be smaller than the angle ADB and the angle ABE will be greater than the angle ABD.

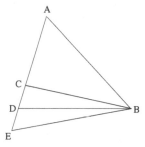

Fig. 4

We repeat this process; then the angles that lie on the side AC continue to decrease with respect to the one which was there at first, and the angles on the side of the straight line AB continue to increase at the point B with respect to the one which was there at first. However, we now need to examine whether their increase and their decrease are coherent in the natural order of things, that is to say compensate one another: for what is added on one side, one takes away something equal from the other. If we find an order as in this example, we shall have found a property for triangles, absolutely:[15] that is to say that the sum of their three angles is the same. In what

---

[14] The word 'sum' has sometimes been added, to conform with English usage.

[15] Lit.: absolute triangles (*al-muthallathāt al-muṭlaqa*). That is, triangles as such.

way do we look for this equality? First of all, as is our custom,[16] we take it
that the sum of the two angles *ABC* and *ACB* is equal to the sum of the two
angles *ABD* and *ADB*, because we have required this step at the beginning
of this. If this is as we have assumed, it necessarily follows that the two
angles *CBD* and *CDB* have a sum equal to the angle *ACB*, and this because,
if this is so, the sum of the two angles *ADB* and *CBD*, to which one adds
the angle *ABC*, is equal to the sum of the angles *CBD* and *CDB*, to which
one adds the angle *ABC*.

In consequence we have what we were looking for here. So if we follow our procedures correctly and obtain a true result, something that is not
impossible, then what we proposed was true. But if what follows is absurd
and impossible, then it necessarily follows that the sum of the angles of the
triangle *ABC* is not equal to the sum of the angles of the triangle *ABD*, nor
to the sum of the angles of any other triangle except one that is similar to it;
we should then need a construction more appropriate to our objective, that
is to say much more closely related to it, or which is of a type which is
associated with it.

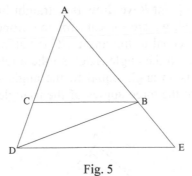

Fig. 5

We draw *DE* parallel to *BC* and we join *AE* and *AD* so that the two
triangles are similar and equal angles are formed, so that one can subtract
them one from another. There necessarily follows for us a result that is
either true or false, because we have first of all supposed it to be true. The
angle *BDE* is equal to the angle *DBC* and the sum of the angles *EDB* and
BDC is equal to the angle *EDC*. Consequently, the sum of the angles *BDC*
and *DBC* is equal to the angle *BCA*. So what we were looking for necessarily follows. But we are investigating the equality of the sum of the
angles of the triangle *ABC* to the sum of the angles of the triangle *ABD*.
Consequently, we have found a property of the angles of a triangle, better,

---

[16] That is, in analysis.

two properties, because at the end of our research we have found that if we extend one of the sides of the triangle, then an exterior angle is formed equal to the sum of the two interior angles which are opposite it in the triangle.

Now we are looking for another property of the angles; that is to say that after having proved that the sum of the angles of any triangle is equal to the sum of the angles of any other triangle, we are looking for the size of the sum of these angles. For this research we need a measure by which to measure these angles and this measure must be of their type, that is to say the right angle. We must take a triangle and make one of its angles a right angle, because if we make two of its angles right angles, our construction will not form a triangle, but the two sides will be parallel and will not meet one another; while a triangle is formed only by the meeting of its three sides. As a consequence, we must suppose the two sides enclosing the right angle to be equal. We suppose we have a triangle *ABC*, right-angled and isosceles, its right angle is the angle *A*; consequently we employ a parallel straight line because it, more than any other, resembles what is appropriate in this place. From the point *B* we draw the straight line *BD* parallel to *AC*, an angle is then formed, we are looking for the properties it has. We then find the angle *DBC* is equal to the angle *BCA*; but we have supposed that the angle *BCA* is equal to the angle *ABC*, so the angles *ABC* and *DBC* are equal. But their sum is an angle equal to the angle *BAC*; consequently it follows that the sum of the three angles of the triangle *ABC* is equal to two right angles.

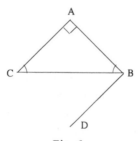

Fig. 6

But we have found that property in a specified triangle, that is to say one in which one of the angles is a right angle and the two sides that enclose it are equal. Now [the sum of] the angles of a specified triangle, and that of the angles of a triangle as such,[17] are equal – as we have men-

---

[17] Lit.: absolute triangles.

tioned. It appears clearly to us, in consequence, that the sum of the three angles of any triangle is equal to two right angle. This is what we wanted to explain.

This is one of the methods for looking for properties. So you must educate your understanding and your mind in this art. In fact, in this process which is the deduction of propositions, educating the understanding and refining the mind is more useful than reading the books on geometry prescribed by the ancients, whose purpose in doing that was to give the reading of geometry precedence over all the books on philosophy of mathematics, as well as to educate the mind.

Let us take another example in connection with another question so that someone who is diligent may exercise himself in this art and see opening up what was enclosed in this question: How to divide a given triangle into three parts in a given ratio?[18]

Let us suppose we have the triangle ABC and the ratios of D, E and G; the shape of the division must be produced by three other straight lines which meet one another in the middle[19] of the triangle.

Let us suppose the triangle has been divided as we wished, that is to say into the triangles ABH, ACH, BCH; and let the ratio of the triangle ABH to the triangle ACH be equal to the ratio of D to E and the ratio of the triangle ACH to the triangle BCH be equal to the ratio of E to G.

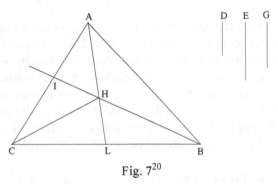

Fig. 7[20]

[18] See mathematical commentary, Section 3.3, pp. 534–7.

[19] That is to say, inside it.

[20] In the figure in the manuscript we have K in place of L, which is different from what we find in the body of the text.

Next we think the matter over to look for a construction that is useful for this question. We extend BH to I so as to show that the ratio of the triangle ABH to the triangle BCH is equal to the ratio of AI to CI. If we divide the side AC in the ratio of D to G, then the division into the two triangles will be along the common straight line BI; this is necessary. We divide AC in I in the ratio of D to G and we join BI. It is necessary that the point of division lies on the straight line BI and that the angle formed from the triangle which is on the side of the straight line AC lies on the straight line BI. So we need to construct a triangle starting from the side AC and two other straight lines drawn from the points A and C and an angle <whose vertex> lies on the straight line BI still with the ratio to one of the two other remaining triangles equal to one of the ratios of E to D or of E to G.

The best of the constructions is the first construction, because its procedure is true: We do with the side BC what we did with the side AC by dividing the side BC, at the point L, in the ratio of D to E. Let us join AL; it is clear that the ratio of the triangle AHB to the triangle AHC is equal to the ratio of D to E.

But we have proved that the ratio of the two triangles whose sides are drawn from the points A and C, and which meet one another on the straight line BI, is equal to the ratio of the triangles ABI and BIC. Consequently, the three triangles are constructed in the triangle ABC in a given ratio. This is what we wanted to prove.

Another method: We suppose the three triangles have been constructed and we extend BH to I. We must attempt to find a triangle AHB; all the same, we imagine it has been constructed, as is our custom, for determining figures by the method of analysis. We reflect on it in mathematical terms and we look for a method whose character is close to that of the first one, that is if we divide BI at the point H so that the ratio of the triangle ABH to the triangle AHI will be known and the triangle ABH will be known. But the ratio of the triangle AHI to the triangle HIC is known because we know the ratio of AI to IC.

Let us consider the two triangles AIH and AHB, both of them, individually, if we can know the ratios;[21] by composition of some of them,[22] the two triangles will be divided according to the known ratio, once we know

_____

[21] We are considering the ratios of AIH to AHC and of AHC to ABH, whose product is the ratio of AIH to ABH.

[22] The ratio of HIC to AIH, which is known; composition gives the ratio of AHC to AIH.

that the ratio of two arbitrary triangles, positioned as the two triangles *ABH* and *CBH* are [positioned], is known to us. We are looking for this method; shall we find it or not? If the ratio of *BH* to *HI* is known to us and if the ratio of *AI* to *IC* is known, once this triangle[23] has been constructed, the ratio of the triangle *HCA* to the triangle *AHB*[24] is known, which is the objective.

Now the triangle has been divided in a ratio we did not intend, we need to divide one of the straight lines in proportion into the same parts as those into which the two triangles *AHI* and *HIC* were divided. We divide *E* into two parts such that the ratio of one to the other is equal to the ratio of *D* to *G* and we put the ratio of *BH* to *HI* equal to the ratio of *D* to one of the two parts of *E*. We join *AH* and *CH*. The ratio of the triangle *ABH* to the triangle *AHI* is equal to the ratio of *D* to one of the parts of *E*, and the ratio of the triangle *AHI* to the triangle *HIC* is equal to the ratio of one of the parts of *E* to the remaining part. So the ratio of the triangle *ABH* to the triangle *AHC* is equal to the ratio of *D* to *E*. Now we have shown that the ratio of the triangle *ABH* to the remaining triangle *BCH* is equal to the ratio of *D* to *G*. This is what we wanted to prove.

Another method for constructing this figure is the following: We divide the side *AB* in the ratio of *D*, *E* and *G* <at the points> *H* and *I*. We join the straight lines *CH* and *CI*. It is clear that each of the triangles we seek is equal to each of the triangles <obtained>.

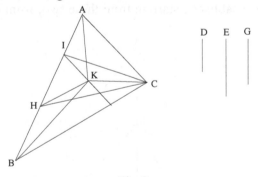

Fig. 8

[23] This is triangle *AHB*.

[24] We have $\dfrac{HCA}{AHB} = \dfrac{HCA}{AIH} \cdot \dfrac{AIH}{AHB} = \dfrac{AIH + HIC}{AIH} \cdot \dfrac{IH}{HB}$, the product of the compound ratio and the ratio in which *BI* has been divided.

This method is foreshadowed in the first approach; we next think this over and search for the point at which the sides of the triangles equal to these constructed triangles meet one another. We draw *IK* parallel to *AC* because we know that any triangle equal to the triangle *AIC*, with base *AC*, meets the straight line parallel to *AC*. In the same way, we draw *HK* parallel to *BC* for the reason we have already mentioned; so they meet one another in *K*. We join *AK*, *BK* and *CK* and we consider that the triangle is divided as we wanted. It is clear how to proceed by these methods, even if we do not explain them completely.

There is another method for this figure, however it reduces to the two methods that we have mentioned. Accordingly we have put it aside and have decided not to describe it.

As an example of what we have said: if there is a lemma or a theorem among the lemmas and the theorems, and if this lemma or this theorem has a lemma, then this lemma also has a lemma, then one can prove the lemma or the theorem starting from the lemma of its lemma; we suppose we have a circle *AB* with centre the point *D*; on the arc *BAC* we put an angle *BAC*; we join *BD* and *CD*. I say that the angle *BDC* is double the angle *BAC*.

Euclid proves this using the property of the exterior angle of a triangle in which one extends one of the sides; it is proposition thirty-two of the first book of his work the *Elements*. But propositions twenty-nine and thirty-one are two lemmas for this proposition. So we must check to see whether it can be established starting from these two, from one of the two, or not.

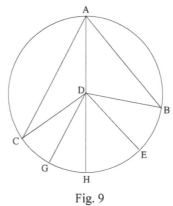

Fig. 9

Through the point *D* we cause to pass a straight line parallel to *AB*, which is *ED*, and another straight line parallel to *AC*, which is *DG*; we extend *AD* to *H*. This is to use proposition thirty-one which is a lemma to its lemma. But the exterior angle *EDH* is equal to the interior angle *BAD* and the angle *EDB* is equal to the angle *DBA*, which is alternate to it. Now the angle *DBA* is equal to the angle *BAD*, because the sides are equal. Now the sides being equal which appears in this figure does not derive from the lemma, it is a property of the figure that it imposes on this proposition.[25] So let us bear this meaning in mind.

Consequently, each of the angles *BDE* and *EDH* is equal to the angle *BAD*, and the angle *BDH* is double the angle *BAD*. It is also clear, for the same reason, that the angle *HDC* is double the angle *DAC*. Consequently, the whole angle *BDC* is double the whole angle *BAC*.

This is to use proposition twenty-nine. So we have used the lemmas of their lemmas; which has shown it is true. This is what we wanted to prove.

We give examples to illustrate how propositions contribute to one another, with the help of propositions built up by starting from the division of a straight line in mean and extreme ratio.[26] In fact, propositions built up from that commonly involve [the number] five. The construction of the regular pentagon in fact involves the division of a straight line in mean and extreme ratio, starting from composition[27] of the semidiameter and the side of the decagon which has a ratio to the side of the pentagon, because it is the chord of half its arc; this gives a straight line divided in mean and extreme ratio. Of the two chords in the circle circumscribed about the pentagon, that is to say the chords drawn from angles of the pentagon inscribed in the circle, one [of these chords] divides the other in mean and extreme ratio. For any straight line is divided into two parts in mean and extreme ratio if, when we add to its greater part half of the whole straight line, then the square of the sum is five times the square of half the straight line. If we divide a straight line into two parts in this ratio and if to the greater part we add double the smaller part, then the square of the whole straight line is five times the square of the first part. If we divide a straight line in mean and extreme ratio and if to the smaller part we add half of the greater part, the square of the sum is five times the square of half the greater part.

---

[25] The word *shakl* is translated here successively as 'figure' and as 'proposition', which are the two senses in which the term is used.

[26] In his *Barāhīn Kitāb Uqlīdis fī al-uṣūl 'alā sabīl al-tawassu' wa-al-irtiyāḍ* (*Proofs from Euclid's Book on the* Elements *according to the Path of Development and of Practice*), al-Sijzī comments on the propositions at the beginning of Book XIII of the *Elements* on division in mean and extreme ratio, as well as on the pentagon (see ms. Dublin, Chester Beatty, 3652, fols 28ʳ–29ᵛ).

[27] Arabic: *tarkīb*.

Starting from the composition of the sides of a square divided into five equal parts and from their separation, there arises a straight line divided in mean and extreme ratio. By composition, I mean: adding certain straight lines to others and joining them so that they become a single straight line; and by separation: dividing the greatest into two parts so that one of them is equal to the smallest part.

Example: We suppose we have the square of *AB* in such a way that the angle *E* is a right angle, so that the sum of the squares of *AE* and *EB* is equal to the square of *AB*. To find another straight line, say *AD*, such that double its square is equal to the square of *AB*; this straight line will be equal to *AG*. Finding the straight line *AD* is easy, it is done by drawing a semicircle *ADB*, which we divide into two halves at *D*; we join *AD*. Double the square of *AD* is equal to the square of *AB*. In consequence, we need to find a straight line *AE* such that if we draw *EB*, *EB* will be equal to *EG* and *GA* to *AD*, and which thus leads to our objective.

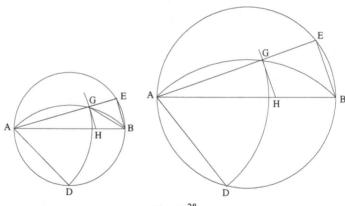

Fig. 10[28]

To obtain this, we work out what is required to determine the straight line. In fact, finding *GE* equal to *EB* arises from the equality of the angle *EBG* and the angle *EGB*. So it is clear that if we draw *AE*, if at the point *B* of the straight line *EB* we construct an angle equal to half a right angle and if we join *BG*, this gives a straight line *GE* equal to the straight line *EB*.

After that we have to make *AG* and *AD* equal; so we must imagine that the straight line *AE* rotates about the point *A* – with centre *A* and distance *AD* let us draw the circle *DG* – and another straight line which passes through point *G* of a circle *DG*; it is necessary that this other straight line

---

[28] These figures do not appear in the manuscript.

meets the circle *DG*. So we must construct an arc that subtends an angle equal to a right angle plus half a right angle, as the arc *AGB*, it being given that, if the circle *DG* cuts it and if we draw *AG* to *E*[29] and we join *BG*, the exterior angle *AGB* is equal to the sum of the interior angles *E* and *B*. But it is clear to us that the angle *E* is a right angle so that it necessarily follows that the angle *B* is equal to half a right angle. As a result, the angle *B* is equal to the angle *G* of the triangle *GEB*, so that the straight line *EG* is equal to the straight line *EB* and the straight line *AG* is equal to the straight line *AD*. Accordingly, we have thus divided *AE* at the point *G* in [such] a way that the square of *AG* is equal to double the product of *AE* and *EG*. So the square of the sum of *AE* and *AG* is equal to triple the square of *AE*. So we suppose that the square of *AB* is equal to triple the square of *AD*, which is equal to *AG*, and as before we prove that the square of *AG* is equal to the product of *AE* and *EG*. So *AE* has been divided as we wanted.

Now by transformation, if we draw *GH* parallel to *EB*, *AB* will be divided as we wanted. The proof of this is easy. This is what we required to prove.

Let us now look for how to prove the proposition that Ptolemy presents in his book *The Almagest*: If we have two unequal arcs of a given circle, then the ratio of the chord of the greater arc to the chord of the smaller arc is less than the ratio of the greater arc to the smaller arc.[30]

In this problem we need to take thought, to work out composite constructions and joining up figures; nevertheless, this proposition and its analogues are easy, given that we know the truth of the matter, and that, moreover, the constructions by which he proved it are shown [by Ptolemy]. In these two respects, this problem and its analogues will be easy. But since it was difficult for us to prove what is sought without adding another construction to it, we are obliged to have recourse to another construction such that, if we add it to it, [then] by starting from the combination of the two, it will be easy to prove it. Employing the construction that Ptolemy gave, our construction will be easy, since we know how he adopted an approach and which things he added to it to prove it. He added triangles made up of straight lines and of arcs; then he proved it by using these triangles as intermediaries, from their angles, from their chords and from their arcs.

---

[29] *E* is on the circle *ADB*.

[30] Claudius Ptolemy, *Almagest*, English trans. G. Toomer, London, 1984, Book 1, Chapter 9, p. 47.

Here we say something that is not concerned with this question. We do, however, need it. In fact, we have derived our procedure in this proposition from the procedure given by the ancients, insofar as the figures have ratios and properties such that, if a skilled man thinks about them, they[31] appear to him to be connected and mingled with one another, as if they had become a single matter and were all in a single state; because, if we imagine them as differing in species and agreeing in type, they have links and connections that are necessarily attached to the essences of their properties, which are common to all things of this type.

For example: two chords cutting one another in a circle: in fact, those of one set are proportional to the others, this is an absolute statement about them with regard to their type. Next comes the mode within the species and the state in which the two chords one of which cuts the other lie within the circle, from which there necessarily follows its property, which is the essence of the proportion. If we examine what that state is, using as intermediary the propositions that provide a grasp on it as well as on the essence of the property – like the necessity of the proportionality of the lines which enclose areas to the surfaces [they enclose]; like the fact that an arc intercepted by equal angles; and like triangles constructed on equal bases and between two parallel straight lines being equal – if we examine these and their analogues, we find their properties and their essences, God willing. It is for this reason and ones like it, starting from properties of the figures and the ordering of them, that we in some way concentrate our attention on their nature, right at the beginning, before finding what we are looking for.

We then return to what we said: we suppose we have the arc *BAC* and we divide it into two unequal parts at *A*, the greater part was *AC*. We draw two chords *AB* and *AC*. I say that the ratio of the arc *AC* to the arc *AB* is greater than the ratio of the chord *AC* to the chord *AB*.

Proof: We join *BC*, we extend *BA* to *K* and we put *AK* equal to *AC*.

We have proceeded in this order while we are adding to this figure constructions that bring about this configuration; no other construction is possible for us.

We then join *CK*; so we have first of all added two triangles to the figure for the proposition (*ṣūrat al-shakl*) that we are looking for, one is the triangle *ABC* and the other is the triangle *AKC*; but the objective is not attained by these two triangles. We draw *AD* parallel to *KC*. If we draw *AD*

---

[31] The figures.

parallel to *KC*, it is because of the order that this entails: the equality of the angle *DAC* or *DAB* with the angle *AKC*.

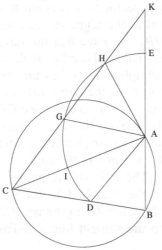

Fig. 11

Here we then exercise our ingenuity, by drawing *AG* parallel to *BC*; so we need to construct segments of circles so as to know the angles themselves and to find the magnitude of the proportion between the sides of the triangles and of the angles of the arcs and then we push on to find the proportion between the arcs *BA* and *AC* and the angles of the sectors. With centre *A* and distance *AG* we draw the arc *IGHE*. We have constructed this circle with centre *A* because the things we are looking for concerning the arc *IGHE* are proportional to the angles that are at the point *A*. We then seek to attain the objective of the construction: that the angles that are at the point *A* and the angles enclosed by the sides *AB*, *AC* and *BC* shall be proportional so that what we are aiming for necessarily follows.

It being given that the straight line *GH* is smaller than the straight line *GK*, we have that the arc *HE* is equal to the arc *GI*.[32] We join *AH*; we have *AH* equal to *AG* and the segment *GIC* equal to the segment *HEK*.[33]

We then try to attain our end by considering the proportionality between the segments, the arcs, the triangles and the sides. Here we first of all need to imagine the results [accomplished] and by analysis go from the

---

[32] See p. 544, n. 24.
[33] *Ibid.*

endpoint to our point of departure, then to set out from our point of departure to [arrive at] the end. It is here that we use intuition.

It being given that the sector *AGI* is equal to the sector *AEH* and that the segment *EHK* is equal to the segment *GIC*, we add together the sector *HGA* and the triangle *AHG*; we have that the ratio of the triangle *AGC* to the triangle *AGK* is greater than the ratio of the sector *AGI* to the sector *AGE*. The ratio of the straight line *GC* to the straight line *GK* is consequently greater than the ratio of the angle *IAG* to the angle *GAK*. But the ratio of the straight line *GC* to the straight line *GK* is equal to the ratio of the straight line *BA* to the straight line *AC*, because *AC* is equal to *AK*, so the ratio of the angle *CAG* to the angle *GAK* is smaller than the ratio of *BA* to *AC*. But the angle *CAG* is equal to the angle *ACD* and the angle *KAG* is equal to the angle *ABC*. The ratio of the angle *C* to the angle *B* is thus smaller than the ratio of the straight line *BA* to the straight line *AC*. In consequence, the ratio of the arc *AC* to the arc *AB* is greater than the ratio of the straight line *AC* to the straight line *AB*. This is what we wanted to prove.

We look for the proof in a different way: we suppose we have the circle *ABH* and the two unequal arcs *AC* and *CB* – the greater is *CB*; we repeat the procedure we have already described. We join *AB* and we divide the angle *C* into two halves with the straight line *CH*. The division of the angle into two halves occurs because it is given that the straight line *AB* is divided at [the point] *D*, such that the ratio of *AD* to *DB* is equal to the ratio of *AC* to *CB*, so the straight line *AB* will be an intermediate element in the constructions we need.

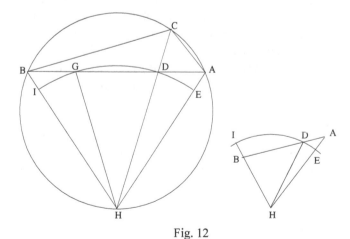

Fig. 12

Next we need, in this circle or outside it, a construction that will allow us to have [an angle equal to] the sum of the angles A and B at a single point, and that so that, if we take this point as a centre and if we draw an arc with a certain distance, that will accomplish our purpose. Now these constructions are at first not clear to us, however this procedure is a valid procedure.

We join AH and BH; the two angles AHC and BHC thus come together at the point H; now they are equal to the angles A and B. The fact of joining the two straight lines AH and BH at the point H – and of not drawing from the points A, C, B three straight lines that meet in another point, whatever point it might be, on the arc AHB – arises from the fact that if we have found what we are looking for by this construction, then it will be easier to find what we are looking for if we draw them to the mid point of the arc AHB, given that the straight lines AD and DB are proportional to the straight lines AC and CB. Now what is more accessible to proportionality and to order is easier to find. Next we have to look for an arc with centre H and an arbitrary distance. I do not as yet know what this distance should be so that from it there necessarily results, through the segments, in the arcs and the angles which are at the point H, through the straight lines AD and DB and through the triangles ADH and DHB, the excess of the ratio of the arc BC to the arc CA over the ratio of the chord BC to the chord CA. This is where we have a fallacy: when someone says 'with centre H and distance HD, we draw the arc EDI and we draw HB as far as I in his figure' and then proves it in accordance with this [figure], we say to him that it is impossible because the straight line AH is equal to the straight line BH and the endpoint of the arc falls at the point E, so the other endpoint falls at the point I, opposite the point E.

But, it being given that, in this proposition, the boundary of the segments and the triangles pass through the point D, in every case, as they did at first, with centre H and distance DH we draw the arc $EDGI$ so as to find, or not find, what we are looking for. We draw HG; the ratio of the sector GDH to the sector DEH is thus greater than the ratio of the triangle GDH to the triangle HDA. By composition, we have that the ratio of the sector DIH to the sector DHE is greater than the ratio of the triangle DHB to the triangle DHA; consequently, the ratio of the arc DI to the arc DE is greater than the ratio of the straight line DB to the straight line DA. But the ratio of the arc DI to the arc DE is equal to the ratio of the angle BAC to the angle ABC, which is equal to the ratio of the arc BC to the arc CA, and the ratio of the straight line DB to the straight line DA is equal to the ratio of the chord BC to the chord CA. This is what we wanted to prove.

But since Ptolemy's purpose is <to find> the part and half the part, the two arcs on which the proof is based must be smaller than a semicircle. For this problem we need this condition, because we are going to give a proof that is different from the earlier proof. For this we employ a different method, which is as follows:

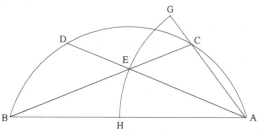

Fig. 13

We take the arc *AB* smaller than a semicircle, and we divide it into two unequal parts at *C*, the greater part being *CB*. We join *AC* and *CB*, we put *BD* equal to *AC*, we join *AD*, with centre *A* and distance *AE* we draw the arc *HEG* and we extend *AC* to *G*. It is clear that this line lies outside the arc <*ACB*>; it is also clear that *AE* is equal to *EB*, so the ratio of the sector *AGE* to the sector *AEH* is greater than the ratio of the triangle *ACE* to the triangle *AEB*. By composition, the ratio of the sector *AGH* to the sector *AEH* is greater than the ratio of the triangle *ACB* to the triangle *AEB*. But the ratio of the arc *GH* to the arc *EH* is equal to the ratio of the arc *CB* to the arc *DB*. Consequently, the ratio of the arc *CB* to the arc *DB* is greater than the ratio of the straight line *CB* to the straight line *EB*. But the straight line *EB* is equal to the straight line *AE* and the straight line *AE* is greater than the straight line *AC*, because we have taken the arc *ACB* to be smaller than a semicircle. Consequently, the ratio of the arc *CB* to the arc *CA* is much greater than the ratio of the chord *CB* to the chord *AC*. This is what we wanted to prove.

Let *ABC* be a given circle. The two arcs *AC* and *CB* are unequal, *AC* is greater than *CB*. We draw *AB* and from the point *C* we drop a perpendicular to *AB*; <let it be *CD*>. I say that the ratio of the straight line *AD* to the straight line *DB* is greater than the ratio of the arc *AC* to the arc *CB*.

Proof: We extend the perpendicular *CD* to *E* and we join *EB* and *EA*. We draw *EI* equal to *EB* and with centre *E* and distance *ED* we draw the circle *GDKH*. The ratio of the triangle *ADE* to the triangle *DBE* is greater than the ratio of the arc *HKD* to the arc *DG*, because the triangle *ADE* is larger than the sector <*EDH*> by the curvilinear quadrilateral *AIKH* and by the segment *IKD*, the arc *HD* is intercepted by the angle *AED* and the arc *DG* is intercepted by the angle *DEG*. In the same way, the arcs *AC* and *CB* are intercepted by these same angles. So the arcs are proportional. The ratio of *AD* to *DB* is thus greater than the ratio of the arc *AC* to the arc *CB*. This is what we wanted to prove.

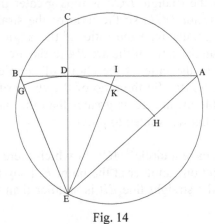

Fig. 14

The arc *AC* is greater than the arc *AB*. I say that the ratio of the chord of double the greater arc to the chord of double the smaller arc is smaller[34] than the ratio of the greater arc to the smaller arc.

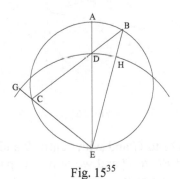

Fig. 15[35]

[34] The statement in the manuscript gives 'is greater than'. But in the course of the proof, al-Sijzī shows we are in fact dealing with 'is smaller than'. So we probably have an error by the copyist.

[35] See the figures for the two other cases in the commentary.

Proof: We draw the diameter *AE*, we draw *EB* and *EC* and we draw *BC*; let it cut *AE* at the point *D*. With centre *E* and distance *ED* we draw an arc *HDG*; so the point *G* falls either at the point *C* of the straight line *CD* or above this line, because if it falls below the straight line *CD*, then the angle *ADC* will be either a right angle or acute;[36] now this is not so. The ratio of the triangle *BDE* to the triangle *DCE* is thus greater than the ratio of the sector *HDE* to the sector *DGE*, so the ratio of the straight line *BD* to the straight line *DC* is greater than the ratio of the angle *HED* to the angle *DEG* and greater than the ratio of the arc *BA* to the arc *AC*. But the ratio of *BD* to *DC* is equal to the ratio of the chord of double the arc *BA* to the chord of double the arc *AC*. So the ratio of the chord of double the arc *AB* to the chord of double the arc *AC* is greater than the ratio of the arc *BA* to the arc *AC*. This is what we wanted to prove.

We suppose we have a circle[37] *ACB* in which there are the two chords *AC* and *BD* which cut one another at the point *E*. I say that the ratio of the straight line *DE* to the straight line *EB* is smaller than the ratio of the arc *AD* to the arc *CB*.

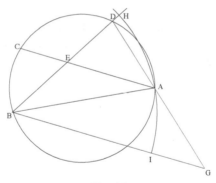

Fig. 16

Proof: We extend *DA* to *G* and with centre *B* and distance *BA* we draw a circle *HAI*; we extend *BD* to *H* and we draw *BG* parallel to *AC*. The ratio of the triangle *ADB* to the triangle *AGB* is smaller than the ratio of the sector *AHB* to the sector *AIB*. In the same way, the ratio of the straight line *AD* to the straight line *AG* is smaller than the ratio of the angle *DBA* to the

---

[36] See p. 551, n. 26.

[37] The figure in the manuscript seems to start with a semicircle *ACB*. The reasoning is the same for a circle: we can have arc *ACB* ≤ π or arc *ACB* > π.

angle *ABG*. But the angle *ABG* is equal to the angle *CAB* and they[38] intercept the arcs *AD* and *CB*, so the ratio of *AD* to *AG* is smaller than the ratio of the arc *AD* to the arc *CB*. But the ratio of the straight line *AD* to the straight line *AG* is equal to the ratio of the straight line *DE* to the straight line *EB*, so the ratio of the straight line *DE* to the straight line *EB* is smaller than the ratio of the arc *AD* to the arc *CB*. This is what we wanted to prove.

How to divide a straight line *AB* into two parts such that the product of the whole straight line *AB* and one of its two parts, to which we add the square of the remaining part, plus the given square *C*, is equal to the square of *AB*?

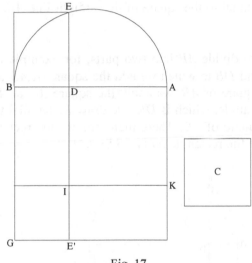

Fig. 17

So we need to apply a square to *AB*, because that is obvious and the final part of the construction depends on it. Next we suppose *AB* has been divided as we wished at the point *D*. If this is so, we have to draw *DE'* parallel to *BG* so that we know that the area *DG* is that enclosed by *AB* and *DB*; we then have to construct a square on the straight line *AD*. We construct the square *AI*. If the square *AI* plus the area *DG* plus the square *C* is equal to the square *AB*, then it is necessary that the area *KE'* is equal to the square *C*. But the two additional pieces are equal, so the area *KE'* is equal to the area *BI*, because they are the two additional pieces. But the surface *BI* is enclosed by the straight lines *AD* and *DB*. So if we draw a semicircle *AEB*, with diameter *AB*, and if we draw *DE* perpendicular to *AB*, then [the

---

[38] That is, the angles *DBA* and *CAB*.

sum of] the squares of the two straight lines *AD* and *DB* is equal to the square of *DE*, so the straight line *DE* will be the side of the square *C*. Consequently the side of the square *C* must not be greater than half the straight line *AB*, because we would not be able to construct that. So we have added another condition to the condition.

By synthesis, on *AB* we draw the semicircle *AEB* and in it we drop a perpendicular to *AB* equal to the side of the square *C*, which is *DE*. We extend it to *E'* and we apply the square *AI* to *AD*. But the product of *AB* and *BD* is <the area> *DG*, the square *C* is equal to *AD* by *DB* which is the area *KE'* and the square of *AD* is *AI*. So if we divide *AB* into two parts at *D*, we have that the product of *AB* and *BD* to which we add the square of *DA* and the square *C*, is equal to the square of the straight line *AB*. This is what we wanted to prove.

If we want to divide *AB* into two parts, for example at *D*, so that the product of *AD* and *DB* to which we add the square of *AD* and the square *C* is equal to the square of *AB*, we apply the square *AG* to *AB* and we make *AD* by *BD* a rectangle, which is *DK*; we draw *KI* parallel to *AD*. It is clear that *AH* is the square of *AD*. There then remains the rectangle *IG* equal to the square *C*. But the rectangle *IG* is *AB* by *BD*.

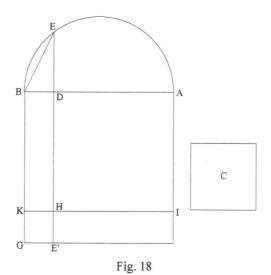

Fig. 18

By synthesis, we need to apply a semicircle *AEB* to the diameter *AB* and to place the side of the square *C* in it as a chord starting from the end-point *B*; which is *BE*. We drop the perpendicular *ED* to *AB*. It is clear that the latter is divided as we wished, because the straight line *EB* squared is equal to the rectangle *IG*, and the rectangle *DK* is the product of *AD* and *DB*; and the square of *AD* is the square *AH*. So we have divided *AB* at *D* such that *AD* by *DB*, added to the square of *AD* and to the square *C*, is equal to the square of *AB*. This is what we wanted to prove.

Having finished with these things, let us now end this book so as not to prolong the discourse, so as not to tire the understanding of the reader nor to make the work inaccessible to him.

Since an investigation of the nature of propositions and of their individual properties is always carried out in one of these two ways: either we bring to mind the necessity of their properties by considering a variety of species, [an act of] imagination that draws on sensation or on what is perceived by the senses; or we assume these properties and also the lemmas they make necessary, in succession, by geometrical necessity; [accordingly] I shall now give an example of this to attract attention from those concerned with this art.

As for imagining their properties being necessary by [considering] variation of their species in the light of what is common to the senses, it is like what we gave an example of earlier: For all triangles, the sums of the angles are equal to one another. And like: let there be two isosceles triangles *ABC* and *ADE*, the side *AD* being equal to the side *AB* and the angle *DAE* being greater than the angle *BAC*, then the base *DE* is greater than the base *BC*. This property is equally imagined by what is common to the senses. For the scholar, the primary properties required are required in this way.

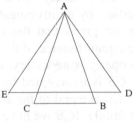

Fig. 19

For the second way, the scholar must enquire into them in a thorough comprehensive geometrical investigation to use them as an exercise and so as to arrive at a conception of the properties of the propositions in reality and this becomes a resource for him. I now give an example of this, which is the following:

We suppose we have a circle *ACB*. In it we draw the two chords *AB* and *CD* which cut one another at the point *H* and we wish to know in what way it necessarily follows that the product of *AH* and *HB* is equal to the product of *CH* and *HD*. We join *CA* and *BD*, there result two similar triangles *ACH* and *BHD*, it being given that equal angles inscribed in the circumference of the same circle intercept the same arc. We have that the ratio of *CH* to *AH* is equal to the ratio of *BH* to *HD*. In the same way, if we draw the straight line *EHG* and if we join *CE* and *DG*, we obtain the two triangles *CEH* and *DHG*, which are likewise similar; so their sides are in proportion.

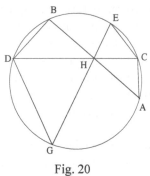

Fig. 20

As for investigating the fact that an arc[39] of a circle is intercepted by an angle equal to the angle formed by the chord of that arc[40] and the straight line which is a tangent to it,[41] we draw the circle *ABC* with diameter *AB* and we draw *BD* tangent to the circle; then it becomes clear to us that the semicircle *ACB* is intercepted by an angle equal to the angle *ABD*. So we draw *BD*, we have to look for the variety of the species of that figure, and the necessity of their properties, by an investigation of their nature. So we join *BC*, *AC* and *CE*. It being given that the variation of the angle *B* is intrinsically common to the circumference of the circle and the two straight lines *AB* and *DG*, such a property is necessary. But since the triangle *ACB* is right-angled and the arc *CB* is intercepted by equal angles, accordingly the angle *E* of the triangle *CEB* is equal to the angle *A* of the triangle *ACB*. Now to the angle *B* of the triangle *ACB* we have added an angle – which is the angle *ABE* – and we have to take away from the angle *ACB* an angle equal to the one we added to the angle *B*, by the same measure. The angle formed by the chord of an arc of a circle and the tangent is equal to the angle which intercepts that arc, which is the angle *BAC*. Consequently, it

---

[39] Lit.: segment (*qiṭ'a*).
[40] *Ibid.*
[41] A tangent at an endpoint of the arc.

appeared to us, in reality and geometrically, how the species of this figure
vary, that their properties are necessary and that *EBD* and the angles which
intercept the arc *EACB* are equal. This is what we wanted to prove.

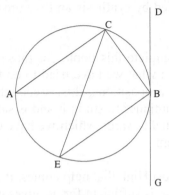

Fig. 21

Let us now start with a figure, by using analysis, so that this will be an
exercise for beginners: Let there be a point *A* and a straight line *BC*; we
wish to draw from the point *A* to the straight line *BC* two straight lines,
such as the straight lines *AB* and *AC*, which enclose the known angle *A*,
that is an angle equal to a given angle, and such that *AB* by *AC* is known,
that is to say equal to a given area.

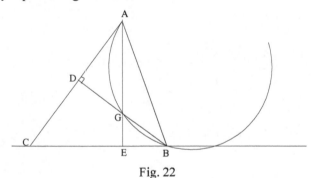

Fig. 22

By analysis, we take *AB* by *AC* to enclose a known area and a known
angle, I mean the angle *A*. Let us draw the two perpendiculars *AE* and *BD*.
Since *AB* by *AC* is known and the triangle *ABD* is of known form, because
the two angles *A* and *D* are known, accordingly the ratio of *AB* to *AD* is
known, so the ratio of *AB* by *AC* to *AD* by *AC* is known, because *AC* is the
common term. So *AD* by *AC* is known; now *AE* by *AG* is equal to *AD* by
AC on account of the similarity of the triangles *AGD* and *ACE*; but *AE* is

known, so *AG* is known. So if we apply to *AG* an arc which is intercepted by an angle equal to the angle *ABD*, then it is to its intersection with the given straight line that we draw *AB*. Now *AC* encloses a known angle with the latter. We thus proceed by synthesis and we prove it by the method of synthesis.

It is here that we complete this book: for those working in this area, these examples suffice. In what we have tried to do to smooth the way for determining geometrical propositions, this is enough for someone who is thinking about it, who continues to study it and is satisfied with accustoming himself to pursuing that towards which we have directed him and what we have pointed out to him.

It is from God the Most High that help comes. It is to Him that we submit ourselves. It is He who is sufficient for us, gives us fulfilment and helps us.

The book is completed, Thanks be to God and to the grace of His assistance.

On the Construction of an Acute-angled Triangle
from Two Unequal Straight Lines

You have asked – may God give you continuing happiness – to construct an acute-angled triangle from two unequal straight lines; and you mentioned that Abū Saʿd al-ʿAlāʾ ibn Sahl constructed this by using an ellipse with the help of proposition fifty-two of Book III of Apollonius' *Conics*,[1] by the method of division and discussion. I remember that I had a solution to the problem, and it is in our book on *Triangles*. But what we have presented in our book was not [done] by the method of discussion. So I have found it by the method of discussion and division starting from the first and third books of Euclid's *Elements* so as to prove to you – may God sustain you – that the propositions established by simple methods and principles that can be found from the books of Euclid's *Elements* are preferable to adopting difficult methods, and in particular those of the *Conics*. For problems whose solutions cannot be established from the book of the *Elements*, one is allowed to have recourse to unusual and obscure methods; but we do not need to show what they can do, because it is obvious; and from God comes help.

Question: Starting from two given unequal straight lines, we wish to construct an acute-angled triangle.

Answer: For this triangle there are three cases. Let the two given straight lines be the straight lines *AB*, *AC*. We want what we have said. We draw a circle on the chord *AB* such that the arc *AB* is intercepted by half a right angle; which is the circle *ADB*. We draw *BD* perpendicular to *AB* to meet the circumference of the circle. We join *AD*. It is clear that *AD* is a diameter of the circle. With centre *A* and distance *AC*, we draw the circular arc *CK*. Either it will touch the circle *ADB* at the point *D*, as in the first case; or it will cut it in two points – in *I* and *K* – as in the second case; or it will lie outside the circle, as in the third case. If it lies outside the circle, let

---

[1] Lit.: the cone.

us extend *BD* to *I* and let us join *AI*.[2] If it cuts it as in the second case, let it cut it in two points on either side of the diameter *AD*, at the points *I* and *K*. Let us then join *AI*. If it touches it, as in the first case, it is necessary that it touches it at the point *D*.

I say that the angle subtended by the straight line *AB*[3] always lies inside the sector *CAD*, in the first case, inside the sector *CAI*, in the second case and the third case. Now, from the two straight lines drawn from the points *A* and *B* to the arc *CED*,[4] and if we draw a straight line from the point *B* to the straight line drawn from *A* to the arc *CED*, such that, with the straight line drawn from *B* to the arc *CED*, this straight line encloses an angle equal to the angle formed on the arc *CED* by the two straight lines drawn from the points *A* and *B*, there is formed[5] an acute-angled triangle. But it is not possible to construct an acute-angled triangle from these two given straight lines by anything other than this.

Let us draw *AE* to meet the arc *CED* and let us join *BE*. On the point *B* of the straight line *AB* we construct an angle equal to the angle *AEB*; let it be <the angle> *EBG*.

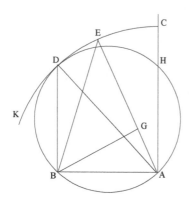 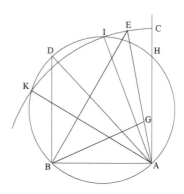

Fig. 1   Fig. 2

[2] In the last two cases of the figure, al-Sijzī uses the same letter, *I*, to designate two different points.

[3] He implies that *E*, the vertex of the angle *AEB* subtended by the straight line *AB*, always lies on the arc *CD* (excluding its endpoints) in the first case, and on the arc *CI* (excluding its endpoints) in the second case and the third case.

[4] Al-Sijzī considers the first case of the figure where *E* is on the arc *CD*. In the second and third cases, we need *E* to lie on the arc *CI*; the arc *CED* exists only in the first case of the figure, and in what follows al-Sijzī deals with the first case.

[5] Lit.: closes up.

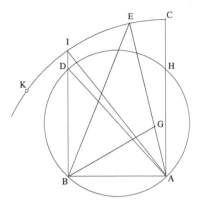

Fig. 3

I say that the triangle *AGB* is acute-angled.

Proof: Since the angle *AEB* is smaller than half a right angle, and the angle *AGB* is double the angle *AEB*, because it is equal to the sum of the two angles *AEB* and *EBG* which are equal to one another, accordingly the angle *AGB* is smaller than a right angle; so it is acute; but since the two straight lines drawn from the points *A* and *B* to meet the arc *CED* enclose with the straight line *AB* two angles each smaller than a right angle, that is to say acute, accordingly each of them[6] is acute; so the triangle *AGB* is acute-angled. It is clear that the straight line drawn from the point *A* in the direction of *H* encloses an obtuse angle with *AB*. Similarly, the straight line drawn from *B* in the direction of *D* encloses an obtuse angle[7] with *AB*. So take the triangle inside the straight lines *AC* and *BD* which are parallel. This is what we wanted to prove.

This is what we have presented by means of division and discussion, employing a universal method, whose easy access, simple course and brevity of expression make it agreeable to your mind and to your understanding. So take profit from it, may God make you happy through it and let there be peace.

The treatise is completed, thanks to God and to His blessings.

I wrote it on Thursday, the day *day* of the month of Ābān in the year 339 in the era of Yazdegerd.[8]

---

[6] Here, 'them' refers to the three angles.
[7] He means: in separating from *AH* and *BD*.
[8] See History of the texts, p. 577, n. 46.

Book by Archimedes on the Foundations of Geometry, translated from Greek into Arabic by Thābit ibn Qurra, for Abū al-Ḥasan ʿAlī ibn Yaḥyā, protected by the Prince of Believers

– I – Let us take an equilateral triangle *ABC*, in it let us draw the height *AD* and on the straight line *BD* let us mark an arbitrary point, which is the point *E*. From the point *E* let us draw two perpendiculars to the straight lines *CA* and *AB*, which are the straight lines *GE* and *EH*.

I say that the straight line AD *is equal to the sum[1] of the straight lines* GE *and* EH.

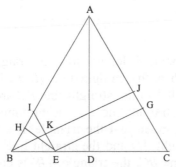

Proof: From the point *E* let us draw a straight line parallel to *AC*, which is the straight line *EI*. From the point *B* let us draw a straight line perpendicular to the straight line *AC*, which is the straight line *BJ*. It being given that the triangle *ABC* is equilateral and that the straight line *AC* is parallel to the straight line *IE*, the triangle *BIE* is equilateral; and it being given that the straight line *BJ* is perpendicular to the straight line *AC* and that the straight line *AC* is parallel to the straight line *IE*, the straight line *BK* is perpendicular to the straight line *IE*. Now the straight line *KJ* is equal to the straight line *EG*, because the surface *KEGJ* is a parallelogram. So the whole straight line *BJ* is equal to the sum of the straight lines *EH* and *GE*;

[1] We sometimes add the word 'sum', to conform with the standard usage in English.

but the straight line BJ is equal to the straight line AD, so the straight line AD is equal to the sum of the straight lines EG and EH. This is what we wanted to prove.

– II – Let us take an equilateral triangle ABC, let us draw inside it the height AD and let us mark inside it an arbitrary point, which is the point E. From this point let us draw perpendiculars to the sides of the triangle, which are the straight lines GE, EH and EI.

I say that AD *is equal to the sum of the straight lines* EG, EH, EI.

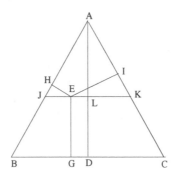

Proof: From the point E let us draw a straight line parallel to the straight line BC, which is the straight line $JELK$. It being given that the straight line JK is parallel to the straight line BC and that the straight line EG is parallel to the straight line DL, accordingly the surface ED is a parallelogram. It being given that the triangle ABC is equilateral and that we have drawn the height AD and the straight line JK parallel to its base, which is the straight line BC, the triangle AJK is equilateral. And it being given that the triangle AJK is equilateral, that we have drawn the height AL, that on the straight line BK we have marked an arbitrary point – which is the point E – and that from this point we have drawn two perpendiculars to the straight lines JA, AK which are the straight lines EH, EI, accordingly the straight line AL is equal to the sum of the straight lines EH, EI. Now we have proved that the straight line LD is equal to the straight line EG, so the straight line AD is equal to the sum of the straight lines EG, EH, EI. This is what we wanted to prove.

Book of Hypotheses by Aqāṭun

– I – Let us take an equilateral triangle *ABC* in which we draw the height *AD* and mark an arbitrary point on the straight line *CB*, which is the point *E*. From this point let us draw two perpendiculars to the straight lines *CA* and *AB*, which are the straight lines *GE* and *EH*.

I say that the straight line AD is equal to the sum of the straight lines GE and EH.

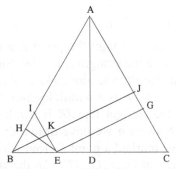

Proof: From the point *E* we draw a straight line parallel to the straight line *AC*; let the straight line be *IE*. From the point *B* let us draw a straight line perpendicular to the straight line *CA*; let the straight line be *BJ*. Since the triangle *ABC* is equilateral and since the straight line *AC* is parallel to the straight line *IE*, the triangle *IBE* is equilateral. Since the straight line *BJ* is perpendicular to the straight line *AC*, which is parallel to the straight line *IE*, accordingly *BK* is perpendicular to the straight line *IE*. Now the straight line *EH* is perpendicular to the straight line *IE*, so the straight line *BK* is equal to the straight line *EH*; but the straight line *KJ* is parallel to the straight line *EG*, so it is equal to it. So the whole straight line *BJ* is equal to the sum of the straight lines *GE* and *EH*. But the straight line *BJ* is equal to the straight line *AD*, so the straight line *AD* is equal to the sum of the straight lines *EH* and *EG*. This is what we wanted to prove.

– II – Let us take an equilateral triangle *ABC*, let us draw inside it the height *AD* and inside it let us take an arbitrary point, which is the point *E*. From this point let us draw three perpendiculars to the sides of the triangle; which are the straight lines *EG*, *EH* and *EI*.

I say that AD is equal to their sum.

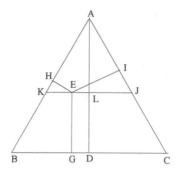

Proof: We cause to pass through the point *E* a straight line parallel to the straight line *BC*; let it be the straight line *JEK*. Since the straight line *JK* is parallel to the straight line *BC* and since *EG* is parallel to the straight line *DL*, accordingly the surface *DE* is a parallelogram. Since the triangle *ABC* is equilateral and since the straight line *JK* is parallel to its base, the triangle *AJK* is equilateral. Now in it we have drawn the height *AL* and on the straight line *KLJ* we have marked a point *E* from which we have drawn two perpendiculars to the straight lines *AJ*, *AK* – let the straight lines be *EH* and *EI* –, accordingly the straight line *AL* is equal to their sum. But the straight line *LD* is equal to the straight line *EG*, so the straight line *AD* is equal to the sum of the straight lines *EG*, *EH*, *EI*. This is what we wanted to prove.

Treatise by Aḥmad ibn Muḥammad ʿAbd al-Jalīl al-Sijzī on the Properties of the Perpendiculars Dropped from a Given Point onto a Given Equilateral Triangle by the Method of Discussion

– I – If the point falls on one of the sides of the triangle, like the point *E*, let us draw the two perpendiculars, *EM* and *EN*, to the sides *AB* and *AC*. We cannot draw a third perpendicular, apart from these two. Since *EM* is the height of an equilateral triangle, the side of the triangle is *BE* and since *EN* is the height of an equilateral triangle, the side of the triangle is *EC*. So the ratio of *EM* to *BE* is equal to the ratio of *EN* to *EC* and is equal to the ratio of *AD* to *BC*. By composition, the ratio of the sum of *EM* and *EN* to the sum of *BE* and *EC* is equal to the ratio of *AD* to *BC*, so the sum of the two straight lines *EM* and *EN* is equal to the straight line *AD*.

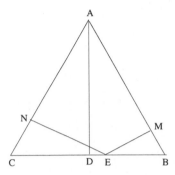

– II – If the point falls on the straight line *AD*, like the point *G*, let us then draw the two perpendiculars *GS* and *GO* to the sides *AB* and *AC* and let us draw the straight line *ZGU* parallel to the straight line *BC*; the sum of the straight lines *GS*, *GO* is thus equal to the straight line *AG* and the sum of the straight lines *GS*, *GO*, *GD* is equal to the sum of *AG* and *GD*, that is to say the straight line *AD*.

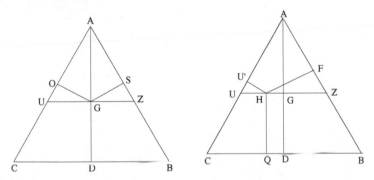

If the point falls in the surface *ABC*, like the point *H*, let us then draw the straight line *ZHU* parallel to the straight line *BC* and let us draw the perpendiculars *HP*, *HU'*, *HQ* to the sides *AB*, *AC*, *BC*. From what we have proved, the sum of the straight lines *HP*, *HU'* is equal to the perpendicular *AG* and the straight line *HQ* is parallel to the straight line *GD* and is equal to it, so the sum of the straight lines *HP*, *HU'*, *HQ* is equal to the sum of the straight lines *AG*, *GD*, that is to say the straight line *AD*.

I. FAKHR AL-DĪN AL-RĀZĪ: IBN AL-HAYTHAM'S CRITIQUE OF THE NOTION OF PLACE AS ENVELOPE

'Ibn al-Haytham marshalled arguments to demolish the assertion that place is a surface; he said: if place were a surface, then it would be able to increase while the thing whose place it determines remained as it was, in two cases. A) If we divide a parallelepiped by parallel surfaces that are parallel to the two original surfaces, the surfaces which enclose this body before it is divided up are, without any doubt, smaller than those that enclose it after it has been divided up into many parts, whereas the thing whose place it determines remains as it was. B) If wax is shaped into a sphere, the surface that encloses the wax is smaller than the surface that encloses it when it is shaped into a cube. Since the sphere is the greatest of figures,[1] the thing whose place it determines, when it is transformed into a cube, thus remains exactly the same whereas the place increases.

It is possible for the place to remain the same whereas the thing whose place it determines decreases: indeed, the water which is in a goatskin has as its place the inner surface of the goatskin. If we press on the goatskin so that the water overflows through its spout, the surface of the goatskin continues to surround what remains of the water; so the thing whose place is determined has decreased and the place is as it was.

It is possible for the thing whose place is determined to decrease and for the place to increase, as when we make a deep hollow in one of the sides of a cube; its concave surface is then necessarily greater than its plane face, and what remains of the body once it has been hollowed out is much smaller than it was at first. Here the thing whose place is determined has decreased whereas the place has increased. And since the consequences are manifestly inadmissible, the antecedents must be so too.'[2]

---

[1] That is, the sphere is the greatest among solid figures that have equal surface areas.

[2] Fakhr al-Dīn al-Rāzī (1149-1210), *Al-Mulakhkhaṣ*, ms. Teheran, Majlis Shūrā, no. 827, fols 92–93. Cf. above, Ibn al-Haytham's treatise *On Place*, pp. 499–500, 511; and the French edition, pp. 670–1 and 955.

II. AL-ḤASAN IBN AL-HAYTHAM AND MUḤAMMAD IBN AL-HAYTHAM: THE MATHEMATICIAN AND THE PHILOSOPHER

ON PLACE

In Volumes II and III we drew attention, under this same title, to the confusion caused by biobibliographers and many historians, from the thirteenth century onwards, regarding the mathematician and the philosopher. We put forward many historical arguments, both regarding technical content and bibliographic, which we still consider to be unanswerable.[3] In Volume III, we called upon two important witnesses: ʿAbd al-Laṭīf al-Baghdādī and Fakhr al-Dīn al-Rāzī, both from the end of the twelfth century.

But old habits die hard. Thus, in what is no doubt a desperate attempt to support the identification of al-Ḥasan with his namesake Muḥammad, it has been considered possible to assert that al-Ḥasan's treatise on place is a revised version of a treatise by Muḥammad, called *Treatise on Place and Time according to what he* [Muḥammad] *Found Following Aristotle's Opinion on them* (*Kitāb fī al-makān wa-al-zamān ʿalā mā wajadahu yalzamu rayʾ Arisṭūṭālīs fīhimā*).

This conjecture is arbitrary in the literal sense, since it is not supported by any argument, whether historical, technical or textual (Muḥammad's treatise is lost, we have only its title), and it is fraught with implications that are, to say the least, implausible.

1. This title by Muḥammad ibn al-Haytham, reported by the biobibliographer Ibn Abī Uṣaybiʿa from the author's autobiography, is from a late composition. From the information supplied by Ibn Abī Uṣaybiʿa himself, it appears that this text was composed after January 1027 and before July 1028, that is after Dhū al-Ḥijja 417 of the Hegira and before the end of Jumādā al-ākhira 419 of the Hegira.[4] Now, in 417 of the Hegira,

---

[3] *Les Mathématiques infinitésimales du IXᵉ au XIᵉ siècle*, vol. II: *Ibn al-Haytham*, London, 1993, pp. 8–19; English trans. *Ibn al-Haytham and Analytical Mathematics*. A History of Arabic Sciences and Mathematics, vol. 2, Culture and Civilization in the Middle East, London, 2012, pp. 11–25. *Les Mathématiques infinitésimales du IXᵉ au XIᵉ siècle*, vol. III: *Ibn al-Haytham. Théorie des coniques, constructions géométriques et géométrie pratique*, London, 2000, vol. III, pp. 937–41; English trans. *Ibn al-Haytham's Theory of Conics, Geometrical Constructions and Practical Geometry*. A History of Arabic Sciences and Mathematics, vol. 3, Culture and Civilization in the Middle East, London/New York, 2013, pp. 729–34.

[4] Ibn Abī Uṣaybiʿa, *ʿUyūn al-anbāʾ fī ṭabaqāt al-aṭibbāʾ*, ed. N. Riḍā, Beirut, 1965, p. 558.

Muḥammad, again according to Ibn Abī Uṣaybi'a, was in his sixty-third (lunar) year. So he would have composed his treatise *On Place and Time* (now lost, along with the major part of the philosopher's huge output) at the age of about sixty-five lunar years: so we are not considering a work from his early youth.

2. Between January 1027 and July 1028, Muḥammad had also composed, among other things, *A Summary of the* Physics *of Aristotle* (*Talkhīṣ al-Samā' al-ṭabī'ī li-Arisṭūṭālīs*), *A Summary of the* Meteorologica *of Aristotle* (*Talkhīṣ* Kitāb al-āthār al-'ulwiyya *li-Arisṭūṭālīs*), and *A Summary of the Book* On Animals *by Aristotle* (*Talkhīṣ Kitāb Arisṭūṭālīs fī al-Ḥayawān*). To which we must add numerous writings on philosophy, theology, medicine and optics. Moreover, before 1027, Muḥammad ibn al-Haytham had also written a *Summary of the Problems of Aristotle's Physics* (*Talkhīṣ al-Masā'il al-ṭabī'iyya li-Arisṭūṭālīs*). So it is indeed a philosopher immersed in Aristotelian learning who conceived on this book *On Place and Time*. It is, moreover, sufficient to run through the titles of many others of his writings on metaphysics, logic and physics for us to register his deep engagement with the work of Aristotle. To confine ourselves to the subject of logic, Muḥammad ibn al-Haytham wrote a summary of Porphyry's *Isagoge*, as well as of seven of Aristotle's logical works; a two-chapter book on the syllogism, a book on proof, and so on. He has also left us a *Book to Refute Philoponus' Criticisms of Aristotle in connection with* The Heavens and the World (*Fī al-radd 'alā Yaḥyā al-Naḥwī wa-mā naqaḍahu 'alā Arisṭūṭālīs ... fī al-Samā' wa-al-'ālam*).

3. We can see clearly the philosophical framework within which Muḥammad ibn al-Haytham was working before and during the period of composition of *On Place and Time*. The combination of place and time suggests, moreover, that he intended to deal with the ideas of the *Physics*; and it does not take a great philologist to recognise, simply from the title he gave the work, that Muḥammad had shaped his *On Place and Time* in accordance with the teachings of Aristotle, or what 'follows' from this teaching.

4. If we now return to al-Ḥasan ibn al-Haytham, we have shown that his treatise is resolutely and explicitly anti-Aristotelian. Moreover, in this treatise he formulated the first geometrical theory of place. His anti-Aristotelianism and the originality of his theory did not, indeed, escape his critics, for instance al-Baghdādī at the end of the twelfth century.

Also, in this treatise on place, al-Ḥasan draws extensively on one of his most original and sophisticated mathematical works on isoperimetric and isepiphanic figures and the solid angle (*On the Sphere which is the Largest of all Solid Figures having Equal Perimeters and on the Circle which is the*

Largest of all the Plane Figures having Equal Perimeters — *Fī anna al-kura awsaʿ al-ashkāl al-mujassama allatī iḥāṭatuhā mutasāwiya wa-anna al-dāʾira awsaʿ al-ashkāl al-musaṭṭaḥa allatī iḥāṭatuhā mutasāwiya*).[5] This same treatise is mentioned in *On Place*, as well as in another book by al-Ḥasan: *For Resolving Doubts in the* Almagest.

Finally, again according to Ibn Abī Uṣaybiʿa and from a list he had found of the writings of al-Ḥasan,[6] the text on place (like the majority of al-Ḥasan's writings) was composed before 1038.

In conclusion, if we accept that Muḥammad and al-Ḥasan are one and the same person, and that the treatise by al-Ḥasan *On Place* is a revised version of the *Treatise on Place and Time according to what he* [Muḥammad] *Found Following Aristotle's Opinion on them*, we have to accept:

1) that, at the age of sixty-five, al-Ḥasan wrote a treatise on place according to the theory of Aristotle, together with a commentary on the *Physics*, before changing his opinion completely and turning against Aristotle's theory. But what event could have led to such a revolution in his thinking, since that is indeed what we are seeing here? Might the composition of his book on isoperimetric and isepiphanic figures and the solid angle (*On the Sphere which is the Largest of all Solid Figures having Equal Perimeters*) have instigated this conversion? However, a careful reading of this book does not permit us to draw any such conclusion, because the theorem Ibn al-Haytham uses in the treatise *On Place* can be derived directly from the treatise by al-Khāzin,[7] since the mathematician has no need to investigate the solid angle which is the central topic of this book, and since he need not have waited until he was in his sixty-fifth year to turn against Aristotle. On the contrary, as we have shown, the geometrisation of the notion of place can be understood through the advances in geometrical understanding that are accumulated in the other geometrical treatises written by al-Ḥasan ibn al-Haytham. So we should not see the geometer's anti-Aristotelianism as a sudden change of opinion and still less as simply a philosophical choice; his attitude is, much rather, derived from various works that involve geometrical transformations and movement. In short, al-Ḥasan ibn al-Haytham's geometrisation of place is an effect of the

---

[5] *Les Mathématiques infinitésimales*, vol. II, Chap. III.

[6] This list is also known through the manuscript in Lahore.

[7] *Les Mathématiques infinitésimales du IXᵉ au XIᵉ siècle*, vol. I: *Fondateurs et commentateurs: Banū Mūsā, Thābit ibn Qurra, Ibn Sinān, al-Khāzin, al-Qūhī, Ibn al-Samḥ, Ibn Hūd*, London, 1996; English trans. *Founding Figures and Commentators in Arabic Mathematics*, A History of Arabic Sciences and Mathematics, vol. 1, Culture and Civilization in the Middle East, London, 2012, Chap. IV.

emergence of geometrical transformations, as operations as well as objects of geometry. In the circumstances, it is hard to see how one could maintain that a mathematician steeped in Aristotelian theory would change his mind because of a result showing that, among solids, it is the sphere that has the minimum surface area, a result known for a long time, that he changed his mind to the point of criticising Aristotle and working out an entirely new theory.

2) that this revolution was not noticed by the man himself, to the point that he did not even refer to it in his second version of the text. This would be the more surprising because it often happens that Ibn al-Haytham returns to an old problem he has already considered, to explore it in a revised text, often a longer one. But in these cases he never failed to refer back to the first version. This is exactly what he did in his treatise *On the Figures of Lunes*,[8] in his treatise *On the Construction of the Regular Heptagon*[9] and in his treatise *On the Principles of Measurement*,[10] among others.

3) that his successors and above all his critics, such as al-Baghdādī, who were conversant with the writings of al-Ḥasan ibn al-Haytham's times, never even noticed this radical change of position. Surely it is implausible that al-Baghdādī, in particular, would have been unaware of 'his' writings on logic to the point of reproaching him for his ignorance of logic; and that because he did not know this first treatise *On Place and Time* he made no mention of it in his critique of the treatise *On Place*.

In the absence of historical and textual arguments, any conjectures are possibly correct and there is no limit imposed on implausibility.[11] Only an intimate understanding of al-Ḥasan ibn al-Haytham's mathematical writings can protect one from the temptation to make conjectures like that of there having been a 'revised version', a conjecture designed to defend an error made by biobibliographers, and one that has had too long a life.

[8] *Les Mathématiques infinitésimales*, vol. II, p. 102.

[9] *Ibid.*, vol. III, p. 454.

[10] *Ibid.*, vol. III, p. 538.

[11] It is with conjectures of this type, and with as little argument for them, that A. Sabra has dedicated himself to defending the identification of the mathematician with the philosopher. The reader will understand why we do not undertake to discuss the conjectures one by one. See A. Sabra, 'One Ibn al-Haytham or Two? An Exercise in Reading the Bio-Bibliographical Sources', *Zeitschrift für Geschichte der arabisch-islamischen Wissenschaften*, Band 12, 1998, pp. 1–50.

BIBLIOGRAPHY

I. MANUSCRIPTS

Aqāṭun, *Kitāb al-Mafrūḍāt*
    Istanbul, Süleymaniye, Aya Sofya, 4830, fols 91^v–92^r.

[Archimedes]
Kitāb fī uṣūl al-handasiyya
    Patna, Khuda Bakhsh, 2519, fols 142^v–143^r.

Ibn al-Haytham
Fī khawāṣṣ al-dawā'ir
    St Petersburg, 600 (formerly Kuibychev, Library V.I. Lenin), fols 421^r–431^r.
Fī khawāṣṣ al-muthallath min jihat al-'amūd
    Patna, Khuda Bakhsh, 2519, fols 189^r–191^r.
Fī al-ma'lūmāt
    St Petersburg, 600 (formerly Kuibychev, Library V.I. Lenin), fols 335^r–347^v.
    Paris, BNF, 2458, fols 11^v–26^r.
Fī al-makān
    Cairo, Dār al-Kutub, 3823, fols 1^v–5^v.
    London, India Office, 1270, fols 25^v–27^v.
    Hyderabad, Salar Jung Mus., 2196, fols 19^v–22^r.
    Istanbul, Süleymaniye, Fātiḥ, 3439, fols 136^v–138^r.
    Teheran, Majlis Shūrā Millī, 2998, fols 166–174.
Fī mas'ala handasiyya
    St Petersburg, B 1030, fols 102^r–110^v.
    Oxford, Seld. A32, fols 115^v–120^v.
Fī al-taḥlīl wa-al-tarkīb
    Cairo, Dār al-Kutub, Taymūr, Riyāḍa 323, 68 p.
    Dublin, Chester Beatty, 3652, fols 69^v–86^r.
    Istanbul, Süleymaniye, Reshit, 1191, fols 1^v–30^v.
    St Petersburg, 600 (formerly Kuibychev, Library V.I. Lenin), fols 348^r–368^r.

Al-Rāzī, Fakhr al-Dīn, *al-Mulakhkhaṣ*
    Teheran, Majlis Shūrā, 827.

Al-Sijzī
Jawāb al-Sijzī 'an masā'il handasiyya sa'ala 'anhu Ahl Khūrāsān
    Dublin, Chester Beatty, 3652, fols 53^r–61^r.
    Istanbul, Süleymaniye, Reshit, 1191, fols 110^v–123^v.
Kitāb fī tashīl al-subul li-istikhrāj al-ashkāl al-handasiyya
    Lahore, coll. Nabī Khān, fols 2–27.

Qawl fī khawāṣṣ al-aʿmida al-wāqiʿa min al-nuqṭa al-muʿṭā ilā al-muthallath al-mutasāwī al-aḍlā ʿ
    Dublin, Chester Beatty, 3652, fols 66ᵛ–67ʳ.
    Istanbul, Süleymaniye, Reshit, 1191, fols 124ᵛ–125ʳ.
Risāla ilā Abī ʿAlī Naẓīf ibn Yumn fī ʿamal muthallath ḥādd al-zawāyā
    Paris, Bibliothèque nationale, 2457, fols 136ᵛ–137ʳ.
    Lahore, coll. Nabī Khān, fols 28–30.
Taʿlīqāt handasiyya min Kitāb al-Sijzī
    Dublin, Chester Beatty, 3045/14, fols 74ʳ–89ᵛ.
    Cairo, Dār al-Kutub, riyāḍa 699, 35 p.

Thābit ibn Qurra
Kitāb Thābit ibn Qurra ilā Ibn Wahb fī al-taʾattī li-istikhrāj ʿamal al-masāʾil al-handasiyya
    Istanbul, Aya Sofya, 4832, fols 1ᵛ–4ʳ (under the title *Risāla fī kayfa yanbaghī an yuslaka ilā nayl al-maṭlūb min al-maʿānī al-handasiyya*).
    Leiden, Or. 14/21, fols 380–388 (under the title *Fī al-ʿilla allatī lahā rattaba Uqlīdis ashkāl kitābihi*).
    Cairo, Riyāḍa 40/11, fols 155ᵛ–159ᵛ (under the title *Risāla fī kayfa yanbaghī an yuslaka ilā nayl al-maṭlūb min al-maʿānī al-handasiyya*).
    Paris, Bibliothèque nationale, 2457, fols 188ᵛ–191ʳ.
    Tunis, Aḥmadiyya, 16167, fols 86ᵛ–90ᵛ (under the title *Fī al-ʿilla allatī lahā rataba Uqlīdis ashkāl kitābihi*).

II. OTHER MANUSCRIPTS CONSULTED FOR THE ANALYSIS

ʿAbd al-Laṭīf al-Baghdādī
Kitāb al-Naṣīḥatayn, Bursa, Hüseyin Çelebi, 823, fols 88ᵛ–93ʳ.
Fī al-makān, Bursa, Hüseyin Çelebi, 823, fols 23ᵛ–52ʳ.

Al-Farghānī, *al-Kāmil*, Kastamonu, 794, fols 89–117.

Ibn al-Haytham
Fī ḥall shukūk Kitāb Uqlīdis fī al-uṣūl, Istanbul, University, 800.
Sharḥ muṣādarāt Kitāb Uqlīdis, Istanbul, Feyzullah, 1359, fols 150ʳ–237ᵛ.

Ibn Hūd, *al-Istikmāl*
    Copenhagen, Or. 82.
    Leiden, Or. 123.

Ibn Sinān, *Maqāla fī ṭarīq al-taḥlīl wa-al-tarkīb fī al-masāʾil al-handasiyya*
    Paris, Bibliothèque nationale, 2457, fols Aᵛ–18ᵛ.

Al-Qūhī
Marākiz al-dawāʾir al-mutamāssa ʿalā al-khuṭūṭ bi-ṭarīq al-taḥlīl
    Paris, Bibliothèque nationale, 2457, fols 19ʳ–21ʳ.

Mas'alatayn handasiyyatayn
  Cairo, Dār al-Kutub, 40, fols 206ᵛ–208ʳ.
  Istanbul, Aya Sofya, 4830, fols 171ʳ–173ʳ.
  Istanbul, Aya Sofya, 4832, fols 123ᵛ–125ᵛ.

Al-Sijzī
Barāhīn Kitāb Uqlīdis fī al-uṣūl 'alā sabīl al-tawassu' wa-al-irtiyāḍ
  Dublin, Chester Beatty, 3652, fols 18ʳ–29ᵛ.
  Istanbul, Süleymaniye, Reshit, 1191, fols 84ᵛ–105ᵛ.
Fī al-masā'il al-mukhtāra allatī jarrat baynahu wa-bayna muhandisī Shīrāz wa-Khurāsān wa-ta'līqātihā
  Dublin, Chester Beatty, 3652, fols 35ʳ–52ᵛ.
  Istanbul, Süleymaniye, Reshit, 1191, fols 31ᵛ–62ʳ.
Fī taḥṣīl al-qawānīn al-handasiyya al-maḥdūda
  Istanbul, Süleymaniye, Reshit, 1191, fols 70ʳ–72ᵛ.
  Paris, Bibliothèque nationale, 2458, fols 3–4.

Thābit ibn Qurra
Fī anna al-khaṭṭayn idhā ukhrijā 'alā aqall min zāwiyatayn qā'imatayn iltaqayā,
  Paris, Bibliothèque nationale, 2457, fols 156–160.

III. BOOKS AND ARTICLES

P. Abgrall, 'Les cercles tangents d'al-Qūhī', *Arabic Sciences and Philosophy*, 5.2, 1995, pp. 263–95.

A. Anbouba, 'Un traité d'Abū Ja'far al-Khāzin sur les triangles rectangles numériques', *Journal for the History of Arabic Science*, 3.1, 1979, pp. 134–78.

Apollonius de Perge, *La section des droites selon des rapports*, commentaire historique et mathématique, édition et traduction du texte arabe par Roshdi Rashed et Hélène Bellosta, Berlin/New York, Walter de Gruyter, 2009.

Apollonius, *Les Coniques*, Tome 2.1: *Livres II et III*, commentaire historique et mathématique, édition et traduction du texte arabe par Roshdi Rashed, Berlin/New York, Walter de Gruyter, 2010.

Aristote, *Physique*, texte établi et traduit par H. Carteron, Collection des Universités de France, Paris, 1961; French trans. P. Pellegrin, *Aristote: Physique*, Paris, Garnier-Flammarion, 2000.

Aristotle, *Physics*, A Revised Text with Introduction and Commentary by W. D. Ross, Oxford, 1936; English trans. by P. H. Wicksteed and F. M. Cornford, London/Cambridge (Mass.), Heinemann and Harvard University Press, 1970; by E. Hussey, *Aristotle Physics*, Books III and IV, Clarendon Aristotle Series, Oxford, 1983.

Arisṭūṭālīs, al-Ṭabī'a, ed. 'A. Badawi, vol. I, Cairo, 1964; vol. II, Cairo, 1965.

A. Arnauld and P. Nicole, *La Logique ou l'art de penser, contenant, outre les règles communes, plusieurs observations nouvelles, propres à former le jugement*, édition critique présentée par Pierre Clair et François Girbal, Le mouvement des idées au XVIIᵉ siècle, Paris, PUF, 1965.

A. Behhoud, 'Greek Geometrical Analysis', *Centaurus*, 37, 1994, pp. 52–86.

Al-Bīrūnī, *Risāla fī istikhrāj al-awtār fī al-dā'ira*, Hyderabad, 1948; ed. A. S. Demerdash, Cairo, 1965.

Al-Biruni and Ibn Sina, *al-As'ilah wa'l-Ajwibah*, ed. S. H. Nasr and M. Mohaghegh, Teheran, 1973.

M. Chasles, *Aperçu historique sur l'origine et le développement des méthodes en géométrie*, Paris, Gauthier Villars, 1889.

J. L. Coolidge, *A History of Geometrical Methods*, Oxford, 1940; repr. Dover, 1963.

P. Crozet
'Al-Sijzī et les *Éléments* d'Euclide: Commentaires et autres démonstrations des propositions', in A. Hasnawi, A. Elamrani-Jamal and M. Aouad (eds), *Perspectives arabes et médiévales sur la tradition scientifique et philosophique grecque*, Paris, Peeters, 1997, pp. 61–77.
'À propos des figures dans les manuscrits arabes de géométrie: l'exemple de Siğzī', in Y. Ibish (ed.), *Editing Islamic Manuscripts on Science*, Proceedings of the Fourth Conference of al-Furqān Islamic Heritage Foundation, 29th–30th November 1997, London, al-Furqān, 1999, pp. 131–63.
'Geometria: La tradizione euclidea rivisitata', in R. Rashed (ed.), *Storia della scienza*, vol. III: *La civiltà islamica*, Rome, Istituto della Enciclopedia Italiana, 2002, pp. 326–41.

R. Deltheil and D. Caire, *Géométrie et compléments*, Paris, ed. Jacques Gabay, 1989.

Girard Desargues, *Exemple de l'une des manieres universelles du S. G. D. L. touchant la pratique de la perspective sans emploier aucun tiers point, de distance ni d'autre nature, qui soit hors du champ de l'ouvrage*, Paris, 1636, pp. 11–12; trans. J. V. Field and J. J. Gray, *The Geometrical Work of Girard Desargues*, London and New York, 1987, pp. 158–60 (French text, pp. 200–1).

Descartes, *Œuvres de Descartes*, publiées par Ch. Adam et P. Tannery, Paris, 1965, t. VI.

A. Dhanani, *The Physical Theory of Kalām: Atoms, Space, and Void in Basrian Mu'tazili Cosmology*, Leiden, E. J. Brill, 1994.

A. Dietrich, 'Die arabische Version einer unbekannten Schrift des Alexander von Aphrodisias über die Differentia specifica', *Nachrichten der Akademie der*

Wissenschaften in Göttingen, I. Philologisch-historische Klasse, 2, 1964, pp. 88–148.

Y. Dold-Samplonius, *Book of Assumptions by Aqāṭun*, Thesis, University of Amsterdam, 1977.

Euclide
Les Œuvres d'Euclide, traduites littéralement par F. Peyrard, Paris, 1819; nouveau tirage, augmenté d'une importante introduction par M. Jean Itard, Paris, Librairie A. Blanchard, 1966.
Les Éléments, French trans. and commentary by Bernard Vitrac, 4 vols, Paris, 1990–2001.

Fakhr al-Dīn al-Rāzī, *Kitāb al-Mabaḥīth al-mashriqiyya*, Teheran, 1966.

Al-Fārābī
Iḥṣā' al-'ulūm, ed. 'Uthmān Amīn, 3rd ed., Cairo, 1968.
Kitāb al-mūsīqā al-kabīr, edited by Ghattās 'Abd al-Malik Khashaba, revised and introduced by Maḥmūd Aḥmad al-Hifnī, Cairo, n. d.
Risāla fī al-khalā', edited and translated by Necati Lugal and Aydin Sayili in *Türk tarikh yayinlarindan*, XV.1, Ankara, 1951, pp. 21–36.
Al-Manṭiqiyyāt li-al-Fārābī, ed. Muḥammad Taqī Dānish Pajūh, Qom, 1310 H, vol. III: *al-Shurūḥ 'alā al-nuṣūṣ al-manṭiqiyya*.

M. Federspiel, 'Sur la définition euclidienne de la droite', in *Mathématiques et philosophie de l'Antiquité à l'âge classique*. Études en hommage à Jules Vuillemin, éditées par R. Rashed, Paris, Éditions du CNRS, 1991, pp. 115–30.

Fermat, *Œuvres de Fermat*, publiées par les soins de MM. Paul Tannery et Charles Henry, Paris, Gauthier-Villars, 1896.

J. V. Field
'When is a proof not a proof? Some reflections on Piero della Francesca and Guidobaldo del Monte', in R. Sinisgalli (ed.), *La Prospettiva: Fondamenti teorici ed esperienze figurative dall'Antichità al mondo moderno*, Florence, 1998, pp. 120–32.
Piero della Francesca: A Mathematician's Art, New Haven and London, 2005.

E. Giannakis, 'Yaḥyā ibn 'Adī against John Philoponus on Place and Void', *Zeitschrift für Geschichte der arabisch-islamischen Wissenschaften*, Band 12, 1998, pp. 245–302.

V. Goldschmidt, *Écrits*, Paris, Vrin, 1984, t. I: Études de philosophie ancienne.

M. Gueroult, *Spinoza*, vol. II: *L'âme*, Paris, Aubier, 1974.

A. Heinen, 'Ibn al-Haitams Autobiographie in einer Handschrift aus dem Jahr 556 H/1161 A.D.', *Die islamische Welt zwischen Mittelalter und Neuzeit, Festschrift für Hans Robert zum 65*, Beirut, 1979, pp. 254–79.

H. Hermelink, 'Zur Geschichte des Satzes von der Lotsumme im Dreieck', *Sudhoffs Archiv für Geschichte der Medizin und der Naturwissenschaften*, Band 48, 1964, pp. 240–7.

J. Hintikka, 'Kant and the Tradition of Analysis', in Paul Weingartner (ed.), *Deskription, Analytizität und Existenz*, Salzburg-München, 1966.

J. Hintikka and U. Remes, *The Method of Analysis*, Dordrecht, 1974.

W. Hinz, *Islamische Masse und Gewichte umgerechnet ins metrische System*, Leiden, 1955.

Hobbes
Elementorum philosophiae sectio prima de corpore, in *Opera philosophica quae latine scripsit omnia...*, ed. G. Molesworth, vol. II, London, 1839.
Examinatio et emendatio mathematicae hodiernae, in *Opera philosophica quae latine scripsit omnia...*, ed. G. Molesworth, vol. IV, London, 1865.

J. P. Hogendijk
'The Geometrical Parts of the *Istikmāl* of Yūsuf al-Mu'taman ibn Hūd (11th century). An Analytical Table of Contents', *Archives internationales d'histoire des sciences*, 41.127, 1991, pp. 207–81.
Al-Sijzī's Treatise on Geometrical Problem Solving (Kitāb fī Tashīl al-Subul li-Istikhrāj al-Ashkāl al-Handasiyya), translated and annotated by Jan P. Hogendijk, with the Arabic text and a Persian translation by Mohammad Bagheri, Teheran, Fatemi Publishing Company, 1996; review by P. Crozet in *Isis*, 90.1, 1999, pp. 110–11.
'Traces of the Lost *Geometrical Elements* of Menelaus in Two Texts of al-Sijzī', *Zeitschrift für Geschichte der arabisch-islamischen Wissenschaften*, Band 13, 1999–2000, pp. 129–64.

C. Houzel, 'Histoire de la théorie des parallèles', in *Mathématiques et philosophie de l'Antiquité à l'âge classique. Études en hommage à Jules Vuillemin*, éditées par R. Rashed, Paris, Éditions du CNRS, 1991, pp. 163–79.

Ibn Abī Uṣaybi'a, *'Uyūn al-anbā' fī ṭabaqāt al-aṭibbā'*, ed. N. Riḍā, Beirut, 1965.

Ibn al-Haytham, *Majmū' Rasā'il Ibn al-Haytham*, Osmania Oriental Publications Bureau, Hyderabad, 1947.

Ibn Mattawayh, *al-Tadhkira*, ed. Samīr Naṣr Luṭf and Faysal Badīr 'Aūn, Cairo, 1975.

Ibn Sīnā, *al-Shifā'*: *al-Ṭabī'iyyāt*, 1. *al-Samā' al-ṭabī'ī*, ed. S. Zayed, rev. I. Madkour Cairo, 1983; ed. Ja'far Āl-Yāsīn, Beirut, 1996. *Al-Najāt*, ed. M. S. al-Kurdī, Cairo, 1938.

A. M. Legendre, 'Réflexions sur les différentes manières de démontrer la théorie des parallèles ou le théorème sur la somme des trois angles du triangle', *Mémoires de l'Académie des sciences*, 12, 1833, pp. 367–410.

G. W. Leibniz, *La Caractéristique géométrique*, texte établi, introduit et annoté par Javier Echeverría; traduit, annoté et postfacé par Marc Parmentier, Mathesis, Paris, Vrin, 1995.

M. Mahoney, 'Another Look at Geometrical Analysis', *Archive for History of Exact Sciences*, vol. V, nos 3–4, 1968, pp. 318–48.

I. Mueller, 'Aristotle's Doctrine of Abstraction in the Commentators', in R. Sorabji (ed.), *Aristotle Transformed: the Ancient Commentators and their Influence*, London, 1990, pp. 463–84.

Al-Nadīm, *Kitāb al-fihrist*, ed. R. Tajaddud, Teheran, 1971.

O. Neugebauer and R. Rashed, 'Sur une construction du miroir parabolique par Abū al-Wafā' al-Būzjānī', *Arabic Sciences and Philosophy*, 9.2, 1999, pp. 261–77.

Abū Rashīd al-Nīsābūrī, *Kitāb al-Tawḥīd*, ed. Muḥammad 'Abd al-Hādī Abū Rīda, Cairo, 1965.

Pappus
Pappi Alexandrini Collectionis quae supersunt e libris manu scriptis edidit latina interpretatione et commentariis instruxit F. Hultsch, 3 vols, Berlin, 1876–1878.
Pappus d'Alexandrie, *La Collection mathématique*, French trans. Paul Ver Eecke, Paris/Bruges, 1933.
Pappus of Alexandria, Book 7 of the Collection. Part 1. Introduction, Text, and Translation; Part 2. *Commentary, Index, and Figures*, Edited with Translation and Commentary by Alexander Jones, Sources in the History of Mathematics and Physical Sciences, 8, New York/Berlin/Heidelberg/Tokyo, Springer-Verlag, 1986.

Philopon, *Ioannis Philoponi in Aristotelis Physicorum libros quinque posteriores commentaria*, ed. H. Vitelli, *CAG* XVII, Berlin, Reimer Verlag, 1888.

Proclus, *In Primum Euclidis Elementorum librum Commentarii*, ed. G. Friedlein, Leipzig, 1873; repr. Olms, 1967; French trans. by P. Ver Eecke, *Proclus: Les Commentaires sur le premier livre des Éléments d'Euclide*, Bruges, 1948.

Claudius Ptolemy, *Almagest*, English trans. G. Toomer, London, Duckworth, 1984.

M. Rashed
'Alexandre et la "magna quaestio"', *Les Études classiques*, 63, 1995, pp. 295–351.

'Kalām e filosofia naturale', in R. Rashed (ed.), *Storia della scienza*, vol. III: *La civiltà islamica*, Rome, Istituto della Enciclopedia Italiana, 2002, pp. 49–72.

R. Rashed

'Ibn al-Haytham et le théorème de Wilson', *Archive for History of Exact Sciences*, 22.4, 1980, pp. 305–21; repr. in *Entre arithmétique et algèbre: Recherches sur l'histoire des mathématiques arabes*, Paris, Les Belles Lettres, 1984, pp. 227–43.

'Nombres amiables, parties aliquotes et nombres figurés aux XIIIe et XIVe siècles', *Archive for History of Exact Sciences*, 28, 1983, pp. 107–47; repr. in *Entre arithmétique et algèbre: Recherches sur l'histoire des mathématiques arabes*, Paris, Les Belles Lettres, 1984, pp. 259–99.

'Mathématiques et philosophie chez Avicenne', in *Études sur Avicenne*, dirigées par J. Jolivet et R. Rashed, Paris, Les Belles Lettres, 1984, pp. 29–39.

Entre arithmétique et algèbre: Recherches sur l'histoire des mathématiques arabes, Paris, Les Belles Lettres, 1984; English trans. *The Development of Arabic Mathematics Between Arithmetic and Algebra*, Boston Studies in Philosophy of Science 156, Dordrecht/Boston/London, Kluwer Academic Publishers, 1994.

Sharaf al-Dīn al-Ṭūsī, Œuvres mathématiques. Algèbre et Géométrie au XIIe siècle, Collection Sciences et philosophie arabes – textes et études, 2 vols, Paris, Les Belles Lettres, 1986.

'Al-Sijzī et Maïmonide: Commentaire mathématique et philosophique de la proposition II-14 des *Coniques* d'Apollonius', *Archives internationales d'histoire des sciences*, vol. 37, no. 119, 1987, pp. 263–96.

'Ibn al-Haytham et les nombres parfaits', *Historia Mathematica*, 16, 1989, pp. 343–52; repr. in *Optique et mathématiques: Recherches sur l'histoire de la pensée scientifique en arabe*, Variorum CS388, Aldershot, 1992, XI.

'La philosophie mathématique d'Ibn al-Haytham. I: L'analyse et la synthèse', *MIDEO*, 20, 1991, pp. 31–231.

'L'analyse et la synthèse selon Ibn al-Haytham', in *Mathématiques et philosophie de l'Antiquité à l'âge classique*. Études en hommage à Jules Vuillemin, éditées par R. Rashed, Paris, Éditions du CNRS, 1991, pp. 131–62; repr. in *Optique et mathématiques: recherches sur l'histoire de la pensée scientifique en arabe*, Variorum Reprints CS388, Aldershot, 1992, XIV.

Optique et mathématiques: Recherches sur l'histoire de la pensée scientifique en arabe, Variorum CS388, Aldershot, 1992.

'La philosophie mathématique d'Ibn al-Haytham. II: Les Connus', *MIDEO*, 21, 1993, pp. 87–275.

Géométrie et dioptrique au Xe siècle. Ibn Sahl, al-Qūhī, Ibn al-Haytham, Paris, Les Belles Lettres, 1993; English version: *Geometry and Dioptrics in Classical Islam*, London, al-Furqān, 2005.

Les Mathématiques infinitésimales du IXe au XIe siècle, vol. II: *Ibn al-Haytham*, London, al-Furqān, 1993; English trans. *Ibn al-Haytham and Analytical Mathematics*. A History of Arabic Sciences and Mathematics, vol. 2, Culture and Civilization in the Middle East, London, Centre for Arab Unity Studies, Routledge, 2012.

Les Mathématiques infinitésimales du IXe au XIe siècle, vol. I: *Fondateurs et commentateurs: Banū Mūsā, Thābit ibn Qurra, Ibn Sinān, al-Khāzin, al-Qūhī,*

Ibn al-Samh, Ibn Hūd, London, al-Furqān, 1996; English trans. *Founding Figures and Commentators in Arabic Mathematics*, A History of Arabic Sciences and Mathematics, vol. 1, Culture and Civilization in the Middle East, London, Centre for Arab Unity Studies, Routledge, 2012.

Les Mathématiques infinitésimales du IXᵉ au XIᵉ siècle, vol. III: *Ibn al-Haytham. Théorie des coniques, constructions géométriques et géométrie pratique*, London, al-Furqān, 2000; English translation by J. V. Field: *Ibn al-Haytham's Theory of Conics, Geometrical Constructions and Practical Geometry*. A History of Arabic Sciences and Mathematics, vol. 3, Culture and Civilization in the Middle East, London/New York, Centre for Arab Unity Studies, Routledge, 2013.

'Ibn Sahl et al-Qūhī: Les projections. Addenda & corrigenda', *Arabic Sciences and Philosophy*, vol. 10.1, 2000, pp. 79–100.

'Fermat and Algebraic Geometry', *Historia Scientiarum*, 11.1, 2001, pp. 24–47.

Les Mathématiques infinitésimales du IXᵉ au XIᵉ siècle, vol. IV: *Méthodes géométriques, transformations ponctuelles et philosophie des mathématiques*, London, al-Furqān, 2002.

'Les mathématiques de la terre', in G. Marchetti, O. Rignani and V. Sorge (eds), *Ratio et superstitio*, Essays in Honor of Graziella Federici Vescovini, Textes et études du Moyen Âge, 24, Louvain-la-Neuve, FIDEM, 2003, pp. 285–318.

Les Mathématiques infinitésimales du IXᵉ au XIᵉ siècle, vol. V: *Ibn al-Haytham: Astronomie, géométrie sphérique et trigonométrie*, London, al-Furqān, 2006; English translation by J. V. Field: *Ibn al-Haytham. New Spherical Geometry and Astronomy*, A History of Arabic Sciences and Mathematics, vol. 4, Culture and Civilization in the Middle East, London/New York, Centre for Arab Unity Studies, Routledge, 2014.

R. Rashed and H. Bellosta, *Ibrāhīm ibn Sinān: Logique et géométrie au Xᵉ siècle*, Leiden, E. J. Brill, 2000.

R. Rashed and C. Houzel, 'Thābit ibn Qurra et la théorie des parallèles', in R. Rashed (ed.), *Thābit ibn Qurra. Science and Philosophy in Ninth-Century Baghdad*, Scientia Graeco-Arabica, vol. 4, Berlin/New York, Walter de Gruyter, 2009, pp. 27–73.

R. Rashed and B. Vahabzadeh, *Al-Khayyām mathématicien*, Paris, A. Blanchard, 1999; English transl. (without the Arabic texts): *Omar Khayyam. The Mathematician*, Persian Heritage Series no. 40, New York, Bibliotheca Persica Press, 2000.

B. A. Rosenfeld, *A History of Non-Euclidean Geometry. Evolution of the Concept of a Geometric Space*, Studies in the History of Mathematics and Physical Sciences, 12, New York, Springer-Verlag, 1988, pp. 49–56.

A. Sabra, 'One Ibn al-Haytham or Two? An Exercise in Reading the Bio-Bibliographical Sources', *Zeitschrift für Geschichte der arabisch-islamischen Wissenschaften*, Band 12, 1998, pp. 1–50.

Ṣāʿid al-Andalusī, *Ṭabaqāt al-umam*, ed. H. Būʿalwān, Beirut, 1985.

A. S. Saidan, *The Works of Ibrāhīm ibn Sinān*, Kuwait, 1983.

Al-Samaw'al, *al-Bāhir*, ed. S. Ahmad and R. Rashed, Damascus, 1972.

D. Sedley, 'Philoponus' Conception of Space', in R. Sorabji, *Philoponus and the Rejection of Aristotelian Science*, Ithaca, 1987, pp. 140–53.

L. A. Sédillot, 'Du *Traité* des Connus géométriques de Hassan ben Haithem', *Journal asiatique*, 13, 1834, pp. 435–58.

F. A. Shamsi, 'Properties of Triangles in Respect of Perpendiculars', in Hakim Mohammed Said (ed.), *Ibn al-Haytham*, *Proceeding of the Celebrations of 1000th Anniversary*, Karachi, Times Press, Sadar, n.d., pp. 228–46.

D. Sourdel, *Le Vizirat abbaside*, Institut Français de Damas, Damascus, 1959–60.

Thābit ibn Qurra, *Rasā'il Ibn Qurra*, Osmania Oriental Publications Bureau, Hyderabad, 1947.

R. B. Todd, *Alexander of Aphrodisias on Stoic Physics*, Leiden, 1976.

R. Taton, 'La géométrie projective en France de Desargues à Poncelet', Conférence faite au Paris de la Découverte le 17 février 1951, pp. 1–21.

F. Woepcke, 'Notice sur une théorie ajoutée par Thābit Ben Qorrah à l'arithmétique spéculative des grecs', *Journal Asiatique*, IV, 2, 1852, pp. 420–9.

Yāqūt, *Mu'jam al-Udabā'*, ed. Būlāq, Cairo, n. d., vol. III.

INDEX OF NAMES

SUBJECT INDEX

INDEX OF WORKS

INDEX OF MANUSCRIPTS